HOLT PHYSICAL SCIENCE

William L. Ramsey
Lucretia A. Gabriel
James F. McGuirk
Clifford R. Phillips
Frank M. Watenpaugh

Holt, Rinehart and Winston, Publishers
New York Toronto London Sydney

THE AUTHORS

William L. Ramsey
Former Head of the Science Department
Helix High School
La Mesa, California

Lucretia A. Gabriel
Former Teacher of Science
Koda Junior High School
Shenendehowa Central School District
Clifton Park, New York

James F. McGuirk
Head of the Science Department
South High Community School
Worcester, Massachusetts

Clifford R. Phillips
Head of the Science Department
Mount Miguel High School
Spring Valley, California

Frank M. Watenpaugh
Science Department
Helix High School
La Mesa, California

About the Cover

The photograph on the cover shows a gyroscope. This instrument was originally used to illustrate the rotation of the earth. A gyroscope can spin on its axis while at the same time rotating around an axis perpendicular to its spin axis.

Editoral Development: William N. Moore, Daniel C. Wasp, Lorraine Smith-Phelan
Editorial Processing: Margaret M. Byrne, Regina Chilcoat, Barbara Russiello
Art, Production, and Photo Resources: Vivian Fenster, Fred C. Pusterla, Robin Swenson, Susan Gombocz, Lois Safrani, Ira A. Goldner, Joan Marinelli, Frances Saracco, Angel Borrero
Product Manager: Laura Zuckerman
Consultant: John Matejowsky
Researchers: James R. George, Erica Felman
Advisory Board: Rhenida Bennett, John W. Griffiths, David J. Miller, William Paul, George Salinger, L. Jean Slankard, John Taggert

Picture Credits appear on page 495.
Cover photograph by Fundamental Photographs

Printed in the United States of America
ISBN: 0-03-056867-6
1234-071-987654321

TO THE STUDENT

Science is like a tree. As scientists discover more about nature, the tree grows new branches. Two of the most important branches of the tree of science have grown from the study of energy and of matter. These two main branches of science together make up the study of physical science. One branch is called physics. Physics is the study of energy. The second branch is called chemistry. Chemistry is the study of matter.

How to Use This Book

This book will help you to explore physical science. There are five main parts to this book. You will begin your study of physical science with the First Steps that must be taken to understand how science works. These First Steps are important in understanding why science has been so successful in explaining the natural world. The next two parts, or units, of this book are devoted to physics, the study of energy. The remaining two units deal with chemistry, the study of matter. Each of these units is divided into chapters. A review at the end of each chapter will help you to remember what you have learned. Each chapter is divided into several lessons. At the beginning of each lesson, the objectives for that lesson are set off by • or ○. These objectives tell you what you will learn as you study the lesson. You will also have a chance to perform an activity in each lesson. This activity will help you to better understand the main idea of the lesson. You will find many interesting and helpful photographs and diagrams in each lesson. Science words that may be new to you are printed in **boldface** type. Their definitions will be found alongside in the margin. Each lesson ends with a short summary of the important ideas of the lesson. A few questions following the summary will help you to check that you have accomplished the objectives of the lesson.

Most chapters contain an optional laboratory exercise. In a scientific laboratory, scientists observe what happens in nature. Your classroom can become a laboratory in which you can observe and begin to understand matter and energy.

Throughout the book, special features on careers in science and science-related fields will tell you about various jobs that you may find interesting. A list of addresses to which you may write for further information is also included.

Safety

In your study of physical science, you will learn many fascinating things about the natural world. However, the study of science can also involve many potential dangers. Science classrooms and laboratories contain equipment and chemicals that can be dangerous if not handled properly. You should always follow the directions and cautions in the book when doing an activity or laboratory exercise. In addition, your teacher will explain to you the precautions and rules that must be followed to insure the safety of you and your classmates. Your responsibility is to follow these rules carefully and to demonstrate a positive attitude toward laboratory safety.

ACKNOWLEDGMENTS

Special thanks are due to the following teachers who taught *Holt Physical Science* in their classes and who made many helpful suggestions and criticisms of the program:

Cheryl H. Cook, Medway High School, Medway, Massachusetts; June Gallagher, Alta Vista Junior High School, Phoenix, Arizona; and Richard G. McKinstry, Mad River Junior High School, Dayton, Ohio.

The authors would also like to thank Dr. Jerry Faughn, East Kentucky State University, Richmond, Kentucky, for his helpful critique of the manuscript; Franklin D. Kizer, Executive Secretary of the Council of State Science Supervisors, Lancaster, Virginia, for his recommendations on safety; and Judith Linscott Martin, Reading Specialist, formerly of the Granville County Public Schools, Granville County, North Carolina, for her assistance with vocabulary.

CONTENTS

To the Student iii
Acknowledgments iv

First Steps 1

1. Investigating 1
2. The Metric System 7
Laboratory Experimenting 12
Review 14

motion and energy — UNIT 1

CHAPTER 1 Motion 18

1–1. Speed and Accelerated Motion 18
1–2. Forces 25
1–3. Mass and Weight 30
1–4. Balanced Forces 37
Laboratory Force and Acceleration 44
Chapter Review 46

CHAPTER 2 Energy 49

2–1. Spending Energy 49
2–2. Simple Machines 55
2–3. Conservation of Energy 63
2–4. Uses of Energy 69
Laboratory Power 76
Chapter Review 78

CHAPTER 3 Waves 81

3–1. Energy and Waves 81
3–2. Wave Motion 88
3–3. Sound Waves 94
3–4. Sound and Music 100
Laboratory Characteristics of Sound 106
Chapter Review 108

CHAPTER 4 Light 111

4–1. The Electromagnetic Spectrum 111
4–2. Movement of Light Waves 118
4–3. Color 125
4–4. Bending Light Rays 134
Laboratory Reflection and Refraction 143

CAREERS IN SCIENCE-RELATED FIELDS 146

Chapter Review 148

UNIT 2 electricity, magnetism, and heat

CHAPTER 5 Electric Charge 152

5–1. Kinds of Electric Charge 152
5–2. Electric Force 159
5–3. Electric Current 165
5–4. Electric Circuits 172
5–5. Measuring Electricity 178
5–6. Electric Power 185
Laboratory Electric Current 194
Chapter Review 196

CHAPTER 6 Magnetism 199

6–1. Nature of a Magnet 199
6–2. Earth as a Magnet 205

CAREERS IN ELECTRONICS AND HOME HEATING 210

6–3. Magnetism and Electricity 212
6–4. Electromagnetic Induction 217
Chapter Review 223

CHAPTER 7 Heat 225

7–1. Heat Energy 225
7–2. Heat Transfer 232
7–3. Temperature and Heat 241
7–4. Behavior of Gases 248
7–5. Liquids and Solids 254

Laboratory Heat and Fusion 250
Chapter Review 263

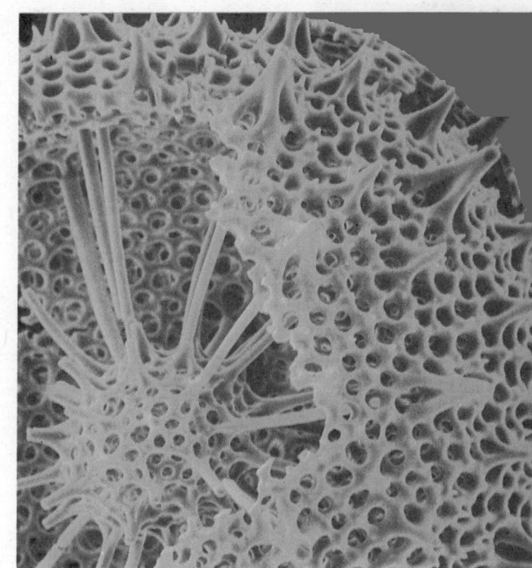

CHAPTER 8 Matter **266**

8–1. Molecules 266
8–2. Mixtures, Compounds, and Solutions 272
8–3. Elements and Atoms 277
Laboratory Analysis of Water 282
Chapter Review 285

CHAPTER 9 Atoms **287**

9–1. Atomic Particles 287
9–2. Atomic Structure 292
9–3. Electron Shells 297
9–4. Atomic Mass 303
Chapter Review 308

CHAPTER 10 Kinds of Atoms **310**

10–1. Ionization 310
10–2. Chemical Activity 318
10–3. Chemical Families 324
Laboratory The Voltaic Cell 330
Chapter Review 332

CHAPTER 11 Atomic Bonds **334**

11–1. Chemical Bonding 334
11–2. Kinds of Chemical Bonds 342
11–3. Chemical Reactions 348
11–4. Speed of Reactions 353
11–5. Energy in Chemical Reactions 359
Laboratory Chemical Reactions 367
Chapter Review 369

CAREERS IN CHEMISTRY 372

CHAPTER 12 Chemical Changes **374**

12–1. Metals and Nonmetals 374
12–2. Water 380
12–3. Ions 385
Laboratory Electrical Conductivity of Solutions 390
Chapter Review 392

the reactions of matter

CHAPTER 13 Acids, Bases, and Salts **396**

13–1. Acids and Their Properties 396
13–2. Bases and Their Properties 401
13–3. An Acid + a Base = a Salt 406
Laboratory Neutralization of $Ba(OH)_2$ with H_2SO_4 411
Chapter Review 414

CHAPTER 14 Organic Chemistry **417**

14–1. Carbon and Its Compounds 417
14–2. Hydrocarbons 422
14–3. Substituted Hydrocarbons 428
14–4. Molecules Necessary for Life 434
Laboratory Chemical Tests for Organic Compounds 440
Chapter Review 442

CHAPTER 15 The Nucleus **445**

15–1. Radioactivity 445
15–2. Using Radioactivity 451
15–3. Nuclear Energy 457
15–4. Controlled Nuclear Reactions 465
Chapter Review 471

PHYSICAL SCIENCE—THE FUTURE 474

GLOSSARY **476**
APPENDIX **483**
PICTURE CREDITS **495**
INDEX **496**

First Steps

1. INVESTIGATING

Notice that the bicycle racers shown above are leaning sideways as they round a sharp corner. You may also have discovered that you may tumble off your bicycle if you do not lean into a turn. Why does a bicycle act this way? A bicycle rider who asks this question has already taken the first step in scientific thinking. Scientists always begin their work by asking a question. Finding an answer depends upon observing all details that might have a relationship to the question. A possible answer is then developed and tested. Scientists observe, think, predict, and prove.

When you finish lesson 1, you will be able to:

- Find and describe a pattern in a given set of *observations*.

- Given a particular *hypothesis*, describe an *experiment* you could set up to test it.

- Explain the relationship between a *theory* and a *scientific law*.

- ○ Perform an experiment to test a hypothesis.

If you have watched bicycle riders turning a corner, you may have noticed a pattern in the way they ride. The rider must lean more when rounding a corner at a high speed than at a slow speed. There must be some relationship between the speed of the bicycle and the need to lean into a turn.

Observation
Anything that we can learn by using our senses, such as sight or hearing.

This discovery of a pattern is a good example of scientific thinking. Scientific thinking begins with **observations** (ob-ser-**vay**-shuns). *Observations* include any information gained by using your senses. The study of physical science, for example, begins with observations of the physical world around you: seeing, feeling, and hearing what goes on around you.

Speed
Distance divided by the time needed to go that distance

Figuring out the **speed** of bicycle riders is an example of an important kind of scientific observation. To calculate *speed*, it is necessary to divide the distance traveled by the time taken to go that distance. For example, to measure the speed of the skateboard rider shown below you might express speed as "pickets per second" (pickets/sec).

If six pickets were passed in 2 sec, the speed is 6 pickets/2 sec = 3 pickets/sec. An automobile traveling

1. a.

1. b.

75 miles (120 km) in 3 hr has a speed of 75 mi/3 hr = 25 mi/hr (120 km/3 hr = 40 km/hr). Observations of speed always require that distance and time be measured. Making measurements is an important part of scientific observation. Many times it is helpful to use graphs to show a relationship between two things that change. For example, the graph in Fig. 2 shows that the speed of a skateboard rider was increasing. As you can see from the graph, after 3 sec the rider has passed 3 pickets. The speed is thus 3 pickets/3 sec = 1 picket/sec. But after 4 sec the rider has passed 6 pickets and the speed is 6 pickets/4 sec = 1.5 pickets/sec.

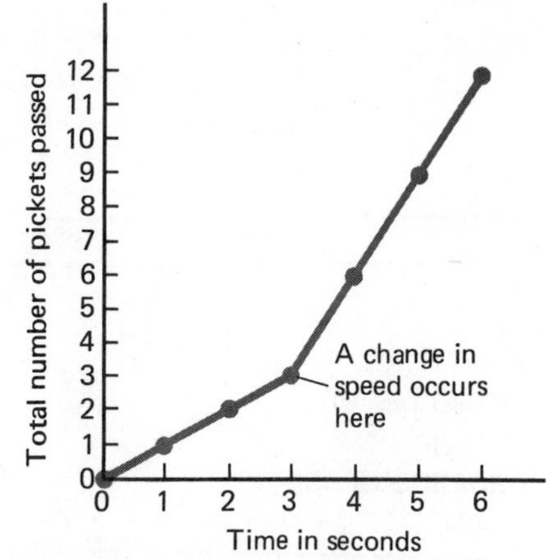

2. *This graph is one way of showing that the speed of a skateboard rider has changed.*

Making observations is only the beginning of scientific thinking. The next step is to look for a pattern in the observations. For example, observations of many skateboard riders might show a pattern in their falls: If the riders raised their arms over their heads, they fell.

The discovery of such a pattern among a set of observations makes it possible to predict what will happen in similar situations. You might predict that you would fall off a moving skateboard if you raised your arms. Scientists call such a prediction, based on observations, a **hypothesis** (hie-**poth**-ih-sis). In order to make a good *hypothesis*, you must find a relationship among the observations.

Hypothesis
A prediction or intelligent guess based on patterns in observations.

The last step in scientific thinking is testing your hypothesis. You might try riding a skateboard and raising your arms. You should try this several times. What happened each time? A test of a hypothesis is called an **experiment** (ik-**sper**-uh-ment). A scientific *experiment* must be done in such a way that everything can be observed and tested. Consider testing the hypothesis about riding a skateboard. The same person must ride the same skateboard in every test. This kind of experiment is called a *controlled experiment*.

Experiment
The testing of a hypothesis.

What happens when the observations are complete and an experiment is finished? You will know if the hypothesis tested is correct. A correct hypothesis may be used to build a **theory**. A scientific *theory* explains what we know about some part of the natural world. Like hypotheses, theories in science must be tested by further experiments. If a theory is found to be incorrect in any way, the theory must be changed or discarded. Some scientific theories have been tested many times and always found to be correct. After a

Theory
An explanation of what we know about some part of nature.

Scientific law
A theory that has been tested many times and always found to be true.

theory has passed many tests, it becomes a **scientific law**. Your study of physical science will make you aware of many important scientific *theories* and *laws*.

ACTIVITY

Materials
waxed paper
vinegar
medicine dropper
flour
baking powder
unknown

Observation, Hypothesis, Experiment

Observation, hypothesis, and experiments are parts of the methods used by scientists to solve problems. You will use these steps to find what is in an unknown substance.

A. Obtain the materials listed in the margin.

B. Place a piece of waxed paper on a flat surface (*not a tilting desk top*). Place a small amount of flour on the waxed paper. Next to the flour, put the same amount of baking powder.

1. Observation: Are there any differences in the appearances of the flour and the baking powder?

C. Record your observations in a table like the one shown.

	Appearance	Reaction to vinegar
Flour		
Baking powder		
Unknown		

D. Add a drop or two of vinegar to both the flour and the baking powder. In your table, record your observation of what happens.

2. Hypothesis: Suppose you had two unlabeled cans of flour and baking powder, so you do not know what is in each can. Write a sentence or two describing a method that could be used to identify the substance in each can.

E. Obtain an unknown white powder. Based on your hypothesis, carry out an experiment to identify the unknown substance as either flour or baking powder. Record your results.

3. As a result of your experiment, can you say that the unknown contains baking powder?

4. Based on the results of your experiment, can you be certain the unknown contains only one material? Other tests can be done to make the identification more complete, but your simple observation, hypothesis, and experiment show how these steps can give a result.

SUMMARY

Scientific investigation begins with curiosity: asking questions. The questions investigated by scientists have to do with the world we live in. Scientific investigation has three important parts: (1) making observations which often are in the form of measurements; (2) looking for some kind of pattern in the observations in order to make a hypothesis; and (3) testing the hypothesis by experiment to develop a theory.

QUESTIONS

Unless otherwise indicated, use complete sentences to write your answers.

1. List the letters of the following observations about an object that could be made without testing or using a measuring tool: **a.** It is yellow. **b.** It is round. **c.** It is 3.0 cm long. **d.** It is made of steel. **e.** It is hard. **f.** It is shiny.
2. Select one of the patterns listed below that best fits the following observations.
   Observations:
   - The red skateboard had a speed of 6 pickets/sec when student A coasted on it.
   - The red skateboard had a speed of 9 pickets/sec when student B coasted on it.
   - The green skateboard had a speed of 9 pickets/sec when student B coasted on it.
   - The green skateboard had a speed of 6 pickets/sec when student A coasted on it.

   Patterns:
   a. The green skateboard goes faster than the red one.
   b. The skateboards can go only 6 pickets/sec or 9 pickets/sec.

c. Student A is heavier than student B.
d. Student B goes faster than student A on either skateboard.

3. One hypothesis that could be made about the skateboard event in question 2 is: "The way student B stands on the skateboard makes it go faster." Which of the following ways would best test whether this hypothesis is true or not?
   a. Ask student B if he can ride faster than A at other times.
   b. Look at the skateboards to see if they were changed after student B rode them.
   c. Perform an experiment to measure the speeds of students A and B after student A is taught to ride the same way as B.
   d. Ask students A and B how much they weigh.
4. In your own words, predict the speed of the red skateboard if a student heavier than student A coasts on it in the same way students A and B coasted on it.
5. In your own words, describe the relationship between a theory and a scientific law.

2. THE METRIC SYSTEM

"Please pick up a half-kilo of hamburger and a liter of milk on your way home." If you were asked to do this errand, would you know how much hamburger and milk to bring home? Kilograms of hamburger and liters of milk might be common in the future. Pounds and quarts may slowly disappear as this country changes to the metric system of measurement. In almost every other country in the world, the metric system is already used. The metric system is always used in scientific work.

When you finish lesson 2, you will be able to:

- Explain the importance of using units that everyone can understand in making *measurements.*
- Identify common metric units used to measure length, volume, and mass.
- Use a metric ruler to make measurements.

Observations made during scientific experiments often involve **measurements.** It is important that a *measurement* means the same thing to everyone who uses it. The speed of the skateboard was expressed as "pickets per second." Do you think that this unit of speed would mean the same thing to different people? Fences do not all have pickets of the same size. What about someone who had not seen the photographs of the skateboard? In science, as in other fields, we use units of measurement that everyone can understand. In this lesson, you will learn some of the units of measurement used in science.

Measurement
Any observation that is done by counting something. Often an instrument, such as a ruler, is used in measuring.

When you make measurements in science, you will use metric units. Metric units are the most commonly

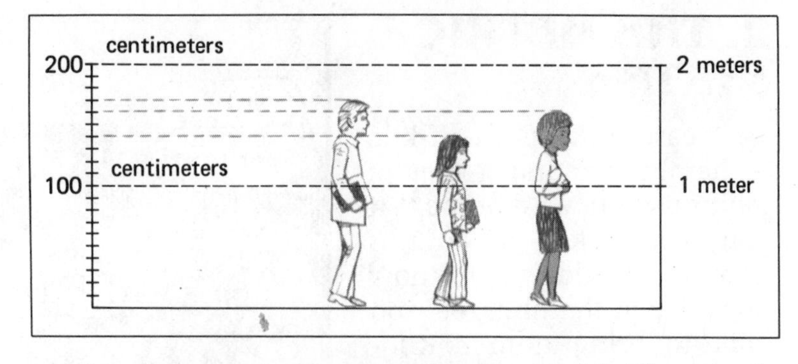

3. These students are between 1 and 2 meters tall. Which units would be easier to use to show that their heights are not the same?

used units of measurement in the world. The metric system is a decimal system. Metric units are related to each other in multiples of ten.

Meter (m)
The basic unit of length in the metric system. A little more than 1 yard in length.

Centimeter (cm)
0.01 of a meter. Try this: How many centimeters are there in 2.5 meters?

You will measure length in **meters** (m) or parts of a *meter*. A meter is slightly more than a yard. You are between 1 and 2 m tall. Doorknobs are usually about 1 m from the floor. In Fig. 3 you can see that the meter is divided into 100 equal parts called **centimeters** (**sent**-uh-meet-urz). Shorter lengths can easily be measured in *centimeters* (cm). The prefix *centi* before the name of any metric unit means 0.01 (1/100) of the unit. A centimeter is 0.01 of a meter. There are 100 centimeters in a meter just as there are 100 cents in a dollar. If you have 176 cents, you have $1.76. If you are 176 centimeters tall, you are also 1.76 meters tall. Notice that only the position of the decimal point changes in converting from centimeters to meters.

Millimeter (mm)
0.001 of a meter (also 0.1 of a centimeter). Try this: How many meters are there in 1,500 millimeters?

By dividing the meter into 1,000 equal parts called **millimeters** (**mil**-uh-meet-urs), you can easily measure lengths even shorter than centimeters. The prefix *milli* before the name of any metric unit means 0.001 (1/1,000) of the unit. A *millimeter* (mm) is 0.001 of a meter.

4. Study the figures in this drawing. Using a metric ruler, measure the height of each figure. How tall is each figure in centimeters? in millimeters? Is your ability to judge size improved by measuring?

	1 cm				
1 cm	1	2	3	4	5
	6	7	8	9	10
	11	12	13	14	15
	16	17	18	19	20

5. *The area of this box is 20 cm^2.*

Area
The amount of surface within a given set of lines; measured in squared metric units (m^2, cm^2, mm^2, etc).

In addition to making measurements directly with a metric ruler, you can also make calculated measurements. An example of a calculated measurement is **area.** *Area* is the amount of surface of an object. The area of a flat square or rectangle can be found by multiplying its length by its width. For example, the length of a piece of paper is 5 cm and its width is 4 cm. The area of this piece of paper is 5 cm × 4 cm = 20 cm^2. The area is expressed as cm^2 or "centimeters squared" or "square centimeters."

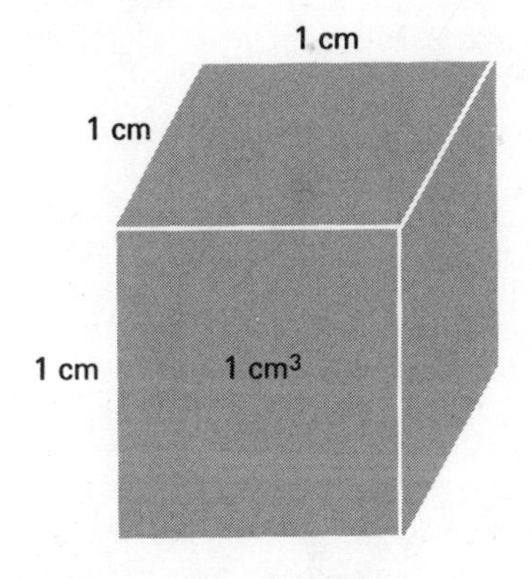

6. *A sugar cube that measures 1 cm × 1 cm × 1 cm has a volume of 1 cm^3.*

Another example of a calculated measurement is **volume.** *Volume* describes the amount of space an object takes up. The volume (V) of a box-shaped object is found by multiplying length (l) by width (w) by height (h): $V=lwh$. For example, a sugar cube might have a length, width, and height of 1 cm. The volume of the sugar cube is 1 cm × 1 cm × 1 cm = 1 cm^3. See Fig. 6. The volume can be expressed as cm^3 or "centimeters cubed" or "cubic centimeters." However, a more commonly used unit is the **liter** (L). A *liter* is equal to 1,000 cm^3 and is slightly more than one quart. Soft drinks are often sold in 1-L bottles.

Another unit used in the metric system measures mass. Mass is measured in **grams** (g). There are 454 g in 1 lb. A common unit used to measure larger weights in the metric system is the *kilogram. Kilo* added in front of a metric unit, such as a gram, means one thousand times the unit. A kilogram means 1,000 g and is equal to 2.2 lbs.

Volume
The amount of space occupied by an object, measured in cubed metric units.

Liter (L)
A commonly used unit of volume in the metric system. A little less than 1 quart.

Gram
A small unit of weight in the metric system. One pound contains 454 grams.

7. The reading at:
A is 2.6 cm
B is ? cm
C is ? cm
D is ? cm

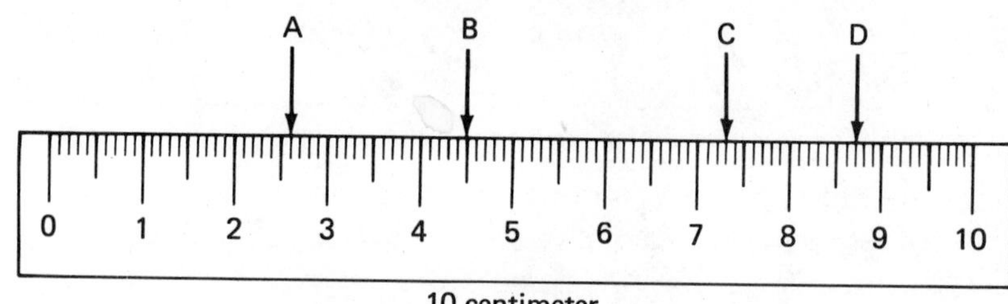

10 centimeter metric ruler

Materials
metric ruler
index card
pencil

Using the Metric System

A. Obtain the materials listed in the margin.

B. If metric rulers are not available, carefully trace the metric ruler above onto a piece of paper. See Fig. 7. This ruler is 10 cm long. The units that are numbered are centimeters. There are 10 spaces between each numbered centimeter line. Each of these spaces is 1 mm. See Fig. 8.

pencil

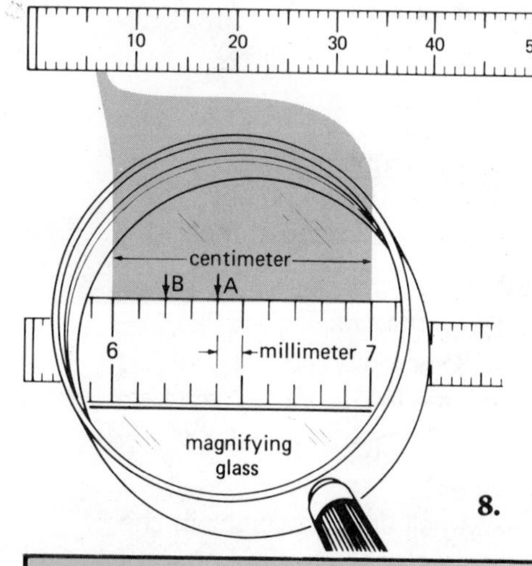

8.

C. Using the ruler that you traced or a metric ruler, answer the following questions.

1. What is the length of your pencil in centimeters?
2. What is the length of your eraser in centimeters? in millimeters?

D. Have someone help you measure your height using the metric ruler.

3. What is your height in millimeters? in centimeters? in meters?

E. Using the metric ruler, measure the length and width of an index card.

4. What is the length of the card in centimeters?
5. What is the width of the card in centimeters?
6. What is the area of the index card in square centimeters?

F. Measure the length and width of the classroom in meters and calculate its area in square meters.

7. What is the area of the room in square meters?

G. Measure the length and width of your desk top in centimeters and calculate its area in square centimeters.

8. What is the area in square centimeters?

SUMMARY

Most experiments include observations in the form of measurements. It is important that the units used in measuring mean the same for everyone using those measurements. The metric system of measurement is used in scientific work. The basic unit for measuring length in the metric system is the meter. A hypothesis shown to be correct by experiment can help to make a theory. A scientific law is a theory shown to be correct by many experiments.

QUESTIONS

Unless otherwise indicated, use complete sentences to write your answers.

1. The most commonly used units of measurement in the world today measure length in **a.** meters **b.** centimeters **c.** millimeters **d.** all of these are metric units.
2. Give an important reason for using the metric system of measurement.
3. If your height is 1.53 m, what is it in centimeters? in millimeters?
4. What is the area of a room, in square meters, that measures 4.5 m by 5.4 m?
5. How many milligrams are in 1 g? in 1 kg?

Materials
Place to hang pendulum
string 1.5 m
weight (eraser or other)
clock or watch with sweep second hand
metric ruler

9.

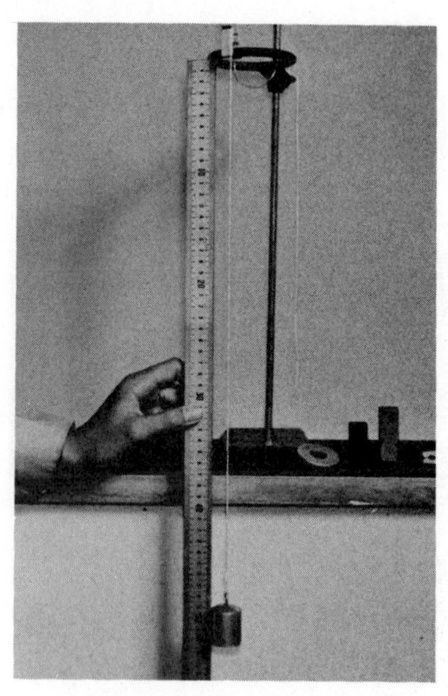

10.

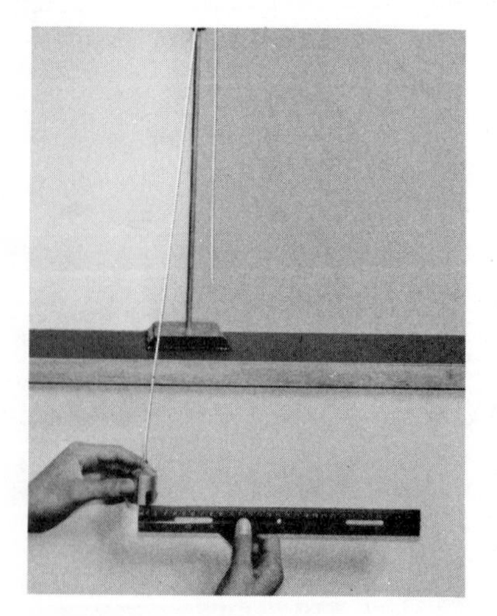

Experimenting

Purpose

In this laboratory, you will make measurements of length and time and also count the number of swings of a pendulum. Measurement gives an observation a number value and makes it easy to keep a record of the observation. You will record your observations in a table.

Procedure

A. Obtain the materials listed in the margin. A simple pendulum is just an object hanging on a string. Make a simple pendulum by tying one end of a string to an eraser (or other light object). Tie the other end of the string to a stand (or tape it to the edge of a table).

B. The length of a pendulum is the distance from the middle of the hanging weight to the point from which it is hanging. Make the length of your pendulum at least 50 cm.

C. Copy the following table in your notebook and record the length of your pendulum.

Trial	*Length (cm)*	*Time (sec)*	*Swings*	*Weight (light or heavy)*	*Distance Pulled Back (cm)*

D. Pull the pendulum weight 5 cm to one side and release it. Count the number of full swings the weight completes in 60 sec. Record the number.

E. Pull the weight 10 cm to one side and release it. Count the number of complete swings in 60 sec and record.

1. Copy and complete the following sentence: "I think the pendulum will swing ______ times in 60 sec when pulled 15 cm to one side and released." This prediction is your hypothesis.

F. To test your hypothesis, pull the weight 15 cm to one side and release it. Record the results in your data table.

G. Put a heavier weight on the string. Using the heavier weight, make a pendulum that is the same length as the first pendulum. Make a hypothesis about the number of complete, back and forth swings the pendulum will make in 60 sec.

2. In your own words, using complete sentences, state your hypothesis. (Will the pendulum complete more, the same, or less swings in 60 sec?)

H. To test your hypothesis, pull the heavier weight 10 cm to the side and release it. Record your observations in your data table.

I. Make the pendulum one-half as long as it was. Make a hypothesis about the number of swings this pendulum will complete in 60 sec.

3. State your hypothesis in your own words. ("I think the shorter pendulum will ______.")

J. Test your hypothesis and record your observations in your data table.

If other people were to use your measurements and the same equipment, they should get the same results. For an experiment to be worthwhile, it should be repeatable.

4. Would someone else get the same results as you if they used your measurements and equipment?

Summary

A pendulum clock that is running slow must be swinging too few times each minute. In your own words, using complete sentences, tell how you would adjust the pendulum on a pendulum clock that is running too slowly.

VOCABULARY REVIEW

Match the number of the word with the letter of the phrase that best explains it.

1. observations
2. speed
3. hypothesis
4. experiment
5. measurement
6. meter
7. area
8. liter
9. theory
10. scientific law

a. Testing a hypothesis.
b. Commonly used unit of volume.
c. The amount of surface.
d. The basic unit of length in the metric system.
e. Learned by using your senses.
f. Distance divided by time.
g. A theory tested many times and always found true.
h. An observation done by counting something.
i. Prediction based on patterns in observations.
j. An explanation of what we know of some part of nature.

REVIEW QUESTIONS

Choose the letter of the answer that best completes the statement or answers the question.

1. The fact that bike riders lean more into a turn at high speeds is an example of **a.** a calculation **b.** an observation **c.** a theory **d.** a scientific law.
2. The speed of an automobile can be determined by **a.** distance ÷ time **b.** distance × time **c.** distance + time **d.** distance − time.
3. A bike rider travels 25 mi in 2 hr. The speed of the rider is **a.** 12.5 mi/hr **b.** 50 mi/hr **c.** 27 mi/hr **d.** 23 mi/hr.
4. A prediction, based on observation, is tested in what is called a **a.** theory **b.** hypothesis **c.** experiment **d.** scientific law.
5. The system of measurement most commonly used in the world is the **a.** English system **b.** customary system **c.** world system **d.** metric system.
6. The unit that is 1/100 of a meter is called a **a.** millimeter **b.** centimeter **c.** kilometer **d.** micrometer.

7. A milliliter is what part of a liter? **a.** 1/10 **b.** 1/100 **c.** 1/1000 **d.** 1/10,000.
8. A rectangle whose sides are 3.0 cm and 7.0 cm has an area of **a.** 21 cm^2 **b.** 10 cm^2 **c.** 2.3 cm^2 **d.** 4.0 cm^2.
9. The number of cubic centimeters in one liter is **a.** 10 **b.** 100 **c.** 1000 **d.** 10,000.
10. The number of grams in one pound is **a.** 0.454 **b.** 4.54 **c.** 45.4 **d.** 454.

REVIEW EXERCISES

Give brief but complete answers to each of the following. Unless otherwise indicated, use complete sentences to write your answers.

1. How can the speed of a bike rider be determined?
2. Look for the pattern in the following series of numbers. Then, assuming that the pattern continues, write the next three numbers. 1, 2, 4, 8, 16, ____, ____, ____.
3. What is the relation between a hypothesis and an experiment?
4. How is a theory related to a scientific law?
5. List the following words in the order they would occur in establishing a scientific law: hypothesis, theory, observation, experiment.
6. Why is the metric system important in making measurements?
7. Name the basic metric units of length, volume, and weight.
8. How are cubic centimeters related to liters?
9. Explain the meaning of each of the following prefixes as used in the metric system: milli, centi, and kilo.
10. Show how you would change 8.8 lb into kilograms.

EXTENSIONS

1. Write to the National Bureau of Standards in Washington, D.C., for information on how standard units are determined.
2. Report to the class on the conversion to the metric system in the United States. Write to the American National Metric Council in Washington, D.C., or the National Bureau of Standards.

3. Use the diagram to complete the following table:

Metric Units	Customary Units
4 kg	______
______ g	44 oz
1,200 g	______ lbs
40 km	______ mi
520 cm	______ in
______ mL	5 cups
12 L	______ gal
______ mL	2 tsp
______ L	5 qt
700 mL	______ pints

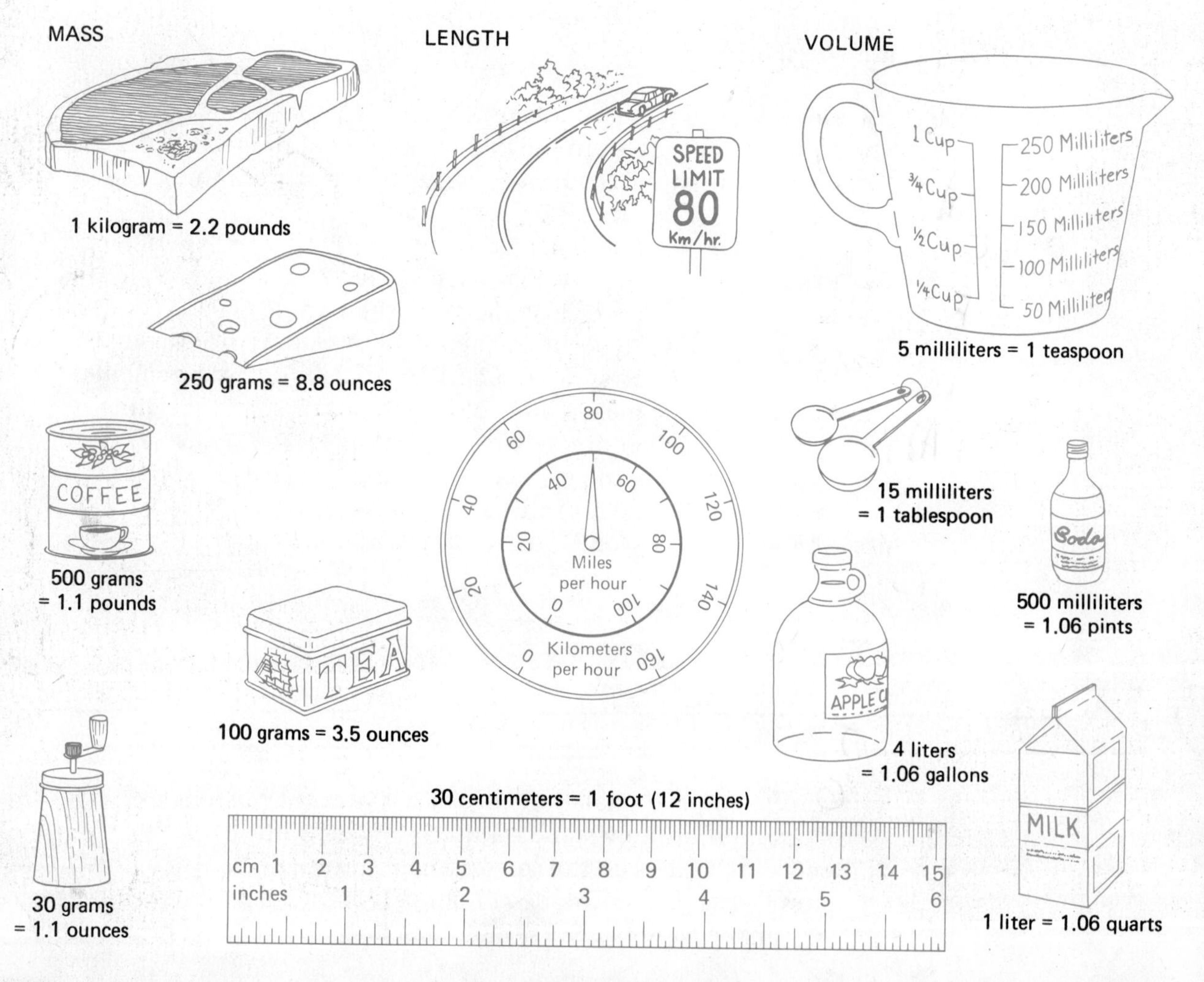

motion and energy

UNIT 1

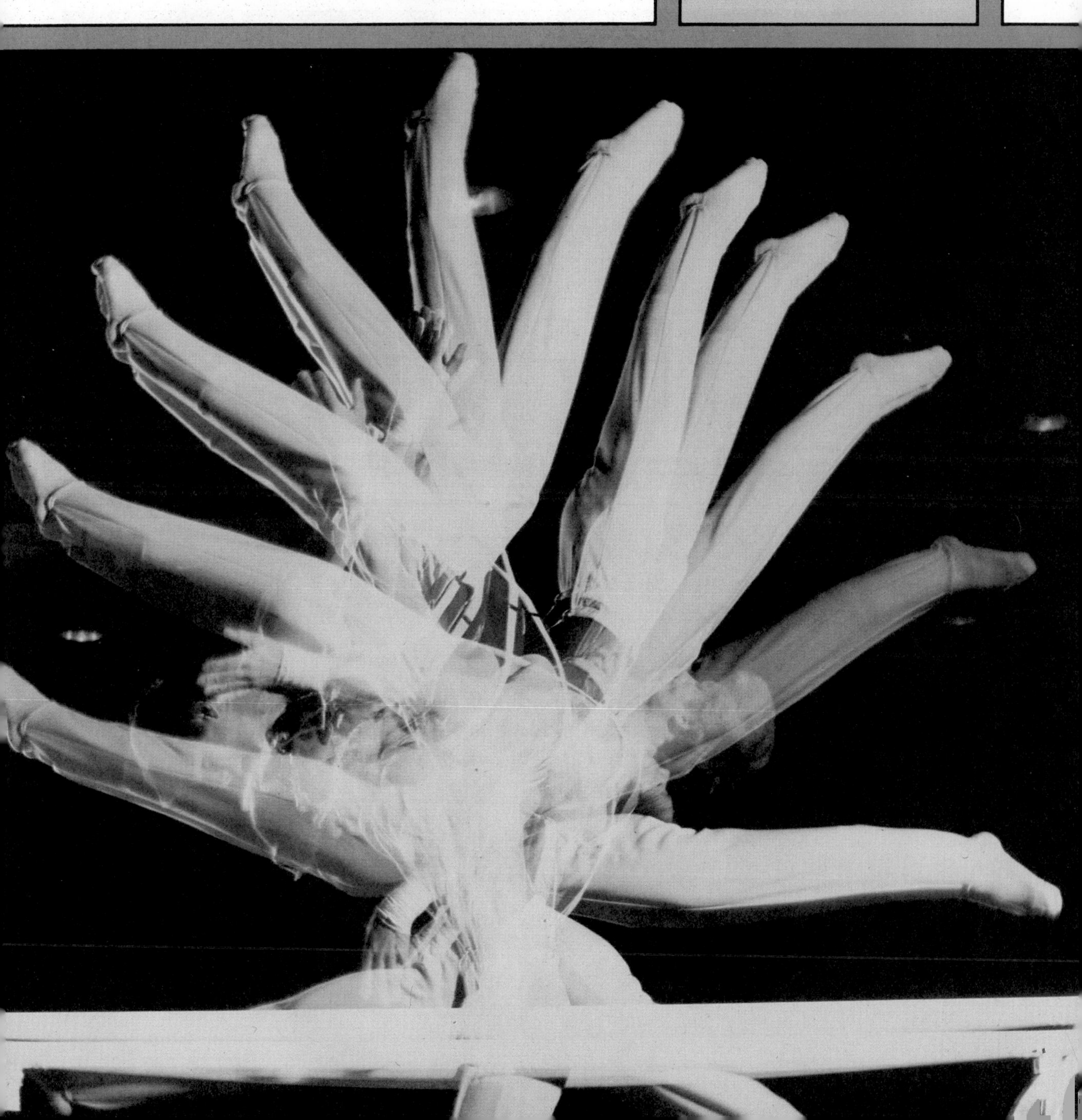

CHAPTER 1

Motion

1-1. SPEED AND ACCELERATED MOTION

While you are reading these words, you are moving faster than a jet plane. Do you feel as if you are moving that fast? Very likely you feel as if you are sitting in one place. The motion is the result of being carried around with the earth as it makes one complete turn each day.

It is hard for you to make observations about the speed of the earth. You know that something is moving only by comparing it to something that is not moving.

When you finish lesson 1, you will be able to:

- Using examples, explain what you mean when you say something is moving.

- Explain how to find the average speed of a moving object.
- Give examples to show that the speed or direction of a moving object may change from one period of time to another.
- ○ Demonstrate some ways in which the speed of a moving object can be measured.

Motion is always observed by comparing the moving object to another object that appears to stay in place. The object that appears to stay in place is a *reference point*. Look at the photos on p. 2. What reference points do you see in these pictures? Every moving thing covers a certain distance in a certain period of time. For example, in 1 min a skateboard rider would move past a certain number of fence pickets of uniform width. Speed is expressed as a measurement of distance moved during a period of time. When the speed of a moving object is measured, distance and time are measured in convenient units. The speed of a running person, for example, might be measured in meters covered each second or meters per second (m/sec). A fast runner might have a speed of 6 m/sec. In metric units, the speed of a car is measured in kilometers per hour (km/hr). The prefix *kilo* added in front of a metric unit, such as a meter, means one thousand times the unit. Therefore, a kilometer means 1,000 m and is equal to 0.621 mi. If a car had a speed of 89 kilometers per hour (km/hr) its speed in customary units would be 55 miles per hour (89 km × 0.621 mi/km). Since the time taken to travel a certain distance in a car is usually not exactly one hour, only the *average speed* is generally given. Average speed is equal to the total distance traveled divided by the time taken. For example, the average speed for a trip of 240 kilometers that lasted 3 hours is $\frac{240 \text{ km}}{3 \text{ hr}} = 80$ km/hr.

Average speed is always indicated by:

$$\frac{\text{distance traveled } (d)}{\text{time taken } (t)} = \text{average speed } (v) \text{ or } \frac{d}{t} = v$$

Motion
A change in position of an object when compared to a reference point.

Find the speed in the following examples:

Distance	Time	Speed
26 m	8 sec	?
96.3 m	3 sec	?

1-1. *Why could you say that the velocity of a moving object is changing, even if it is moving at a constant speed?*

What would be the average speed for a trip of 97.5 km taking 1.5 hr? What is the average speed of a marble that rolls 45 cm in 5 sec?

Any object that is in motion must be moving in a certain *direction* as well as at a certain speed. Direction of motion is just as important as how fast the object is moving. The term **velocity** expresses both speed and direction. Expressing the *velocity* of a moving car requires that both speed and direction be given; for example, 80 km/hr east. See Fig. 1-1. In non-scientific language, the words "speed" and "velocity" are used as if they have the same meaning. You can talk about speed without mentioning direction. However, you should include both speed and direction when talking about velocity.

Velocity
The speed and direction of a moving object.

Most of the time moving objects do not move with a steady speed. For example, when you ride a bicycle you usually pedal for a time, then coast. Your speed increases while pedaling and then drops as you coast. If you pedal uphill, your speed will be slower than when you pedal on level ground. On the other hand, the speed will increase when you go downhill. Because of these changes in speed, your motion on a bicycle trip would be called **accelerated** (ik-**sel**-uh-rate-ud). If the speed of a moving object is found to change from one time period to another, its motion is *accelerated*.

Acceleration
The change in velocity during a given time interval; either speed or velocity—or both—may change.

The word "accelerated" is most commonly used to mean an increase in speed. However, scientists use "accelerated" to mean any change in the velocity of a moving object. Remember that velocity includes both speed and direction of motion. Therefore, a scientist would say that when a bicycle slows down or turns a corner, its motion is accelerated. As you will see in the next lesson, thinking of acceleration as either a change of speed or direction helps you to understand the behavior of moving objects.

When acceleration is measured, often only the change in speed is included. In this case, acceleration is measured as the change in speed divided by the time taken for the change to happen. Then acceleration is given by:

$$\frac{\text{change in speed}}{\text{time taken for change to happen}} = \text{acceleration}$$

For example, suppose a car starts out from a stop sign. It reaches a speed of 36 km/hr in 12 sec. Its acceleration is:

$$\frac{36 \text{ km/hr}}{12 \text{ sec}} = 3 \text{ km/hr each second}$$

This means that the speed of the car increased by 3 km/hr during each second. This acceleration may be written as 3 km/hr/sec. This acceleration unit could be read as 3 kilometers per hour each second.

A common example of accelerated motion is a falling object. Gravity will give a freely falling object an acceleration of 9.8 m/sec/sec. That is, a falling object will increase its speed by 9.8 m/sec during each second it is falling. See Fig. 1-2.

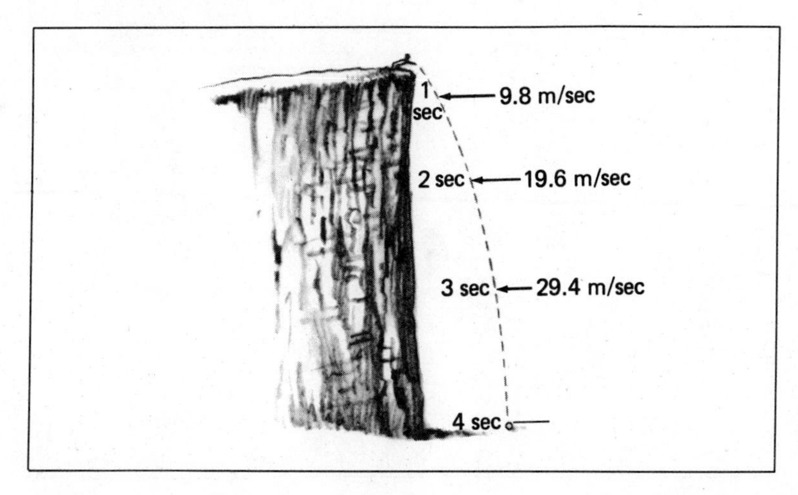

1-2. *What is the speed of the ball at 4 sec?*

1-3. *How does the acceleration of a falling skydiver change when the parachute opens?*

Something falling through the air will pick up speed or accelerate only for a definite time. The resistance of the air causes the acceleration to stop once a certain speed is reached. A skydiver like the one shown in Fig. 1-3 will reach a top speed of about 193 km/hr (about 120 mi/hr). The greatest speed reached by an object falling through air is called its *terminal speed*. All objects falling through the air reach a certain terminal speed.

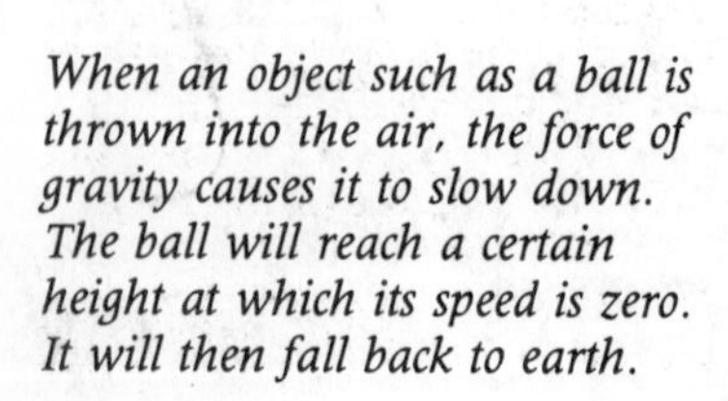

When an object such as a ball is thrown into the air, the force of gravity causes it to slow down. The ball will reach a certain height at which its speed is zero. It will then fall back to earth.

ACTIVITY

Average Speed

A. Obtain the materials listed in the margin.

B. Find a table top or floor space that is smooth and level for a distance of about 1.5 m. You must be able to see the second hand on a watch or clock from the place you choose.

1. Using a marble, how can you tell if the table is level?

C. Place a ruler so that one end is raised about 1.5 cm and the other end is on the table top. See Fig. 1-4.

D. From the upper end of the ruler, roll a marble down its groove onto the table.

E. Determine how far the marble rolls from the bottom end of the ruler in 2 sec. Record this measurement. Do not stop the marble before the 2 sec is up.

F. Repeat this procedure at least three times. Record the distance each time. Finding the distances are your observations.

To find the average distance rolled in 2 sec, add together all the distances measured. Divide this total by the number of measurements.

2. What was the average distance the marble rolled in 2 sec?

To find the average speed of the marble, divide the distance traveled by the time to travel that distance. For instance, if the marble rolled 30 cm in 2 secs its average speed would be $\frac{30 \text{ cm}}{2 \text{ sec}}$ or 15 cm per second $\left(15 \frac{\text{cm}}{\text{sec}}\right)$.

3. What was the average speed of the marble that you rolled?

G. Determine how far the marble rolls from the bottom end of the ruler in 3 sec. Make several more measurements. Record your observations.

4. What was the average distance in 3 sec?

5. What was the average speed of the marble?

6. Copy and complete the following sentence using one of the choices provided in parenthesis. "The average speed during the 3-sec runs was __________ (greater than, the same as, less than) the average speed during the 2-sec runs."

Materials
metric ruler
marble

1-4.

SUMMARY

Motion is observed by comparing the moving object to some reference point that is not moving. Speed is a calculated measure of motion. Average speed is calculated by measuring the distance an object moves and dividing the distance by the measured time it took to go that distance. Speed, distance, and time are measured in convenient units. Velocity expresses both speed and direction of a moving object. Objects in motion do not always move at the same speed. Many factors such as moving up and down hills change the speed of an object. An object that changes from one speed to another is accelerating.

QUESTIONS

Complete the paragraph below by using the following list of terms to fill in the numbered blanks. A term may be used more than once. You will not use all of the terms.

a. accelerating
b. motion
c. reference point
d. speed
e. time
f. 2 m/sec
g. 5 m/sec

In order to observe __(1)__ of an object, its position must be compared from time to time with a __(2)__ that is not moving. The distance moved by an object divided by the __(3)__ it took to move that distance is called __(4)__. For example, a skateboard rider who moved 10 m in 5 sec would have a speed of __(5)__. When an object's speed changes it is __(6)__. A marble rolling downhill would roll faster and faster and thus is __(7)__. The slowing down of an object also means it is __(8)__.

9. In your own words, using complete sentences, describe the difference between velocity and speed.

1-2. FORCES

The time is many years in the future. A spaceship is taking off for a star beyond the sun. It is a journey of many years. The distance is too great even for your imagination. Once the spaceship is clear of the earth, the rocket motors are turned off. The ship then continues to move at the same speed until it reaches its goal.

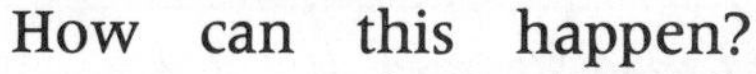

How can this happen? Does a car continue to coast along for years after the engine is shut off? Certain natural rules control the motion of spaceships and automobiles. To understand these rules, you must remember the observations you have already made of moving objects.

When you finish lesson 2, you will be able to:

- ● Predict the behavior of a moving object when a *force* acts upon it.
- ● Explain the effect of the earth's gravitational force on an object.
- ○ Demonstrate that a force is needed to cause a moving object to change its speed or direction.

Your experience tells you that moving objects finally stop. A bicycle coasting on level ground will lose speed unless the rider pedals. All moving objects feel a resistance called **friction.** *Friction* is a force caused by two surfaces rubbing together. Because of friction, a push or a pull is needed to move a book resting on a tabletop. The push or pull must overcome the friction between the book and the tabletop. Wheels can reduce the amount of friction. Friction can also be decreased by using a slippery substance like oil between the rubbing surfaces. But some friction is always present to resist motion. How would moving objects behave if there were no resistance caused by friction?

Friction
The force that slows the motion between two objects in contact with each other.

1-5. *Sir Isaac Newton (1642–1727) was one of the greatest scientific thinkers of all time.*

On earth it is very hard to escape friction. Most of your common observations of moving objects show that it is natural for them to slow and then come to a stop. A spaceship, however, would feel almost no friction. Once started, it would continue to move in a straight path. To change its direction, a push would be necessary. Any push or pull is called a **force.** A *force* can cause a moving object to change its speed or direction. Because friction is everywhere in our natural environment, it took many centuries for scientists to discover this principle. It was first put in the form of a scientific law by one of the greatest scientists, Sir Isaac Newton. See Fig. 1-5. Newton stated this principle as his first law of motion: *Every object remains at rest or moves with a constant speed in a straight line unless acted upon by some outside force.* See Fig. 1-6.

Force
Any push or pull that causes an object to move or to change its speed or direction of motion.

Anyone familiar with Newton's first law of motion knows that a force is present when a change in motion is observed. If you want to coast at a constant speed on a bicycle, no force is needed. Accelerating to a higher speed will require a force supplied by your muscles. To slow down, the brakes must supply a force by creating friction against the moving wheels. Every change in motion requires that some force be supplied. This resis-

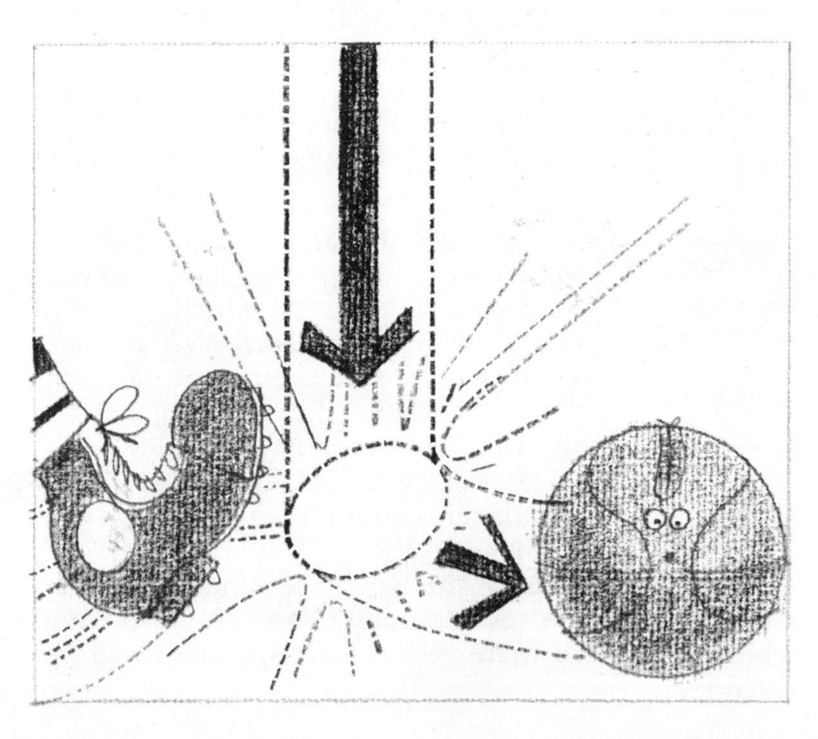

1-6. *An object remains at rest or moves with constant speed in a straight line unless acted on by an outside force.*

tance of objects to any change in motion is called *inertia*. A moving object, in the absence of friction, continues to move because of its inertia.

A knowledge of the first law of motion should remind you always to use a seat belt when riding in a car. Suppose that you are without a seat belt in a car that stops suddenly. You keep moving forward until you feel a force needed to stop your motion. That force will be supplied when you crash into some part of the car such as the windshield or dashboard. A seat belt will make it less likely that you will be injured. It will supply the force needed to hold you firmly in the seat.

The world is filled with forces that cause changes in motion. Some of the most common forces around us are **gravity** (**grav**-it-tee) forces. *Gravity* acts on all objects and pulls them toward the earth's center. It is the force of gravity acting on falling objects that causes them to pick up speed or accelerate. Gravity force acting on a particular object is called **weight.** For example, a bathroom scale measures the gravity force acting on a person in units of pounds. See Fig. 1-7. A pound is a measure of a certain amount of force. In the metric system, force is measured in units called *newtons* (N). A bathroom scale could also measure gravity force (*weight*) in newtons. On such a scale, a 111-lb person would register ~~25 N.~~ 490 N.

Gravity
The force that pulls an object toward the center of the earth.

Weight
The measurement of the force of gravity. In the metric system, gravity force is measured in newtons (N). The customary unit for measuring gravity force is the pound (lb).

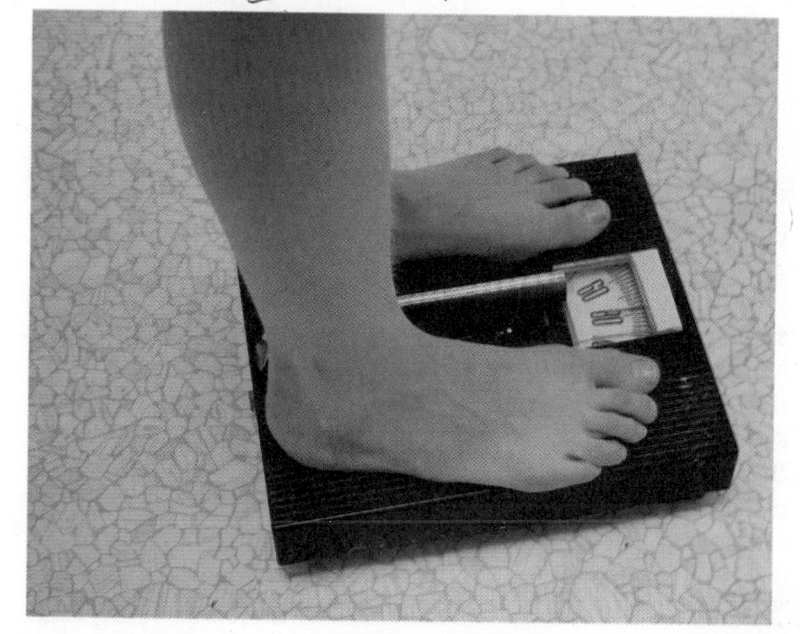

1-7. *A bathroom scale measures the gravity force acting on the person being weighed.*

ACTIVITY

Materials
marble
metric ruler

Forces
In this activity, you will use a set-up similar to that used in the last activity.

A. Obtain the materials listed in the margin.

B. Raise the end of a ruler about 1.5 cm, as before.

C. Starting at the top, roll a marble down the groove in the ruler.
From the bottom of the ruler, measure the distance that the marble rolls in 2 sec. Repeat at least three times. Record.
1. What was the average distance the marble rolled in 2 sec?
2. What was the average speed of the marble?

D. Repeat Step C but raise the end of the ruler about 1 cm higher. Keep everything else the same. Roll the marble several times and record your measurements.
3. What was the average distance the marble moved in 2 sec with this setup?
4. What was the average speed of the marble?
5. The average speed of the marble in step D was (smaller than, the same as, greater than) the speed in step C.
The difference in speed between these two setups shows that a force was acting on the marble?
6. What force do you think was acting on the marble?

SUMMARY

Friction is present in most cases of motion we observe on earth. Without friction, an object like a spaceship would continue its motion in a straight line until it hit something. Sir Isaac Newton called friction a force. Newton's first law of motion explains that every object remains at rest or moves with a constant speed in a straight line unless acted upon by some outside force. If there is a change in the motion of an object, then a force must have been applied. A falling object's motion changes since its speed increases as it falls. The force causing this change is gravity.

QUESTIONS

Unless otherwise indicated, use complete sentences to write your answers.

1. Why does a spaceship moving in deep space continue to move in a straight line with no change in speed after its engines are shut down?

2. In which of the following examples can you be sure that a force is acting? Describe the force in each case.
   a. A book sliding across the table slows to a stop.
   b. An object falls to the floor.
   c. A marble rolling down a ruler speeds up.
   d. A model airplane goes higher as it flies.
   e. An object is whirled around in a circle.

3. Copy and complete in your own words the following incomplete sentence. "A falling object will pick up speed because __________."

1-3. MASS AND WEIGHT

In April 1976, the Apollo 16 spacecraft was on its way to the moon. The three astronauts discovered as had others before them, that anything in the spacecraft that was not tied down had a tendency to float about. Astronauts in a spacecraft experience a kind of "weightlessness" although gravitational forces are still present. When you finish lesson 3, you will be able to:

- Compare mass with weight.
- Give examples to show that the size of the force acting on an object determines how much the object's motion will change.
- Demonstrate that a force is required to cause motion other than in a straight path.

Suppose you were standing on a scale inside an elevator. The scale registers the gravitational force acting on you as your weight, that is, a person with a mass of 50 kg weighs 409 N. As the elevator drops from a high floor, your weight—as registered on the scale—seems to fall to zero. You seem to be "weightless." However, your mass (50 kg) remains the same. In a spacecraft, your weight seems to decrease as you get farther from the earth. For example, if you were twice as far from the earth as you are now, your weight would be one-quarter what it is now. (Remember, the force of gravity decreases with distance.) No matter how far you were from earth, however, your mass would always remain the same. The amount of matter (mass) in an object is the same whether the object is "weightless" or not. For scientific purposes it is necessary to think of the matter in objects apart from their weight. The weight of an object changes somewhat

1-8. *You can do many unusual things when you are "weightless!" What happens to your mass in this case?*

from one location to another, but the **mass** is always the same. The amount of matter in any object is called its *mass*. A book, for example, has the same mass at all times. Even when "weightless" in a spaceship, the book would have the same mass as on the earth. See Fig. 1-8. If you want to think about the weight of an object, it is necessary to say where the weight is measured. On the moon, for example, gravity force on a particular object is much less than on the earth. Your weight on the moon would be only about 0.17 as much as your weight on the earth. However, this would not mean that your mass has changed. Thus, scientists are always careful to distinguish between mass and weight.

Mass
A measure of the amount of matter contained in an object.

In his study of motion, Newton discovered that the size of gravity force is related to mass. This means that

1-9. *The mass of the car is greater than the mass of the bicycle. Therefore, a larger force is required to accelerate the car.*

if you always stay in the same place, objects with more mass will have more weight. As long as you stay at the same place on the earth, you can measure the amount of mass in an object by measuring the gravity force pulling on it. This is the reason that in everyday language mass and weight are used as if they mean the same thing.

Common observations will show that the mass of an object is important when it is moving. For example, think about the force required to cause a bicycle and an automobile to accelerate to 8 km/hr. See Fig. 1-9. A much larger force would be needed for the heavy car than for the lighter bicycle. The mass of an object determines the size of the force needed to produce a certain change in speed.

Suppose you drop a heavy and a light rock at the same time. Will the heavy one fall faster? Aristotle, a famous scientist in ancient Greece, said that is was natural for heavier objects to fall faster. For two thousand years Aristotle's teachings about falling objects were accepted. But in the sixteenth century, the Italian scientist Galileo found the correct answer to the problem. Galileo said that if the effects of air resistance were removed, a feather and a rock would both fall at the same rate. Both would accelerate at the same rate and reach the ground at the same time. The rock with the larger mass does require more force to produce its acceleration. But that larger force is supplied by its greater weight. Thus, gravity force can cause all falling objects to have the same acceleration in the absence of air resistance.

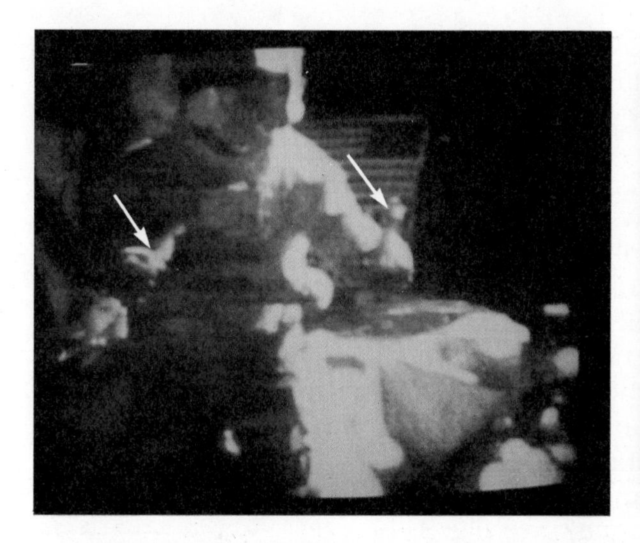

1-10. *An astronaut on the surface of the airless moon drops a hammer and a feather at the same time. Which one will hit the ground first?*

Galileo was not able to test his hypothesis. There was no way for him to remove the effect of air since vacuum pumps were not invented until much later. When vacuum pumps were available and the experiment was tried in the absence of air resistance, Galileo was proven correct. Three hundred years later, an astronaut on the moon also proved that Galileo was right. The astronaut shown in Fig. 1-10 dropped a feather and a hammer on the surface of the moon. Since there is no air on the moon, both the feather and the hammer hit the surface at the same time.

Another example of how the mass of an object is connected to its motion can be discovered by making something move in a curving path. A force is needed to make a moving object follow a curving path. (Remember Newton's first law: An object will continue to move in a straight line unless acted on by an outside force.) The size of the force is related to the mass of the moving object. As the mass increases, the amount of force needed to push an object out of a straight path

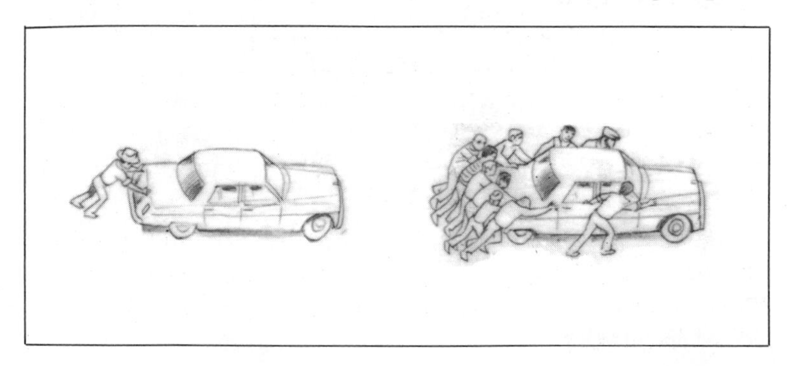

1-11. *The acceleration of an object depends on the size and direction of the force applied.*

also increases. The entire relationship between the mass of an object and the forces acting on it was summed up by Newton in his second law of motion: *The acceleration of an object of certain mass is determined by the size of the force acting and the direction in which it acts.*

To show how Newton's second law of motion works, think about pushing a stalled car to get it started again. If only one person pushes, the force applied to the car will probably not be enough to make the car start. If ten people push, the car accelerates to the required speed. The size of the force determines the acceleration. Remember also that the direction of the force is important. If five people were pushing the car one way and five were pushing the other way, what would happen? See Fig. 1-11.

ACTIVITY

Materials
book
rubber band
metric ruler
string loop, 60 cm around
2 rubber stoppers
heavy string, 1 m

Acceleration

A. Obtain the materials listed in the margin.

B. Measure the length of the folded rubber band loop. Stretch it until it measures 5 cm more than its relaxed length. Notice the feel of the force (pull) you must apply. Now stretch it until it measures 10 cm more than its relaxed length.

1. Which stretch (5 cm or 10 cm) required the greater force?

C. Attach the rubber band to the string loop and fasten the string to your book by looping it over the cover as shown in Fig. 1-12.

D. While measuring its stretched length, pull on the rubber band until the book slides across the level desk. After the book is moving slowly, pull only hard enough to keep the book moving at a constant, slow speed.

2. What is the length of the rubber band while the book moves slowly?

E. Apply a force to the book by stretching the rubber band 3 cm more than your answer to question 2. Notice about how long it takes the book to move 50 cm. Do not try to measure the time.

F. Apply a force to the book by stretching the rubber band 6 cm more than your answer to question 2. Notice about how long it takes the book to move 50 cm.

3. Which force (pull) takes the least time to pull the book 50 cm: 3 cm or 6 cm?

4. Since the book is speeding up while it is being pulled, which force (pull) (3 cm or 6 cm) produces the greatest change in speed?

5. Which force produces the greatest acceleration?

G. Tie the string on the stopper (or eraser). Make several knots. Starting at the center of the stopper measure 25 cm along the string. Tie a knot in the string. Tie another knot at 50 cm.

H. Hold the string at the 50-cm knot.

I. Swing the stopper around your head so that it goes around as near to 10 times in 5 sec as you can make it. Notice how hard it pulls on your hand. We will call this the "standard pull."
Caution: Be very careful you have room so you do not hit your classmates.

J. Add another identical stopper to the string.

K. Again, hold the string 50 cm from the stoppers and swing them around your head 10 times in 5 sec.

6. Is the pull on your hand more than, the same as, or less than the "standard pull"?

L. Remove the second stopper.

M. Hold the string at the 50-cm knot. Swing the stopper 10 times in 5 sec to get the "feel" of it again.

N. Now swing the stopper faster until it goes twice as fast (20 times in 5 sec).

7. Is the pull at the faster speed more than, the same as, or less than the "standard pull"?

If you are not sure of your answer, try it again.

O. Swing the stopper 10 times in 5 sec again to get the "feel."

P. Now hold the string at 25 cm.

In order to have the stopper go the same speed as before, it must go around 20 times in 5 sec. Think about it. To go the same speed in a circle one-half as big, it must go around twice as many times in 5 sec. It must go the same distance in the same time.

Q. Swing the stopper at 25 cm so that it goes around 20 times in 5 sec.

8. Is the pull more than, the same as, or less than the "standard pull"?

9. What three things have you tried in this activity which affect the pull needed to make a stopper move in a circle?

10. Using your own words, write one or more complete sentences telling the effect of changes in mass, speed, and radius on the pull needed to make an object move in a circle.

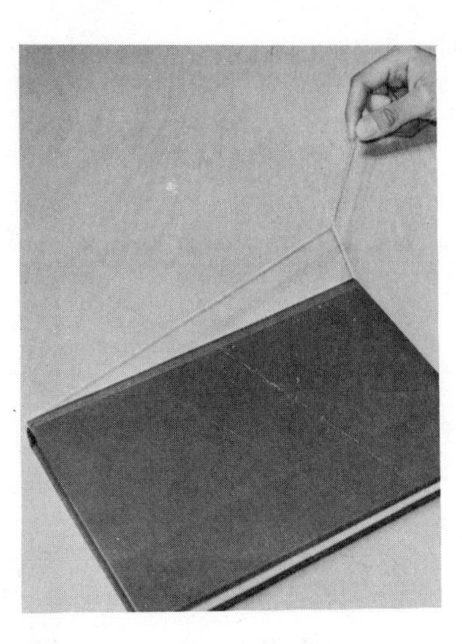

1-12. *Applying a force to a book.*

SUMMARY

Mass describes the amount of matter contained in an object. Mass is usually measured by the gravity force or weight acting on the object. In a spaceship or a falling elevator, you could be "weightless." Anything with a large mass requires a large force to accelerate it to a certain speed. A large force is also required to cause an object with a large mass to follow a curving path. Both of these ideas are expressed as Newton's second law of motion: The acceleration of an object is determined by the size of the force acting and the direction in which it acts.

QUESTIONS

Unless otherwise indicated, use complete sentences to write your answers.

1. The weight of an object may change from place to place; however, its mass stays the same. Explain why this is true.
2. A certain amount of force is applied to each of three objects, one at a time, with the following results: object A did not move, object B accelerated 2 m/sec^2, object C accelerated 5 m/sec^2.
   a. Which object has the greatest mass?
   b. Which object has the least mass?
3. Copy and complete in your own words the following sentence. "A force is needed to make an object move in a curving path because ________."
4. Describe how you could show that a force is required to make an object travel in a curved path.

1-4. BALANCED FORCES

One of the most popular rides in amusement parks shows how the natural laws of motion work. People sit on a large, smooth turntable. The turntable then begins to rotate. Moving slowly at first, the rotating platform picks up speed. As it turns faster and faster, the people riding on it begin to slide off. The idea is to stay on the turntable as long as possible. But the speed soon becomes too fast. Finally, everyone is thrown off. When the turntable stops, everyone climbs on again for another ride.

When you finish lesson 4, you will be able to:

- Use examples to show that an unbalanced force will cause an acceleration (change in speed or direction).
- Identify gravity as the force that causes planets to circle the sun.
- Explain why a force never exists alone.
- ○ Use a force gauge to show that a force cannot exist by itself.

Newton's second law of motion shows that a force is needed to make something move in a curving path. If you sit on a rotating turntable, a force is required to cause you to move in a circular path. This force is supplied by friction against the surface of the turntable. On the smooth surface, the friction force is not large. You will be thrown off when the force of friction is not great enough to cause you to move in a circular path.

1-13. *Can you tell where the force comes from to keep the object moving in a circle?*

Whirling an object around on the end of a string is similar to the turntable ride. See Fig. 1-13. Your pull on the string supplies the force needed to pull the object out of a straight-line path and into a curving path. Your hand also feels another force pulling outward on the string. This test shows an important fact about all forces: No force exists by itself. Forces always exist in pairs.

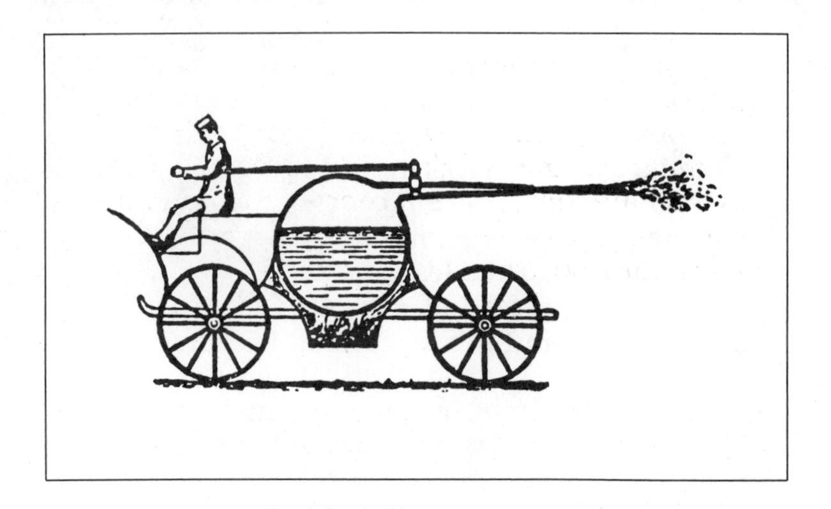

1-14. *This carriage was powered by a jet of steam. Although this practical illustration of Newton's third law of motion was never built, it could have worked.*

The fact that forces never exist alone was also discovered by Newton. It is the principle of his third law of motion: *For every action, there is an equal and opposite reaction.* See Fig. 1-14. The weight of a book pushes down on a table with a certain force. This force could be called the action force. The table pushes up on the book with an equal but opposite force. The force supplied by the table is the reaction force. See Fig. 1-15. If you held the book up by your hand you would feel the reaction force because your muscles would supply it.

1-15. *Newton's third law of motion says that for every force there is an equal but opposite force.*

A jet or rocket engine works on the principle of action and reaction forces. In both cases, hot gases are forced out to the rear of the engine. The engine exerts a force on the gases to push them out; the gases, in turn, exert an equal and opposite force on the engine. This force moves the jet or rocket ahead. Figure 1-16 shows how the same principle can cause you to fall into the water when trying to step out of a canoe.

Newton also studied gravity forces. A force is required to make the planets follow a curving path around the sun. See Fig. 1-17. This force is supplied by the sun's gravity pulling the planets toward it. Newton determined that if the sun attracts a planet, the planet must also attract the sun. Two objects, such as the sun and a planet like the earth, must pull on each other with a gravity force. The size of the gravity force depends on the mass of the planet, the mass of the sun, and the distance between them. Gravity must pull two objects toward each other with a force whose size is related to the masses of the two objects. The sun and earth pull on each other with equal but opposite forces. Gravity force also becomes smaller as the distance between objects becomes greater. See Fig. 1-18. Gravity force becomes larger as the distance between the objects becomes smaller. The planets that are far away from the sun feel less attraction toward the sun. These planets follow less curving paths than they would if they were closer to the sun.

1-16. *Why does the boat move backward when the passenger tries to step out?*

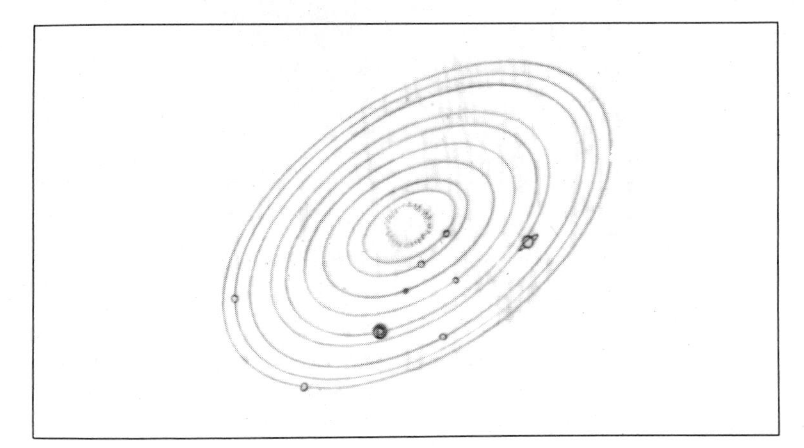

1-17. *The planets follow a curving path around the sun because of the pull of the sun's gravity.*

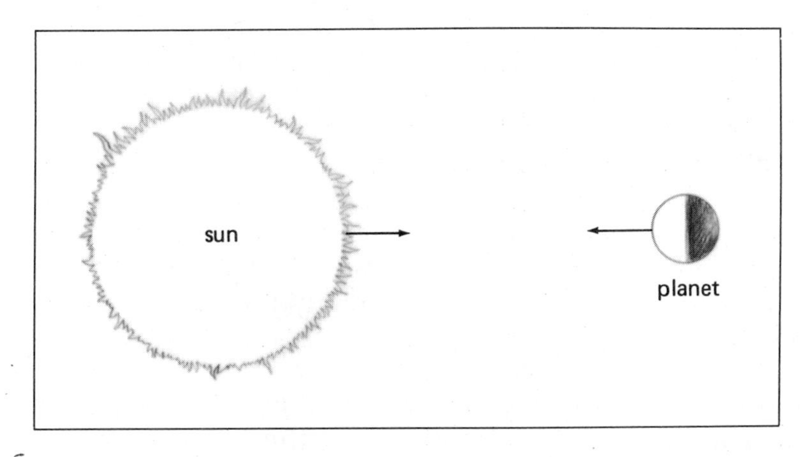

1-18. *The gravitational force on a planet, caused by the sun's pull, is equal and opposite to the gravitational force on the sun by the planet.*

For example, consider the two planets Mars and Venus. Mars is farther away from the sun than Venus. Because of the difference in distance, the sun pulls

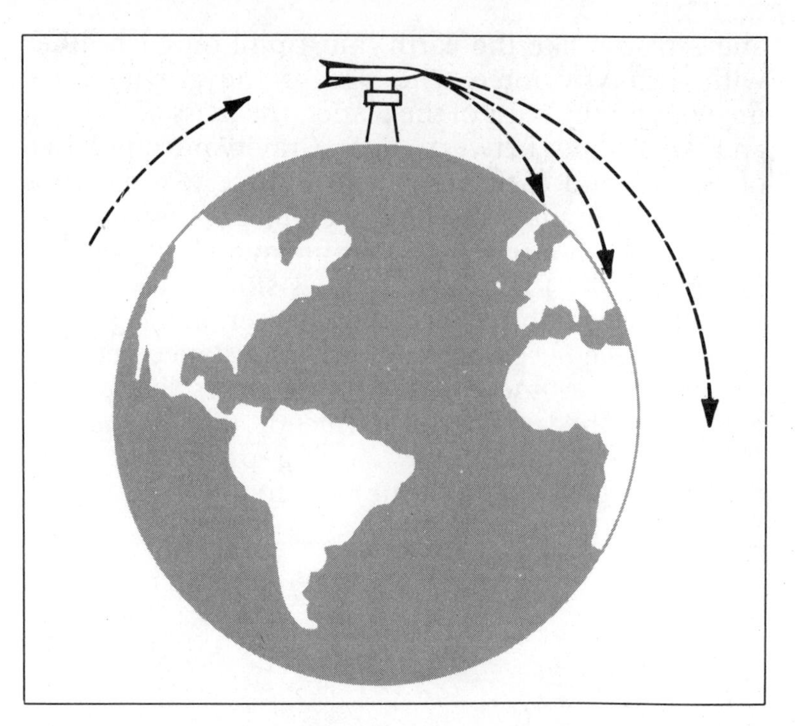

1-19. *If fired at a high enough speed, a cannon shell fired from a mountaintop could become a satellite of the earth.*

more strongly on Venus than on Mars. The path Mars follows around the sun is therefore less curved than the path of Venus.

Gravity force causes the planets to follow curving paths around the sun. In the same way, an object can be made to circle the earth. Suppose a shell was fired from a cannon on a mountaintop as shown in Fig. 1-19. At low speeds, the shell would follow a curving path bringing it back to earth at different distances from the cannon. But at a certain very high speed, the shell would follow a curved path equal to the curved shape of the earth. Then, the shell would always be falling in a path carrying it around, but never hitting the earth. The shell would become a *satellite* of the earth. It would continue to follow an orbit around the earth. Many such satellites have been put into orbit around the earth. They have many purposes, such as helping to relay TV signals over great distances.

Earth satellites put into orbit around the earth are launched by rockets. The satellite must be launched to an altitude of at least 160 km. At this height, the satellite is above the effects of air resistance. At lower altitudes, the high speed needed to keep the satellite in orbit would cause the satellite to burn up in the earth's atmosphere because of air resistance.

ACTIVITY

Force Gauge

This activity will give you the opportunity to make some observations about groups of forces.

Materials
index card
rubber band
paperclip
stapler

A. Obtain the materials listed in the margin.

B. Using an index card, a rubber band, a staple, and a paperclip make a *force gauge* as shown in Fig. 1-20. As you can see, the force gauge is just a rubber band stapled to the card with a paperclip hanging from it.

C. With the gauge hanging as shown, mark the place on the card where the rubber

1-20.

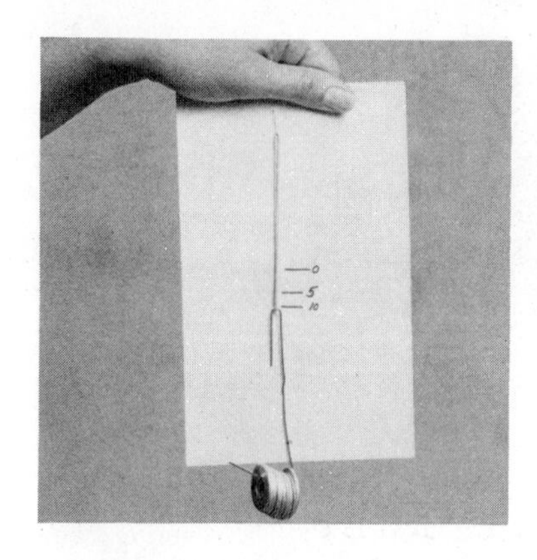

1-21.

band is attached to the paperclip. Write a zero by this mark.

D. Hang one washer on the paperclip. Mark the card again where the rubber band and paperclip join.

E. Add another washer. Mark the card the same as before.

F. Continue adding one washer at a time and marking the card until 30 washers have been added. To make the scale easier to read, write 5, 10, 15, 20, 25, and 30 at the marks. See Fig. 1-21.

G. Hook the paperclip of your gauge onto a table leg or top. See Fig. 1-22.

H. Pull the gauge until it shows a reading of 10. You are going to measure the force on the table leg by using another force gauge.

I. Hook a second gauge onto yours and pull on it. The first gauge should just barely release from the table leg.

1. What is the reading on the second gauge that was necessary to release the first gauge?

J. Reverse the gauges. Use the first gauge to release the second gauge.

2. How do the gauge readings compare this time?

3. In your own words, describe the effect on an object if the forces acting on it add up to zero.

Save the force gauge for future use. You will use it in a later chapter.

1-22.

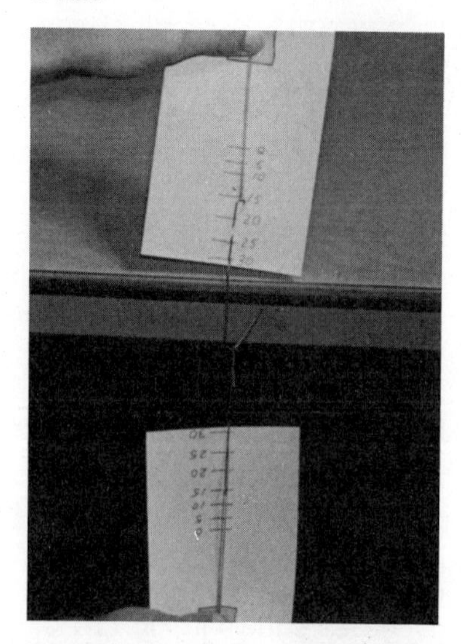

SUMMARY

A turntable ride or a carousel at an amusement park illustrate an important law of motion. A force is needed to make an object follow a curving path. Investigation of forces shows that they do not exist by themselves. Newton's third law of motion describes these combinations of forces. Combinations of forces may be balanced. Unbalanced forces always cause acceleration. Gravity describes a pair of forces that attracts two objects according to their masses and distance apart.

QUESTIONS

Unless otherwise indicated, use complete sentences to write your answers.

1. Are any of the following examples of unbalanced forces? **a.** You whirl an object on a string in a circle and notice the pull on the object through the string. **b.** You calibrate a rubber band force gauge and notice that heavier objects stretch the rubber band more than lighter ones. **c.** You hold a rock in your hand and notice its weight pressing downward. **d.** A spaceship moves in a straight line while accelerating.
2. Newton's strong feeling that forces come in opposite and equal pairs was shown when he reasoned that: **a.** A force is needed to make planets circle the sun. **b.** If the sun attracts a planet, then the planet attracts the sun. **c.** Planets closer to the sun have more attraction for the sun. **d.** Larger planets have a greater attraction for the sun.
3. Describe the force that keeps the planets in their curving paths about the sun.
4. Complete the following sentence: "Satellites above 160 km altitude can remain in orbit because they are above the effects of ______."

Force and Acceleration

Materials
pencil
clock with sweep secondhand
paper tape
index card
force gauge
metric ruler

Purpose

In this laboratory, you will use a skateboard and force gauge to demonstrate the relationship between force and acceleration.

Procedure

A. Obtain the materials listed in the margin. To do this exercise, you will need to practice tapping your pencil on a piece of paper with the point straight down. Time the taps with a sweep secondhand clock. Tap your pencil at as regular a rate as you can. Try to get 4 taps per second. This would mean 20 taps in 5 sec. Try it several times until you can do it. You will be working in pairs on this exercise. You will each need to be able to do all the operations.

B. Set up a card with a hole in it and a strip of paper tape under it. Tape the card to the table top as shown in Fig. 1-23.

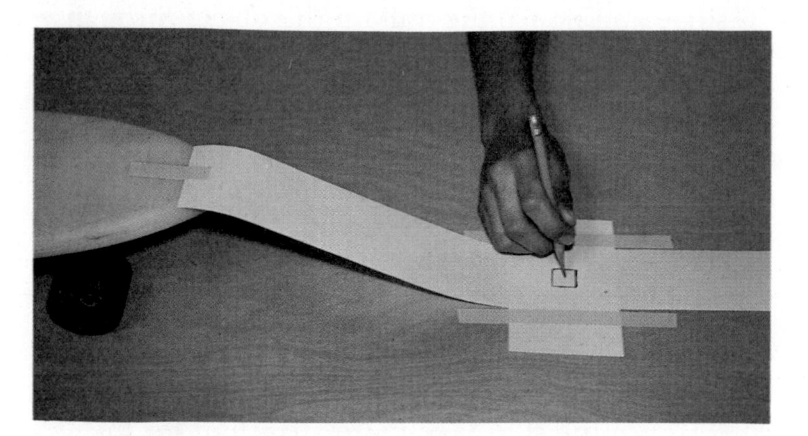

1-23.

C. One person pulls the tape. A second person should tap a pencil in the hole at a regular rate of 4 times per second. Try it so you will know that it works and what it is like. The marks on the tape also show the speed the tape was moving (the speed that the hand pulling the tape was moving).

D. Hook the force gauge you made in lesson 1-4 to one end of a skateboard. Pull on the skateboard with a force equal to 10 washers. Try this 3 or 4 times. With a little practice, you can keep the force quite steady.

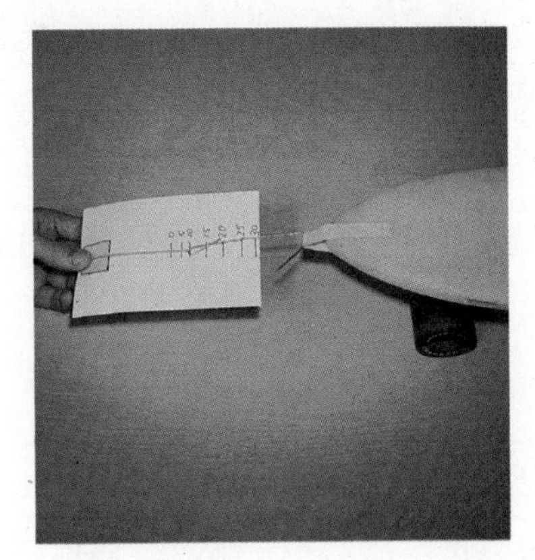
1-24.

E. Now fasten a piece of paper tape to the other end of the skateboard. Keep the skateboard on a smooth, level surface. Put the paper tape under the card with a hole as shown in Fig. 1-23.

F. Have one person tap a pencil in the card hole at a rate of 4 times/sec. The second person pulls on the skateboard with a force equal to 10 washers. Label the tape "10." See Fig. 1-24.

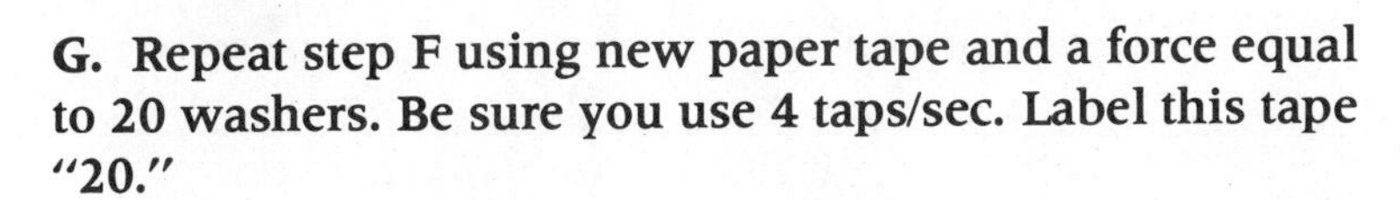

G. Repeat step F using new paper tape and a force equal to 20 washers. Be sure you use 4 taps/sec. Label this tape "20."

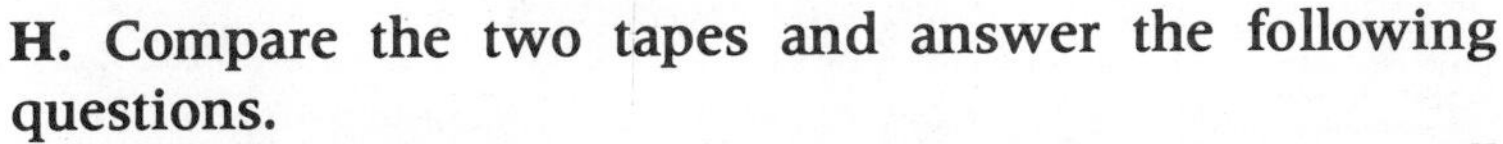

H. Compare the two tapes and answer the following questions.

1. Is the distance moved between marks the same all along each tape?

2. How can you tell when the speed is greatest?

3. What is the distance in centimeters between the seventh and eighth marks on the tape labeled "10"? Record.

4. What is the distance in centimeters between the seventh and eighth marks on the tape labeled "20"? Record.

5. From your answers to questions 3 and 4, which tape shows the greatest acceleration?

Summary

Complete the following sentence with the choice that agrees with your results. "When the force on a skateboard is made greater, the acceleration will be ______." (greater, the same, smaller)

Write a statement predicting what you believe the distance would be between the seventh and eighth marks on a tape if the skateboard were pulled with a force of 30 washers. If you have the time available, check your prediction by doing the experiment.

VOCABULARY REVIEW

Match the number of the term with the letter of the phrase that best explains it.

1. force
2. acceleration
3. speed
4. friction
5. motion
6. gram
7. gravity
8. mass
9. weight
10. satellite
11. velocity
12. inertia

a. A change in position of an object when compared to a reference point.
b. The distance moved by an object divided by the time it took to move that distance.
c. Any push or pull.
d. That which slows the motion between two objects in contact with each other.
e. The change of speed during a given interval of time.
f. A measure of the amount of matter contained in an object.
g. The force which pulls an object toward the center of the earth.
h. The measurement of the force of gravity.
i. A small unit of mass in the metric system.
j. Speed and direction of a moving object.
k. Continues to follow an orbit around the earth.
l. Tendency to stay stationary.

REVIEW QUESTIONS

Complete each statement by choosing the best word or phrase, or by filling in the blank.

1. In order to know that an object is moving, it is necessary to **a.** measure its speed **b.** compare it to some reference point over a period of time **c.** measure its acceleration **d.** show that a force is acting on it.
2. If a runner covered 10 m in 2 sec, the average speed would be ________.
3. Since the speed of a falling object changes as it falls, the motion is called **a.** acceleration **b.** gravity **c.** free fall **d.** constant motion.
4. Using Newton's ideas, you know that a book sliding across the table slows down because frictional ________ between book and table top are present.

5. Objects accelerate as they fall to earth since **a.** there is no force being applied **b.** they are heavier than air **c.** a gravity force is present **d.** there is no frictional force present.
6. The weight of a spaceship as it leaves the earth changes; however, its mass ___________.
7. An object on a string moves in a curving path as you whirl it around your head because **a.** the pull on it is straight ahead **b.** the pull on it is from the side **c.** there is no pull on it **d.** gravity pulls downward.
8. When you push on a skateboard, it moves in the direction you pushed because **a.** it has little friction **b.** it pushes back on you **c.** forces come in pairs **d.** all of these are true.
9. The two things that determine the strength of a gravity force are ___________.
10. Planets move in curved paths about the sun because **a.** a force is pulling them from the side **b.** there is an attraction between the planets and the sun **c.** of gravity force **d.** all of these are true.

REVIEW EXERCISES

Answer the following as briefly and completely as you can.

1. The following measurements were taken as a marble rolled across the floor.

	Distance (*d*)	Time (*t*)
Trial 1	100 cm	2 sec
Trial 2	95 cm	2 sec
Trial 3	105 cm	2 sec

   a. What was the average distance rolled in 2 sec?
   b. What was the average speed?
   c. If the time for a trial was increased to 4 sec, with all else staying the same, how would the average speed change?

2. The average distance a marble rolled in 3 sec after being rolled down a ruler was 63 cm. The ruler was raised 2 cm and the average distance the marble rolled in 3 sec was 84 cm.
   a. What was the average speed for the 63-cm roll?
   b. What was the average speed for the 84-cm roll?
   c. Use Newton's first law of motion to explain why the average speed changed.
3. **a.** Why is it that here on earth a car weighs more than you?
   b. Would a car weigh more than you on the moon?

4. In your own words, describe Newton's second law of motion.
5. When your force gauge was hooked to the table leg and pulled out to a reading of 10 units on the scale, you were applying force on the table leg.
   a. With what force was the table leg pulling on you and the force gauge?
   b. How much would the table leg pull if you were to increase the reading on the gauge to 20?
   c. What would happen if the leg was too weak to withstand that force?
6. a. What kind of path would the earth, or any other planet, have if there were no gravity force between it and the sun?
   b. Is the gravity force a push or pull? Write a complete sentence telling why you think gravity is a push or pull.

EXTENSIONS

1. Look at each figure in chapter 1. List each figure by number in which you think that a force is being applied. For each of these figures describe the force you believe is being applied.
2. Find out from your teacher how to calculate the acceleration shown on the tapes you produced in the laboratory. Calculate the acceleration shown on each tape. Explain why the acceleration is different for the two tapes.
3. In recent years, NASA has sponsored many explorations to other planets. Some of the data gathered suggest that Jupiter and Saturn have more satellites (moons) than previously thought. Find out how many moons each is now known to have. How were the recent satellites discovered?

CHAPTER 2

Energy

2-1. SPENDING ENERGY

When the pyramids were built, massive blocks of stone were raised hundreds of meters from the ground. Huge amounts of energy must have been spent in building these monuments. This lesson deals with the forces involved in moving an object over a distance.

When you finish lesson 1, you will be able to:

- Explain what is meant by *work* and give several examples.
- Describe the relationship among *energy,* force, and work.
- Explain how unbalanced forces result in changes in motion.

○ Demonstrate by using a force gauge that changes taking place in the physical world involve forces.

2-1. *The weight of the student is balanced by the force of the chair pushing up.*

2-2. *The force necessary to lift the student's weight up the stairs is supplied by her leg muscles.*

At least one force is pulling on you this very minute. It is the force called gravity. The size of the gravity force is equal to your weight. If you are sitting still and not moving, an equal force is pushing on you in the direction opposite to that of the gravity force. This equal but opposite force is probably supplied by a chair. See Fig. 2-1. Now think about climbing a flight of stairs. Your muscles must supply a force to lift your body up each step. See Fig. 2-2. Sitting in a chair and climbing stairs are examples of situations in which forces are acting on your body. The force provided by the chair does not cause you to move. You will not be concerned with this type of force in this lesson. Instead, you will be looking at the forces that cause motion.

Many changes that take place in the physical world are the result of forces. What is the difference between forces that cause changes in motion and those that do not?

Even if you pushed very hard against a wall, you would probably not be able to move the wall. On the other hand, if you pushed a book just as hard, you certainly would expect to see the book move. See Fig. 2-3. The difference between the two forces lies in the fact that the force moving the book is said to accomplish some **work.** When a force moves an object, the amount of *work* done is equal to the size of the force multiplied by the distance the object was moved.

Work
The work done by a force is equal to the size of the force multiplied by the distance through which the force acts.

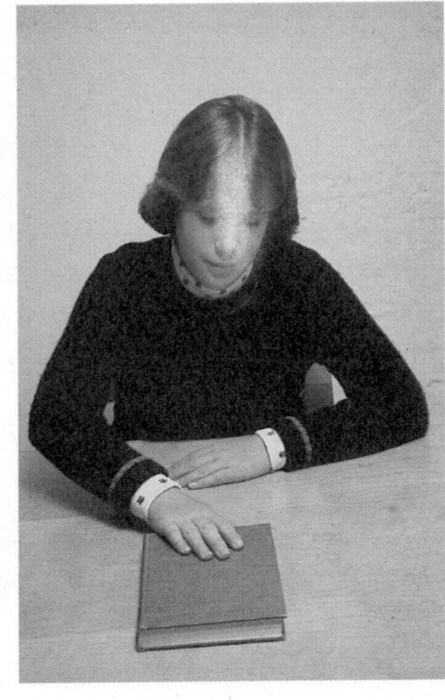

2-3. *In moving the book, some work was done.*

In the metric system, force is measured in units called *newtons* (N). A newton is not a large amount of force. If you push down on a bathroom scale with a force of 1 N, the scales would register about 1/4 lb. Suppose that a force of 2 N is needed to push a book 1 m across a table. The amount of work done is:

$$W = F \times d$$
$$= 2\text{N} \times 1 \text{ m}$$
$$= 2 \text{ newton-meters } (\text{N} \bullet \text{m})$$

Work equals force multiplied by distance: $W = F \times d$.

If a force does 6 N• m of work in moving an object 3 m, how much work would the same force do in moving the object 12 m? Show how you got your answer.

In climbing stairs, your muscles do more work if they must act over a very long distance. Also, a heavy person does more work in climbing stairs than a lighter person going the same distance. This is because of the larger force required to lift the heavier person. Each kilogram of mass is attracted toward the earth with a gravity force of 9.8 N. This means that for each extra kilogram, a person must supply an additional 9.8 N of force to go a certain distance up the stairs.

What do you do to an object when you do work on it? The answer to this question is the cause of all changes that take place in the physical world. The answer is that when you do work on an object you change its **energy.** For example, if you lift a book, you are doing work on it and, therefore, you are adding *energy* to the book. The amount of energy added can be determined by measuring the amount of work done.

Energy
The property of matter that enables it to do work.

Sometimes, we need to know how fast work is being done. For example, almost anyone can lift a brick. See Fig. 2-4. Given enough time, you could lift—one at a time—enough bricks to make a building. The speed with which work is done often determines what kind of job you can do. The rate at which work is done is called **power.** For example, the amount of *power* you would use to move a pile of bricks one brick at a time would be very small. If work is done at a rate of 1 newton-meter per second (1 N•m/sec), the power produced is one *watt* (W). The watt is the unit for power in the metric system. A larger unit of power is a *kilowatt* (kW); 1 kW = 1,000 W.

Power
The rate at which work is done. Power = work ÷ time.

2-4. *When you move an object such as a brick, you do work. The faster you work, the more power you produce.*

The rate at which an electric appliance uses energy is usually expressed in watts (W). For example, a light bulb might be labeled 60 W or a television labeled 250 W. The total amount of electric energy used by an appliance depends on how long it operates. A 60 W light bulb burning for 1 hr uses 60 W × 1 hr = 60 watt-hour (W–hr) of electricity.

The total electrical energy used in a house is measured in kilowatt-hours by a meter such as that shown in Fig. 2-5. The meter reading from left to right is 5856. Each month the electric power company reads the meter. Subtracting each monthly reading from that of the previous month tells how many kilowatt-hours of electricity were used during the month.

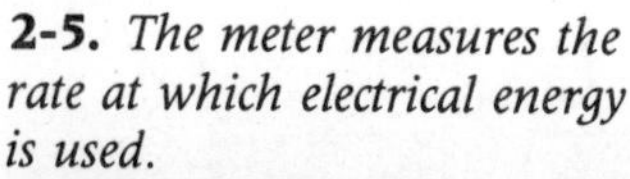

2-5. *The meter measures the rate at which electrical energy is used.*

Work and Energy

A. Obtain the materials listed in the margin.

B. Hook a force gauge onto the leg of a table and pull with a force of 10 units. See Fig. 2-6. The amount of work done is the force acting and the distance the object moves ($W = F \times d$).

1. Does the table move?
2. How much work did you do? You used energy in your own body while you were pulling on the force gauge. However, if you did not move the table, you did not do any work on the table.
3. Does the work done tell you how much energy you have used?

C. Now hook the force gauge onto a small book. See Fig. 2-7.

D. Move the book 10 cm by applying the needed force.

4. Did you do any work?
5. Was a measurable amount of energy used in moving the book?

E. Now move the book 20 cm using the same amount of force, as in step D.

6. How does the amount of work done to move the book 20 cm compare to the amount of work needed to move the book 10 cm?
7. Copy and complete the following sentence by supplying the missing words: "The energy needed to move the book 20 cm is ______ (greater, less) than the energy needed to move it 10 cm because while the force is ______ (greater, the same, less) the distance is ______ (greater, less)."

If you moved two books 10 cm each, you would do twice as much work as you would to move one book the same distance. Review your answers to questions 1–7.

F. Copy and complete the following statements:

8. The amount of work done on a book ______ (increases, decreases) as the force needed to move it increases. The work ______ (increases, decreases) as the distance the book is moved increases.

Materials

force gauge (from Chap. 1) or spring scale
metric ruler

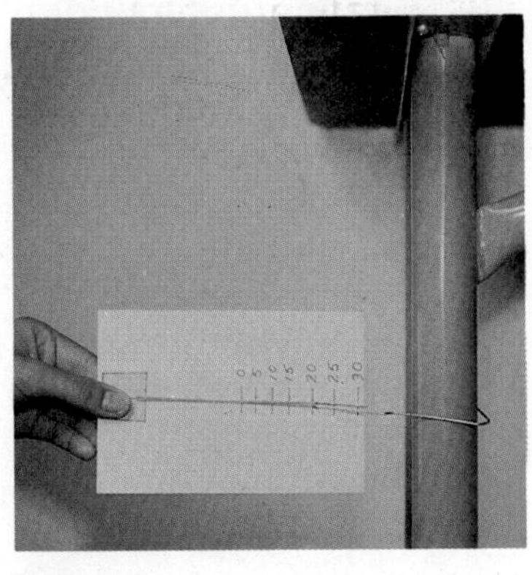

2-6.

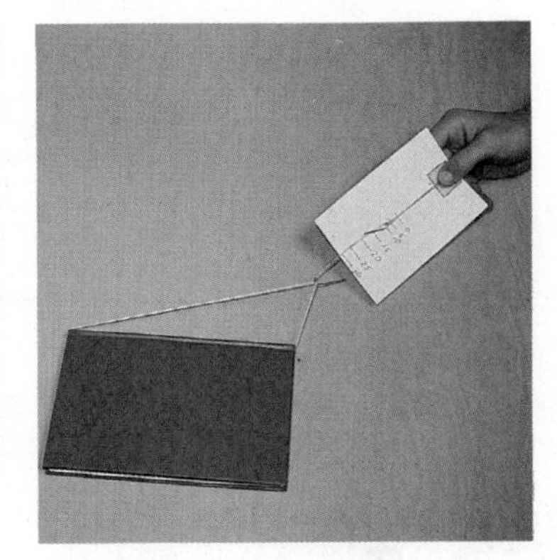

2-7.

SUMMARY

The ancient Egyptians must have done a great deal of work in building the pyramids. When a force moves an object through a distance, work is done. The amount of work depends on the size of the force and the distance over which the force acts. Energy is always spent when work is done. Power describes the rate at which work is done.

QUESTIONS

Use complete sentences to write your answers.

1. A force gauge attached to a book lying on the table shows a reading of 10 units.
   a. Is force being applied to the book?
   b. What additional information do you need to know to tell whether work is being done on the book?
   c. If the force applied by the gauge causes the book to move, how can you tell the amount of work being done?

2. A force is present in each of the following examples. In which of them is measurable work being done due to the force?
   a. You push as hard as you can on the wall of the room.
   b. You pull the door open.
   c. A nail refuses to come out of a piece of wood as you pull on it.
   d. You lift a book from the floor and place it on the table.

3. In your own words, explain the relationship of work, force, energy, and power.

2-2. SIMPLE MACHINES

Suppose all the machines in the world went on strike. At one moment all machines would stop working. City streets would be filled with stalled vehicles. You would not be able to watch TV, tell the correct time, or even open a can of food. A great deal of energy would still be available, but you could not make much use of it. Machines are an essential part of our lives. In this lesson, you will learn how machines are used to apply energy.

When you finish lesson 2, you will be able to:

- Explain what a *simple machine* is and give three or four examples.
- Demonstrate by specific examples that a simple machine is a way of transferring energy.
- Explain how a machine can produce only as much work as is put into it.
- ○ Using a ruler and a pencil, build a simple lever.

You could think of yourself as a storage place for energy. Your body is like a bank that receives energy from food and later spends this energy in many activities. Each time your muscles deliver a force that causes an object to move, you are spending some of your stored energy.

Sometimes your muscles are not able to give enough force to do the work that you want. For example, suppose you want to move a very heavy rock. You might not be able to move it at all by pushing on it. However, by applying the same force to a bar placed a certain way, you can move the rock. Can you explain why? See Fig. 2-8.

2-8. *Could these fishermen move their boat as easily without using the pole?*

Lever
A rigid bar that turns on a fixed point and changes the direction and size of a force applied to it.

Simple machine
A device that can be used to change the direction and size of forces.

The rigid pole or bar used to move a load is called a **lever.** It might seem that *levers* are able to supply more energy than is spent. This is not true. The work done in moving one end of a lever is always equal to the work done on the load at the other end. This means that a lever only transfers energy. A lever does not increase energy. For example, suppose you wanted to move a heavy rock. You could either try to lift the rock by hand or use a lever. In either case, you would do the same amount of work. (Remember, work depends only on the force applied and the distance moved.) The advantage in using a lever is that you do not have to use as much force. By using a lever, you can do the same amount of work using less force. When a hammer is used to pull a nail, the long handle makes a lever. The head of the hammer applies a stronger force to pull the nail than you apply on the handle. A lever is an example of a **simple machine.** There are different kinds of *simple machines.* All simple machines are alike in that they change the size or direction of forces applied to them. However, no machine is able to deliver more energy than is put into it.

A pole or bar that is used as a lever always turns on some fixed point called a *fulcrum.* The *fulcrum* is the place where the lever is supported as it turns. The position of the fulcrum determines how the lever works. For example, the pole used as a lever shown in Fig. 2-8 has the fulcrum placed between the effort applied and the resistance to be moved. This kind of lever is called a *first-class lever.* Another example of a first-class lever is shown in Fig. 2-9. A *second-class lever* has the resistance located between the effort and the fulcrum. A nutcracker is an example of a second-class lever. Another second-class lever is shown in Fig. 2-9. A lever in

2-9. *Many examples of all three classes of levers are used everyday. Classify each example shown here as first-, second-, or third-class lever.*

which the effort is applied between the resistance and the fulcrum is called a *third-class lever*. A shovel is a third-class lever. Figure 2-9 shows another example of a third-class lever.

All simple machines like levers change forces in some way. For example, a lever can change the *size* of a force. The extent to which a lever can change a force depends upon two relationships: (1) the distance between the effort and the fulcrum; and (2) the distance between the fulcrum and the resistance. A first-class lever with the fulcrum placed close to the resistance magnifies the effort force applied. See Fig. 2-10. Such a lever requires an effort force that is less than the force applied to the resistance. However, the effort force must move farther than the distance the load is lifted. This means that the input work (effort force × effort distance) is equal to the output work (resistance

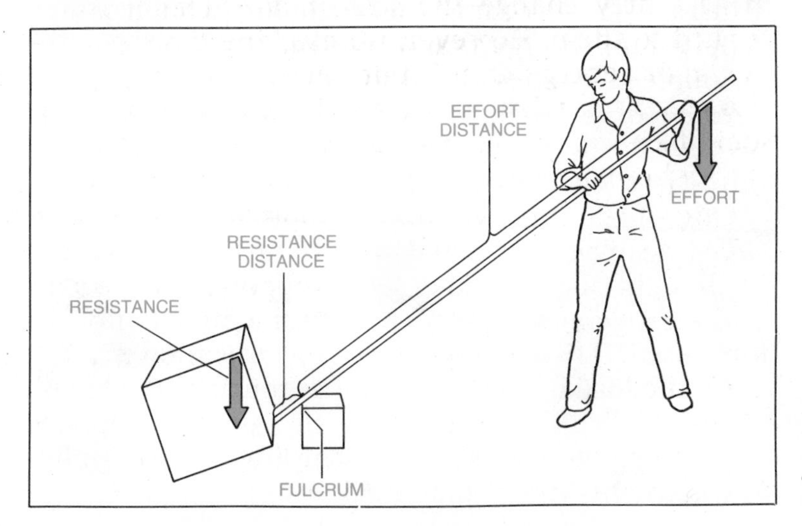

2-10. *A lever used to lift a heavy object must have its fulcrum located close to the resistance.*

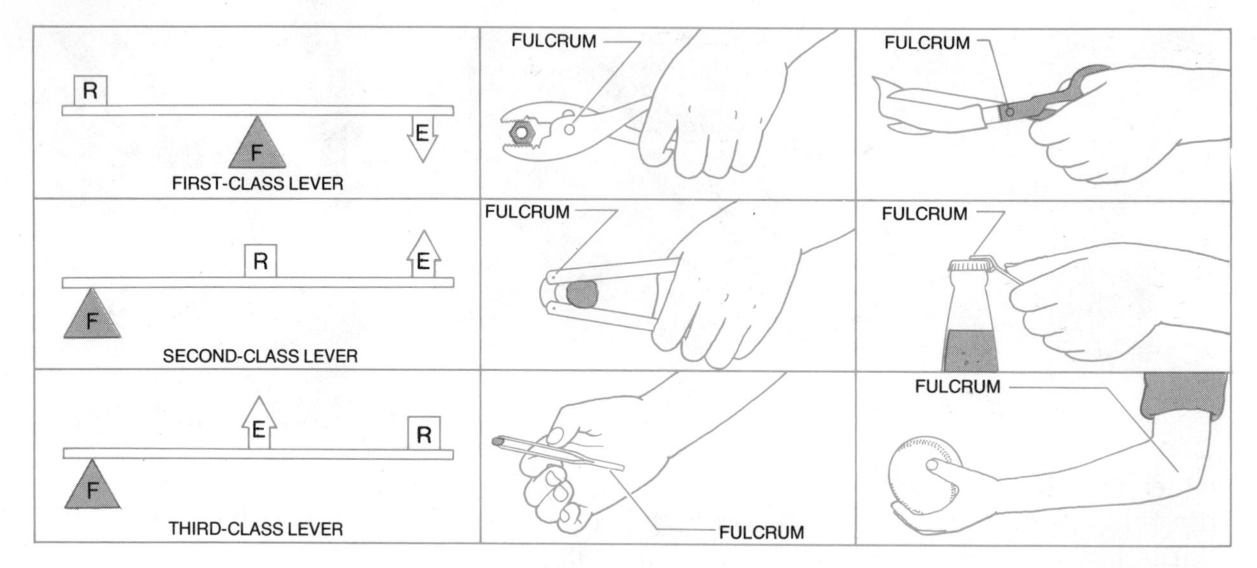

2-11. *Each class of lever changes the direction of the applied force in a different way. F = fulcrum, R = resistance, E = effort.*

force × resistance distance). Suppose, for example, that you use a lever with the fulcrum placed 1 m away from the load. Your part of the lever is 5 m away from the fulcrum. You push down with a force of 2 N. Then the resistance force applied to lifting the load can be found from:

$$\text{resistance force} \times \text{resistance distance} = \text{effort force} \times \text{effort distance}.$$

$$\text{resistance force} \times 1\text{ m} = 2\text{ N} \times 5\text{ m}$$

$$\text{resistance force} = \frac{2\text{ N} \times 5\text{ m}}{1\text{ m}} = 10\text{ N}$$

The resistance force is five times larger than the effort force you applied. However, your effort force had to move five times farther than the load was lifted. Thus, a lever can change the size of forces but it cannot supply more energy than is put in.

Simple machines can also be used to change the *direction* of a force. For example, a first-class lever reverses the direction of the applied force. Second- and third-class levers do not. See Fig. 2-11.

One of the most common simple machines is the *pulley*. A single pulley, like that shown in Fig. 2-12, can be thought of as a kind of lever supported at the axle. This arrangement simply changes the direction of the force used. If two pulleys are arranged as shown in Fig. 2-13, the load is lifted only half as much as the length of rope pulled down. Then, ideally, the force applied must be equal to half the weight lifted. Other pulley arrangements can also be made.

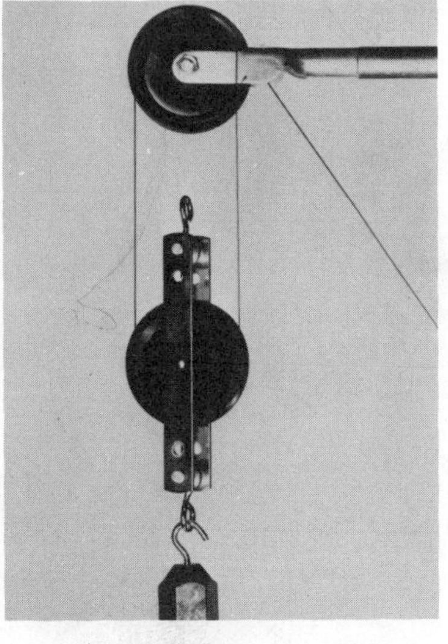

2-12. *A example of a fixed pulley.*

2-13. *One fixed and one movable pulley.*

Another simple machine is the *gear wheel.* If two gears of equal size operate together, energy is transferred. See Fig. 2-14. There is no change in the size of the forces used. If the gear wheels are of different sizes, also shown in Fig. 2-14, the larger wheel always turns more slowly than the smaller. This means that a point on the outside of the larger wheel moves through a shorter distance than does a point on the outside of the smaller wheel. The turning force supplied by the larger wheel is greater than the force applied to the smaller wheel.

2-14. *Gear wheels have many uses in machinery, for example, clocks and watches.*

One kind of simple machine uses liquids to transfer forces. It consists of two pistons of different sizes. See Fig. 2-15. When the smaller piston is pushed down, the liquid pushes up on the larger piston. The larger piston moves up a shorter distance than the piston

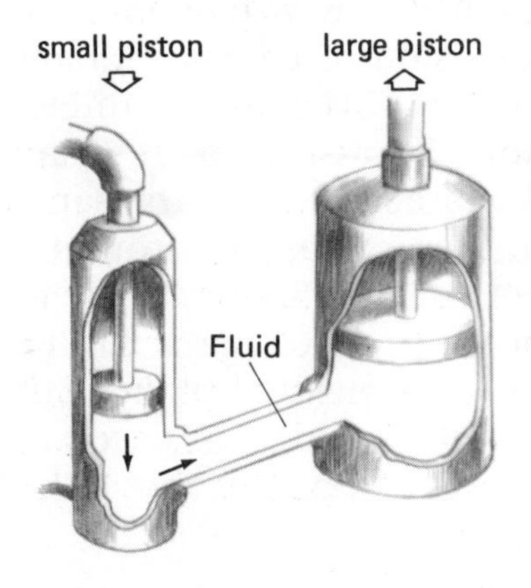

2-15. *How does the downward force on the small piston compare with the upward force on the large piston?*

2-16. *Inclined planes are used in many ways to help people.*

2-17A. *A screw can be made by wrapping a wedge-shaped piece of paper around a pencil.* **B.** *A wedge can be used to split logs.*

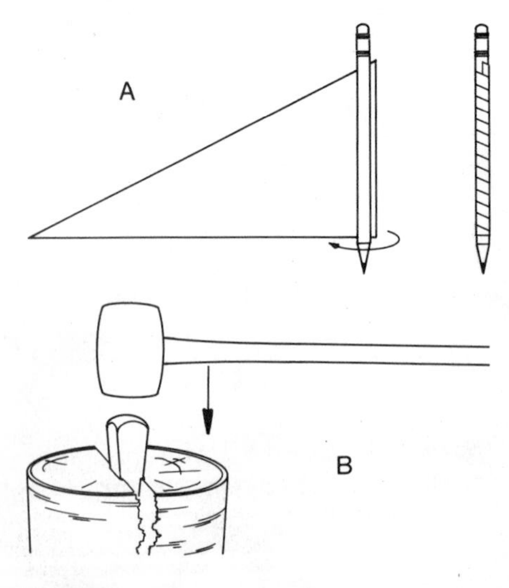

moves down. The force delivered by the large piston is greater than the force applied to the smaller. A machine that works this way is called a *hydraulic* (hie-**draw**-lik) *press.* The brakes on most cars use the hydraulic press.

An *inclined plane* is a simple machine with no moving parts. An inclined plane spreads out the distance over which the applied force must act. This is shown in Fig. 2-16. With an inclined plane, it is possible to lift a load using less force than would be needed to lift the same weight directly. An inclined plane wrapped around a cylinder makes a screw. See Fig. 2-17. Screws are very useful because the length of the inclined plane can be spread over a large distance. Two inclined planes put together make a *wedge.* When a wedge is used, as shown in Fig. 2-17, the two inclined planes move the resistances apart on each side.

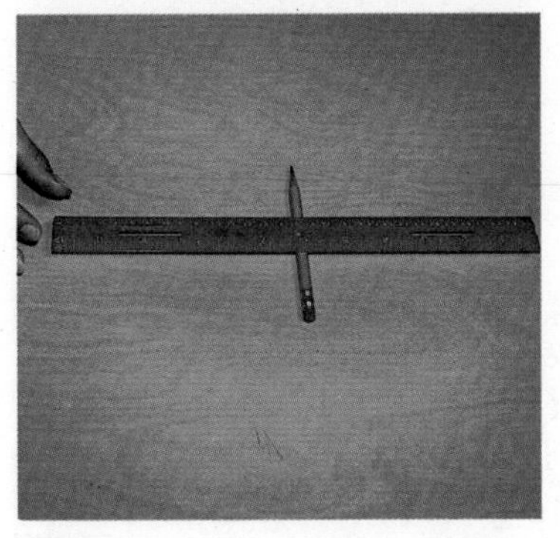

2-18.

Levers

A. Obtain the materials listed in the margin.

B. Balance the metric ruler on a pencil. Be sure to place the pencil on a flat, level surface. See Fig. 2-18.

C. Read the place on the ruler, in centimeters, where the pencil is balanced.

1. What is the reading at which the ruler balances?

D. Very carefully place two washers on the ruler. The centers of the holes should be 7.0 cm to the left of the balance point.

2. What is the reading on the ruler where you put the washers?

E. Now place one washer to the right of the balance point so that it brings the ruler back into balance. Read the location on the ruler at the center of the washer.

3. What is the location on the ruler of the single washer?
4. How far is this location from the balance point?
5. A small weight (one washer) can balance a large weight (two washers) on a ruler if the distance from the balance point to the smaller weight is ______ (larger than, smaller than, the same as) the distance from the balance point to the larger weight.

F. Now tilt the ruler to make the ends move up and down. See Fig. 2-19.

G. Note the distance the single washer moves. Compare this distance with the distance the two washers move.

6. Is there a difference in the two distances?

The amount of work done by the single washer is equal to the work done to lift the two washers. (Remember, $W = F \times d$.) One washer balances two washers on a ruler. You tilt the ruler. The work done by the single washer equals the work done on the two washers.

7. In your own words and using complete sentences, explain why the input work is equal to the output work. (Hint: Are the distances moved the same?)

Materials
metric ruler
3 washers
pencil

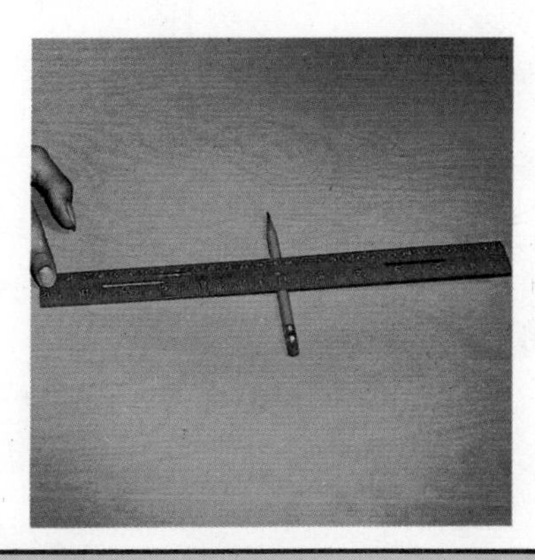

2-19.

SUMMARY

Suppose you have a friend who weighs twice as much as you. You might have trouble lifting that friend off the ground by yourself. If your friend sat on a seesaw, you could easily lift him or her off the ground by pushing down on your end. A seesaw is an example of a lever. The amount of work applied to a lever is equal to the work taken out if you ignore loss of energy due to friction. This means that a lever only transfers energy. It does not change the amount. All simple machines transfer energy. Types of simple machines include the lever, pulleys, gears, hydraulic press, and the inclined plane.

QUESTIONS

Use complete sentences to write your answers.

1. Describe how you could use a lever to demonstrate that a simple machine is a way of transferring energy.
2. A simple machine is used to lift a heavy object 1 m high by applying a downward force. The downward force is applied through a distance of 2 m. Does this mean that twice the work is done in moving the heavy object? Explain your answer.
3. How much work is done to lift a 10-N object 3 m?
4. If a lever was used and the force applied was 5 N, how far would this force have to move to lift the object described in problem 3?
5. In your own words, describe what is meant by work.
6. What is meant by work input equals work output?

2-3. CONSERVATION OF ENERGY

Why will King Kong have more energy when he reaches the top of the building than he had on the ground? After reading this lesson, you will be able to answer this question. Energy is an important part of the world you live in. In this lesson, you will learn about two different forms of energy.

When you finish lesson 3, you will be able to:

- Use examples to explain the difference between *potential* and *kinetic* energy.
- Demonstrate how one form of energy can be changed into another form.
- Use examples to show that energy is neither created nor destroyed.
- Use a diagram to explain the change from potential to kinetic energy.

Try this experiment: Stand with your feet firmly on the floor. Hold a book in both hands. With your eyes closed, slowly raise and lower the book several times. A force pulls the book down. This force is the weight of the book caused by gravity. Suppose that instead of gravity pulling on the book, a spring holds the book to the floor. See Fig. 2-20. If you try hard, you will be able to picture the spring and see it stretch as you raise the book. Now think about the energy added to the spring as it stretches.

A spring that is pulled out of its normal shape has energy stored in it. This stored energy cannot be seen. We know it is there because the stretched spring can apply a force over a distance to another object as it returns to its normal shape. Energy that is stored in an

2-20. *The force of gravity pulling down on the book can be thought of as a spring pulling down on the book.*

2-21. *On which board would the diver have more potential energy?*

Potential energy
Energy stored in an object as a result of a change in its position.
How much more potential energy does a book have if it is raised 2 m rather than 1 m?

2-22. *When the book is allowed to fall, the stored potential energy is changed into kinetic energy.*

Kinetic energy
Energy that moving things have as a result of their motion.

object as a result of a change of position is called **potential energy.** A spring gains *potential energy* when it is stretched. A book that is lifted against the force of gravity stores energy in much the same way as a stretched spring. The energy spent by your muscles in raising the book is stored in the book as potential energy. More energy is needed to lift the book to a higher position. The amount of potential energy stored in the book increases as the book is raised through a greater distance. Also, more energy is needed to lift a heavy book than a light one. The amount of gravitational potential energy stored in an object is a result of its weight and the distance it is lifted. See Fig. 2-21. Thus, King Kong has more energy on top of the building than on the ground.

When relaxed, a stretched spring releases the potential energy stored in it. A book held above the floor releases its potential energy if it is allowed to fall. See Fig. 2-22. The potential energy stored in the book held above the ground is changed into energy of motion as it falls. Energy that an object has as a result of its motion is called **kinetic energy.** An object that is moving in any direction has *kinetic energy*. See Fig. 2-23. The kinetic energy of a falling book can be changed into another form of energy if the book stops moving. For example, the kinetic energy of the falling book might result in possible damage to the book, the floor, or your foot!

2-23. *Which child has the greatest amount of potential energy? kinetic energy?*

2-24. *A roller coaster goes through many energy changes as it moves.*

Where have you seen energy changed from one form to another? See Fig. 2-24. On a roller coaster, potential energy, when the car is on top of a hill, changes into kinetic energy when the car swoops down. The kinetic energy can then be changed back into potential energy as the car climbs another hill. Of course, some energy is used to overcome friction losses.

Suppose you dropped a book on your foot. How many times as much energy would be delivered to your foot if you dropped two books from the same height?

Law of conservation of energy
A natural law that says energy cannot be created or destroyed but may be changed from one form to another.

The gravitational potential energy of an object can be changed into kinetic energy. Kinetic energy can be changed back into potential energy or into some other form of energy. These energy changes occur throughout the natural world. They are summarized in the **law of conservation** (kon-ser-**vay**-shun) **of energy.** This law says that *energy is never created or destroyed but is only changed from one form to another*. For example, the potential energy stored in a book held above the floor is determined by the weight of the book and the distance it was lifted above the floor. The *law of conservation of energy* says that if the book is dropped to the floor, its kinetic energy when it hits the floor is equal to its original potential energy.

A book on a high shelf has a certain amount of potential energy. The book has this energy as a result of its position above the floor. If the book falls from the shelf, the potential energy is changed into kinetic energy. According to the law of conservation of energy, as the book's potential energy decreases, its kinetic energy increases. When the book hits the floor, all of its potential energy has been changed into an equal amount of kinetic energy.

ACTIVITY

Potential and Kinetic Energy

Materials
none

A. In Fig. 2-25, you see a roller coaster car going down a steep hill. Use the lettered positions A, B, C, and D to answer the following questions.

1. At what position does the car have the greatest potential energy?

2. At what position does the car have the least potential energy?

3. At what position does the car have the greatest kinetic energy?

4. At which point, B or C, does the car have more kinetic energy?

5. In your own words, explain the relationship between potential and kinetic energy for a roller coaster.

B. Figure 2-26 shows a pendulum swinging. The photo is a time exposure made with a strobe light flashing on at equal time intervals to show the position of the bob. It is traveling fastest where the spacing is greatest.

6. The pendulum bob is moving fastest at the _______ (center, end) of its swing.

7. The pendulum bob is highest above the table at the _______ (center, end) of its swing.

8. Where does the pendulum have the greatest kinetic energy?

9. Where does the pendulum have the greatest potential energy?

10. Does each increase in kinetic energy occur when there is a decrease in potential energy?

11. Using complete sentences and in your own words, describe the energy changes that take place when a pendulum swings.

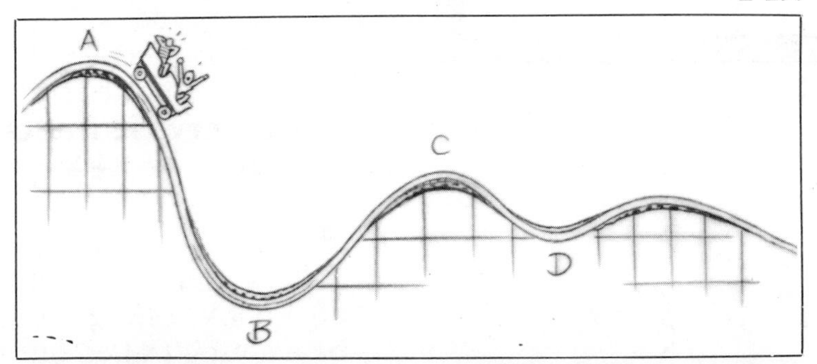

2-25.

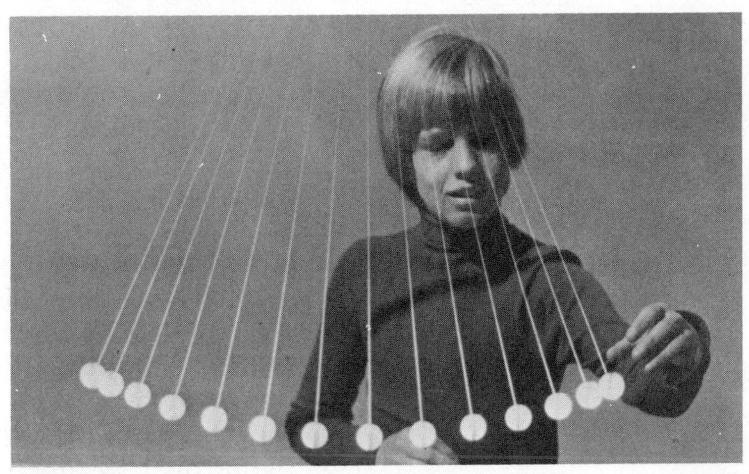
2-26.

SUMMARY

If an object is lifted against the force of gravity, energy is stored in that object in the form of potential energy. That stored energy can be turned into kinetic energy if the object is allowed to fall. When energy changes in form, as from potential to kinetic, the law of conservation of energy states that no energy can be gained or lost.

QUESTIONS

Unless otherwise indicated, use complete sentences to write your answers.

1. Identify each of the following as examples of either potential or kinetic energy.
   a. a book resting on a shelf
   b. a spring pulled out of its normal shape
   c. a book being lifted from the table to a high shelf
   d. a book falling from a shelf to the floor
   e. a roller coaster at the highest point of the track
   f. a roller coaster at the lowest point of the track

2. Describe how you could use a pendulum to demonstrate how potential energy changes to kinetic energy which in turn changes back to potential energy.

3. a. State the law of conservation of energy in your own words.
   b. Give several examples of energy changes in everyday life that illustrate this law.

2-4. USES OF ENERGY

Once in a while the earth gives evidence of a huge amount of energy stored beneath its surface. A volcano such as Mt. St. Helens in Washington erupts with a sudden release of some of the earth's internal heat. Although volcanoes are not common, everywhere on earth there is other evidence of energy changing from one form into another.

When you finish lesson 4 you will be able to:

- Name and describe six forms of energy.
- Describe how energy is used in our daily lives.
- Identify three sources of energy needed by modern civilization.
- ○ Make a chart of the forms of energy you use every day.

We live in an ocean of energy. Everywhere there is evidence of the natural ways by which energy is changed from one form to another. As rivers flow to the sea, falling water releases large amounts of kinetic energy. Energy in wind creates waves on the sea or turns windmills. What forms of energy are needed to describe all these changes? Scientists say that at least six forms of energy are needed to account for all changes from one form of energy to another. See Fig. 2-27. These six forms of energy are described below. You will study them more completely in later chapters.

1. *Mechanical energy.* This is energy due to the position or motion of an object. An example of mechanical energy is the motion of the pistons and other moving parts of an automobile engine. The energy of a hammer hitting a nail is also mechanical.

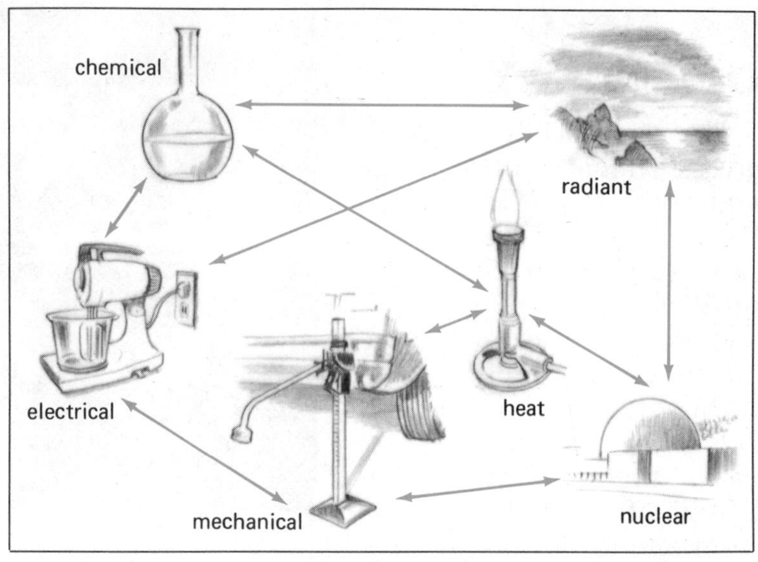

2-27. *Energy can be converted from one form to another.*

2-28. *This photograph shows a diesel automobile engine. Gasoline engines in cars are only about 20–25% efficient. In contrast, diesel engines are about 40% efficient. The efficient use of fuel is an important part of energy conservation.*

2. *Electric energy.* Try combing your hair and then quickly hold the comb close to some small pieces of paper. Often the pieces of paper will jump toward the comb and briefly cling to it. The energy needed to move the paper pieces came from electricity built up in the comb. Electric energy often builds up when two materials are rubbed together. Electrical energy can be carried through many materials, for example, copper wires.

Magnets attract many metal objects just as a comb may attract paper. Magnetic energy is also able to produce forces. Experiments show that magnetic energy and electrical energy are closely related. A kind of magnet can be created by electricity. Magnets are able to make electricity flow through wires.

3. *Chemical energy.* When a piece of paper burns, energy is given off. The paper is also permanently changed into other materials such as the ashes. When one kind of matter is changed into another kind of matter, any energy given off is chemical energy.

4. *Nuclear energy.* Matter may be changed in another way and produce energy. Changes in the cores or nuclei of the atoms in matter release nuclear energy. Splitting the nucleus of some kinds of atoms into two or more parts is the most common way to release nuclear energy. Nuclear energy is the most concentrated form of energy.

2-29. *The concentrated energy of a laser beam can melt metal. Here an industrial carbon dioxide laser is being used to cut metal into patterns.*

5. *Electromagnetic energy.* A laser beam can make a hole in a piece of metal. See Fig. 2-29. A powerful laser beam shows the energy that can be carried by light. Energy that spreads out and passes through space is called electromagnetic energy. This type of energy includes radio waves and X rays as well as visible light. Laser light is a form of electromagnetic energy.
6. *Heat energy.* When your hands are cold you can warm them by holding them against something with a higher temperature. Heat energy always flows from a hot object to a cold object. Heat is the form of energy that causes changes in the temperature of any form of matter.

How many forms of energy do you use in your everyday life? See Fig. 2-30. Even an incomplete answer to this question will quickly show that our lives are built on a foundation of energy. For example, everyone needs the chemical energy supplied by food. We would find it hard to live without heat energy to warm our houses and buildings and to cook our food. We use large amounts of energy to run the machines

2-30. *You use energy in many different ways each day.*

used to produce the necessities of life and for transportation. All over the world, the average person uses about eight times as much energy as could be produced from muscle power alone. In some highly developed countries such as the United States, each person uses ten times more energy than the average person in most other parts of the world.

Most of these energy needs are now met by burning of fuels such as petroleum and coal. Water power, nuclear energy, wind, and heat from the interior of the earth all together supply only a small amount of the energy needs of the world. The supply of petroleum and coal is limited and will be gone some day. Future civilizations must learn to apply the knowledge of natural energy to help meet growing energy needs.

Supplies of energy come from two kinds of sources. First, there are the sources of energy that cannot be replaced, for example, petroleum and coal. Both of these common fuels were formed millions of years ago within the earth's crust. Petroleum and coal are examples of *fossil fuels*. Once they are used up, there will be no way to make new supplies of these fuels. No one knows exactly how much petroleum remains in the earth. But the total reserves now known to exist will be used up within 50 years. Petroleum is already

2-31. *The solar energy collected by the panels on this house can be used to supply electricity as well as heat and hot water.*

becoming more expensive as the world's supply is running out. There seems to be enough coal to last several hundred years.

Other energy sources are replaced as they are used. An example of this kind of energy is the energy that comes from the sun. There is no limit to the amount of solar energy that is available. Homes, schools, and office buildings are now being equipped with panels that collect heat from the sun. See Fig. 2-31. In the future, special collectors called *solar cells* that can change sunlight into electricity may be used.

The wind is also a source of energy that is never used up. Windmills using wind energy to pump water used to be a common sight. Similar modern windmills have been built to generate electricity as they turn. See Fig. 2-32.

Heat stored deep within the earth may help meet future energy needs. Deep wells drilled in certain regions of the earth can bring hot water and steam to the surface. Electric generators can then use that energy to produce electricity.

2-32. *Modern windmills like this one may soon be a common sight as they use wind energy to generate electricity.*

Ocean tides can also be used as an energy source that never runs out. Large dams can be built across the mouths of some bays. Water brought into the bay by the rising and falling tides can then be made to turn electric generators.

Nuclear power plants already supply a part of our electrical energy. These power plants are able to produce a large amount of energy by using a small amount of nuclear fuel. However, nuclear power plants produce radioactive waste material. A safe method of disposal for this dangerous waste is needed.

In the future, our energy needs will probably be met by these sources of energy that are renewed as they are used. It is likely that several different sources will be used together to supply the energy that supports our civilization.

ACTIVITY

Materials
paper
pencil

Uses of Energy

This activity should help make you more aware of the many ways in which you use energy each day and how you can help conserve energy.

A. On a sheet of paper, make a table like the one shown below. In the first column, list as many things as possible that you use everyday that require energy. In the second column, write the kind(s) of energy used by that activity. In the third column, write "cut down" if you could cut down on that energy use or "live without" if you could live without that energy use.

Things I use every day	*Kind(s) of energy*	*Cut down or live without*
1. automobile	chemical, mechanical	
2. typewriter		
3. lamp		
4.		
5.		
etc.		

SUMMARY

Energy exists in at least six different forms. These forms include mechanical, chemical, heat, electrical, electromagnetic, and nuclear. Day-to-day living requires a supply of energy that is presently met almost entirely by burning of fossil fuels. Because the supply of fossil fuels is limited, new sources of energy must be developed for the future.

QUESTIONS

Use the clues given to fill in the following crossword puzzle. DO NOT WRITE IN THIS BOOK.

DOWN

1. ______ can be converted from one form to another.
2. Energy for most of today's living.
3. A hammer hitting a nail.
4. All forms of energy when being stored.
5. Causes a change in temperature.

ACROSS

6. Includes radio waves, X rays, etc.
7. Caused by changes in an atom's core.
8. Energy supplied by food.
9. Most common form of electromagnetic energy.
10. Related to electric energy.

Power

Materials
metric ruler
watch with sweep second hand

Watt
The unit of power used in the metric system.

Horsepower
The unit of power used in the customary (English) system of measurement.

Purpose

You are going to measure how much power, in **watts** and in **horsepower,** you develop in running up a flight of stairs. You will do this by finding the amount of energy you use each second.

Procedure

A. Obtain the materials listed in the margin.

B. You will need to know the height of the whole flight of stairs. If each step is the same height as the others, measure the height of one step in centimeters and multiply by the number of steps. If they seem to differ from each other, measure each step in centimeters and add them together.

1. What is the height of the flight of stairs in centimeters? in meters?

C. Get in position at the bottom of the stairs. Have someone with a watch give a signal for you to start running up the stairs. The person with the watch will measure the time it takes you to go up the stairs. You may want to practice. Be careful not to stumble and fall. See Fig. 2-33.

2-33.

2. How many seconds did it take for you to go up the stairs?

D. To find how many *watts* of power you developed, you will need to know your mass in kilograms. A kilogram weighs 2.2 lbs. If you divide your weight in pounds by 2.2, you will get your mass in kilograms.

3. What is your mass in kilograms?

E. Now multiply your mass in kilograms by the height of the set of stairs in meters.

F. Divide this product by the number of seconds it took for you to go up the stairs.

G. Multiply this result by 9.8 to find out how many watts of power you developed. (This is the same watt you find on light bulbs and appliances.)

4. How many watts of power did you develop?

H. There are 746 W in one *horsepower*. Divide the number of watts you developed by 746 to find the number of horsepower you developed.

5. How many horsepower did you develop?

Summary

In your own words, write a paragraph that would explain to someone how to measure the amount of power, measured in watts and horsepower, that he or she could develop.

VOCABULARY REVIEW

Match the number of the term with the letter of the phrase that best explains it.

1. energy
2. simple machine
3. lever
4. power
5. work
6. kinetic energy
7. hydraulic press
8. potential energy
9. watt
10. inclined plane

a. Size of the force multiplied by the distance through which the force acts.
b. The speed with which work is done.
c. That property of something which makes it able to do work.
d. A device that may change the direction and size of forces.
e. A rigid bar turning on some fixed point which usually changes the direction or size of a force applied to it.
f. Energy that moving things have as a result of their motion.
g. Energy stored in an object as a result of a change in its position.
h. The unit of work in the metric system.
i. A simple machine using liquids.
j. A simple machine with no moving parts.

REVIEW QUESTIONS

Complete each statement by choosing the best word or phrase, or by filling in the blank.

1. When a force moves an object, the amount of work done depends on **a.** the size of the force **b.** the distance the object moves **c.** both the size of the force and the distance the object moves **d.** how fast the object moves.
2. Which one of the following is *not* an example of work? **a.** a student walking up a flight of steps **b.** someone pushing on a wall that does not move **c.** a boy riding a bicycle **d.** a girl riding a horse.
3. That which allows a force to do work is called ___________ .
4. When a force is applied to an object causing it to move, ___________ is done, which produces energy.

5. A lever simply transfers energy since it is used with **a.** less force to move a heavier object a greater distance **b.** more force to move a heavier object a shorter distance **c.** more force to move a heavier object a greater distance **d.** less force to move a heavier object a shorter distance.
6. All of the following are simple machines *except* **a.** lever **b.** pulley **c.** gear wheel **d.** bicycle.
7. Kinetic energy is energy due to motion while potential energy is **a.** stored energy **b.** energy of a spring **c.** energy of a speeding bullet **d.** energy of a book.
8. Write *True* if the following statement is correct; if it is incorrect, change the underlined phrase and write the statement correctly. "A lever, as with all simple machines, can change the direction or size of the applied force as well as give more output energy than input energy."
9. The form of energy that is most concentrated is **a.** electromagnetic **b.** electrical **c.** nuclear **d.** chemical.
10. Day-to-day living requires a supply of energy that is almost entirely supplied by **a.** burning of fuels **b.** water power **c.** nuclear energy **d.** heat from the sun.

REVIEW EXERCISES

Give complete but brief answers to the following exercises.

1. Give one example to show that changes taking place in the physical world are the result of forces.
2. Compare the amount of work done when **a.** a book is moved 20 cm and when it is moved 40 cm **b.** one book is moved 20 cm and two books are moved 20 cm.
3. Explain how a lever cannot produce more work than is put into it.
4. A ruler is balanced on a pencil. Three washers of the same size are stacked on the ruler so that the center of their holes is 3 cm from the pivot point. Where, on the ruler, would you place one washer the same size as the others in order to balance them?
5. List five simple machines and describe how each is used.
6. Given a spring to use, how would you demonstrate to someone else how one form of energy can be changed into another form?
7. State the law of conservation of energy.

8. Use all of the Vocabulary Review words in column I on p. 78 and any additional key words you need to write a short paragraph summarizing each of the four lessons in this chapter.
9. List the six forms of energy described in lesson 3 and give an example of each.
10. Name the present sources of energy used by modern civilization and predict how this energy use will change in the future.

EXTENSIONS

1. Does your family use more electrical energy during the daylight hours or during the night? Read the electric meter at your house when you first get up in the morning and again when the sun goes down. Keep a record of readings for a week. Suggest ways your family can decrease their use of electrical energy.
2. Study the lever (crank), gear (sprocket), and wheel system on a bicycle. Write a report on the purpose of each part. Does it increase or decrease the force needed? Does it increase or decrease speed?
3. In science, units are often named for famous scientists. Look up the following names in the library: **a.** Newton **b.** Joule **c.** Watt. Write a short biography of each scientist. Include an interesting story about each scientist.
4. Report on some historical examples of perpetual motion machines. Include: **a.** a description of each proposed machine **b.** a statement of why each machine would eventually stop **c.** your ideas on whether the machines would work in outer space or on another planet.

CHAPTER 3

Waves

3-1. ENERGY AND WAVES

Have you ever been near the ocean or a large lake during a storm? If you have, you have probably been impressed by the energy of the waves pounding the shore. Where does this energy come from? In this lesson, you will learn what causes waves.

When you finish lesson 1, you will be able to:

- Explain what causes *waves* in water.
- Give examples of the ways waves may be produced when energy moves from one place to another.
- Identify three features of all waves.
- ○ Draw a wave diagram of the movement of a pendulum and label the features of the wave.

3-1. These waves were caused by the energy added to the water by an object falling into the water.

Wave
A disturbance caused by the movement of energy from one place to another.

If you have ever sat by a small lake or pond, or even a puddle of rainwater, and thrown pebbles into the water, you have seen **waves** travel across the surface of the water. The *waves* spread out from the point at which the pebble hit the water. See Fig. 3-1. These waves are caused by energy traveling through the water. Where do you think the energy came from? Remember that an object in motion has kinetic energy. The pebble flying through the air has kinetic energy. When the pebble hits the water, some of the kinetic energy is transferred to the water. This energy causes the individual water particles to start moving in circular paths. This movement, in turn, causes neighboring water particles to move. See Fig. 3-2. As the water particles move in small circles, a wave moves across the

3-2. As the wave moves forward, the individual water particles move in circles.

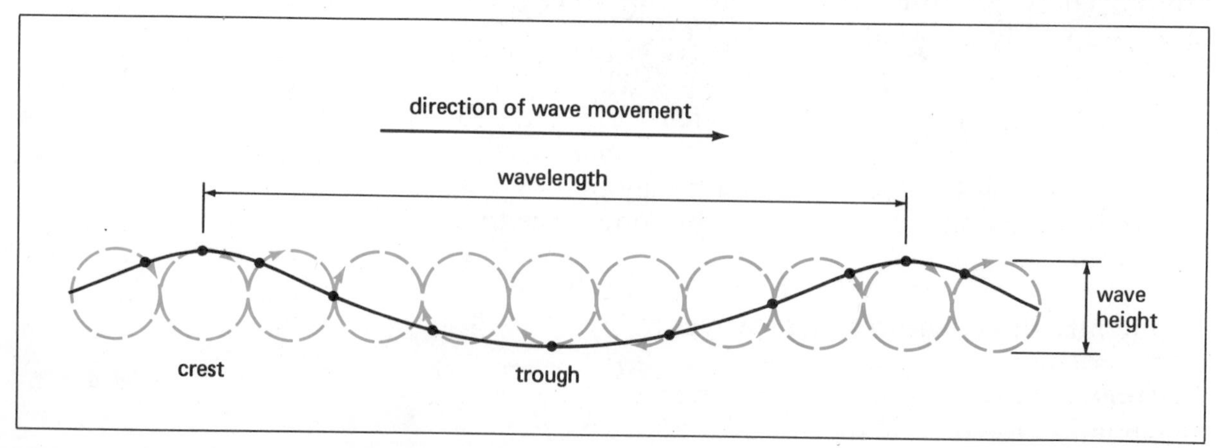

3-3. *This wave motion is the result of up-and-down movement.*

surface of the water. The water itself does not move along with the wave. The energy that causes the wave moves through the water. The huge waves in a storm are the result of the energy of the storm moving through the water.

Waves are always the result of energy moving from one place to another. You can see how this happens if you hold the end of a rope tied to a doorknob. If you move your hand up and down, energy is added to the rope. The energy travels along the rope as a wave. See Fig. 3-3. Notice that the wave moves along the rope while the rope moves up and down. A wave can also be made to move along a spring. See Fig. 3-4. The wave in the rope is the result of up-and-down motion. The wave in the spring is caused by back-and-forth motion.

The kind of wave produced by the up-and-down motion of a rope is called a transverse wave. Transverse means "across." In a transverse wave, the material carrying the wave moves across, or at right angles to, the direction in which the wave is moving. The back-and-forth motion of the wave in a spring is called a *longitudinal wave*. Longitudinal means "lengthwise." A wave in a spring is a longitudinal wave when the motion of the coils of the spring is along its length as shown in Fig. 3-4.

No matter what kinds of waves you are dealing with, you can use the same three properties to describe them. First, a wave can be described by the distance between neighboring crests (high parts) or between neighboring troughs (low parts). This distance is called

3-4. *This wave motion is caused by back-and-forth movement.*

Wavelength
The distance between two neighboring crests or troughs of a wave.

the **wavelength** of a wave. In the case of a wave moving along a spring, the *wavelength* (λ) is the distance between two neighboring squeezed-together or pulled-out coils of the spring. Wavelengths are usually measured in meters for longer waves or in centimeters for shorter waves. The symbol for wavelength is the Greek letter lambda, λ. See Fig. 3-5.

Second, since waves always move, they have a speed. The speed (v) can be determined by measuring how fast a certain point on the wave moves. For example, the speed of a particular wave crest can be measured. Wave speed is usually measured in meters per second.

A third way of describing a particular wave calls for counting the wave crests. This number is called the

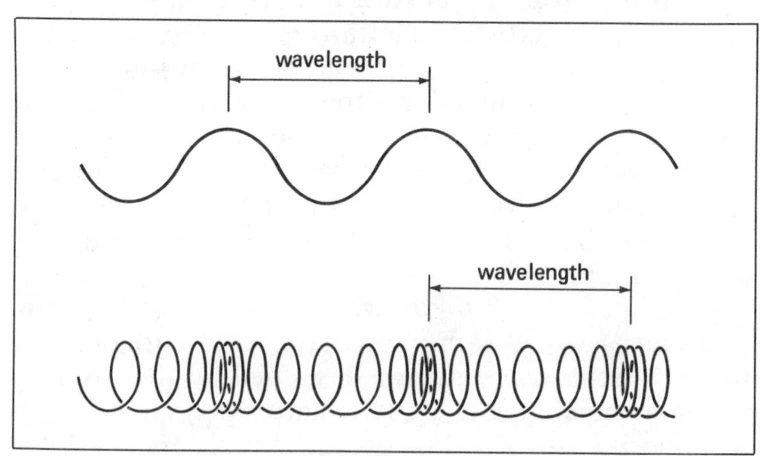

3-5. *This diagram shows the wavelength on a transverse and a longitudinal wave.*

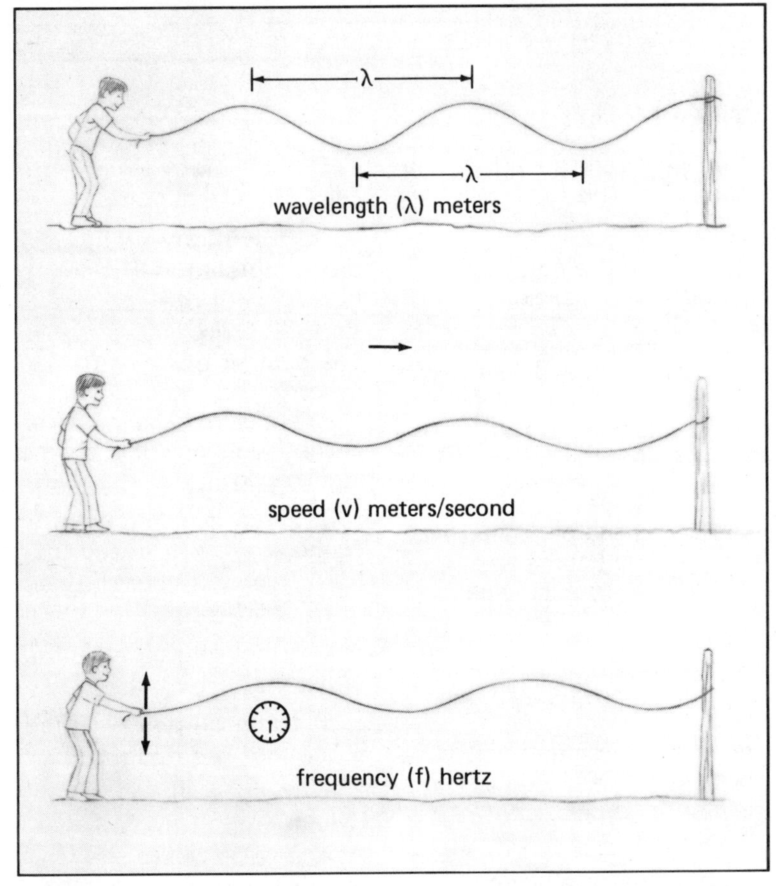

3-6. *The three basic properties of all waves: wavelength, speed, and frequency.*

frequency (free-quen-see) of the wave. The *frequency* (f) of a wave can be measured by counting the number of wave crests that pass by a certain point in 1 sec. Frequency is measured in **hertz** (Hz) (hurts). The unit *hertz* is named after a scientist who was one of the first to study certain kinds of waves. One hertz means a frequency of one complete wave passing a point each second.

Wavelength, speed, and frequency for waves in a rope are illustrated in Fig. 3-6.

Suppose that you were fishing from a stationary boat. Several waves pass by the boat. You could find the speed of these waves by counting the number of waves passing in 1 sec (frequency) and measuring the length of one wave. Their speed is equal to frequency multiplied by wavelength or frequency × wavelength = speed. For example, ($f \times \lambda = v$) if the frequency of a wave is 2 Hz (two waves/sec) and the wavelength is 0.5 m, the speed is 2 waves/sec × 0.5 m/wave = 1.0 m/sec.

Frequency
The number of complete waves that pass by a point each second.

Hertz
A unit used to measure the frequency of a wave. A frequency of 1 Hz means that one complete wave passes a point each second.

Materials
string, 1 m
eraser or rubber stopper
pencil
paper

3-7.

Wave Motion

You will need to work with a partner in this activity.

A. Obtain the materials listed in the margin.

B. Set up a pendulum with a string 50–75 cm in length. Use an eraser or rubber stopper for the bob.

C. Swing the pendulum so that the bob moves 10–15 cm to each side.

D. While the pendulum is swinging, move your finger to follow exactly the back-and-forth motion of the bob. Practice this until you are sure you can do it.

E. Now place a piece of paper under the pendulum so that the bob is centered on the top edge of the paper. The bob should swing over the short end of the paper.

F. With a pencil, follow the motion of the bob over the paper. Have your partner slowly and steadily pull the paper under the pendulum. See Fig. 3-7. The trace you will make on the paper is a wave graph.

G. Swing the pendulum again and count the number of back-and-forth swings the pendulum makes in 1 min.

1. How many swings are completed in 1 min?

2. How many swings are completed in 1 sec? Express your answer in decimal form.

The number of swings in 1 sec is the frequency of the pendulum in hertz. This is the frequency of the wave you traced.

3. What is the frequency of the wave you traced in hertz? Express your answer in decimal form.

4. Label all points on the wave that are crests.

5. Label all points on the wave that are troughs.

6. Indicate the wavelength of the wave.

At the point where the pendulum has the least amount of kinetic energy, the wave has the greatest potential energy.

7. Label all points on the wave that show when the pendulum bob and the wave had maximum potential energy.

8. Label all points on the wave that show when the wave had maximum kinetic energy.

9. In complete sentences, using your own words, describe the following terms as they are related to water waves: **a.** frequency **b.** crest **c.** trough **d.** wavelength

SUMMARY

Waves are the result of energy moving from one location to another. All waves are basically the same. They have wavelength, speed, and frequency. All three properties can be measured for a particular wave.

QUESTIONS

Unless otherwise indicated, use complete sentences to write your answers.

1. Make a labeled drawing to show the parts of a wave.
2. Describe how you could produce waves on a spring.
3. What three features do all waves have in common?
4. Supply the missing values a, b, c, and d in the following table:

v (cm/sec)	f (Hz)	λ (cm)
a	5.0	6.0
50	b	2.5
75.0	13.9	c
d	12.5	5.4

5. You notice that 10 waves reach the shore of a lake in 5.0 sec of time. What is the frequency of the waves?

3-2. WAVE MOTION

You have probably seen films of people surfing on giant waves in the ocean. Maybe some of you have tried it yourselves. Have you ever noticed what happens to a wave when it approaches the beach? Does the wave speed up or slow down? Does the wave change its direction? When a wave hits a barrier in its path, such as a dock, what happens to the wave?

When you finish lesson 2, you will be able to:

- Explain what happens to a wave when its speed changes.
- Describe how a wave changes when it meets a barrier in its path.
- Give examples of the Doppler effect.
- ○ Examine photographs to identify *reflection* and *refraction* of a wave, and the *Doppler effect*.

3-8. *Ocean waves are refracted as they reach the shore as shown in this photo.*

At the shore of an ocean or a lake, you can make an important observation about waves. By watching the waves approach the shore, you can see that the direction in which the waves move changes. Waves almost always approach the shoreline straight on. The direction usually changes as the waves move in closer to the shore. See Fig. 3-8. This change of direction is caused by the slowing of the wave as it reaches shallow water. The wave then meets with friction against the bottom. If the wave approaches the shore at an angle, one side reaches shallow water first and is slowed. This causes the rest of the wave to move ahead until the wave is lined up to come straight into the shore. The behavior of waves in water can be studied by using a special tank called a ripple tank. In a ripple tank, ripples or small waves are made in a shallow layer of water. The tank holding the water has a glass bottom. A light shines downward through the water onto a sheet of white paper. Ripples in the tank cause a shadow of the wave pattern to appear on the white paper. See Fig. 3-9.

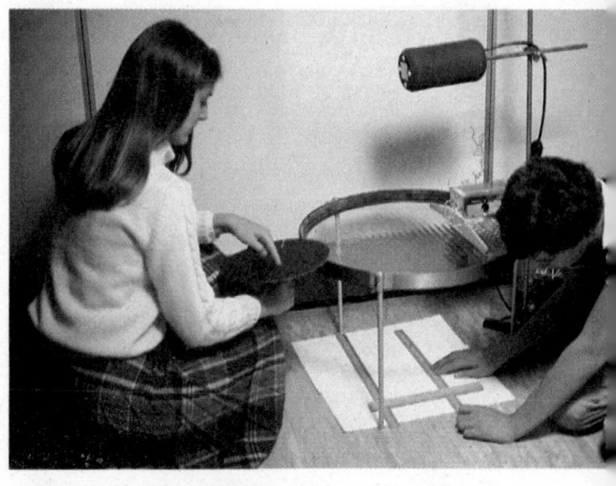

3-9. *In a ripple tank, the pattern of waves in the tank can be seen on the paper below the tank.*

A ripple tank can be used to observe the behavior of waves as they move into shallow water. This is done by putting a flat piece of glass in the ripple tank. Waves moving into the shallow water over the sheet of glass are slowed. They can be seen changing their direction. See Fig. 3-10. When waves change the direction in which they are moving because of a change of speed, the process is called **refraction** (rih-**frak**-shun). Waves in water show *refraction* when the depth of the water changes. The speed of waves is slower in shallow water. See Fig. 3-10.

Refraction
The process in which a wave changes direction because its speed changes.

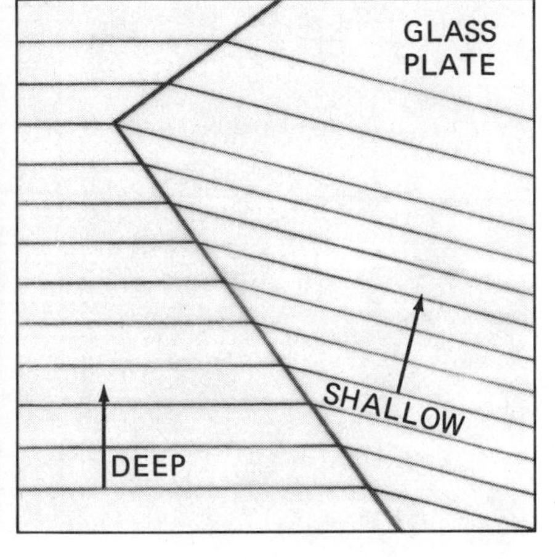

3-10. *The waves in this ripple tank change direction when the depth of the water changes from deep to shallow.*

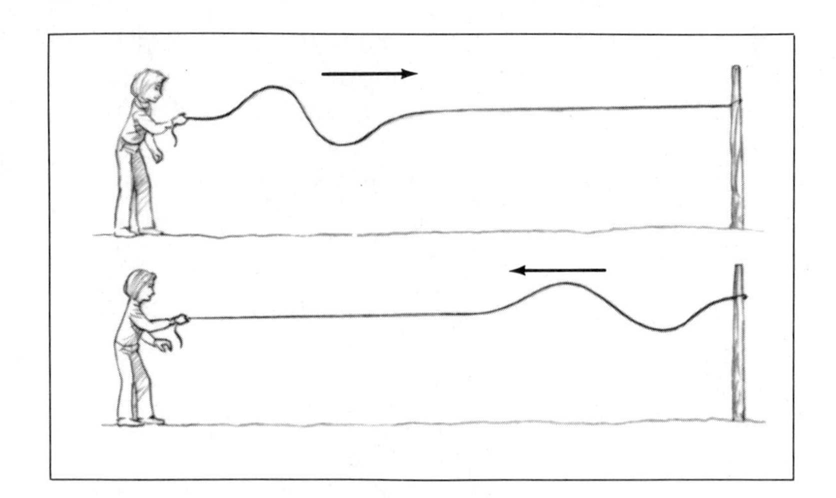

3-11. *Why is the wave reflected back along the rope?*

Reflection
The process in which a wave is thrown back after striking a barrier that does not absorb the energy of the wave.

Waves may also change direction if they meet a barrier in their path. If the barrier does not absorb the energy of a wave, **reflection** (rih-**fleck**-shun) will occur. For example, a wave sent down a rope can be *reflected* back along the rope if the end of the rope is tied down. See Fig. 3-11. In this case, the direction of the wave is reversed as the reflected wave travels back along the rope. *Reflection* can also be observed in water waves. When a wave strikes a barrier straight on, reflection sends the wave back in a directly opposite direction. A wave that strikes a barrier at an angle is reflected back at the same angle. See Fig. 3-12.

Sometimes the frequency of waves can appear to change. This can happen when waves are observed by someone who is moving. An example is the observation of water waves from a moving boat. See Fig. 3-13. If the boat is heading into the waves, the waves strike

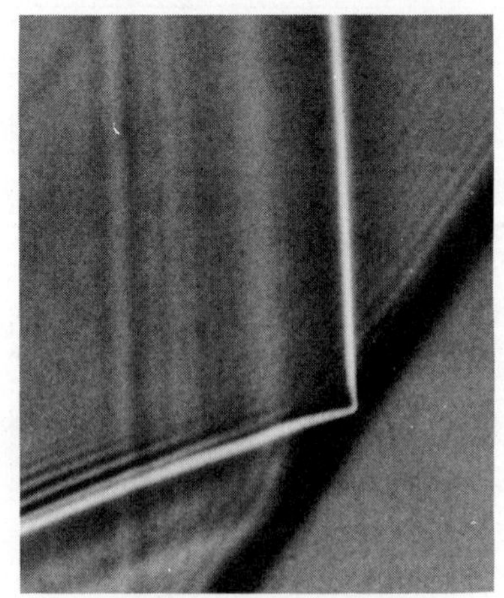

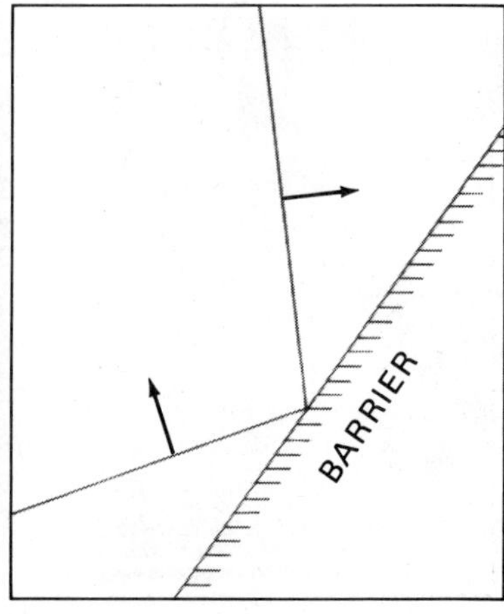

3-12. *Reflected waves are returned at the same angle at which they hit the barrier.*

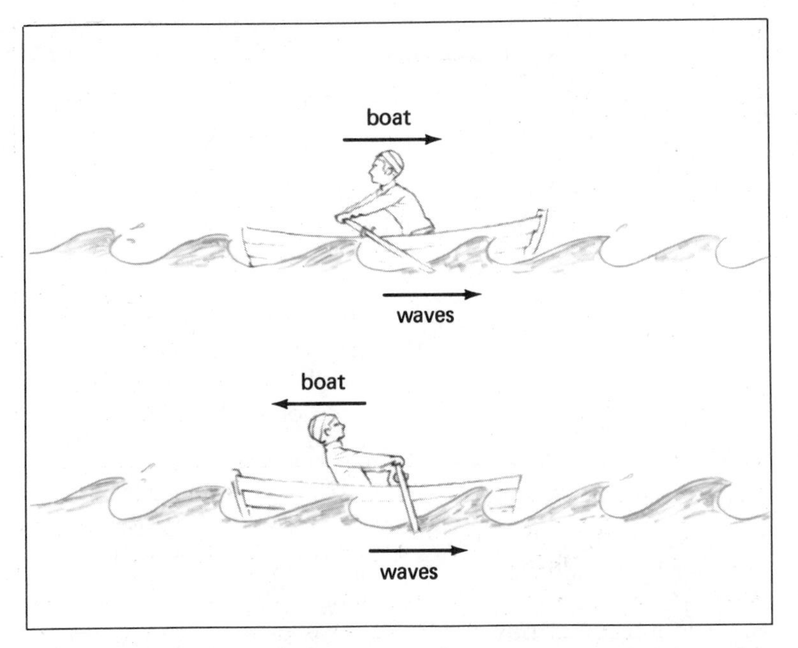

3-13. Which boat would pass through more waves in the same amount of time?

the boat often and the wave frequency appears to be high. If the boat turns around and moves in the same direction as the waves, fewer waves reach the boat in a given time. When the frequency of waves seems to change as a result of the motion of an observer, the result is called the **Doppler effect.** The *Doppler effect* also occurs when an observer is stationary and the source of the waves is moving toward or away from the observer. For example, the sound of an approaching car horn becomes lower as the car goes past you.

Doppler effect
An apparent change in the frequency of waves caused by the fact that the observer or the source of the waves is moving.

Police radar uses the Doppler effect to measure the speed of cars on a highway. The stationary radar sends out pulses that are reflected back from the target car. The Doppler effect on the returning signal can then be used to calculate the speed of the car.

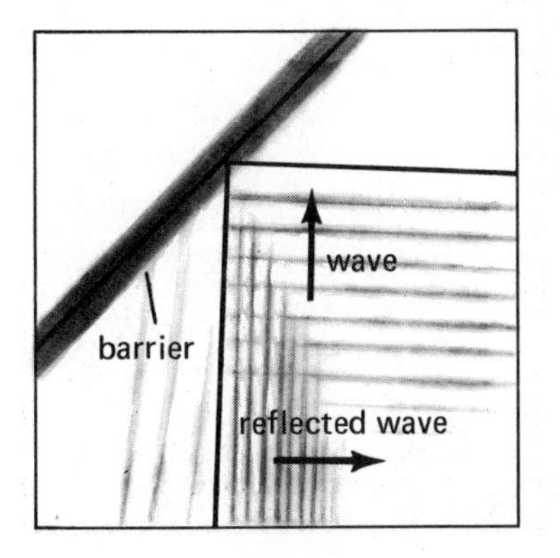

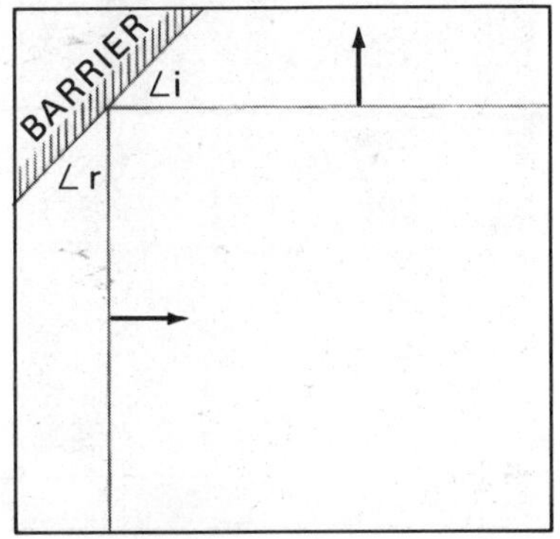

3-14. Before reflecting from the diagonal barrier, the straight waves shown in the photo are moving up. After reflecting, the waves move to the right.

ACTIVITY

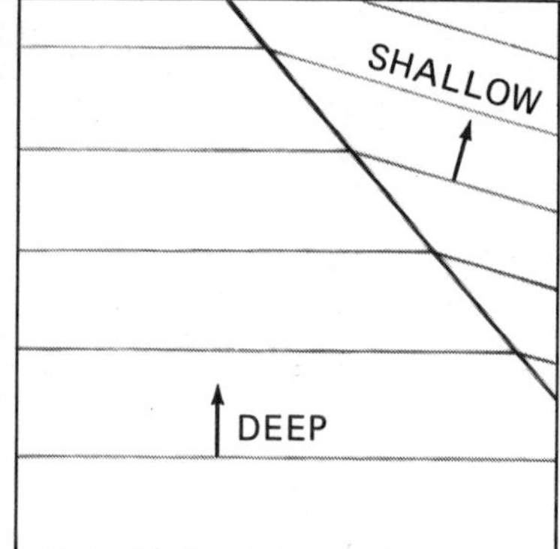

3-15. *These illustrations show the refraction of waves as they move from deep to shallow water. The arrow shows the direction of motion of the waves.*

Materials
paper

Properties of Waves

A. Look carefully at Fig. 3-14 on page 91.

B. Fold a piece of paper so that it forms an angle equal to the angle between the barrier and the wave headed toward the barrier.

C. Compare this angle with the angle between the barrier and the reflected wave.

1. Are these two angles nearly equal?
2. According to the text, should they be equal?

D. Look carefully at Fig. 3-15.

E. From your observations, answer the following questions.

3. Where are the wavelengths shortest (deep or shallow water)?
4. Where is the angle between the line separating deep and shallow water and the wave the smallest (deep or shallow water)?

F. Look carefully at Fig. 3-16. In this photograph, the source of the waves is moving. The direction of motion of the source is shown in the picture. The waves are traveling at the same speed in all directions.

5. Where are the waves the shortest (ahead of or behind the source)?
6. Where would the frequency of the waves be highest (ahead or behind)?
7. In your own words and using complete sentences, explain what is meant by each of the following terms: **a.** reflection **b.** refraction **c.** Doppler effect. Give an example of how each affects wave motion.

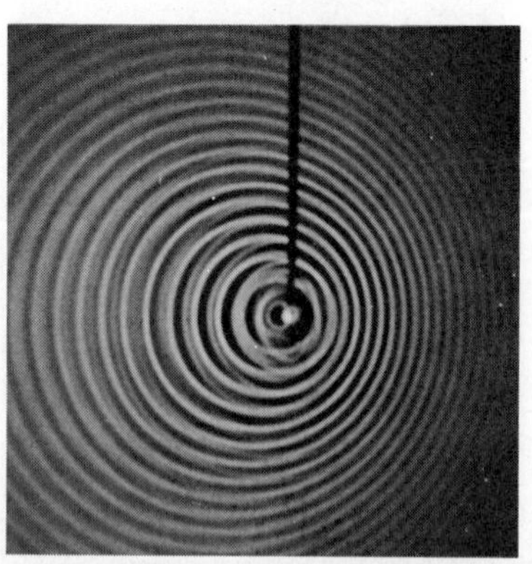

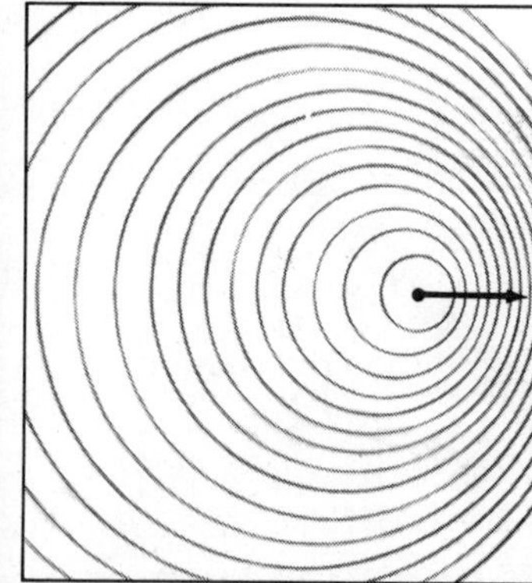

3-16. *The Doppler effect is illustrated in a ripple tank. The arrow shows the direction of motion of the source.*

SUMMARY

The next time you are at the beach, or watching a film of waves hitting a beach, remember what you have learned in this section. As waves approach the beach, they slow down. This change in speed causes the waves to change direction. A barrier, such as a dock or pier, will also change the direction of the waves. Finally, if you were in a boat traveling in the same direction as the waves, the frequency of the waves would appear different from the way it does if you were moving in the opposite direction.

QUESTIONS

Look at the following photos of waves traveling on water. Then answer the questions in complete sentences.

1. Name the wave behavior that each of the photos represents.
2. What is the process called when waves change direction as a result of a change in speed?
3. What happens when a wave on a rope hits a barrier at the end of the rope?

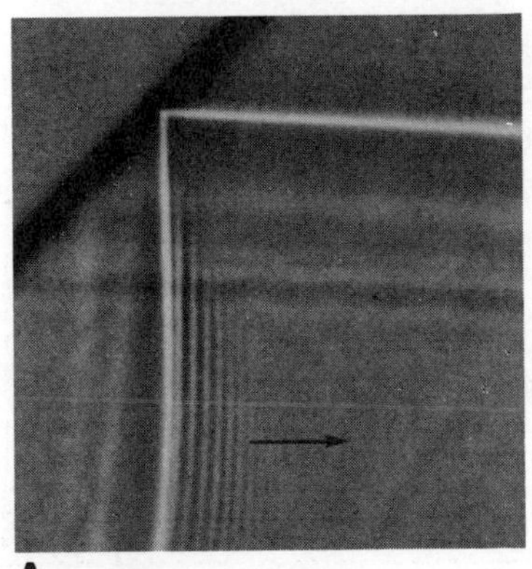

A.

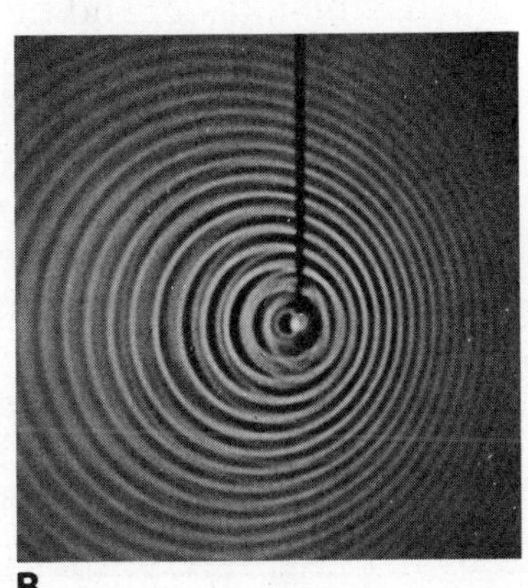

B.

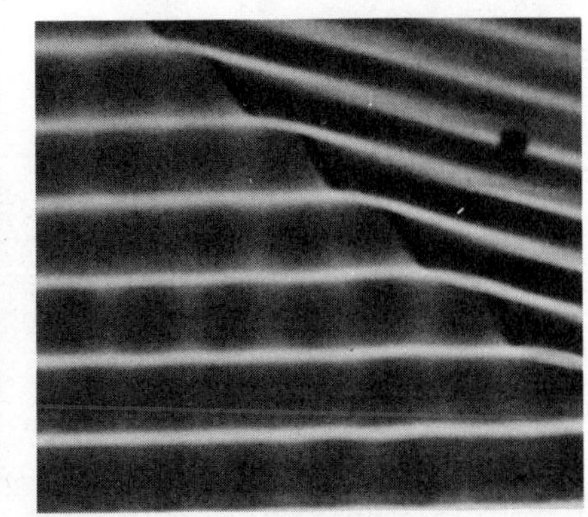

C.

3-3. SOUND WAVES

In 1880, a girl named Helen Keller was born. When she was nineteen months old, Helen had a serious illness. As a result of the illness, she became blind and deaf. She could not hear the sounds or voices of the world around her. Because she could not hear sound, she could not speak. How would it be to be like young Helen? How important is sound in your life? What would the world be like without sound? In this lesson you will learn what sound is.

When you finish lesson 3, you will be able to:

- Explain what is meant by *sound*.
- Show how sound waves are like other kinds of waves.
- Relate the properties of sound waves to the way we hear sounds.
- ○ Identify some properties of sound waves.

3-17. *Deaf students can often be taught to speak using special teaching methods.*

Sound
A form of energy caused by vibrations of the particles of matter through which the sound wave passes.

The only way you can tell what is going on around you is through your senses. These include sight, touch, smell, taste, and hearing. Which of them would you choose as the most important? It is a hard choice to make, but the sense of hearing should be given serious thought. It is very hard to learn to talk without being able to copy the sound of other people speaking. Without the ability to hear and understand speech, a person is cut off from most contact with other people.

Sound can be described as a kind of wave. *Sound* is different in one important way from other kinds of waves. For example, when a wave moves along a piece of rope, the different parts of the rope move up and

3-18. *The vibration of these guitar strings produces a sound.*

down as the wave passes. Sound waves also move through something, but they are the result of back-and-forth motions. A string on a guitar produces a sound by vibrating back and forth. See Figs. 3-18 and 3-19. The sound travels through the air because the motion is passed from one particle of air to the next. Sound waves travel much faster than waves in water. In air at 0°C, sound waves have a speed of 331 m/sec. As air gets warmer, the speed of sound increases. When air is heated, its particles move more rapidly and are able to carry the sound waves faster. The speed of sound in air increases by about 0.6 m/sec for each 1°C increase in temperature. At 20°C, the increase in the speed of sound can be given by:

$$0.6 \text{ m/sec} \times \text{temperature increase above } 0°\text{C}$$
$$0.6 \text{ m/sec} \times 20 = 12 \text{ m/sec}$$

This means that the speed of sound at 20°C is 343 m/sec (331 m/sec + 12 m/sec).

If you assume that the speed of sound at ordinary temperatures is about 350 m/sec, it is easy to find how far away a lightning flash is. Just count the number of seconds between the time you see the flash of light and the moment the sound of the thunder reaches you. Then multiply this time by 350 m/sec. For example, if you hear thunder 3 sec after you see a lightning flash, the lightning was 350 m/sec × 3 sec = 1,050 m (about 1 km) away.

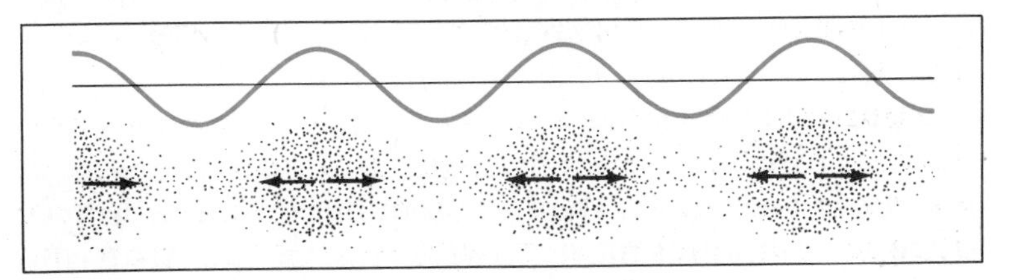

3-19. *As a sound wave moves through the air, the particles of air are crowded together and spread apart.*

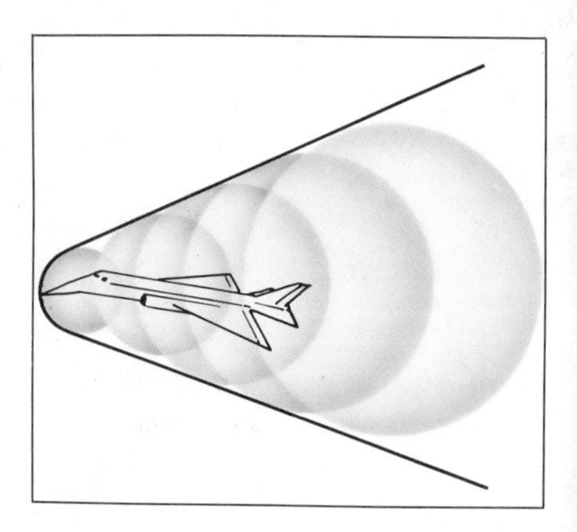

3-20. *An airplane moving faster than the speed of sound causes a sonic boom because it is moving faster than its own sound waves.*

A jet airplane flying at the speed of sound can make powerful sound waves called *sonic booms*. Any object moving through air makes a kind of sound wave by pushing through the air. See Fig. 3-20. Aircraft that move slower than the speed of sound never catch up with this disturbance. When flying at less than the speed of sound, the aircraft moves through a path of air that easily moves away from the aircraft. Above the speed of sound, the aircraft must thrust the air aside. This creates a powerful sound wave that sounds like thunder when it reaches the ground. The material through which sound waves move also has an effect on their speed. For example, in water at 20°C, sound moves at a speed of about 1,460 m/sec. In steel at the same temperature, the speed of sound is about 5,000 m/sec. In general, sound moves faster in solids and liquids whose particles are able to vibrate easily. An approaching train can be heard at a greater distance by listening to the noise made through the rails than the same noise made through the air. How do you account for this?

Like all waves, each sound has a particular frequency. Since sounds are created by something vibrating, the frequency is determined by how fast the source is vibrating. See Fig. 3-21. If the vibration is fast, the resulting frequency is high. A short guitar string produces a sound of higher frequency than a longer string that vibrates more slowly. Most people are easily

3-21. *The vibrations of these guitar strings can produce sounds of higher or lower pitch, depending on their frequencies.*

able to hear sounds with frequencies from about 60 to 10,000 Hz. However, the average ear is most sensitive to sounds with frequencies between 1,000 and 4,000 Hz. If the frequency of a sound is high, the sound is usually said to have a high **pitch.** *Pitch* describes the way the sound is heard. For example, a short guitar string vibrating at high frequency is said to have a high-pitched sound.

Pitch
A property that describes the highness or lowness of a sound and which is determined by the frequency of the sound wave.

Because of the Doppler effect, the pitch of a moving source of sound appears to change. A locomotive horn, for example, will seem to have a higher pitch when approaching you than when moving away from you. The Doppler effect causes the frequency of the sound waves to increase as the source moves toward you. When moving away, the frequency is decreased, causing the pitch to be lowered.

The amount of energy carried by sound waves is very small compared to that carried by most other kinds of waves. The energy of sound waves is transmitted by the vibration of the particles of the substance through which the sound travels. This back-and-forth motion is not very great. The energy involved is not large. Usually this small amount of energy is changed into other forms of energy and the sound quickly dies out. Sound waves sometimes hit a barrier that does not absorb their energy. Then the waves are reflected and you hear an *echo*. See Fig. 3-22.

3-22. *An echo is a reflected sound.*

ACTIVITY

Materials
copy of word game

Words about Waves

In this activity, you are asked to recall some of the words that are used in the study of waves. Follow the directions and clues carefully. Spell the words described by the following numbered clues. Write the letters in order in the blank spaces after the number of the clue. DO NOT WRITE IN THIS BOOK. When you have finished, you will be able to identify the key word in the vertical column shown.

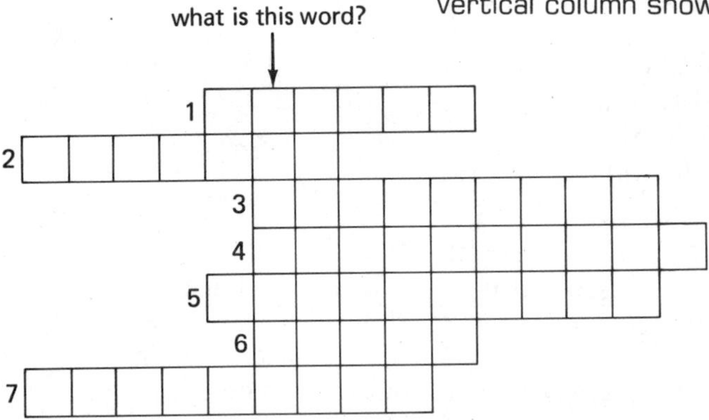

Clues

1. The name given to the bottom of the dip in waves.
2. The name of the effect caused by a source of waves moving.
3. The term that describes the number of waves per second.
4. The process in which waves bounce off a surface.
5. The distance from a point on one wave to the same place on the next wave.
6. The name given to the top of a wave form.
7. The kind of energy that is present in the part of a water wave where the water is piled up.

Now you can answer these questions.

1. What is the key word?
2. What does the key word mean in relation to waves?

SUMMARY

You have learned that sound is made up of waves. These waves consist of back-and-forth motions of the particles of the material through which the sound moves. Like all waves, sound waves carry energy. The pitch of a sound is closely related to the frequency of the waves.

QUESTIONS

Use complete sentences to write your answers.

1. How is sound produced?
2. In what ways are sound waves like waves on a rope?
3. How do sound waves differ from waves on a rope?
4. Describe how the sound produced by a guitar string reaches your ear.
5. You hear an echo from a wall 660 m away in 4.0 sec. What is the speed of sound? (Hint: Did you include both directions?)
6. The speed of sound is 344 m/sec. If you hear an echo in 6.0 sec, how far away is the reflecting surface?

3-4. SOUND AND MUSIC

All people do not necessarily like the same kind of music. One person may like rock, another opera. Luckily, there is a wide variety of musical styles. What is the basic difference between a musical sound and a nonmusical sound? What turns a collection of sound waves into something that we call music? Why are seventy-five or more people playing different instruments in an orchestra able to produce a sound that is pleasing to hear? In this lesson you will study the properties of sound waves that help create music.

When you finish lesson 4, you will be able to:

- Distinguish between a musical and a nonmusical sound.
- Use examples to explain how sounds can be heard around a barrier.
- Describe how two or more sound waves may affect each other.
- ○ Show how two or more sound waves may affect each other.

A book dropped on the floor makes a sound. A musical instrument also makes a sound. A window that is opened only a little can allow a whole room to fill with sounds from outside. These observations about sound can be explained by the fact that sound is a special kind of energy carried by waves.

Sound waves behave like other waves. Consider the effect of waves on each other when they meet. If two or more waves arrive at the same place at the same time, they will affect each other by a process called **interference** (int-er-**fir**-unts). Two waves that are identical will combine to make a single wave that has

Interference
The effect two or more waves have on each other if they overlap.

higher crests and troughs. See Fig. 3-23. Two sound waves with the same wavelength and frequency can be added to make a louder sound with nearly the same pitch. Suppose two waves whose crests and troughs occur at different times come together. If their wavelengths are the same, these waves will tend to cancel each other. See Fig. 3-24. Waves that have different wavelengths can combine to produce a new wave. See Fig. 3-25.

Interference is important in producing the kinds of sounds we hear. Most musical sounds are mixtures of waves whose frequencies are related in some simple way. For example: one musical note is twice the frequency of another note. When both notes are played together, they combine to make a pleasant musical sound. The same note played on different instruments has the same basic frequency.

Other frequencies are added which identify the sound as coming from a particular instrument. A noise is a mixture of sounds whose frequencies have no connection. Such sounds are unpleasant to most people. A stringed instrument such as a guitar produces a musical sound from a vibrating string. If the string vibrates as a whole, a particular tone called the *fundamental* tone results. However, if the string vibrates in two parts, it then produces a higher pitched sound called an *overtone*. See Fig. 3-26. Strings can usually be made to vibrate in ways that produce a number of overtones. Overtones always have frequencies that are the result of multiplying the fundamental frequency by some whole number. For example, if the fundamental fre-

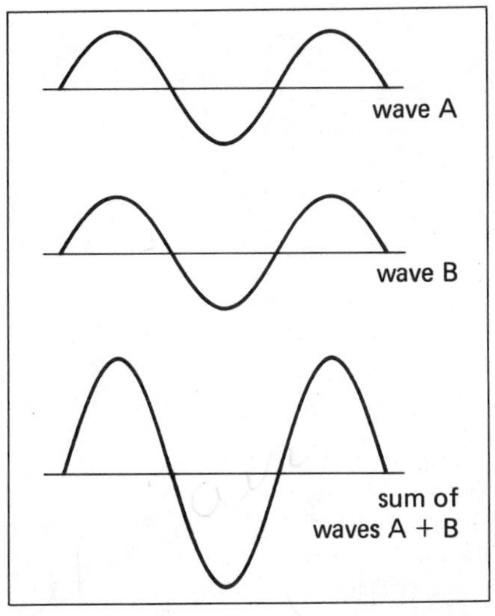

3-23. *If two waves are "in phase," they can combine to produce one wave with greater energy.*

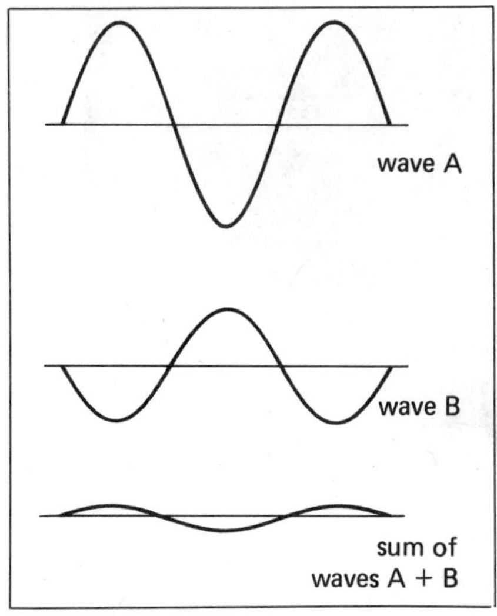

3-24. *If two waves are "out of phase," they will cancel each other.*

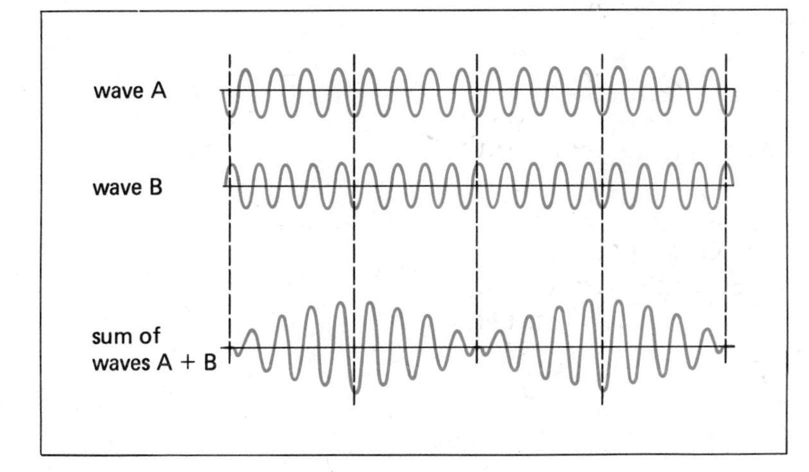

3-25. *Two waves combine to produce a more complicated wave.*

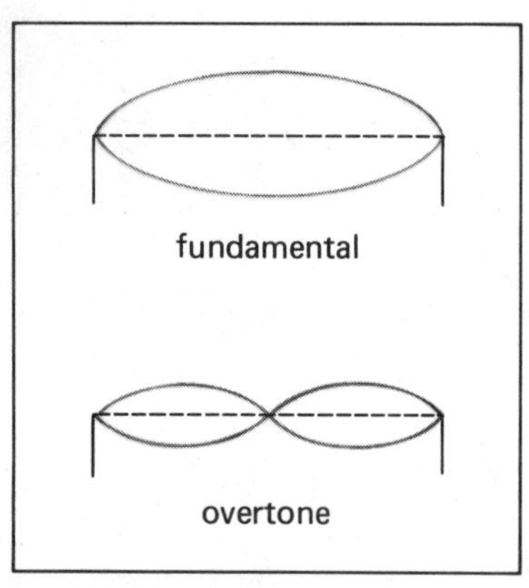

3-26. *This diagram shows a fundamental tone and its first overtone.*

3-27. *How can all these different instruments playing together produce a pleasant musical sound?*

Diffraction
The ability of waves to bend around an obstacle in their path.

quency of a string is 300 vibrations per sec, its overtones will be 1 × 300 = 300 vibrations/sec, 2 × 300 = 600 vibrations/sec, and so on. Because of this simple relationship, the fundamental and its overtones blend together to produce a musical sound. See Fig. 3-27.

In wind instruments, the sounds are produced by causing air inside the instrument to vibrate. You may have used the same principle to produce a sound by blowing across the narrow mouth of a bottle. Some wind instruments, such as a trumpet, use the motion of the player's lips to cause the air to vibrate. Other instruments, such as the clarinet, use vibrations caused by a wooden reed. The pitch of the tone made by a wind instrument depends upon the length of the vibrating column of air. As the length of the air column increases, the pitch decreases. Various wind instruments change the length of the air column by valves covering or uncovering holes (flute), or sliding a tube in and out (trombone).

Percussion instruments, such as drums, are played by being struck with an object. This causes a part of the instrument to vibrate—thus creating the sound.

Music and noise are both made up of sound waves. But noise is a mixture of frequencies with no pattern or connection. However, noise can be more than just an unpleasant sound. Noise can be a form of pollution. Constant exposure to noise from sources such as machines, jet engines, or city traffic may damage hearing and cause other health problems. Loudness of sounds is measured in units called *decibels* (dB). Each increase of 10 dB doubles the loudness of the sound you hear. Exposure to sound levels of above 90 dB may cause hearing damage. Sounds of 160 dB can cause permanent loss of hearing. Table 3-1 shows the typical decibel rating of some sounds. Since you know that waves move in straight paths, how can you hear a sound around a corner? This behavior of sound is so common that you probably never think about it. Yet it is the result of a complicated kind of wave behavior called **diffraction** (dif-**rak**-shun). Although sound waves continue to travel in the direction in which they start, they can bend around a corner by the process of *diffraction*. Diffraction is a result of a new group of waves that are created when the original waves strike a barrier. The edges of a doorway, for example, can act

TABLE 3–1

Transportation and Industry Noise Levels	Loudness (decibels)	Community Noise Levels
Jet plane at takeoff	135	Air-raid siren
	130	Thunder
New York subway	120	
Chain saw	115	Rock concert
Loud motorcycle	110	
Snowmobile	105	Loud conversation
Loud outboard motor	100	Power mower
Air hammer	95	Police siren at 10 m
Farm tractor	90	Loud singing
	85	
Heavy traffic	80	Noisy restaurant
Average factory	75	Vacuum cleaner
Average car accelerating	70	Noisy office
	65	Conversation at 1 m
Near freeway	60	Average office
	55	Background music
Light traffic	50	Average home
	45	Quiet street
	30	Quiet auditorium
	20	Faint whisper
	10	Rustling leaves
	0	

as second sources for sounds passing through the open door. See Fig. 3-28. When a sound is carried around a barrier by diffraction, its frequency and wavelength are unchanged. Diffraction is a property of all other kinds of waves as well as sound.

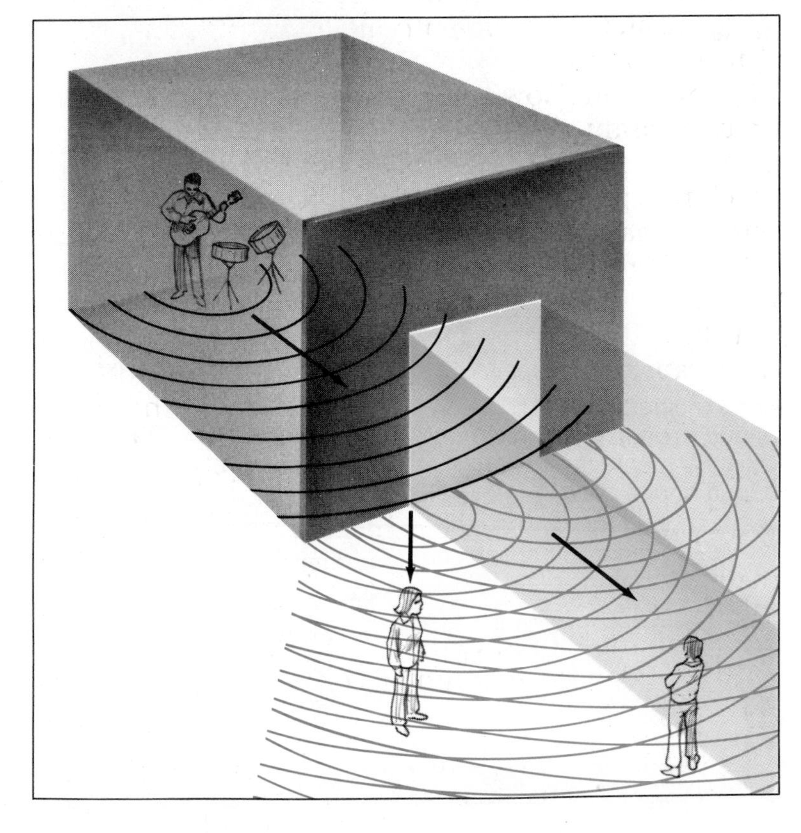

3-28. *Sounds can be heard through an open doorway or around a corner by diffraction.*

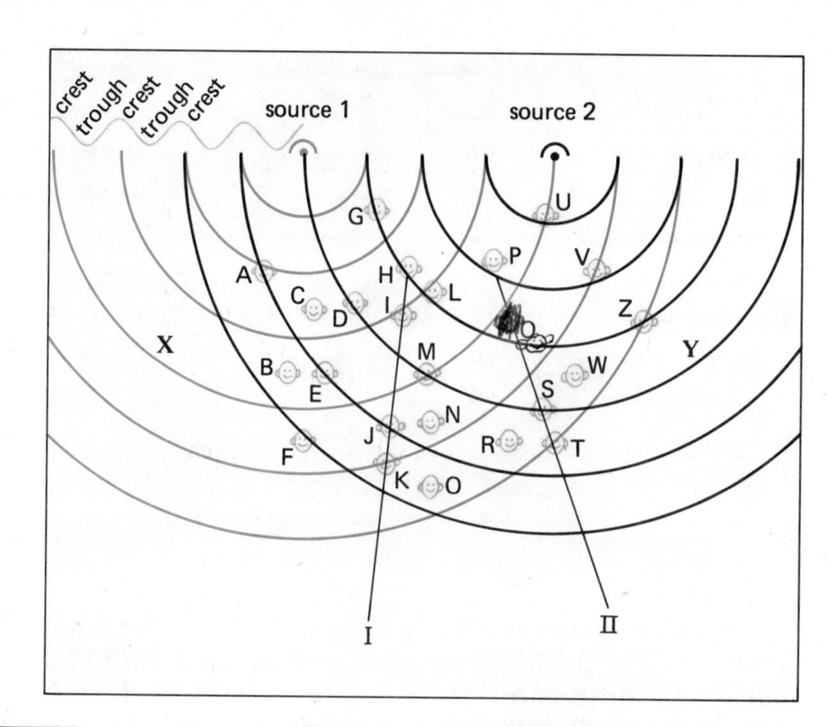

3-29. This diagram shows the interference of sound waves produced by two different sources.

ACTIVITY

Materials
paper
pencil

Interference

Figure 3-29 shows two sources of sound that are broadcasting the same frequency in an auditorium. Students are located as shown, and each student has an identifying letter. Waves are drawn in with a semicircle indicating a crest. Of course, there is a trough between crests. Where two crests (lines) meet, there is a loud sound. Where two troughs meet, there is also a loud sound. Where a crest and a trough meet, the trough cancels the crest and there is little or no sound.

1. What are the letters of the students who will hear a loud sound?

2. Explain why they will hear a loud sound.

3. What are the letters of the students who will hear little or no sound?

4. Explain why they will hear little or no sound. I and II are lines along which little or no sound can be heard. There are also several other lines along which no sound can be heard.

5. Which four letters indicate positions along another line of no sound?

6. What do you think you would hear if you walked along x to y in the auditorium?

SUMMARY

When you listen to music, you hear many different sound waves. Waves that occur together may change each other by the process of interference. The difference between pleasant musical sounds and nonmusical noise is the result of interference. You can hear a radio playing in the next room because waves are able to move around a barrier like a doorway by diffraction.

QUESTIONS

Use complete sentences to write your answers.

1. Explain what happens when sound waves meet, causing an increase in loudness.
2. Explain what happens when sound waves meet, causing a canceling of the sound.
3. What condition must be met for a mixture of sounds to produce music?
4. Explain why you do not have to stand directly in a doorway to hear sound coming from inside a room.

LABORATORY

Characteristics of Sound

Materials
STATION A
tuning fork
wad of aluminum foil
string
beaker of water
balloon
rubber band
STATION B
meter stick
metal rod
ticking wristwatch
STATION C
6 test tubes of water

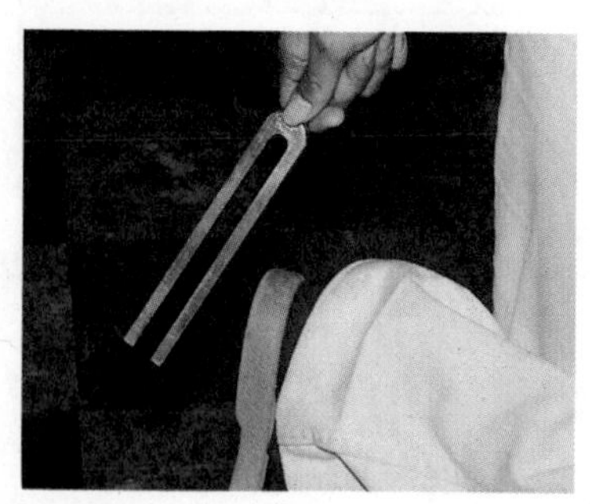

3-30.

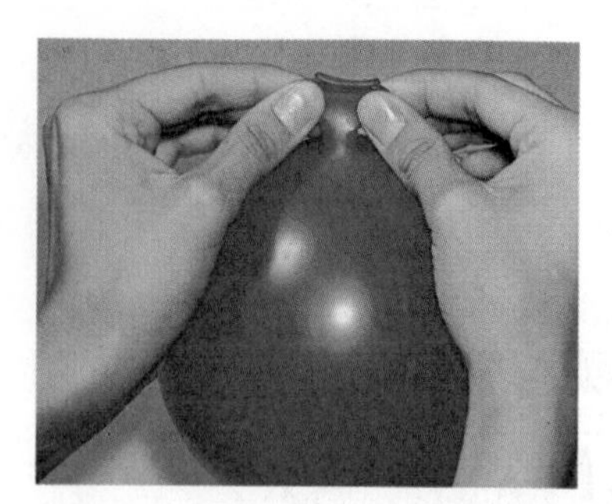

3-31.

Purpose

In this laboratory, you will be studying some of the things that affect the sounds you hear. You will be in a group assigned to one of three stations in the room. The stations are lettered A, B, and C. At each station, you are to follow the procedure suggested below. Use your eyes and fingers as well as your ears for observing in these activities. When you complete the activities at a station, compare your results with others in your group before you move to the next station.

Procedure

STATION A

A. Gently tap the tuning fork on a rubber heel. See Fig. 3-30. Look at the prongs and listen.

1. What do you observe? Record your observations.

B. Tap a tuning fork and let it touch a small ball of aluminum foil hanging from a string.

2. What happens to the ball?

C. Tap the tuning fork again and carefully let the prongs touch the surface of the water in the beaker.

3. What happens to the water? Why?

D. Hold a rubber band tightly and have a partner strum it.

4. What do you observe about the rubber band?

E. Blow up a balloon. Allow the air to escape from it as you stretch the neck of the balloon. See Fig. 3-31.

5. What do you observe at the neck of the balloon?

F. Look at the results of each activity. Prepare a hypothesis (guess) explaining each of your observations. If you observed nothing in common that could cause sound, repeat the activities.

6. State your hypothesis for the cause of the sound in each activity.

STATION B

G. Look at both the meter stick and the metal rod.

7. Are they the same length?

H. If the stick and the rod are not the same length, use a light pencil to mark the length of the shorter one on the longer one.

I. Place the wristwatch on the mark on the longer stick or rod. One partner at a time should try to hear the watch ticking by placing his or her ear at the other end of the stick. See Fig. 3-32. Take turns listening.

8. Can you hear the watch ticking?

3-32.

J. Replace the longer stick with the shorter one and repeat step I.

9. Can you hear the ticking better on the metal rod or the meter stick?

K. Hold the watch in the air and listen from 1 m away.

10. Can you hear the watch as well in the air as on the rod?

11. Why were you able to hear the watch from 1 m away by using a rod?

Loudness
The effect that the energy of sound waves has on the ear.

STATION C

In this activity, you will be studying the causes of pitch and loudness.

L. Fill three test tubes with different amounts of water. Blow across the top of each of them. See Fig. 3-33.

12. Which test tube gave the highest pitch?

13. Which test tube gave the lowest pitch?

M. If you wanted to make the sound from each test tube louder, what would you do? Try it.

14. What did you decide to do to make the sound louder? Did it work?

3-33.

Summary

Write a short paragraph summarizing the information obtained in this laboratory.

VOCABULARY REVIEW

Match the number of the term with the letter of the phrase that best explains it.

1. wave
2. wavelength
3. frequency
4. refraction
5. reflection
6. Doppler effect
7. sound
8. pitch
9. interference
10. diffraction

a. An apparent change in frequency of waves because observer or source is moving.
b. Waves are thrown back after striking a barrier.
c. A wave that causes a back-and-forth vibration of particles through which it travels.
d. The distance between two neighboring crests or troughs of a wave.
e. The ability of a wave to bend around an obstacle in its path.
f. The effect two or more waves have on each other if they overlap.
g. A disturbance caused by energy moving from one place to another.
h. A description of a sound wave that is closely related to frequency.
i. The number of complete waves passing a point each second.
j. A process that causes a wave to change its direction because its speed changes.

REVIEW QUESTIONS

Complete each statement by choosing the best word or phrase, or by filling in the blank.

1. Energy moving across water, a rope, or a coil spring is best seen as ________.
2. The feature of a wave that is measured from crest to neighboring crest is the **a.** wavelength **b.** speed **c.** frequency **d.** direction.
3. A wave graph made of a swinging pendulum will show the **a.** wavelength **b.** crests **c.** troughs **d.** wavelengths, crests, and troughs.
4. A wave changes direction when its speed changes. This is called **a.** the Doppler effect **b.** reflection **c.** refraction **d.** frequency shift.
5. A barrier in the path of a wave may affect the wave by bending the wave as it passes the edge or by causing it to bounce off the barrier. The first effect is called ________; the second is called ________.

6. Of the following, the one that is *not* an example of the Doppler effect is **a.** a musician changing the pitch of an instrument while tuning it **b.** the change in pitch of an approaching auto horn **c.** the change in pitch of the hum of a motorcycle as it passes you.
7. Unlike the up-and-down motion of water waves, sound waves cause the particles to move ________.
8. It is not difficult to recognize that the sound of a guitar is different from that of a piano since **a.** their pitches differ **b.** one is louder than the other **c.** the mixture of frequencies each produces is different **d.** a piano is larger than a guitar.
9. When one sound wave meets another, both are changed by the process known as **a.** diffraction **b.** reflection **c.** interference **d.** refraction.
10. Sound waves that are musical differ from sound waves called noise in that their ________ have a simple mathematical relationship.

REVIEW EXERCISES

Give complete but brief answers to each of the following.

1. Explain how water waves result from energy moving through the water.
2. A 25-cm pendulum is making 60 full swings each minute.
   a. Draw a wave graph representing this motion and label a wavelength, crest, and trough.
   b. What is the frequency in hertz of this pendulum?
3. Describe how a water wave would change when:
   a. It strikes a barrier in its path.
   b. It enters shallow water.
4. Describe the changes in frequency of an auto horn that account for the change in pitch that occurs as the horn approaches and then passes you.
5. Refer to Fig. 3-29, p. 104. Write a statement describing exactly what is happening at each of the dashed vertical lines. Start with the left-hand side and use words from the following list: wave A, wave B, crest, trough, combine, canceled. You will have five statements when finished.
6. Explain how pitch is related to frequency of sound.
7. Describe what generally happens to the energy carried by sound waves as it moves through a substance.
8. In what three general ways may two sound waves affect each other?
9. Compare the sound waves which would be called noise to those which would be called music.

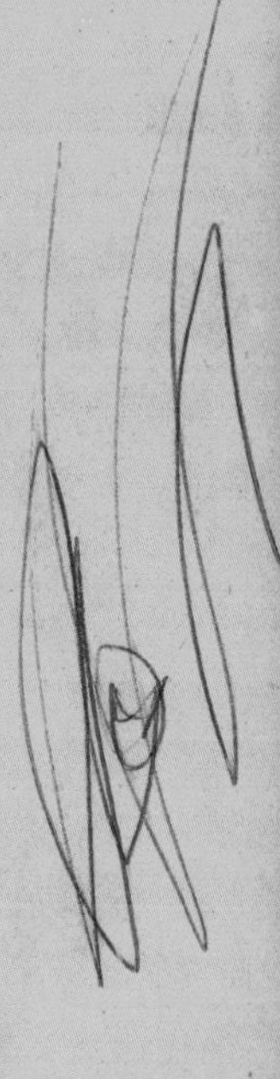

10. Give an example to show that sound waves bend around a barrier.
11. If 24 waves pass an observer in 8.0 sec, what is the frequency of the waves in hertz?
12. The frequency of a wave is 12 Hz and it has a wavelength of 3 cm. What is the speed of the wave?
13. You see a pistol being fired and 4 sec later you hear the sound. Later you measure the distance to the spot where the pistol was fired and find it to be 1,440 m. What is the speed of the sound made by the pistol?
14. Assume that sound travels 350 m/sec. If you see a lightning flash and 5 sec later hear the thunder, how far away was the lightning?

EXTENSIONS

1. Research and report on beat frequencies. Most physics textbooks discuss this phenomenon.
2. Sound bends around a corner when it passes through an open doorway, but light apparently does not. Look up diffraction in an encyclopedia or physics text and find out why this is so.
3. Write a report on the steps necessary to make a phonograph record of a musical performance.

CHAPTER 4

Light

4-1. THE ELECTROMAGNETIC SPECTRUM

The instrument shown in the photograph above is a radio telescope. As the name implies, this is a telescope that gathers radio waves instead of light waves. Using radio telescopes, astronomers are able to collect information about the most distant objects in the universe. Radio telescopes have even been used to search for signals from possible civilizations in outer space. As you will learn in this lesson, radio and light are similar forms of energy.

When you finish lesson 1, you will be able to:

- Use examples to show that light behaves as if it is made of waves.

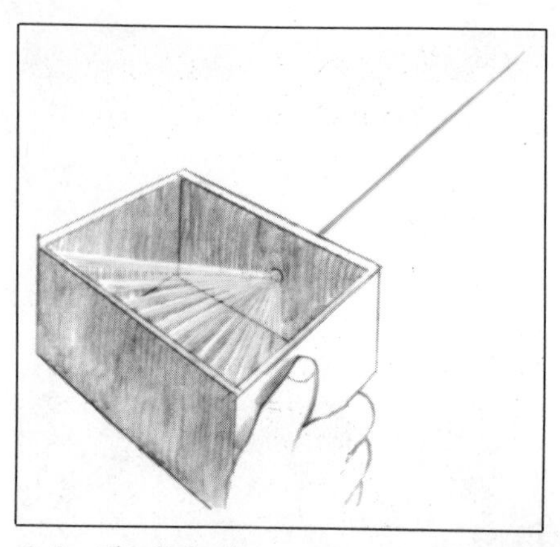

4-1. *The light from a distant source will spread out around the pinhole.*

- Explain what is meant by the *electromagnetic spectrum*.
- Name and give the characteristics of six types of waves in the electromagnetic spectrum.
- ○ Measure the heating effect of the sun's *infrared* waves.

To begin your study of light waves, try this experiment. Make a small pinhole in an index card. Look through the pinhole at a light some distance away. If the hole is very small, you will see the light spread out. That is, the light will appear to be larger than the pinhole through which it is coming. This strange effect can be explained if light is thought of as a series of waves.

The light will appear spread out because the light waves are diffracted when they pass through the pinhole. This is shown in Fig. 4-1. Light seems to be a result of energy moving in the form of waves.

Measurement of the amount of energy carried by light shows that this energy arrives in the form of small packages. These small packages of light energy can be thought of as particles of light. A particle of light energy is called a *photon*. Light can be thought of as a stream of photons guided by waves. However, most of the behavior of light can be explained by thinking only of the light waves.

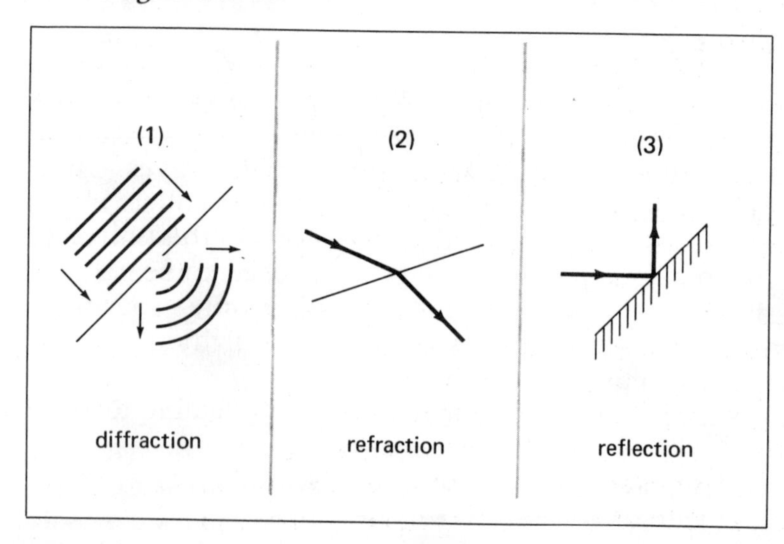

4-2. *Like all waves, light can be diffracted, refracted, and reflected.*

Light has many of the properties of waves, such as diffraction, reflection, and refraction. See Fig. 4-2. Light waves are different from other waves, however, in some important ways. For example, unlike sound, light can move through empty space. See Fig. 4-3. Light waves also move faster than sound waves. Moving with the speed of light, you could make about 7½ trips around the world in 1 sec. Such great speeds are very hard to measure. It took scientists many centuries to find a way to make this measurement.

4-3. *There is no air on the moon. On the moon, you would be able to see by reflected sunlight. However, you would not be able to hear a sound.*

The Italian scientist Galileo tried to measure the speed of light in the sixteenth century. He attempted to measure the time it took for light to travel about 2 km. However, Galileo had no way to measure the very small amount of time it took for light to travel such a short distance. A Danish astronomer named Roemer was more successful. In 1675, he was able to measure the time it took for light from one of the moons of the planet Jupiter to travel across the diameter of the earth's orbit. Dividing the diameter of the earth's orbit by this time gave Roemer a value for the speed of light. His result was not accurate because the actual diameter of the earth's orbit was not known in his day. Modern scientists have developed very accurate ways to measure light's great speed. This speed has been found to be 3×10^8 m/sec in a vacuum (3×10^8 means that 3 is followed by 8 zeros: 300,000,000).

Waves that can move through empty space at the speed of light are called **electromagnetic** (ih-**lek**-troe-mag-**net**-ik) **waves.** Visible light waves are not the only form of *electromagnetic waves*. Radio waves also move through empty space and travel at 3×10^8 m/sec. Experiments show that light and radio waves travel through space at the same speed. So, radio waves are also electromagnetic waves. There are many other electromagnetic waves. For example, microwaves used to cook food are electromagnetic waves.

Electromagnetic waves A form of energy capable of moving through empty space at 3×10^8 m/sec.

Electromagnetic waves are usually described by their frequencies. (Remember, frequency is the number of complete waves passing a point in 1 sec.) All of them have rapid frequencies, ranging from 10^4 to 10^{21} hertz (Hz). Visible light has a frequency of about 10^{14} Hz. The frequency of microwaves is about 10^9 Hz.

All types of electromagnetic waves known to exist make up the **electromagnetic spectrum.** A chart of

Electromagnetic spectrum The series of waves with many properties similar to those of light.

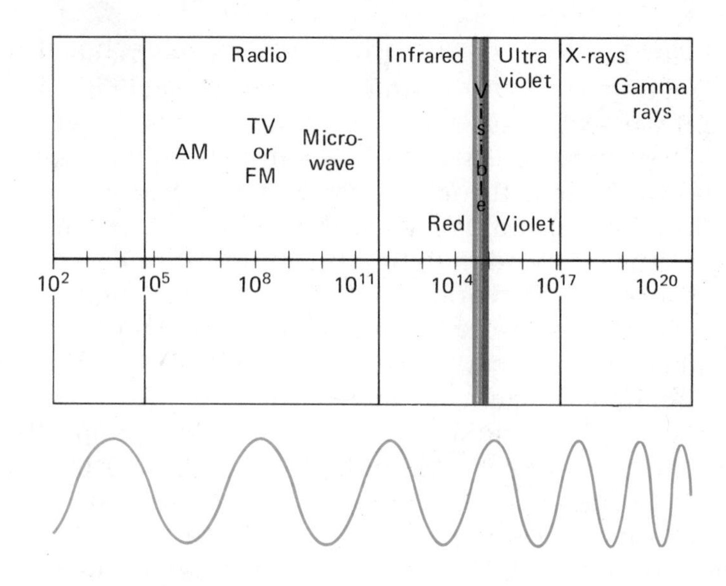

TABLE 4-1

the *electromagnetic spectrum* is shown in Table 4-1. The major parts of the electromagnetic spectrum are:

1. *Radio waves.* These waves include the parts of the electromagnetic spectrum with the lowest frequencies. Various frequencies of radio waves have different uses. For example, ordinary AM radio broadcasts use frequencies between 535 kilohertz (kHz) and 1,605 kHz. A kilohertz is one thousand waves per second. AM radio waves can travel long distances. A part of the atmosphere reflects these waves back to the earth's surface. Thus, AM radio broadcasts can be received far away from the station. FM radio waves have much higher frequencies. These waves have frequencies between 88.1 megahertz (MHz) and 107.9 MHz. One megahertz is one million waves per second. FM radio waves are not reflected from the atmosphere. This means that FM waves cannot be received far away from the station. TV broadcasts also use high-frequency radio waves like FM. For this reason, TV signals do not travel far. Each radio or TV station sends out radio waves of only a certain frequency. When you change from one radio station or TV channel to another, the set is being adjusted to receive a particular frequency.

Radar waves have frequencies above those of FM and TV. Waves at these very high frequencies can be reflected by many objects. The radar waves that are reflected can be detected in the same way that reflected sound waves are heard as an echo. This means that radar can be used to "see" objects through darkness, fog, or at a great distance.

Microwaves have the highest frequencies of any kind of radio wave. Unlike radar, microwaves pass through or are easily absorbed by some materials. Absorption of microwaves can cause many materials to be heated, as food is in a microwave oven.

2. *Infrared waves.* These waves have frequencies greater than all other types of radio waves. Infrared waves are given off by objects when they are heated. Cooler objects absorb infrared waves which then change into heat. The sun, for example, produces infrared waves. Your skin feels warm when exposed to sunlight because of the infrared waves it absorbs. A dark surface absorbs infrared waves better than a light-colored surface. This is the reason that light-colored clothing feels cooler in hot summer sunlight.

3. *Visible light waves.* A very narrow band of frequencies higher than infrared can be seen with the eye. We see the different frequencies making up this narrow part of the electromagnetic spectrum as different colors. The lowest frequencies of visible light are seen as red. The highest frequencies appear as blue or violet. Electromagnetic waves with frequencies above or below the visible part of the spectrum cannot be seen by the human eye.

4. *Ultraviolet waves.* These waves occupy a part of the electromagnetic spectrum at frequencies just above

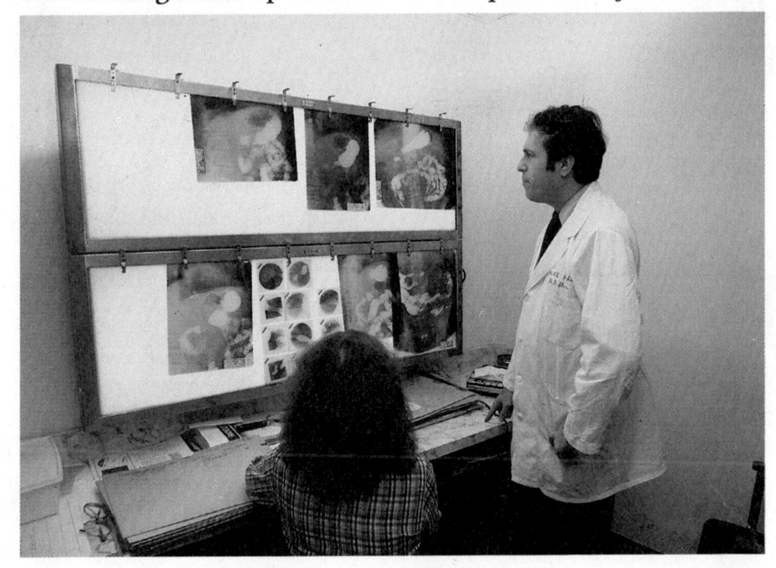

visible light. Ultraviolet waves in sunlight are responsible for *sunburn*. Too much ultraviolet can be dangerous and may cause serious skin damage. But ultraviolet light can also kill germs and it is often used for this purpose in hospitals.

5. *X rays.* These high-frequency electromagnetic waves are very useful. X rays easily pass through most materials. They can also produce a photograph on film or a picture on a special screen. This means that X rays can be used to see inside solid objects. You have probably had X-ray photographs made of parts of your body by a doctor or dentist. However, too much exposure to X rays can kill living cells. Machines that produce X rays must be used only by trained persons.

6. *Gamma rays.* These waves are in the part of the electromagnetic spectrum having the highest frequencies. Gamma rays are similar to X rays but are much more dangerous because they can pass through matter with greater ease. Gamma rays are useful in fighting cancer because of their ability to penetrate matter easily. Beams of gamma rays can be focused on cancerous cells, destroying them but not harming surrounding tissue.

ACTIVITY

Materials

white cloth or paper
black cloth or paper
2 thermometers

Infrared waves

In this activity, you will measure the heating effect of infrared waves from the sun on two types of surfaces.

A. Obtain the materials listed in the margin.

B. On a window sill, place a piece of white cloth in direct sunlight. Next to the white cloth, place a piece of black cloth.

C. Leave both pieces of cloth in direct sunlight for 4–5 min.

D. Place a thermometer under each piece of cloth.

1. Which piece of cloth is warmer?
2. Based on what you know about infrared waves, explain why one piece of cloth is warmer than the other.

SUMMARY

Light has characteristics that cause it to act as if it were made up of waves. Light waves can be diffracted, reflected, and refracted. Light can also move through empty space and travel at high speeds. The electromagnetic spectrum is composed of light waves arranged according to their different frequencies.

QUESTIONS

Unless otherwise indicated, use complete sentences in writing your answers.

1. Describe one experiment that you could do to show that light behaves as waves.
2. For each of the following terms, find the letter in Fig. 4-4 that matches each of the kinds of electromagnetic waves.
   - **a.** visible light
   - **b.** X rays
   - **c.** infrared rays
   - **d.** ultraviolet rays
   - **e.** gamma rays
   - **f.** radio waves
3. Choose three of the types of electromagnetic waves listed in question 2. Write a brief paragraph describing how these types of waves are important in daily life.

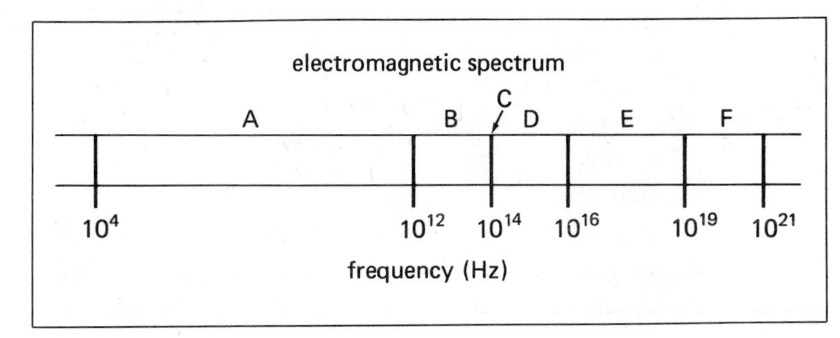

4-4.

4-2. MOVEMENT OF LIGHT WAVES

The picture shows several index cards with holes in them in front of a candle. If all the holes are lined up, you can look through all of them at once and see the candle. If one card is moved to the side, however, you will not be able to see the candle. You might try this experiment yourself. What do you think this experiment proves about light waves?

When you finish lesson 2, you will be able to:

- Show that light waves ordinarily move in straight paths.
- Give two examples to show that light may change its direction when it passes from one material into another.
- Explain how the bending of light waves can produce *mirages*.
- ○ Using a glass of water, demonstrate the fact that light changes its direction when it passes from one material into another.

4-5. *Heliographs were once used to send coded messages for short distances. The heliograph operated by reflecting sunlight from a mirror.*

During the late 1800's, the armies of several countries, including the United States, used a device to send signals by means of sunlight reflected from a mirror. This device was called a *heliograph* (from the Greek words *helios* meaning "sun" and *graphein* meaning "to write"). The heliograph equipment consisted of one or two mirrors mounted on a tripod. A shutter to interrupt the flashes was mounted on another tripod. See Fig. 4-5. With the sun in front of the sender, its rays could be reflected directly to the receiver. The shutter was opened and closed to produce short and long flashes of sunlight. These flashes reproduced the pattern of dots and dashes in Morse code. A message could be received as far away as 48 km under ordinary conditions. Why was it possible to send messages by sunlight?

Light seems to move in a straight path as it travels through the air. See Fig. 4-6. Because light travels in a straight path, you can easily direct reflected light. The motion of light can be shown by straight lines called **rays.** The *rays* show the path that light follows as it moves through a substance such as air.

4-6. *Sunlight would not appear like this unless light traveled in straight lines.*

Rays
Straight lines showing the path followed by light.

Light rays reflected from the surface of a flat mirror all follow the same path. This happens because all the rays are reflected at the same angle with which they strike the mirror. See Fig. 4-7a. As a result, the pattern of the light rays is not changed. Thus, when you look in a flat mirror, you see the reflected rays forming an exact image of everything in front of the mirror. See Fig. 4-7b.

Light rays falling on an uneven surface such as the walls of a room are reflected in many different directions. An uneven surface such as a wall does not look shiny like a mirror. No images are seen since the reflected light rays have no regular pattern. See Fig. 4-7c. Only the uneven surface itself is visible.

When light rays move from one material into another, their speed changes. For example, suppose light moves from air into water. Light waves travel about 25 percent slower in water than in air. This change of speed causes light entering the water at an angle to be *refracted* or bent into a new path. See Fig. 4-8. Refraction of light makes objects underwater appear to be

4-7.

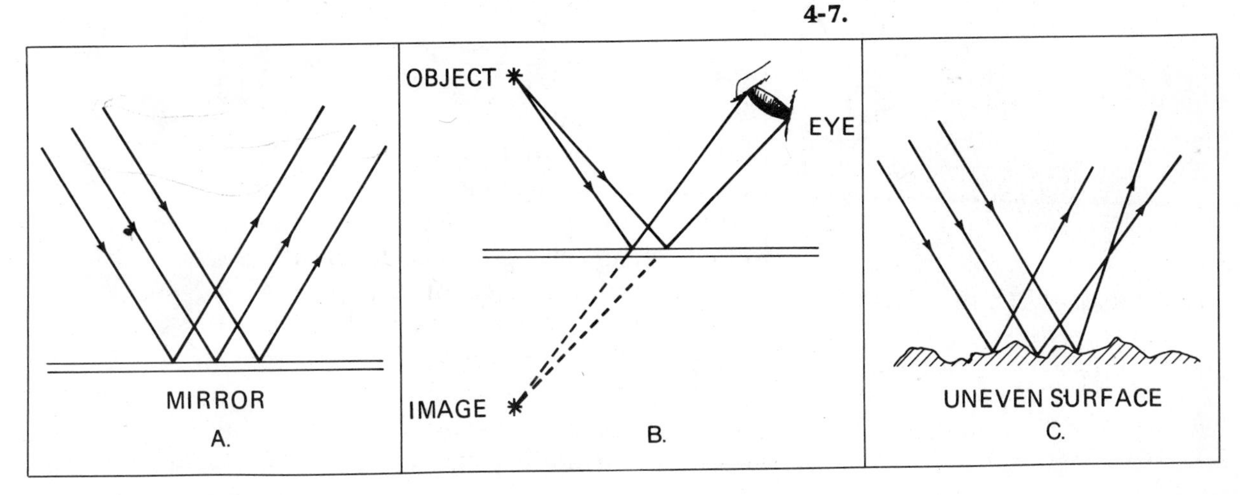

4-8. *The light beams coming from beneath the water change their direction when entering the air. (Courtesy Time/Life Books)*

closer to the surface than they really are. See Fig. 4-9. Refraction of light also makes water seem less deep than it really is.

Some materials slow light more than water. A diamond, for example, slows light to less than 50 percent of its speed in air. As a result, light rays entering a diamond from air are refracted through very large angles. The amount of bending of light rays that go from air into various materials is given by the materials' *index of refraction*. The index of refraction for any material is found by dividing the speed of light in air by the speed in that material. For example, the index of refraction for a diamond is given by:

$$\text{index of refraction} = \frac{\text{speed of light in air}}{\text{speed of light in diamond}}$$

$$= \frac{3.00 \times 10^8 \text{ m/sec}}{1.24 \times 10^8 \text{ m/sec}} = 2.42$$

The index of refraction of some common materials is given in Table 4-2. A large index of refraction indicates the greatest change in the path of light rays.

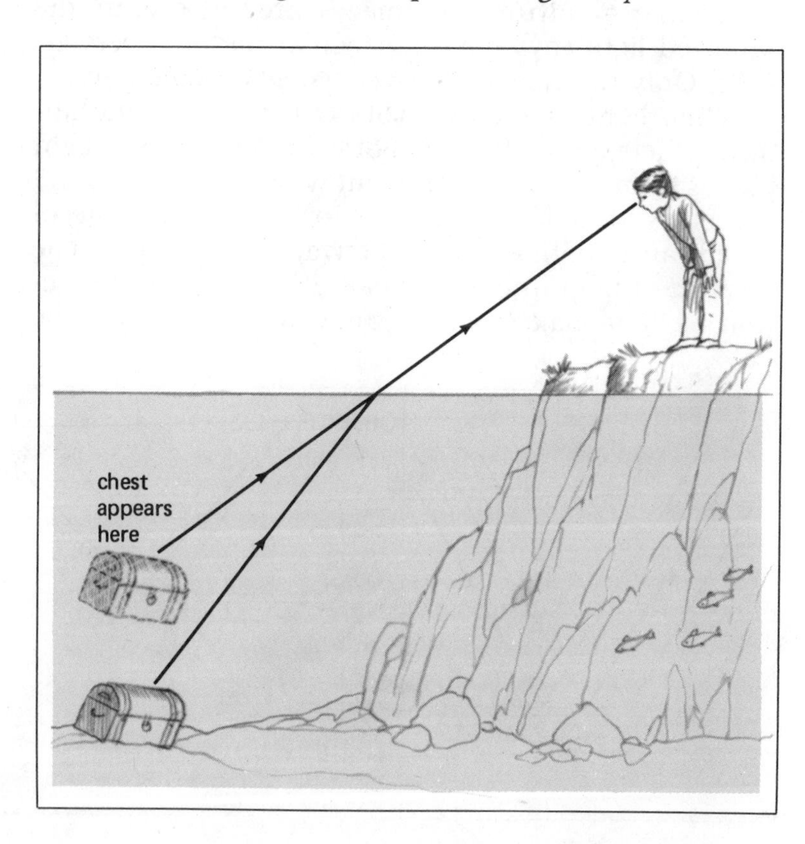

4-9. *An object seen underwater is really lower than it seems because the light rays reflected from the object are refracted as they move from the water into the air.*

TABLE 4-2
Index of Refraction of Some Common Substances

Material	Index of Refraction
Water	1.3
Glass	1.5
Cooking oil	1.5
Diamond	2.4

You may have noticed an unusual example of refraction of light rays while swimming. If you swim underwater and look up, the surface of the water acts like a mirror. This happens because many light rays moving from the water into the air are bent back into the water. This effect is called internal reflection. One of the light rays in Fig. 4-8 shows total internal reflection. These light rays that are reflected back into the water make the water surface look like a mirror when seen from below.

A kind of "light pipe" can be made by using this method of reflecting light rays. A thin fiber of clear glass or plastic is used. A ray of light can be kept inside the fiber as shown in Fig. 4-10. The use of fibers to carry light in this way is called *fiber optics*. Light can be used to carry messages along fiber optic cables. Other devices using fiber optics can be used to look inside the human body.

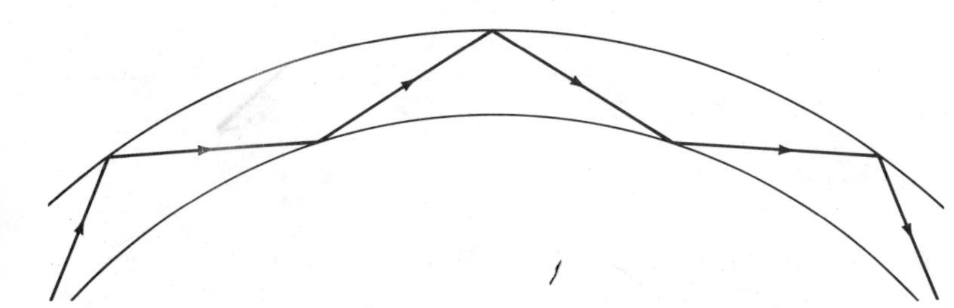

4-10. *A ray of light can travel inside a single clear fiber as thin as a hair. The light ray is totally internally reflected and never escapes from the fiber.*

Refraction of light explains the appearance of **mirages** (muh-**rah**-zhez). *Mirages* are illusions or false impressions. These illusions cause distant objects to seem to be upside down or floating in the air. Mirages occur when light is refracted while passing through air layers of different temperatures. On a warm day, the air close to the ground is warmer than the air at higher levels. Light is refracted by the warm air. This refraction makes distant objects appear to be upside down.

Mirage
An illusion caused by the refraction of light in which distant objects are seen upside down or floating in the air.

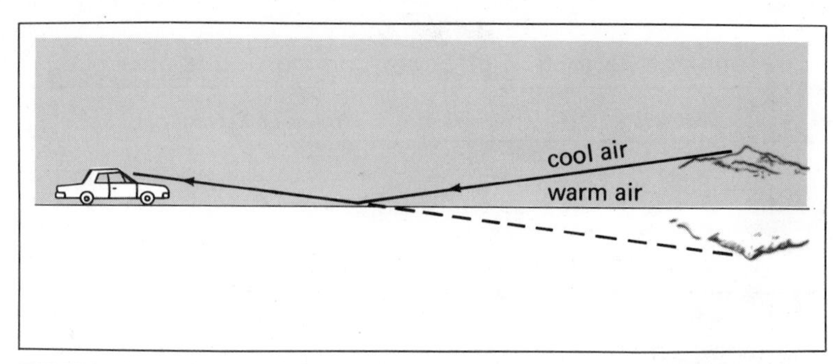

4-11. A mirage in which the image seems to be below the horizon.

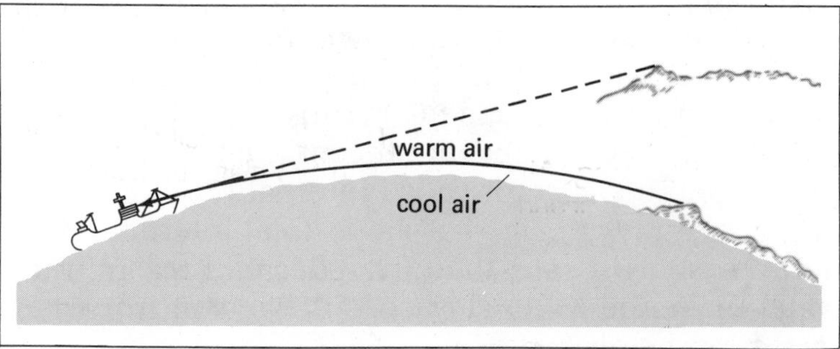

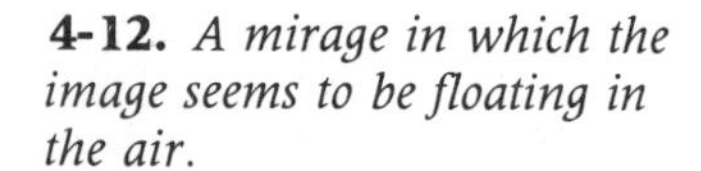

4-12. A mirage in which the image seems to be floating in the air.

See Fig. 4-11. Sometimes the air near the ground is cooler than the air at higher levels. When this happens, a mirage is formed as shown in Fig. 4-12.

ACTIVITY

Materials
2 glasses
rubber band or tape
coin or washer
pencil
paper
water

Refraction of Light

A. Obtain the materials listed in the margin.

B. Fill one of the glasses about 3/4 full of water.

C. Put a rubber band around the other glass about 4 cm from the bottom.

D. Place a coin in the bottom of the empty glass, making sure it is in the center.

E. From about 30 cm away, move your head until you see the edge of the coin lined up with the rubber band. Figure 4-13 shows the path followed by the light from the coin to your eye.

F. Hold the coin with a pencil and watch the coin while you pour the water from the first glass into the glass with the coin. See Fig. 4-14. Do not move your

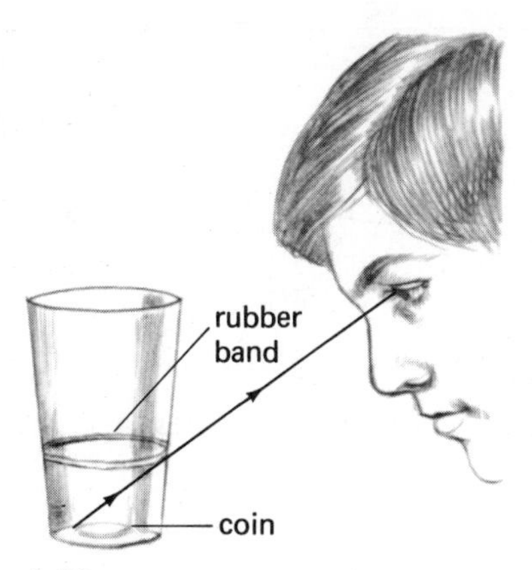

4-13.

head while pouring. Always look at the coin through the top surface of the water.

1. Does the coin appear to move? Part of the path of the light which allows you to see the coin is shown in Fig. 4-15. Only the path from the water surface to your eye is shown.

G. Copy Fig. 4-15 on a sheet of paper. Now draw a line to show the path the light must follow from the coin to the water surface.

2. Does putting water in the glass change the path of the light from the coin to your eye? If so, how is the path changed?

Now try another experiment to see if the direction of light changes in going from water to air.

H. Put a pencil at an angle into the glass of water.

I. Look along the length of the pencil down into the water.

3. Does the pencil look straight?

4. In a complete sentence, explain what happens when light travels from water into air. What is this effect called?

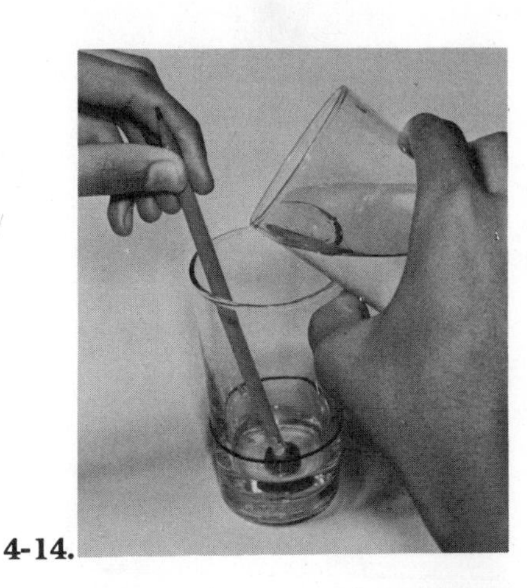

4-14.

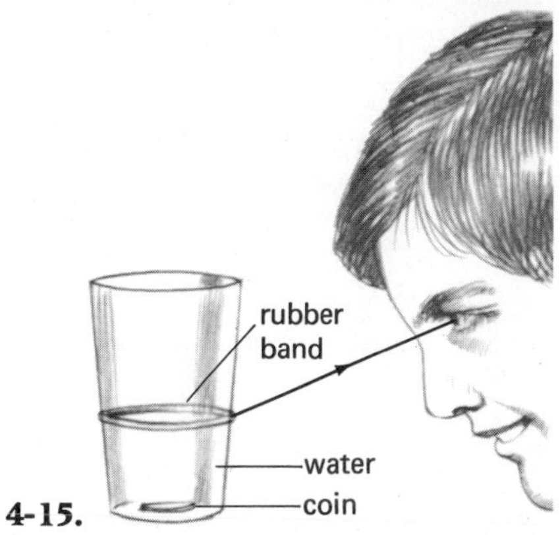

4-15.

SUMMARY

The experiment with the index cards and the candle shows that light moves in a straight path. You can reflect light from a mirror easily because light travels in a straight path. Light moves in straight paths in any particular substance. A ray is a straight line showing the path followed by light. When light passes from one substance to another, its speed changes. This change in speed causes the light to be refracted. Refraction of light explains the appearance of objects under water and mirages.

QUESTIONS

Use complete sentences to write your answers.

1. Give several examples to show that light waves travel in a straight line.
2. Explain why the depth of water seems to be less than it really is.
3. Why do mirages appear?
4. The speed of light in water is 0.75 of the speed of light in air. From this information, explain how to determine the index of refraction of water.
5. Some artificial substances that resemble diamonds have an index of refraction of 1.9. **a.** How does this compare with that of real diamonds? **b.** What decimal fraction of the speed of light in air is the speed in the artificial diamond?

4-3. COLOR

The phenomenon shown in the photograph is a rainbow. Rainbows are seen immediately after a rain shower when the sun comes out. Rainbows are made up of bands of color. The colors of the rainbow range from red, orange, and yellow, through green to blue and violet. The study of colors, such as those in the rainbow, is an important part of the study of light.

When you finish lesson 3, you will be able to:

- Interpret diagrams of the spectrum to show that white light is made up of colors.
- Relate the different colors of light to differences between light waves.
- Explain why objects can have many different colors.
- ○ Show that white light is made up of colors by using a diffraction grating.

4-16. *Can you imagine how this scene would look if the sky were pink or green instead of blue?*

4-17. *Why does the light from the sun appear to be different colors at sunrise and sunset?*

4-18. *Sir Isaac Newton performed many important experiments with light.*

Prism
A specially shaped piece of glass. A prism divides white light into its separate colors.

Visible spectrum
The band of colors produced when white light is divided into its separate colors.

Why is the sky blue? Where do the different colors come from at sunrise and sunset? The answers to these questions involve the properties of light. All color comes from light. The bright blue of the sky and the reds and yellows of sunrise and sunset are the result of white light from the sun being separated into different colors.

A simple experiment can be performed to show that white light is made up of different colors. This experiment was first performed by the great scientist Sir Isaac Newton. See Fig. 4-18.

When a ray of light passes from air into glass, its speed is slowed. This change in speed causes the ray to be bent, or *refracted*. The amount of bending of the ray depends on its frequency. See Fig. 4-19. Light rays with high frequencies are bent the most. A specially shaped glass called a **prism** will spread out the rays according to how much they are bent. White light passed through a *prism* is separated into a band of colors. This band of colors is called the **visible spectrum.** A *visible spectrum* produced by a prism is shown in Fig. 4-20.

All light that you see contains color. As Newton's experiments with the prism showed, white light is made up of all colors added together. Sunlight, for example, ordinarily appears white. Sometimes you can see the colors in sunlight spread out in a rainbow. A rainbow is made when droplets of water from rain act somewhat like natural prisms. Each raindrop spreads out the white sunlight into a tiny spectrum. See Fig. 4-21. As you look up from the ground at a rainbow, you can see only one color from each little spectrum. Looking highest, you can see only the reds. As a result, the top of the rainbow is seen as a band of red. Below the red band you see bands of orange, yellow, green, blue, and violet. See Fig. 4-21. All of these colored bands have a curved shape because of the way the rainbow is seen from the ground.

It would be easy for you to make a real rainbow. You could use a shower of droplets from a garden hose to take the place of the rain. The experiment should be done in the late afternoon when the sun is about half-way down from directly overhead. With your back to the sun, turn on a fine spray from the hose. Point the spray high in the air. The rainbow can be seen best if you look at it toward a patch of dark shade. Your small rainbow will be made in the same way as the giant ones you have seen in the sky.

You know from your study of the electromagnetic spectrum that violet light has a higher frequency than

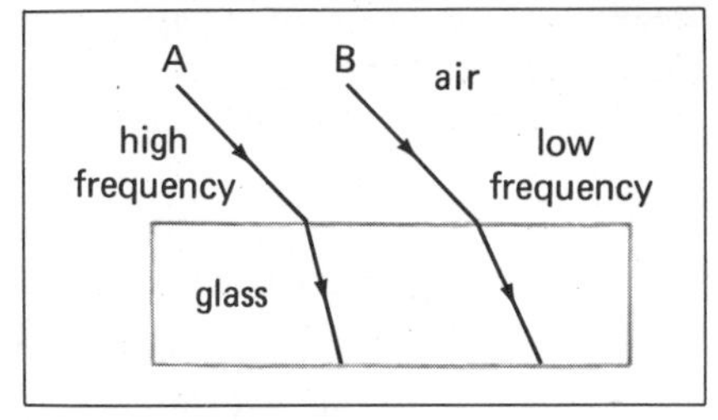

4-19. *If two rays of light, A and B, have different frequencies, they will be refracted by different amounts when passing through glass.*

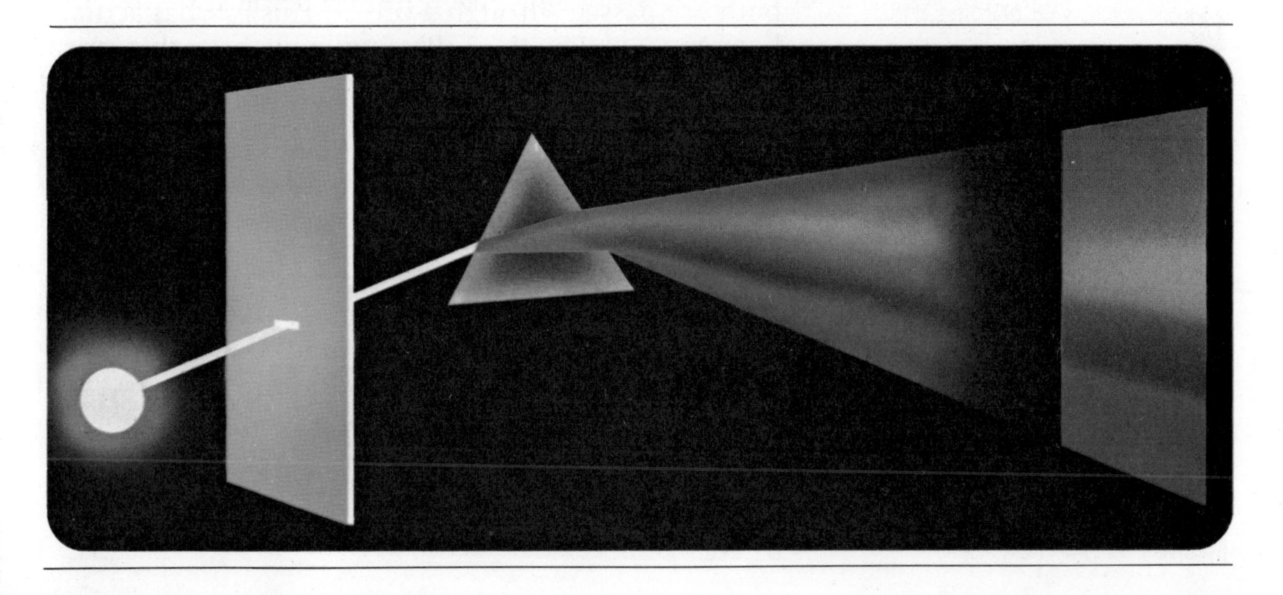

4-20. *A visible spectrum produced by a prism. Does this spectrum remind you of a rainbow?*

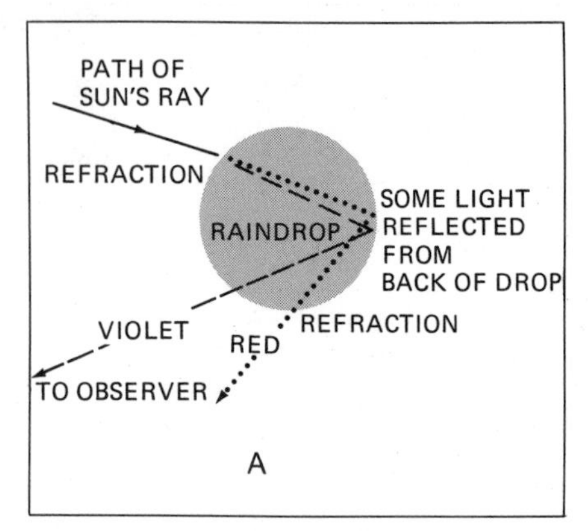

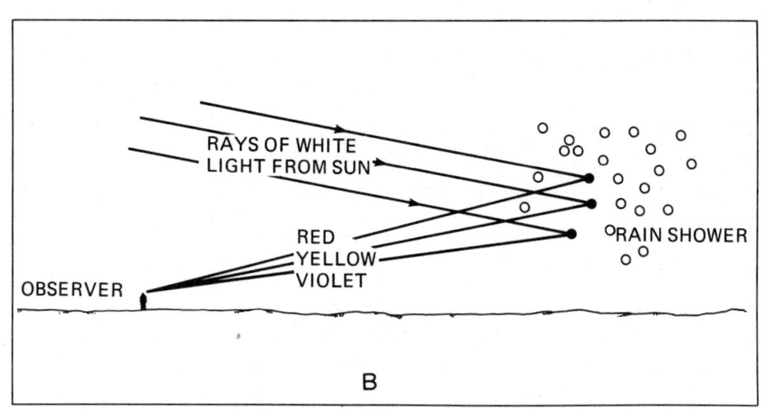

4-21. A. *A ray of sunlight is refracted as it enters and leaves a raindrop. As in a prism, this causes a separation of the white light into different colors.* **B.** *As you look at the sky through a rain shower, you see red refracted from the highest drops, yellow from lower ones, and violet from the lowest.*

red light. See Table 4-3. Rays with high frequencies are bent more than rays with low frequencies when they are refracted by a prism.

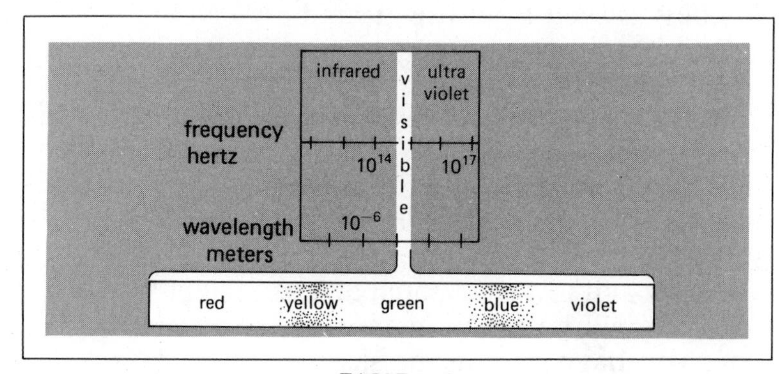

TABLE 4-3

Violet light should be bent more than red light when both are passed through a prism. This is what actually happens. See Fig. 4-22. Passing light through a prism shows that the different colors of light are made up of waves of different frequencies.

Almost all light that you see is a mixture of different frequencies. Sunlight, for example, contains a nearly equal mixture of all visible frequencies. However, a

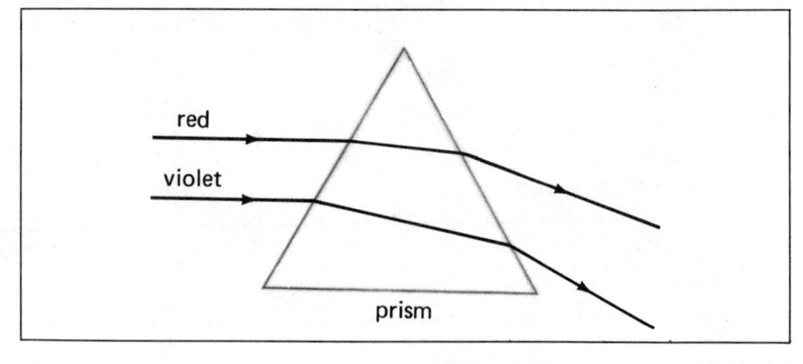

4-22. *Violet light has a higher frequency than red light. Violet light is bent more than red light when both are passed though a prism.*

special kind of light is made up of only a single frequency. This is the light produced by lasers. Laser light has other special properties. A beam of laser light does not spread out very much as it travels over great distances. See Fig. 4-23. By comparison, a flashlight beam may spread out after traveling only a few meters. A laser beam can be used to measure distances with great accuracy. To do this, a laser beam is sent out and reflected back. The time taken for the beam to travel out and back can be measured and the distance determined.

The energy contained in a laser beam sent into the eye can make repairs to the back of the eye. Laser beams can also be used to drill small holes in very hard substances such as diamonds. The unusual properties of laser beams will allow them to be put to many other uses in the future.

You see light waves of different frequencies as different colors. At midday, the sun is high in the sky. The white light from the sun passes through a relatively narrow band of atmosphere. Dust particles in the atmosphere reflect and scatter mostly blue light waves.

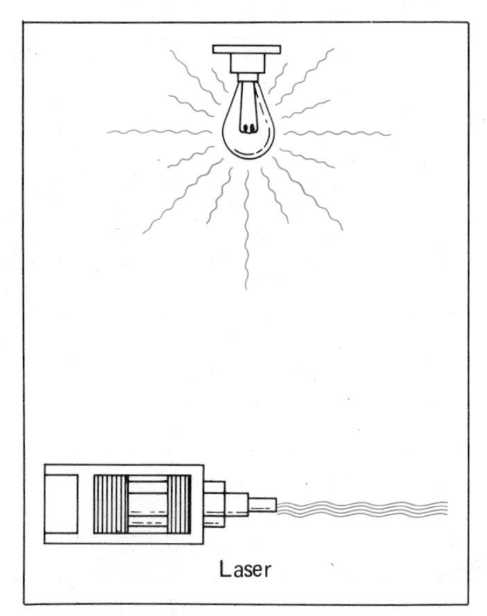

4-23. *The light produced by an ordinary light bulb is made up of many frequencies. Laser light consists of light waves with only one frequency.*

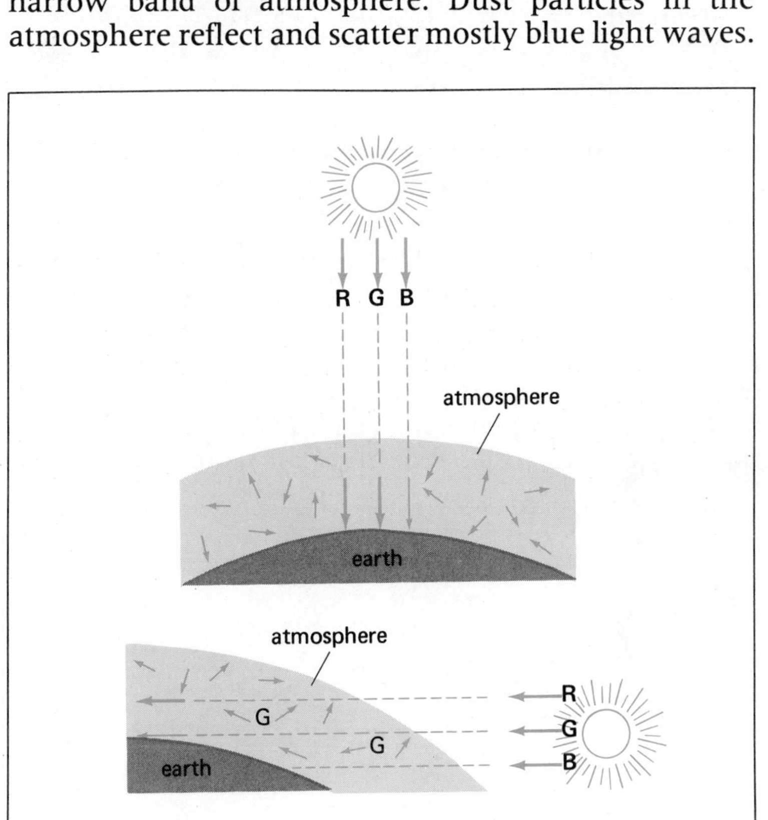

4-24. *The color of the sky is the result of the scattering of different wavelengths of light. At noon, the sky overhead appears blue. At sunset, the sky near the horizon appears reddish.*

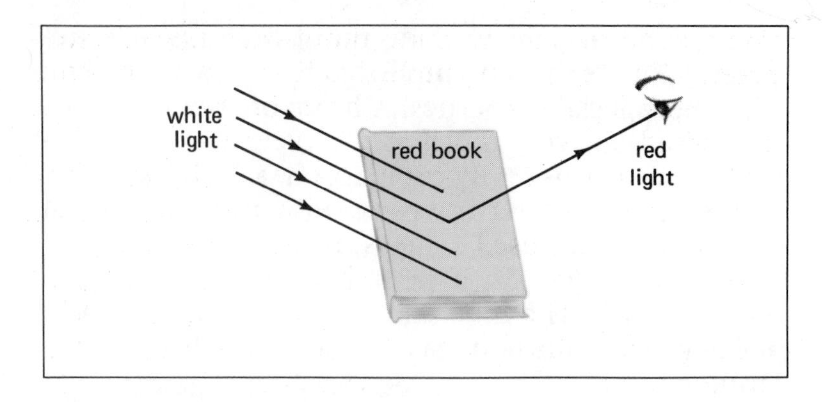

4-25. *A red object such as a book reflects only red light.*

The other frequencies of light are not affected as much. This scattering of blue light waves makes the sky appear blue.

At sunrise and sunset, the sun is lower in the sky. White light from the sun comes from an angle and must pass through more of the atmosphere. See Fig. 4-24. The atmosphere scatters all other frequencies more than red, orange, and yellow so that by the time the light reaches you, you see only reds and oranges. Most light that reaches your eyes has been reflected from the surface of some object. For example, if white light falls upon a red object, all frequencies except red are absorbed. Red light is reflected. See Fig. 4-25. Only the frequency of light called red is reflected to your eyes. You say the object is red. Most objects reflect more than one frequency. The color you see is the combination of those colors that are reflected. If light of a particular color is to be reflected, it must be present in the light falling on the object. White light contains all colors. If only red light shines on a red object, the object appears red. If only green light shines on a red object, the object appears black. See Fig. 4-26. No red light is present to be reflected. For example, suppose

4-26. *If there is no red light falling on the red book, no light is reflected. The book will appear black.*

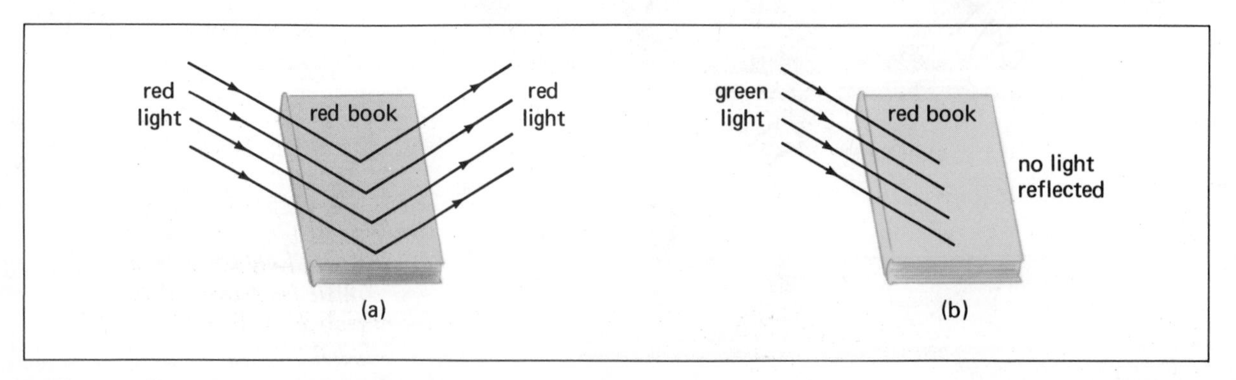

you wear a red coat in a room in which there is only green light. There is no red light for the coat to reflect. The green light is absorbed. The coat does not reflect any light. It will appear black.

The color of an object depends upon the frequencies of light that it reflects. Suppose you want to color an object by painting it. It is possible to produce almost any color by mixing together only three basic colors of paint. These basic colors are red, blue, and yellow. Mixing these paint colors causes specific color frequencies to be reflected. See Fig. 4-27. The color pictures in this book are made by using inks in the three basic colors of red, blue, and yellow, plus black. See Fig. 4-28.

If you could look at a color TV tube with a magnifier when the set is turned off, you would find that the face of the tube is covered with tiny dots. These dots are arranged in groups of three. When the set is on, one dot in each group gives off red light, blue light, or green light. See Fig. 4-29a. All of the colors seen in

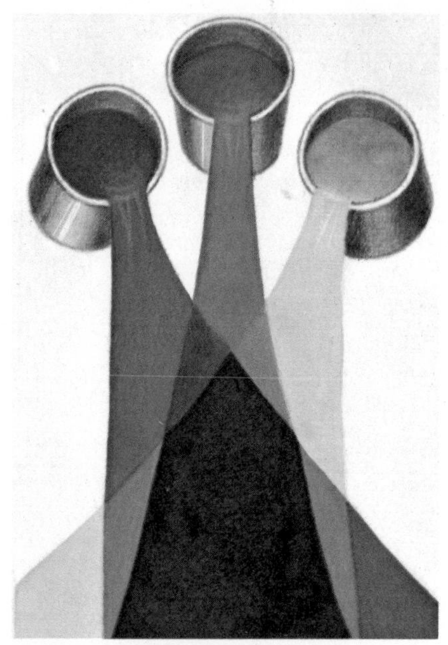

4-27. *When the three primary colors of paint (blue, yellow, red) are mixed, the result is black.*

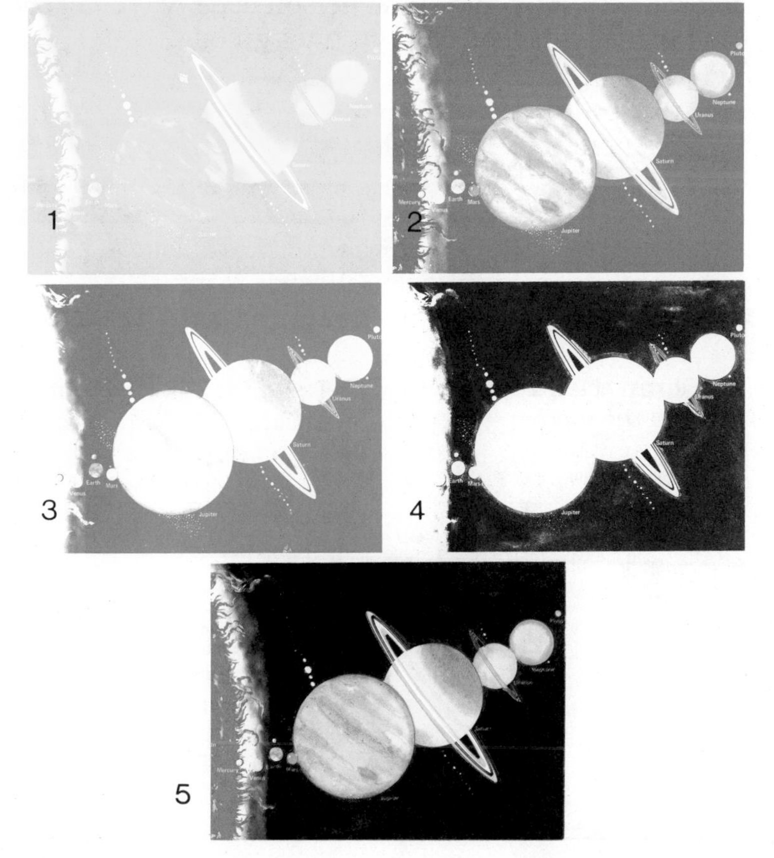

4-28. *A color picture is printed by first separating the picture into the basic colors of blue, yellow, and red. A fourth color (black) is also included. First, the yellow picture is printed (1), then red is added (2), and finally blue (3) and black (4). The finished picture is made up of dots of these four colors combined (5) to reproduce all the colors of the original picture.*

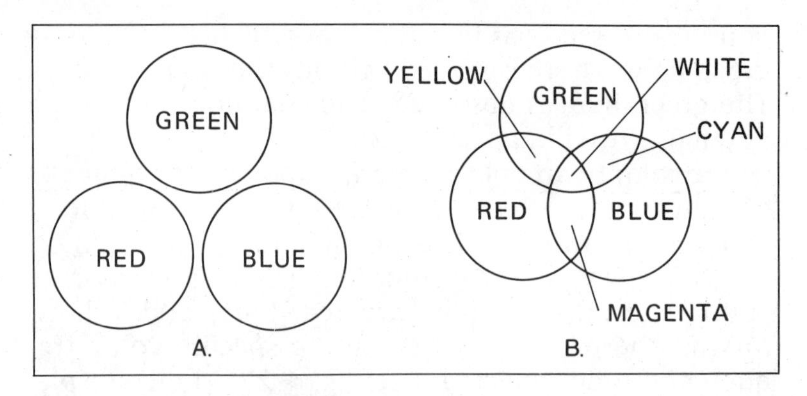

4-29. A. *One of the groups of three tiny colored dots that cover the face of a color TV tube.* **B.** *Various colors are produced by mixing the three colors of the glowing dots together.*

the TV picture come from mixing the three kinds of colored light produced by the separate dots. See Fig. 4-29b. These three colors are called primary colors of light. Notice that the primary colors of light are not the same as the basic colors of reflecting materiais such as paint.

Not all people see colors in the same way. About 8 percent of men and less than 1 percent of women have trouble seeing certain colors. This condition is called colorblindness. A colorblind person usually sees all colors normally except red and blue-greens.

ACTIVITY

Materials
diffraction grating

The Visible Spectrum

A. Obtain the materials listed in the margin.

B. Look through the plastic diffraction grating at a light bulb. The colorful display you see is called the spectrum of visible light.

C. Turn the diffraction grating so that you see all the colors to the left and right of the bulb.

1. Do the colors repeat themselves to the right of the light bulb?
2. What is different about the colors to the left of the light bulb?

D. Look at the colors to the right of the light bulb.

3. Name the colors you see from left to right.

SUMMARY

By using a prism, you can show that white light is made up of different colors. Sometimes you can see many of the different colors of sunlight in a rainbow. Each color of light is made up of light waves with a particular frequency. The colors you see depend on how the different frequencies of light are refracted or reflected.

QUESTIONS

Use the following diagram (Fig. 4-30), which represents light passing through a prism, to answer questions 1–3. Use complete sentences to write your answers.

1. If A is a beam of white light, what color will be found at B?
2. If A is a beam of white light, what color will be found at C?
3. How would the results differ if A were a beam of red light?
4. Explain: **a.** why grass appears green, **b.** why the sky appears blue, **c.** why a red coat appears black when seen in green light.
5. What would the American flag look like when seen in a green light? in a red light?

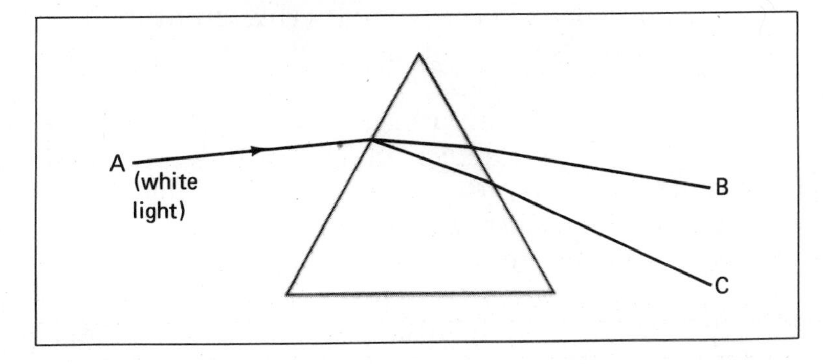

4-30.

4-4. BENDING LIGHT RAYS

Cameras similar to the one shown in the photograph first came into use early in the nineteenth century. Today, modern cameras can produce full color photographs in seconds. Cameras also play an important role in astronomy, geology, medicine, and many other branches of science. In this lesson you will see how cameras use light to make pictures.

When you finish lesson 4, you will be able to:

- Interpret lens diagrams to show that light is refracted in passing through a lens.
- Explain how a *lens* produces an *image*.
- Name two kinds of lenses and describe their properties.
- Use a lens to demonstrate how the refraction of light passing through it forms an image.

Image
The picture formed by a lens.

Lens
A piece of transparent material with curved surfaces that refract light passing through it.

Each of your eyes is like a camera. In a camera, light reflected from, or produced by, an object falls on film that records the picture. In an eye, reflected light is also received. See Fig. 4-31. This light falls on a special nerve that causes the brain to see an **image.** Both a camera and the eye have an important part that makes the formation of an *image* possible. This important part is a **lens.** A *lens* bends light rays. Light rays are *refracted* by the lens. See Fig. 4-31.

Lenses refract light in different ways, depending on the purpose of the device in which they are used. When the lenses in our eyes do not refract light properly, we need eyeglasses. Lenses in eyeglasses help the lenses in our eyes produce clear images of what we see. Telescopes use lenses to help us see distant objects clearly. The lenses in microscopes allow us to see small

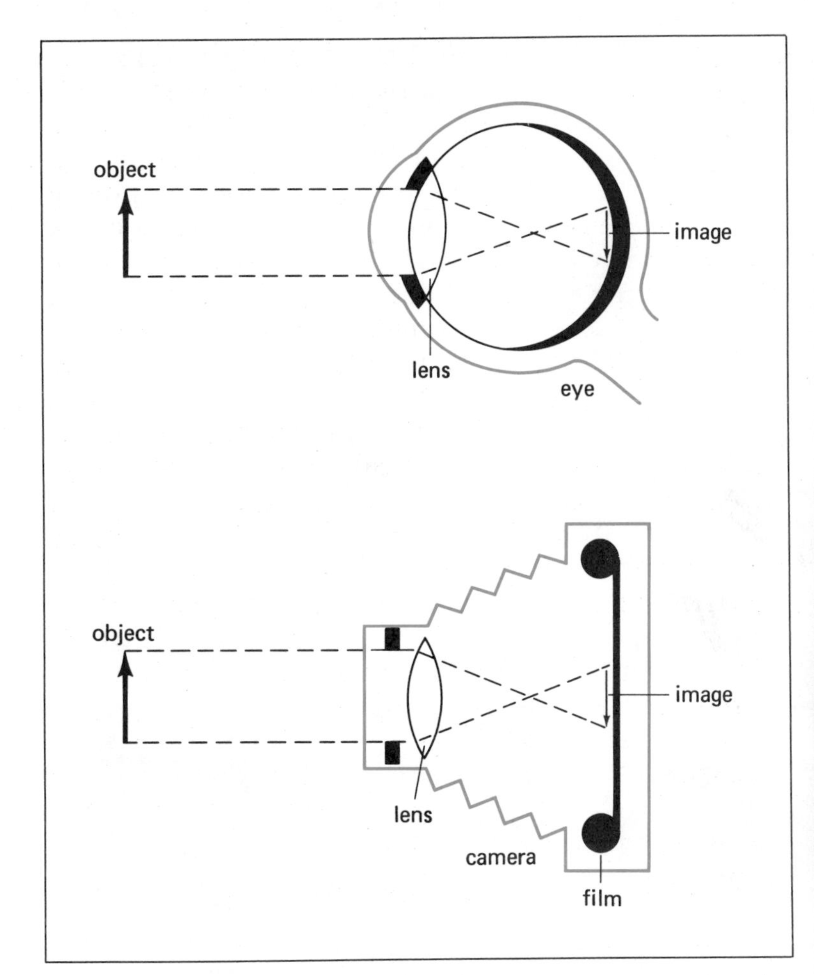

4-31. *The lens in a human eye and the lens in a camera perform similar functions.*

things that are not visible to the eye alone. See Fig. 4-32. We are able to do all this with lenses because they bend light rays.

To see how lenses work, look at a simple magnifying glass. Like all lenses, a magnifying glass has at least one curved surface. See Fig. 4-33. The surface of the lens is curved so that the edges are thinner than the center. This shape is called a **convex** lens. The refraction of light through a *convex* lens can be compared to the way light is refracted by a prism. The convex lens has a shape similar to two prisms combined as shown in Fig. 4-34. The light rays that pass through both prisms meet. The smoothly rounded surface of a convex lens bends light rays in the same way. However, the lens causes the light rays to meet in a much smaller region. See Fig. 4-34. The point at which parallel light rays are brought together after being refracted by a lens is

4-32. *All of these instruments contain lenses. People use eyeglasses, binoculars, telescopes, and microscopes to improve their range of vision.*

Convex
A lens shape in which the edges are thinner than the center.

Focal point
The point at which parallel light rays meet after being refracted.

Focal length
The distance from the center of a lens to the focal point.

4-33. *A magnifying glass has at least one curved surface.*

4-34. *A convex lens bends light rays in the same way as two prisms set base-to-base.*

called the **focal point** of the lens. See Fig. 4-35. The distance from the center of the lens to the *focal point* is called the **focal length.**

Try holding a convex lens close to the printing on this page. Then move the lens closer to your eye. A large image of the printing can be seen. A convex lens held this way can be used as a magnifier. By following the path of the rays through the lens, you can see how a convex lens is able to magnify. See Fig. 4-36a. In order to be used as a magnifier, the lens must be held closer to the object than the focal length of the lens. Rays coming from an object within this distance are still spreading apart after passing through the lens.

Holding the convex lens in front of your eye and looking at a distant object produces a different result. When held close to the eye, only a blur is seen through the lens. Moving the lens away from the eye will finally produce an upside-down image. See Fig. 4-36b.

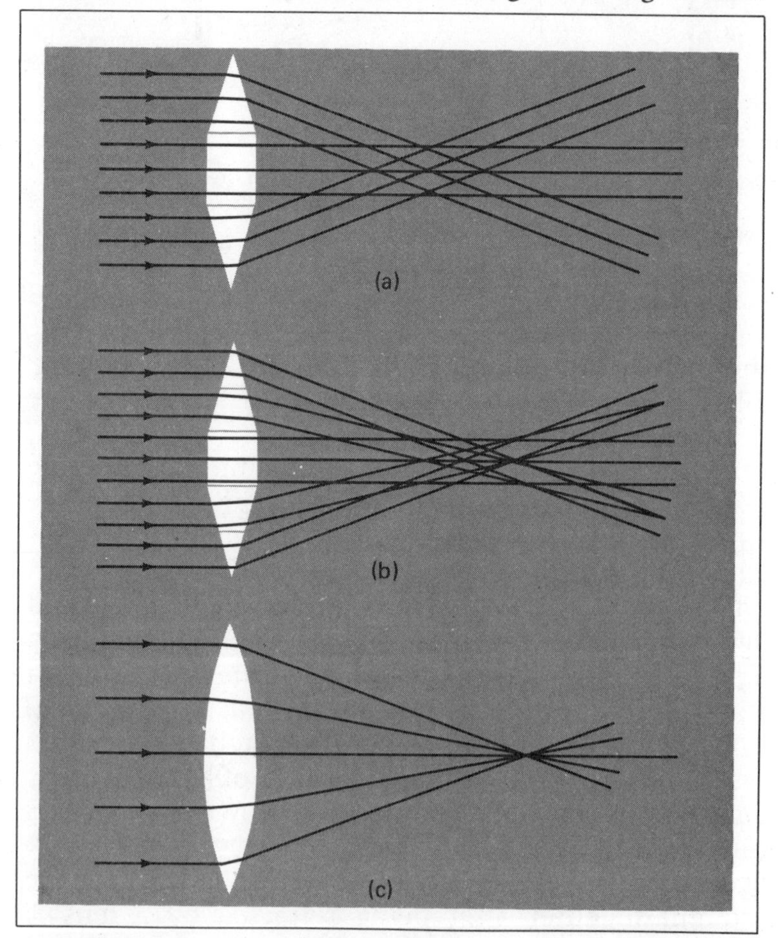

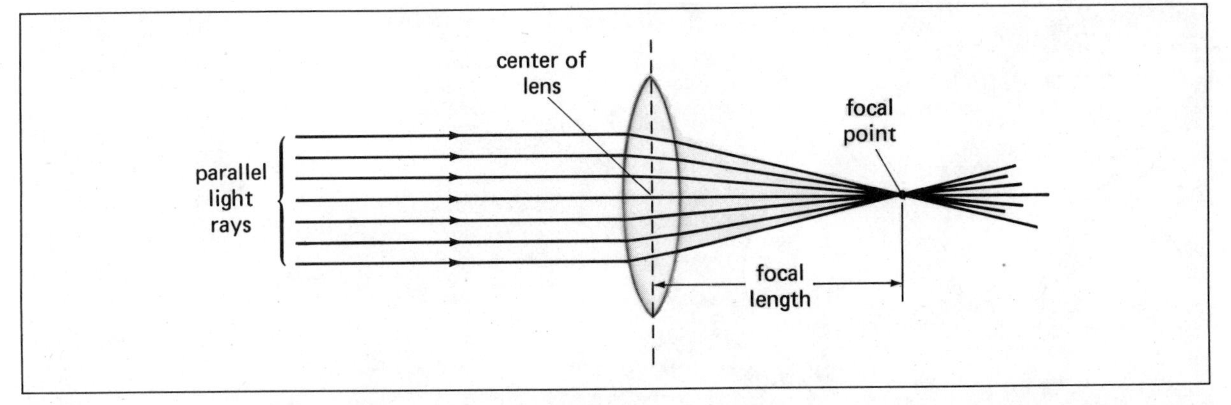

4-35. *The refraction of parallel light rays by a convex lens.*

This type of image is called a *real image* because it can be projected onto a suitable surface behind the lens.

Figure 4-36c shows how a lens can be set up to produce a real image. Comparing the distance between

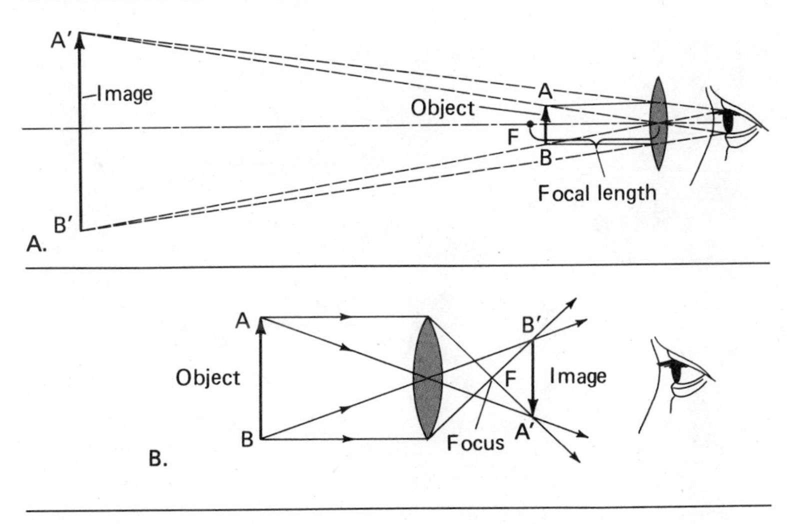

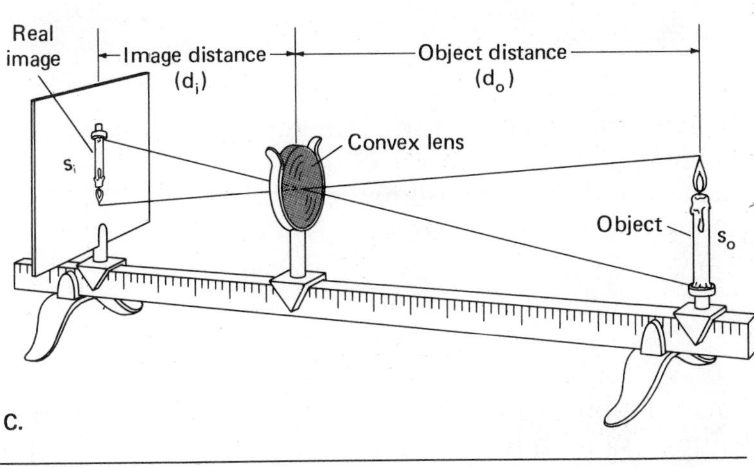

4-36. A. *When a convex lens is used as a magnifier, the object must be within the focal length of the lens.* **B.** *An object beyond the focal length of a convex lens is seen as an upside-down image.* **C.** *A convex lens can also produce a real image as shown.*

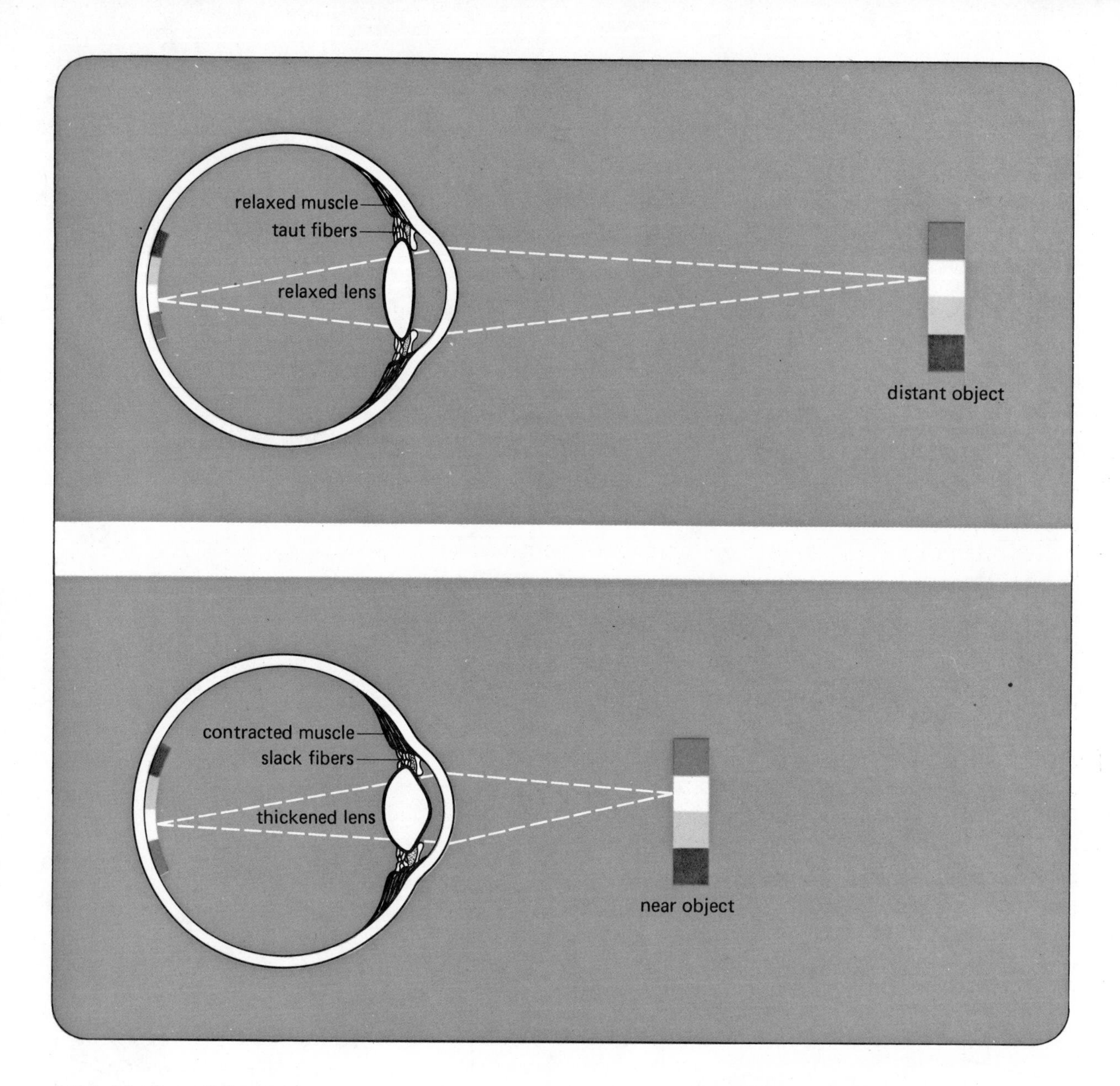

4-37. *The lens of the eye changes shape in order to focus on far away or nearby objects.*

the lens and object (d_o) with the distance between lens and image (d_i) and the size of the object (s_o) and size of image (s_i) shows the following relationship:

$$\frac{\text{distance of object}}{\text{distance of image}} = \frac{\text{size of object}}{\text{size of image}}$$

$$\frac{d_o}{d_i} = \frac{s_o}{s_i}$$

This relationship can be used to find any one of the values if you know the other three. For example, suppose an object 4 cm tall is placed 20 cm from a convex lens. The image is formed 15 cm from the lens. What is the size of the image? Substituting the known values gives:

$$\frac{20\text{ cm}}{15\text{ cm}} = \frac{4\text{ cm}}{s_i}$$

$$s_i = \frac{4\text{ cm} \times 15\text{ cm}}{20\text{ cm}} = 3\text{ cm}$$

The lens of a camera produces a real image that is recorded on the film. Most cameras have lenses that can be moved. The lens is moved closer to the film to produce an image of a distant scene. When photographing a close object, the lens is moved away from the film. Thus, a sharp image can be produced for either close or distant scenes.

How do your eyes produce clear images of both close and distant objects? The lens of your eye is not hard and rigid. It is soft and flexible. Small muscles change the shape of the lens so that reflected light rays from both near and far objects are refracted, thus forming an image on the back of the eye. See Fig. 4-37.

Unlike the eye, most instruments such as telescopes and microscopes have lenses whose shape cannot be changed. Instead, these instruments have combinations of different kinds of lenses. One common type of lens is thicker at the edges than in the middle. This type of lens is called a **concave** lens. Light rays are spread apart by a *concave* lens. See Fig. 4-38.

Sometimes a person's eyes are not able to make clear images of everything that is seen. A person may be able to see distant objects clearly but not nearby objects.

Concave
A lens shape in which the edges are thicker than the center. (The center is *caved* in.)

4-38. *Light rays are spread apart by a concave lens. The image formed by a concave lens is always smaller than the object.*

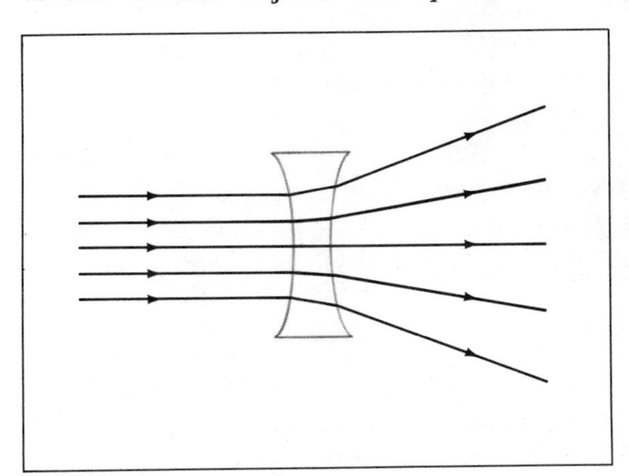

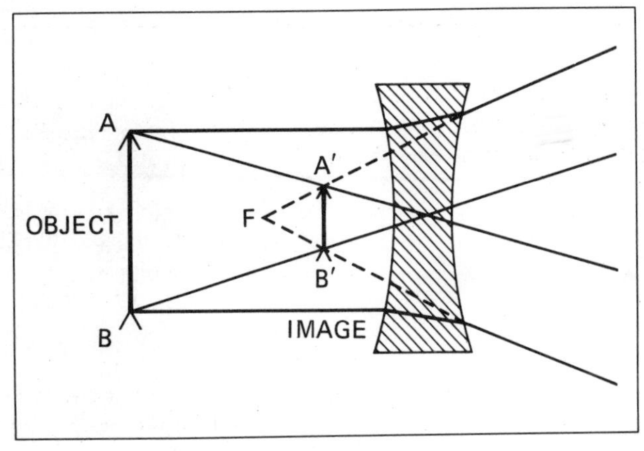

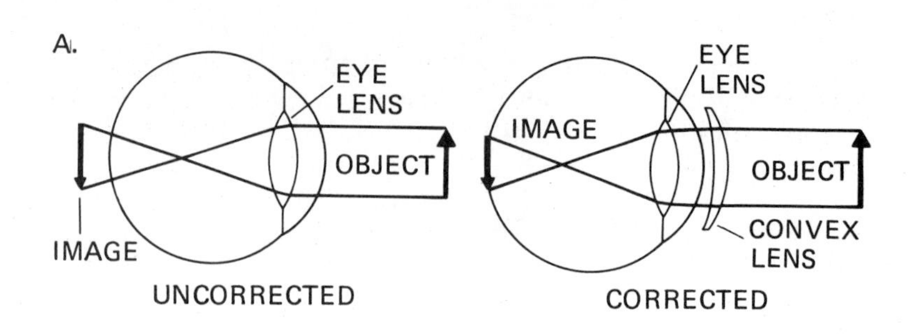

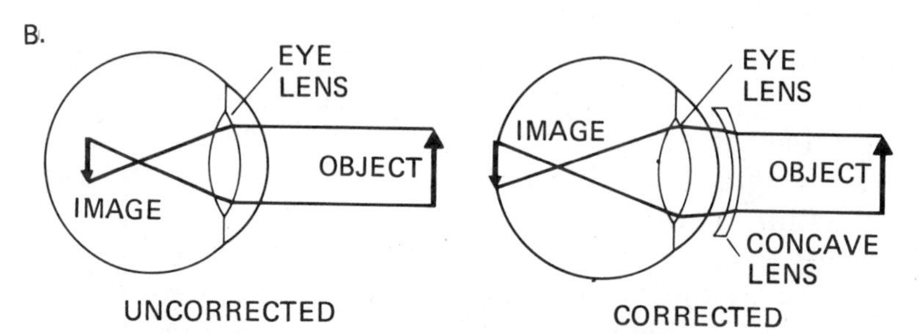

4-39. A. *Farsightedness is caused when the image of a nearby object falls behind the back of the eye. A convex lens corrects this problem.*
B. *Nearsightedness is caused when the image of a distant object falls in front of the back of the eye. A type of concave lens corrects this problem.*

Such people are said to be farsighted. Eyeglasses with convex lenses can correct this problem. See Fig. 4-39a. A nearsighted person can see nearby objects clearly but not distant objects. Concave lenses are used in eyeglasses worn by nearsighted people. See Fig. 39b.

ACTIVITY

Materials
piece of glass
small magnifying glass
white cardboard
paper
ruler

Lenses

A. Obtain the materials listed in the margin.

4-40.

B. Lay a piece of flat window glass on top of several printed lines in your book. While looking through the glass, pick it up and move it slowly away from the printing.

1. What happens as you move the glass away from the page?

C. Lay a small magnifying glass or lens on top of the same printing in your book. While looking through the lens, pick it up and move it slowly away from the printing.

2. What happens as you move the lens away from the page?

D. Hold a piece of white cardboard upright in front of the windows in your room. See Fig. 4-40.

E. Hold the lens between the cardboard and the window, about 1 cm from the cardboard. Now slowly move the lens away from the cardboard.

F. When an image of the window frames is clear, study it carefully.

3. Is the image larger or smaller than the window itself?
4. Is the image rightside up or is it upside down?

G. Move the lens farther from the cardboard.

5. What happens to the image?

H. Now stand on the side of the room opposite the windows.

I. Hold the lens so that the windows are clearly seen on the cardboard. In this case, the distance from the lens to the cardboard is the focal length of the lens.

6. What is the focal length of the lens you are using?

J. Cover one-half of the lens with a piece of paper or cardboard. See Fig. 4-41. Now try to form an image of the windows.

7. Describe the image formed.

4-41.

SUMMARY

One of the most important parts of a camera is the lens. Lenses refract light passing through them. Depending upon the shape of the lens, light rays may be brought together or spread apart. The lens of the eye can see both near and far objects.

QUESTIONS

Unless otherwise indicated, use complete sentences in writing your answers.

1. How does a lens affect light rays in order to form an image?
2. Name two kinds of lenses and describe how they differ.
3. Study Fig. 4-36c. Use the equation $\frac{d_o}{d_i} = \frac{s_o}{s_i}$ to complete the following table:

d_o (cm)	d_i (cm)	s_o (cm)	s_i (cm)
75.0	25.0	6.0	a. ?
b. ?	10.0	5.0	2.5
100.0	c. ?	10.0	2.0
30.0	12.0	d. ?	1.6

4. A person 2.0 m tall is being photographed from a distance of 5.0 m, using a camera that produces an image 0.050 m (5.0 cm) from the lens. Will the image fit a 35 mm film? How big is the image?
5. A lens forms an image of the sun 0.00005 m wide and 0.005 m from the lens. The distance to the sun is about 150,000,000,000 m. From this information, what is the diameter of the sun?

Reflection and Refraction

Materials
Parts I and II
cardboard (23 cm × 30 cm)
4 pins
small mirror
2 pieces of paper
metric ruler
rectangular dish
water

Purpose

In this laboratory you will trace light paths as they reflect from a mirror and refract when going from one material to another.

Procedure

Part I. Reflection

A. Obtain the materials listed in the margin.

B. Draw a line across the center of a piece of paper. Set up a small mirror with its reflecting surface on the line. See Fig. 4-42. The reflecting surface for most mirrors is the back surface. A small piece of clay or a piece of wood can be used to support the mirror.

4-42.

C. Stick a pin in the paper about 5 cm in front of the mirror. Cardboard beneath the paper makes it easy to place the pin. See Fig. 4-43. As shown, draw a small circle around the pin so you can keep track of it. This pin will be called the object pin.

4-43.

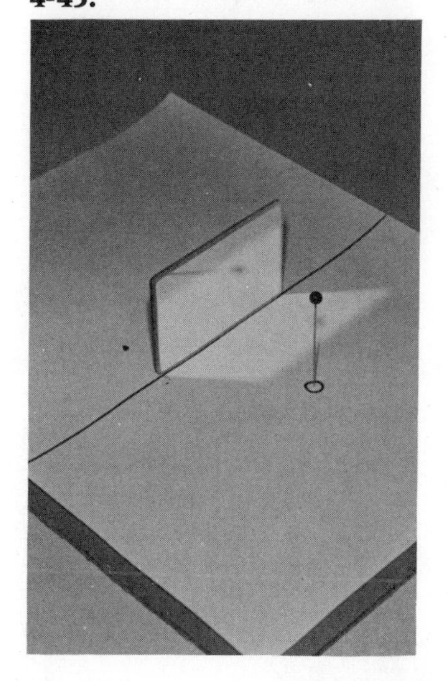

D. Now with your head at the level of the paper, close one eye and look at the image of the pin in the mirror.

E. When your eye is slightly off to the side, place two pins in the paper so that they line up with the image as shown. See Fig. 4-44. These two pins are along the path followed after light from the object pin reflects from the mirror.

F. On the same diagram, draw the line showing the path followed by the reflected light. This path is called a ray. Draw the ray so that it seems to come from the mirror. Light travels from the object pin to the mirror. From the mirror the light travels along the path to your eye. Light always travels in straight lines.

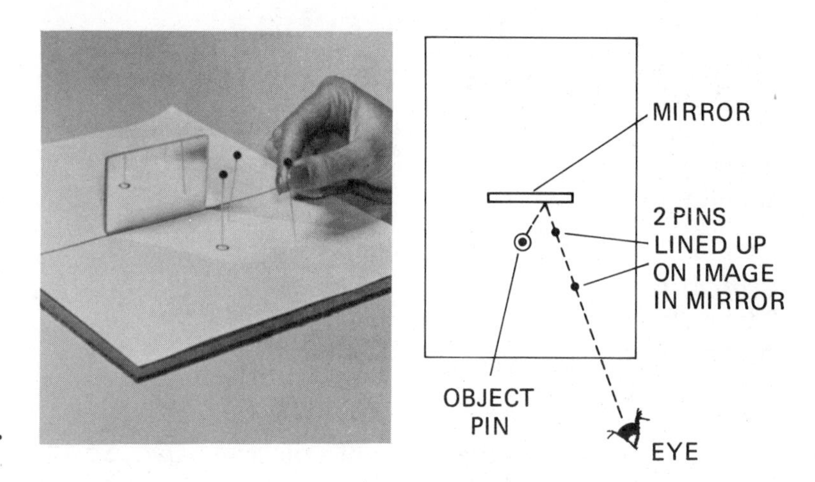

4-44.

G. Draw a line showing the path the light followed from the object pin to the mirror. Compare this angle with the angle between the mirror and the path of the reflected light.

1. Are these angles equal?

Part II. Refraction

A. Place a piece of paper on top of a piece of cardboard as in Part I.

B. Place a rectangular dish about ¾ full of water on top of the paper as shown in Fig. 4-45. Outline the dish with a pencil.

C. Stick two pins in the paper as shown. With your head at the same level as the cardboard, look through the water. Keep only one eye open. Line up the two pins as you see them through the water. See Fig. 4-46.

D. Place two pins in the paper on your side of the box so that they seem to line up with the two pins on the far side of the box. The two pins you have just placed are along the path the light followed after leaving the water. The two pins on the far side are along the path the light followed before it entered the water. Remove the plastic box.

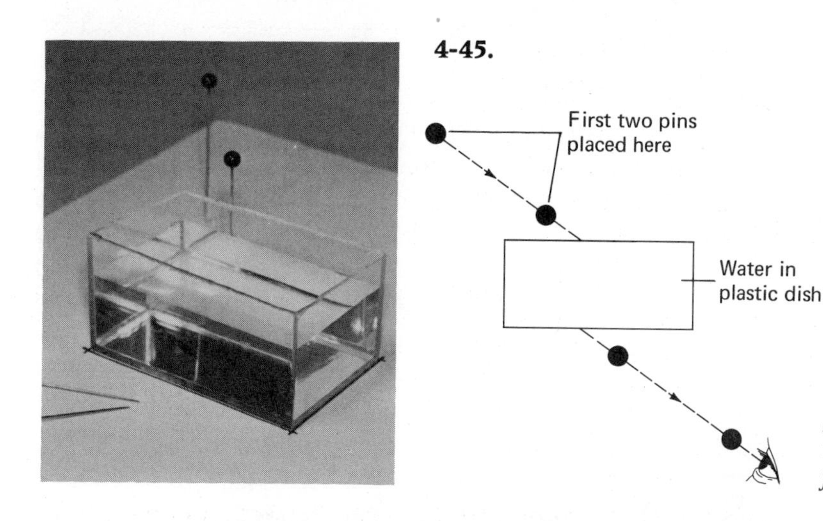

4-45.

4-46. *Looking through the water with one eye, place two pins so that they appear to line up* through the water *with the first two pins.*

E. Draw a line showing the path the light followed before it entered the water. Draw a line showing the path the light followed after it left the water. Draw a line showing the path the light must have followed through the water.

2. Are the paths before entering the water, while in the water, and after leaving the water all in a single straight line?

3. How many times does the light change direction in going from one side of the water to the other?

4. Compare the angle between the side of the box and the direction of the light before entering the water with the angle between the opposite side of the box and the direction of the light after leaving the water. Measure the smaller angle on each side of the box.

Summary

Write a short paragraph describing what you did and the results you obtained for Part I. Title it "Reflection of Light by a Flat Mirror."

Write a short paragraph describing what you did and the results you obtained from Part II. Title it "Refraction of Light by a Container of Water."

CAREERS IN SCIENCE-RELATED FIELDS

As you begin your study of physical science, stop a moment and think about the role science has in your life. Science and its technological applications have had a great impact on our lifestyles and standards of living. The influence of science on your life will be highlighted throughout this book. In addition, special features on careers will concentrate on career opportunities in various areas of science. In this first career spread, you will notice that the careers illustrated are not strictly "science" careers. We chose to begin with these careers in order to emphasize an important fact: Even if you do not choose to make your living as a research scientist or engineer, science will still be an important part of your life. A knowledge of science will be a great asset to you.

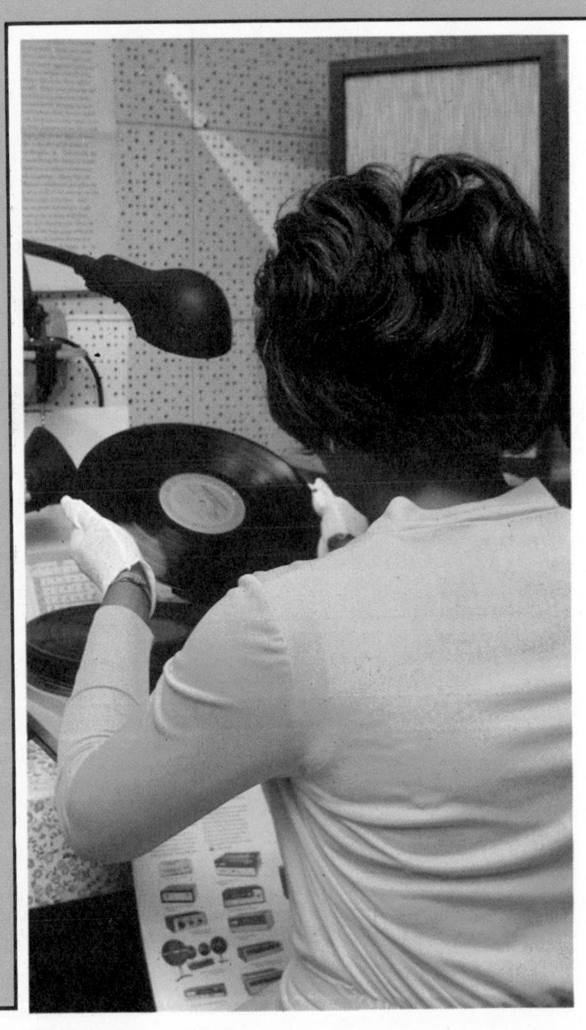

Broadcast Technician

Description: Broadcast technicians operate and maintain the equipment used to record or transmit radio and television programs. They may work with microphones, sound recorders, light and sound effects, television cameras, video tape recorders, and other equipment.

Requirements: A person interested in becoming a broadcast technician should plan to get a First Class Radiotelephone Operator License from the Federal Communications Commission (FCC). Applicants for a license must pass a series of written examinations. High school courses that provide a valuable background for a career in broadcasting include algebra, trigonometry, physics, and electronics. Taking electronics courses in a technical school is also a good way to prepare to become a broadcast technician.

For more information:

Federal Communications Commission, Washington, DC 20554

National Association of Broadcasters, 1771 N St. NW, Washington, DC 20036

Corporation for Public Broadcasting, 1111 16th St. NW, Washington, DC 20036

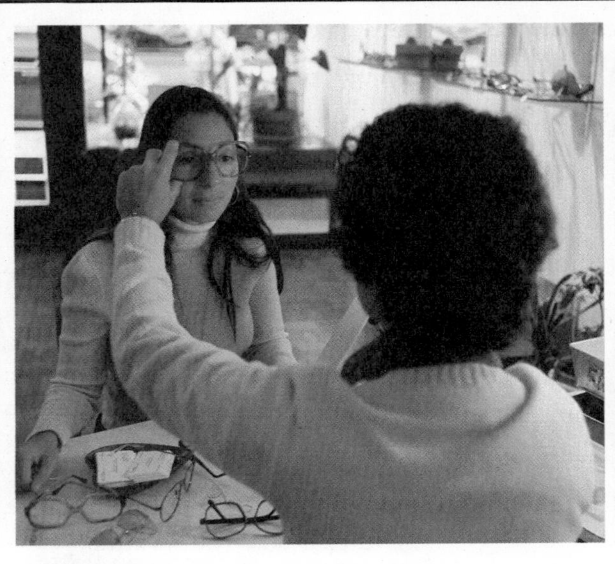

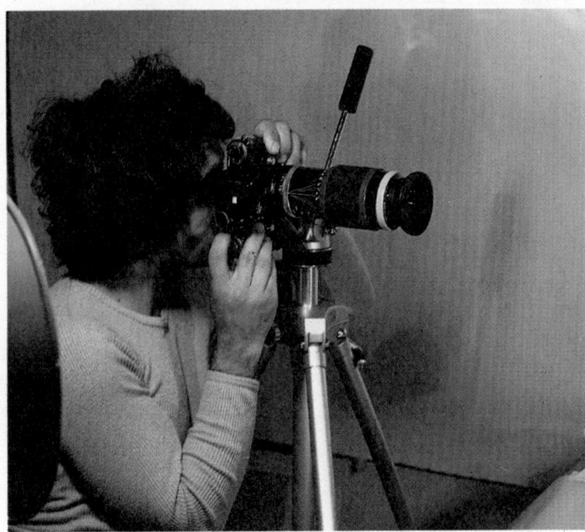

Optician

Description: Millions of people use some form of corrective lenses to improve their vision. Opticians determine where lenses should be placed in relation to the customer's eyes by measuring the distance between the centers of the pupils.

Requirements: Most opticians learn their skills through on-the-job training lasting for several years.

For more information:

National Federation of Opticianry Schools, 300 Jay St., Brooklyn, New York 11201

Opticians Association of America, 1250 Connecticut Ave. NW, Washington, DC 20036

Photographer

Description: Photographers use cameras and film to record people, places and events. Some photographers specialize in a particular field such as scientific, medical, or engineering photography. Photographers work in commercial studios, for newspapers and magazines, or free-lance.

Requirements: Photographic training is available in colleges and art schools. On-the-job training is possible in commercial studios. Amateur experience is often helpful.

For more information:

Photographic Art and Science Foundation, 111 Stratford Rd., Des Plaines, Illinois 60016

Extensions

1. Look at the photograph of the gymnast on p. 17. How do you think a photograph such as this was made? What must a photographer know about motion in order to photograph action scenes such as this?
2. If any of the careers described here interests you, write to one of the addresses given for more information. Write a report on one of these careers and present it to the class.
3. Visit a local photography studio or optician's office and inquire about on-the-job training and educational requirements.

VOCABULARY REVIEW

Match the number of the term with the letter of the phrase that best explains it.

1. electromagnetic waves
2. electromagnetic spectrum
3. visible spectrum
4. image
5. prism
6. mirage
7. light rays
8. focal point
9. lens
10. convex

a. Straight lines showing the path followed by light.
b. A specially shaped piece of glass that separates white light into colors.
c. A lens shape in which the edges are thinner than the center.
d. A piece of transparent material with curved surfaces that refract light.
e. A form of energy able to move through empty space at a very high speed.
f. The point at which parallel light rays meet after being refracted.
g. The picture formed by a lens.
h. The series of waves with properties similar to light.
i. The band of colors produced when white light is separated into colors.
j. An illusion caused by the refraction of light in which distant objects are seen upside down or floating in the air.

REVIEW QUESTIONS

Choose the letter of the answer that best completes the statement or answers the question.

1. Light is said to be energy traveling by means of waves since **a.** we need it to see things by **b.** experiments show it behaves like other waves **c.** it can travel through space **d.** it travels at a very fast speed.
2. Light, infrared, gamma rays, and X rays are all parts of the **a.** electromagnetic spectrum **b.** radio spectrum **c.** microwave spectrum **d.** wave spectrum.

3. The colors of the rainbow make up the spectrum of **a.** radio waves **b.** X rays **c.** gamma rays **d.** white light.
4. Write *True* if the following statement is correct. If it is incorrect, change the underlined phrase and write the statement correctly.
   Rays usually are drawn to show the path of light since light waves travel in straight-line paths.
5. A stick lying partly submerged in water appears bent at the water's edge because of **a.** diffraction **b.** refraction **c.** reflection **d.** warping.
6. A mirage is readily explained by the difference in air layers which causes **a.** diffraction **b.** refraction **c.** reflection **d.** warping.
7. A glass prism separates white light into a spectrum of colors because **a.** each frequency of light is refracted differently **b.** it reflects each color differently **c.** it absorbs each color differently **d.** it speeds up each frequency differently.
8. The lens diagram in Fig. 4-34 shows the formation of a focal point by parallel rays passing through the lens being **a.** reflected **b.** diffracted **c.** magnified **d.** refracted.
9. Write *True* if the following statement is correct. If it is incorrect, change the underlined phrase and write the statement correctly.
   In Fig. 4-3[illegible] the rays passing through the lens of the eye show a focal point located at the rear of the eye where the image is formed.
10. An image formed by a lens can be shown to be the result of **a.** reflection **b.** diffraction **c.** refraction.

REVIEW EXERCISES

Give complete but brief answers to each of the following:

1. Describe an experiment that you could perform to show that light has wave properties.
2. Compare the speed and frequency of sound waves with light waves.
3. Name the six principal parts of the electromagnetic spectrum.
4. Explain why astronomers can point a telescope directly at a distant object in order to see it.
5. Explain why the water always appears shallow when looking down into a clear water lake.
6. Explain mirages using the idea of refraction.
7. White light is composed of all colors.
   a. Give a reason why this must be so based on common everyday observations.
   b. Describe an experiment that proves that this is so.
8. Assuming white light is a mixture of all colors, explain:

a. why the sky is blue.
b. why a green object looks green in white light and black in red light.

9. Explain what happens to light when it passes through a lens to produce an image.
10. Compare the way a camera focuses far away and nearby objects to the way the eye focuses far away and nearby objects. Write one sentence about the camera and then one sentence about the eye to make your comparison.

EXTENSIONS

1. Investigate and report on the advantages and disadvantages of using a microwave oven for cooking. You can begin by asking a dealer for an advertising pamphlet.
2. Studies have been made of the effect of colors on humans. Some colors create a "cool" atmosphere while others create a "warm" atmosphere. Colors may make you feel happy, sad, excited, calm, etc. Write a report on which colors create these moods and suggest ways to improve the color schemes around you.
3. Using the following information, calculate the wavelength for various parts of the electromagnetic spectrum. (Remember, velocity = frequency × wavelength.)

speed of electromagnetic waves = 3×10^8 m/sec
frequency, visible = 5×10^{14} Hz
frequency, gamma = 1×10^{23} Hz
frequency, X rays = 1×10^{17} Hz
frequency, infrared = 1×10^{14} Hz
frequency, radio = 1×10^6 Hz

electricity, magnetism, and heat

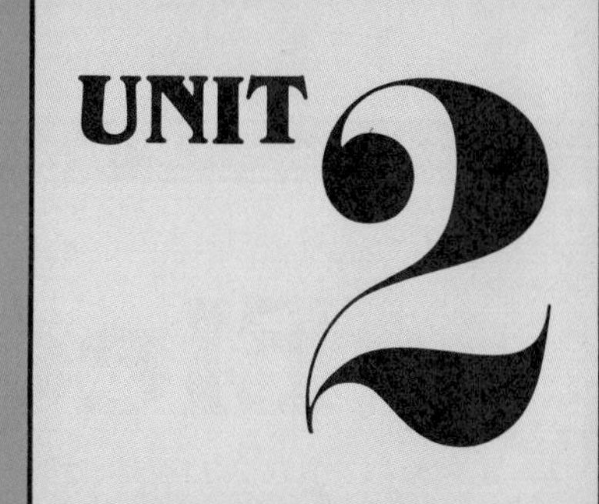

CHAPTER 5

Electric Charge

5-1. KINDS OF ELECTRIC CHARGE

Have you ever walked across a carpet and then felt a shock when you touched a metal doorknob? Have you ever heard a small crackling sound when you rubbed a cat's fur? Did your hair ever crackle and cling as you combed it? If it was dark enough, perhaps you saw tiny sparks jumping between the comb and your hair. In this lesson, you will explore some of these experiences.

When you finish lesson 1, you will be able to:

- Use examples to show how a person can have an electric charge.
- Explain how two objects with electric charges can influence each other.

5-1. *(left) The rubber rod and the ball have the same type of charge. They repel each other.*

5-2. *(right) The glass rod and the ball have different charges. They attract each other.*

- Distinguish between *positive* and *negative* electric charges and explain when an object is *neutral*.

- Demonstrate with the given materials that two charged objects can influence each other.

The experiences described above are the result of electric charges. For example, a comb rubbed briskly with a piece of cloth will attract small bits of paper. The comb can attract the paper because it was given an electric charge by being rubbed. Rubbing two different materials together often causes the materials to become electrically charged. When you walk across a carpet, the friction of your shoes on the carpet may cause your body to be given an electric charge.

Objects having an electric charge affect each other. Two charged objects may pull toward each other. Two other charged objects may push away from each other. For example, a hard rubber rod rubbed with fur is touched to a light-weight ball suspended as shown in Fig. 5-1. After touching the rod, the ball is pushed away. The ball is repelled any time the rubber rod is near. If a glass rod is rubbed with a silk cloth, the ball is attracted to the glass rod as shown in Fig. 5-2.

Observations like these have led scientists to the conclusion that there are two kinds of electric charges.

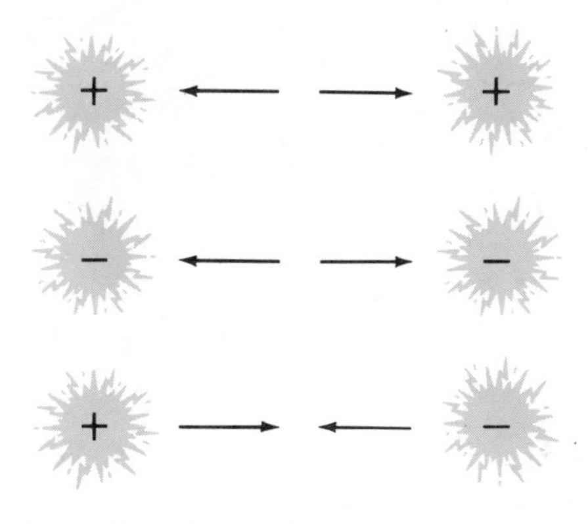

5-3. *Like electric charges repel; unlike charges attract.*

5-4. *An old painting shows Benjamin Franklin flying his kite in a thunderstorm. The young assistant was his son. Why was this a dangerous experiment for Franklin to attempt?*

Experiments show that two objects with the same kind of electric charge will repel or push away from each other. The ball was touched to the rubber rod and given the same charge as the rubber rod. The ball and the rubber rod then repelled each other. Objects with different charges attract each other. The ball was attracted to the different charge on the glass rod. The behavior of objects with electric charges can be described by a simple rule: *Like charges repel each other, while unlike charges attract.* See Fig. 5-3.

One of the scientists who studied electricity in the past was Benjamin Franklin. In addition to being a statesman and author, Franklin was an outstanding scientist. Much of his scientific work was with electricity. One of his most famous electrical experiments showed that lightning is a form of electricity. In 1752, Franklin flew a kite held by a thin wire into a thunderstorm. (**Caution:** Do *not* attempt to repeat this experiment.) The lower end of the wire was attached to a string that held a metal key. Franklin knew that holding directly onto the wire could be dangerous if the full lightning charge traveled down the wire to the key. A spark did jump from the key to his hand. See Fig. 5-4. This proved that the clouds contained an electrical charge. The spark from Franklin's key was similar to the shock you sometimes feel when your body is charged and you touch a metal object. However, the charges causing lightning are very large. Franklin's experiment could easily have killed him in spite of his safety measures.

Like other scientists of his time, Franklin thought that electricity was an invisible fluid that flowed from one object to another. He thought that the two kinds of electrical charges were the result of the amount of this electrical fluid in an object. Although he was wrong about the cause of electrical charges, we still use Franklin's method of naming the two kinds of charges.

Franklin suggested that the two kinds of electric charges be called **positive** (+) and **negative** (−). The charge on the rubber rod rubbed with fur is a *negative* charge. The charge given to a glass rod rubbed with silk cloth is *positive.* When an object has neither a positive nor a negative electric charge, the object is **neutral.** An object that is electrically *neutral* (1) may not have any electric charge, or (2) may have an equal amount of positive and negative charges that cancel each other.

Positive charge
The electric charge given to a glass rod when rubbed with a silk cloth.

Negative charge
The electric charge given to a hard rubber rod when rubbed with fur.

Neutral
The term describing an object that has neither a positive nor a negative charge.

The behavior of electrically charged objects is useful in many ways. For example, most machines that make copies of printed sheets use this electrical principle. First, a flat plate made of special material is given a positive charge. Then a light projects an image of the material to be copied onto the plate. The parts of the plate on which the light falls lose their electrical charge. The dark areas remain charged. A powder carrying negative charges is sprinkled over the plate. This powder sticks to the positively charged parts of the plate. A piece of paper is then laid on the plate and given a positive charge. The powder sticks to the paper, creating a copy of the original printing. Heating the paper melts the powder and permanently attaches it to the paper. These steps are shown in Fig. 5-5. Copying by this method is called *xerography.* Xerography means "dry writing" (from the Greek words *xeros* meaning "dry" and *graphein* "to write").

Electrical charges can also be used to filter unwanted solid materials from the air. Smoke particles, for example, carry electrical charges. The smoke particles can be removed by passing the air over a filter containing electrically charged wires or plates. See Fig.

5-5. *Steps in making a xerographic copy. 1. A special plate is given a positive charge. 2. The material to be copied is projected onto the charged plate. 3. An electrically charged powder is spread over the plate. 4. Paper with an opposite electrical charge is laid on the plate. 5. The powder is held on the paper by electrical charges and is then melted onto the paper.*

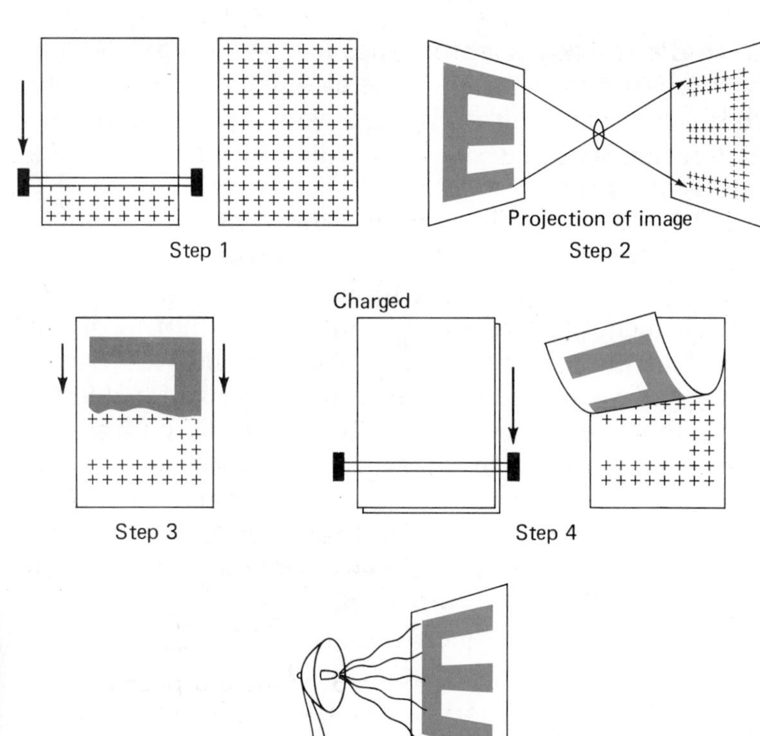

5-6. *An electrostatic device removes polluting dust particles from the gases escaping from smokestacks.*

5-6. The charged particles are attracted to either the positively or negatively charged parts of the filter. Removal of solid particles from smoke greatly reduces the pollution caused by the smoke. However, the gases passing out of the smokestack are not removed by the process.

ACTIVITY

Materials
2 metric rulers (plastic)
nylon thread
waxed paper or wool

Observing Electric Charges

You may need to work with a partner in this activity.

A. Obtain the materials listed in the margin.

B. Tie one end of a nylon thread around the center of a plastic ruler.

C. Hold the ruler at the center. Now rub both ends of the ruler with a piece of waxed paper. Rub fairly hard.

D. After rubbing, hang the ruler by the thread so that the ruler is balanced. See Fig. 5-7.

E. Hold a second plastic ruler by one end. Rub the other end with waxed paper.

F. Now bring the end of the second ruler near one end of the hanging ruler. See Fig. 5-8.

1. Is there a force between the two rulers? What is the evidence for this force?
2. Does the force pull the rulers together or push them apart?

If the rulers pull each other together, repeat steps C through F.

G. Bring the second ruler near the other end of the hanging ruler.

3. Is the reaction at this end different from that at the other end?

H. You need to test only one end of the ruler from now on. Bring your hand near one end of the hanging ruler.

4. Do your hand and the ruler attract each other?

5-7.

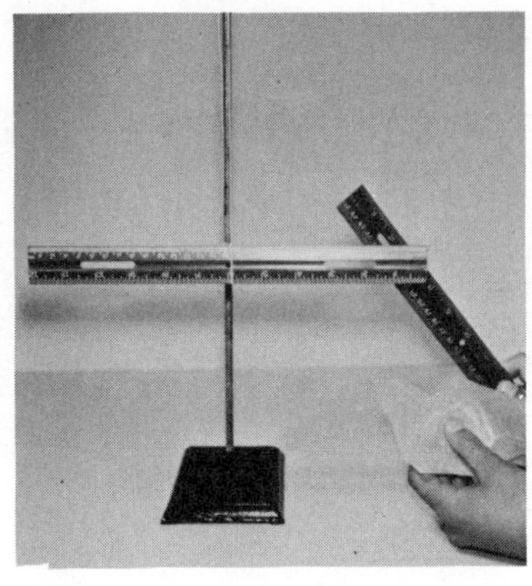

5-8.

I. Rub the second ruler with other materials such as paper, a chalk eraser, your shirt, or anything else handy. Now bring the second ruler close to the hanging ruler.

5. Is there a push or a pull in each case?

J. If there is no reaction, charge the hanging ruler with waxed paper again and repeat the instruction.

6. Explain how you can produce an electric charge on a plastic object. Use complete sentences.

SUMMARY

Ordinary objects like a comb may take on an electric charge when rubbed against some other material, for instance, your hair. Observations show that there are two kinds of electric charges. Objects carrying like charges repel each other. Objects carrying unlike charges attract each other. It is also possible for an object to show no evidence of having an electric charge.

QUESTIONS

Use complete sentences to write your answers.

1. Describe one way in which you may become electrically charged.

2. How might two electrically charged spheres affect each other when brought close together?

3. What kind of electric charge does each of the following have?

   a. Rubber rod having been rubbed with fur.

   b. Glass rod having been rubbed with silk.

4. How can an object be electrically neutral?

5. Given two plastic rulers, a piece of waxed paper, and thread, describe what you could do to demonstrate how charged objects affect each other.

5-2. ELECTRIC FORCE

The machine shown in the photograph is a Van de Graaff generator. The person touching the Van de Graaff generator is given an electric charge. Because all of the hairs on the person's head have the same charge, they repel each other. This causes the hairs to move apart and gives the person the strange appearance you see. The behavior of electrically charged hair and all other electrically charged objects is a result of a particular kind of force.

When you finish lesson 2, you will be able to:

- Explain what effects an *electric force* has on charged objects.
- Describe what affects the size of the electric force between two charged objects.
- Define an electric field.
- ○ Demonstrate, using a ruler and an aluminum ball, what affects the size of the electric force between two charged objects.

If a book resting on a desk top begins to move, you look for a force pushing or pulling it. See Fig. 5-9. On a sloping desk top, the force causing the book to move could be gravity. You have learned that some kind of force is always needed to set an object in motion. When two electrically charged objects cause each other to move, a force must be acting upon them. This force is called the **electric force.** *Electric forces* cause charged objects to move apart or come together according to the kind of charge they have.

Electric force
The force that causes two like-charged objects to repel each other or two unlike-charged objects to attract each other.

In order for the force of your hand to move a book, your hand must touch the book. Electric force works at

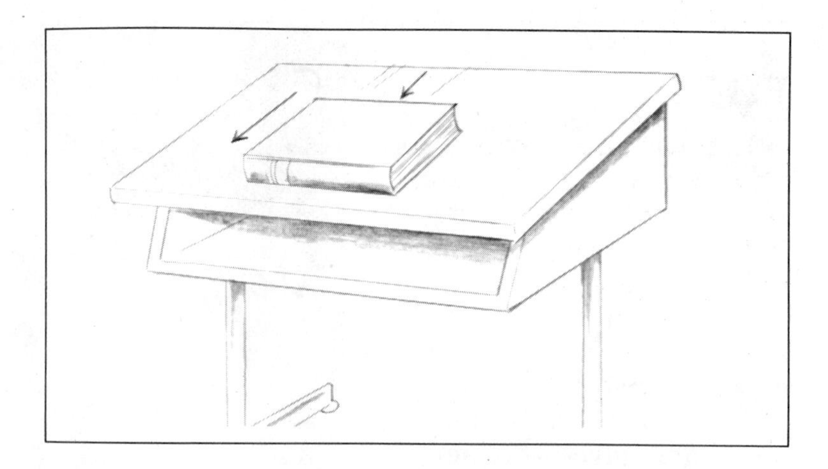

5-9. *The force of gravity could cause the book to slide off the desk and fall to the floor.*

a distance. That is, two electrically charged objects can be affected by an electric force even when they are some distance apart.

Two things seem to affect the size of electric forces. First, the more an object is rubbed to give it an electric charge, the stronger the electric force produced. For example, consider a plastic ruler rubbed once with a cloth and held a certain distance from another charged object. The ruler usually produces only a small attraction or repulsion force. Now consider that same ruler rubbed many times. The amount of charge increases and makes the attraction or repulsion stronger. The strength of the electric force between two objects increases as the amount of electric charge on one or both objects increases.

It is difficult to build up very large electrical charges by rubbing a ruler with a cloth. However, machines that do build up very large electric charges can be made. An example of such a machine called a Van de Graaff generator is shown in the photograph on p. 159. The way the machine works is explained in Fig. 5-10. These machines are used by scientists to help study the structure of atoms. When electric charges build up, they produce *static electricity.* Static electricity is caused by electric charges that are stationary, or fixed in one place. The behavior of two objects carrying static electrical charges can show an important fact about electric forces. To discover what this behavior is, slowly bring the charged objects closer together.

Charged objects have a greater effect on each other as they come closer together. If the two charged objects are 100 cm apart, reducing the distance to 50 cm apart

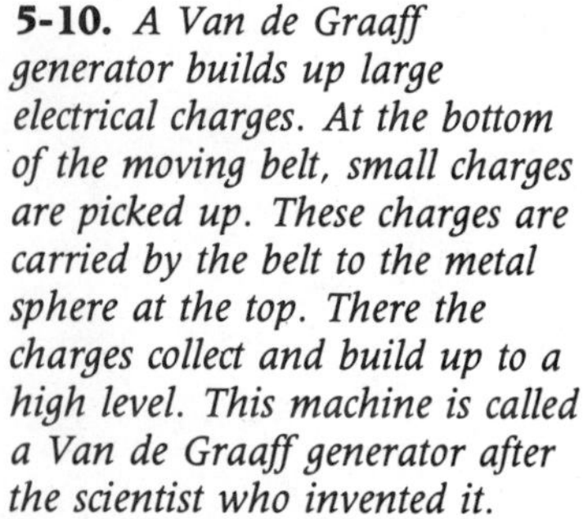

5-10. *A Van de Graaff generator builds up large electrical charges. At the bottom of the moving belt, small charges are picked up. These charges are carried by the belt to the metal sphere at the top. There the charges collect and build up to a high level. This machine is called a Van de Graaff generator after the scientist who invented it.*

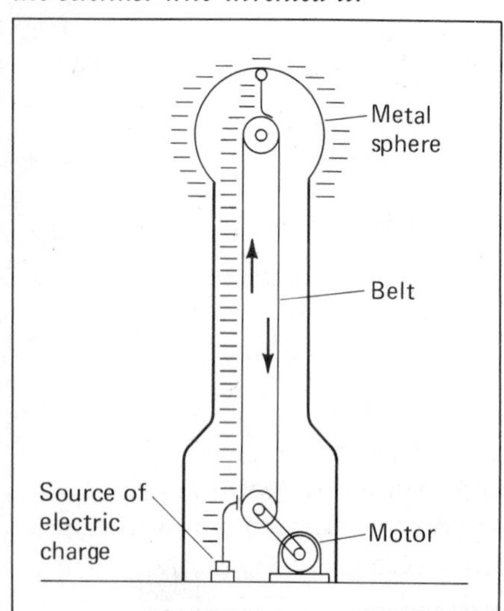

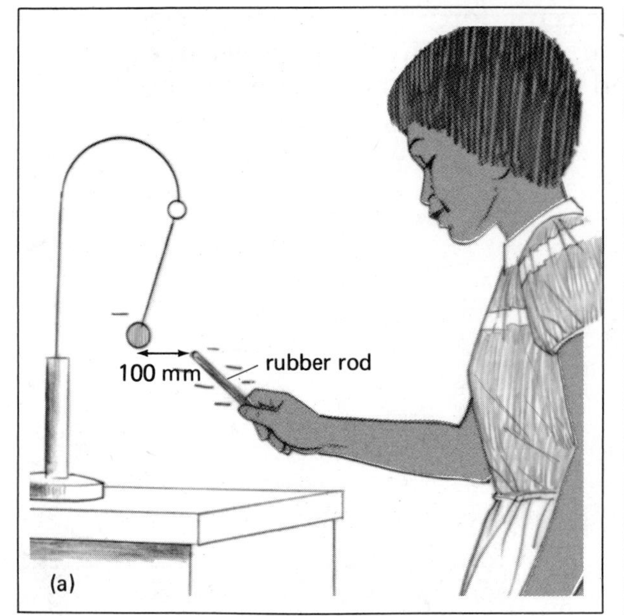

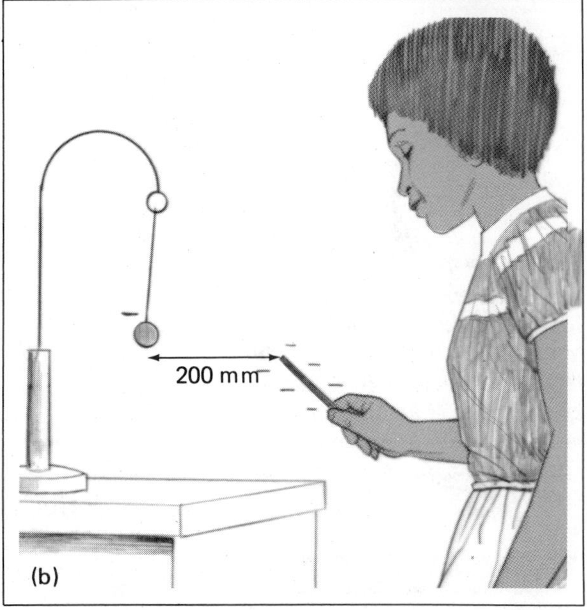

5-11. *This experiment shows that the size of the electric force becomes less with greater distance between the charged objects.*

causes the electric force to become four times greater. In other words, cutting the distance in half causes the force to become four times greater. In the same way, moving the objects 200 cm apart reduces the force to one-quarter of what it was. The relationship between electric force and distance between charged objects is illustrated in Fig. 5-11.

An instrument that shows the presence of static electrical charges makes use of the electric force between charged objects. This instrument is called an *electroscope.* An electroscope contains two thin metal leaves that are supported in a glass container. The metal leaves are often made of gold and are very thin. When the electroscope has no charge and is neutral, the two metal leaves hang down. When a charged object is touched to the top of the metal rod supporting the leaves, the leaves pick up part of that charge. The electric force between the two charged metal leaves pushes them apart. See Fig. 5-12.

Electric forces work at a distance. Around every charged object, there is a space in which the electric force is noticeable. For example, a ruler given an electric charge is surrounded by a space in which the charge is effective. Such a region of space is called an **electric field.** The *electric field* surrounds the charged object in all directions. As you have already seen, the

Electric field
A region of space around an electrically charged object in which electric forces on other charged objects are noticeable.

size of the electric force becomes smaller as the distance becomes larger. Finally, at great distance, the electric field grows too weak to be noticeable.

5-12. *When an electrically charged object is touched to the top of an electroscope, the metal foil leaves move apart. Can you explain why either a negatively or a positively charged object will produce the same reaction on the leaves?*

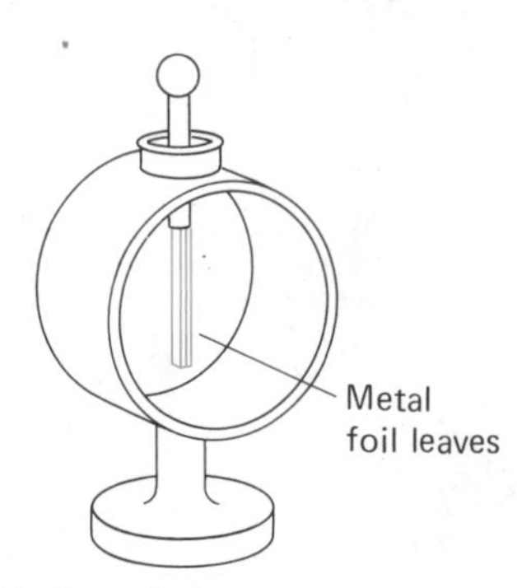

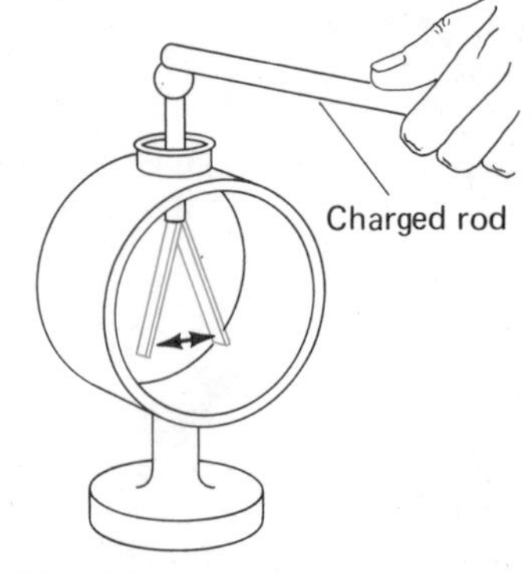

ACTIVITY

Materials
piece of aluminum foil
nylon thread
transparent tape
metric ruler (plastic)
waxed paper or wool

Measuring Electric Forces

A. Obtain the materials listed in the margin.

B. Crush a piece of aluminum foil into a ball about 0.5 cm in diameter.

C. Tie one end of a piece of thread around the ball.

D. Tape the end of the thread to the edge of your desk. See Fig. 5-13.

5-13.

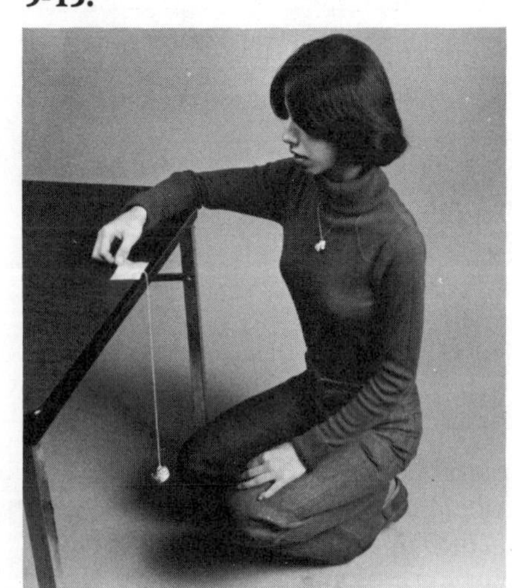

E. Rub a ruler briskly with the waxed paper. Now bring the ruler near the hanging ball. See Fig. 5-14.

1. When you first bring the ruler near the ball, is the ball attracted or repelled by the ruler?

F. Move the ruler back and forth so that the ball touches the ruler at several places. See Fig. 5-15.

G. Continue rubbing the ruler and touching it to the ball until they repel each other forcibly.

H. After the ball and the ruler repel each other, bring the ruler to about 5 cm from the ball. Slowly move it closer.

5-14.

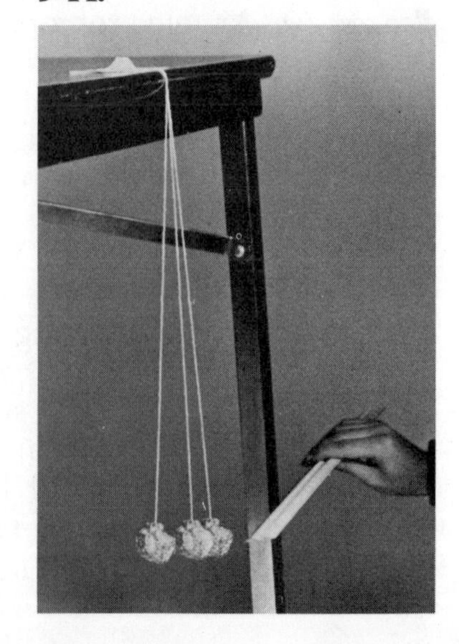

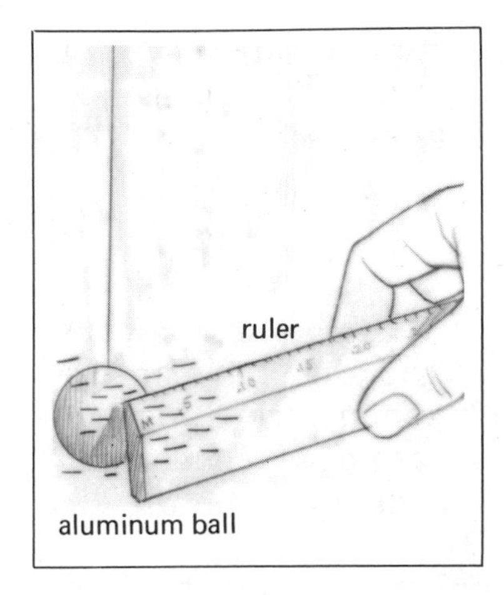

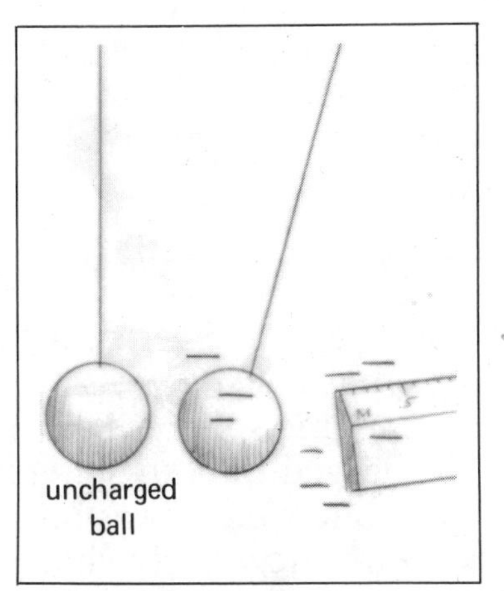

5-15. *Touch the ball with the ruler.*

5-16. *Touch the charged ball with an uncharged ball.*

2. Describe the effect on the ball as the ruler gets closer.

3. What does this effect tell you about the force acting on the ball?

You might try taking some of the charge from the aluminum ball to see if that affects the force.

I. While the charged ball and the ruler are repelling each other, touch the charged ball with an uncharged aluminum ball. The uncharged ball should be hanging from a thread. See Fig. 5-16. Two people can work together.

4. Does changing the charge on the balls change the force?

5. In your own words, explain how the amount of charge and the separation of two charged objects affects the force between them.

SUMMARY

Two electrically charged objects attract or repel each other as a result of electric forces. The electric force increases as the size of the charge increases. The electric force changes in size very quickly as the distance between the charged objects is changed. Any object with an electric charge is surrounded by an electric field.

QUESTIONS

Use complete sentences to write your answers.

1. What effect would the electric force have on two hanging spheres if **a.** both spheres were positively charged **b.** both spheres were negatively charged **c.** one sphere was positively charged and the other negatively charged?
2. How can you change the size of the electric force between two charged objects?
3. Describe what is meant by an electric field.
4. You have charged an aluminum ball by touching it with an electrically charged ruler.
   a. What would you do to show the effect of the amount of charge on the electric force?
   b. What would you do to show the effect of distance on the electric force?

5-3. ELECTRIC CURRENT

How does a vending machine work? You put in your money, push or pull a button, and perhaps a candy bar comes out. What happens inside the machine? Maybe you imagine complex runways, tubes, or springs, or little gates activated by the coins. Maybe you even imagine there is a very tiny person inside the machine. Because you cannot see inside, a mental picture helps you explain to yourself how the machine works. Scientists do something similar when they work with subject matter too small to see or not visible at all. Electricity cannot be seen. Therefore, scientists use a mental picture to help explain how electricity moves from one place to another.

When you finish lesson 3, you will be able to:

- Explain what is meant by a *scientific model*.
- Explain the behavior of charged objects by using the *electron* model.
- Describe what causes an *electric current*.
- List three *conductors* and three *insulators*.
- ○ Use the given materials to demonstrate that an electric current can move easily through some materials but not others.

Rubbing your feet across a carpet may give your body an electric charge. You feel nothing until you touch a metal object such as a doorknob. Then you suddenly feel a small electric "shock." See Fig. 5-17. The shock you feel is the result of the electric charge moving from your body to the doorknob. The electric charge cannot be seen as it moves. When scientists want to study something that cannot be seen, they

5-17. *If you touch a metal doorknob after walking across a rug, you might receive a slight electrical shock. Why?*

Scientific model
A kind of mental picture used by scientists to describe something that cannot be seen directly.

Electron
A negatively charged particle of matter so small as to be invisible.

often use a **scientific model.** A *scientific model* is a sort of mental picture of something that cannot be seen directly.

To help explain how objects can become electrically charged and how electric charges can move from one place to another, scientists use a model called the **electron** (ih-**lek**-tron). An *electron* is a negatively charged particle of matter so small as to be invisible. The fact that many objects can be given an electric charge by being rubbed with another material can be explained by the presence of electrons. Rubbing two materials together can cause electrons to be torn away from one

5-18. *When you pet a cat, electrons move from the cat's fur to your hand. The cat then has a positive charge. Why?*

5-19. *Lightning is caused by the movement of excess electrons.*

material and added to the other. For example, running a comb through your hair may cause electrons to move from the hair to the comb. In this case, the comb takes on a negative electric charge because of the extra electrons. See Fig. 5-18.

Objects do not ordinarily have an electric charge even if electrons are present. Those objects are neutral because they also contain particles similar to electrons but with positive charges. Such particles are actually found in all matter. These particles are called **protons** (**proe**-tons). A *proton* is a very small particle of matter with a positive electric charge. Matter, then, is usually neutral because it contains an equal number of protons and electrons. Normally you do not have an electric charge. You are neutral because the electrons present in your body are canceled by positive charges (protons). An object, such as a comb, has a negative electric charge only when extra electrons are added.

Proton
A very small particle of matter with a positive electric charge.

In some ways, electrons behave like water. Water can flow from one place to another. Usually, gravity causes the water to flow downhill. Electrons can also "flow" from one place to another. It is not gravity, however, that causes electrons to move. Electrons move from a place where there are a greater number of them to a place where there are fewer. This movement of electrons causes an **electric current** to flow. An *electric current* is the result of electrons moving from one place to another.

Electric current
The result of electrons moving from one place to another.

A common example of the movement of electrons is lightning. See Fig. 5-19. Electrons build up in one part

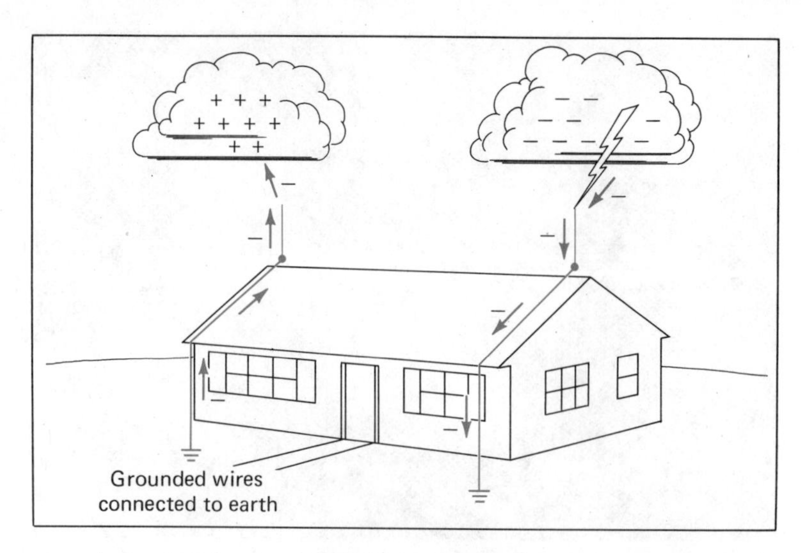

5-20. *The lightning rod on the right carries electrons, moving the lightning from the cloud to the earth. On the left, the lightning rod prevents lightning by supplying electrons from the earth to the cloud.*

of a cloud. These electrons suddenly move from one part of the cloud to another or jump from the cloud to the ground. When this happens, the electrons moving through the air cause it to become very hot. A brilliant light is given off by the heated air. At the same time, the heated air rushes outward to make sound waves that we hear as thunder.

Buildings can be protected from lightning by lightning rods. Lightning rods are pointed metal rods on the highest part of a building. These rods are connected to the earth by a heavy wire. A lightning rod can protect the building in two ways: (1) If electrons move from the cloud to strike the building, they strike the lightning rod and are carried from the lightning rod to the earth. The building itself is not struck by the lightning. (2) Some parts of a cloud may lose electrons to become positively charged. In this case, electrons can flow from the earth to the cloud through the lightning rod. See Fig. 5-20. The charge on the cloud is neutralized and lightning is prevented. In a house not protected by lightning rods, any high part such as a TV antenna may be struck by lightning. This can cause a fire and seriously damage the electrical wires in the house. A person standing close to any part of a house struck by lightning can be injured or killed. For the same reason, when outdoors in a thunderstorm, you should stay away from trees and tall objects. If you are standing in an open field, you may be struck directly. A large building with a steel frame is safe from lightning. The metal in the building acts like a lightning rod. The

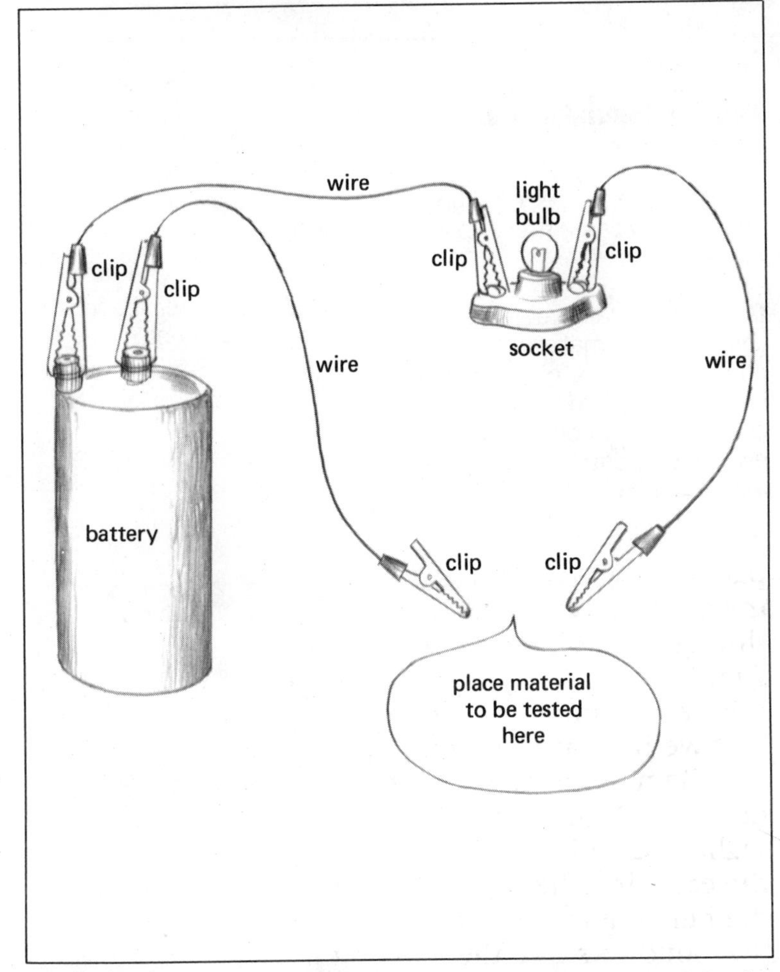

5-21. *A setup such as this can be used to find out if a particular material is an insulator or a conductor. What would happen to the light bulb if the material tested is a conductor?*

inside of an automobile is also safe. The metal body keeps the charges outside of the car.

Lightning would not happen if electrons could move easily through the air. Air is an example of an **insulator** (**in**-suh-late-ur). Any substance that does not allow electrons to flow easily through it is called an *insulator*. If air were not an insulator, electrons could not build up in clouds but would flow away through the air.

Insulator
A material that does not allow electrons to flow through it easily.

Some materials such as metal doorknobs allow electrons to flow freely. These materials are called **conductors** (kun-**duk**-turs). Any *conductor* carries electric current because electrons can easily move through it. For example, see Fig. 5-21. When the two free wires touch an object and the material in it is a good conductor, the bulb will light. The bulb will not light if the material is an insulator.

Conductor
A material that can carry an electric current because electrons can move through it easily.

ACTIVITY

Materials
battery (6 V)
light bulb
3 wires
paperclip
wood
glass
plastic
chalk
staple
eraser
pencil

Testing Insulators and Conductors

A. Set up a battery, a light bulb, and wires as shown in Fig. 5-21.

B. Make a data table listing the following materials: air, paperclip, wood, glass, pencil lead, chalk, staple, paint, eraser, metal holder for eraser on pencil, plastic. Indicate which materials are conductors and which are insulators.

C. Test any other materials you find around you.

1. Which materials are good conductors of electric currents?
2. Which materials are insulators?

SUMMARY

If you build up an electric charge on your body and then touch a metal object, the charge moves from you to the object. Scientists use a model called an electron to help explain how electric charges move. All matter contains both electrons and protons. Objects can become electrically charged when electrons are added or removed. Electrons may move from a place where they are abundant to a place where they are scarce. The movement of electrons produces an electric current. Some materials allow electrons to pass through them freely while other materials do not.

QUESTIONS

Fill in the blanks with the term described in the clues given for each numbered item. DO NOT WRITE IN THIS BOOK.

```
1.                S _ _ _ _
                  C
2.      _ _ _ _ _ I _ _
3.            _ _ E _ _ _ _ _
4.              _ N _ _ _ _ _ _ _
5.    _ _ _ _ _ _ T
                  I
                  F
6.          _ _ _ I _ _ _ _
                  C
                  M
7.        _ _ _ _ O _
8.          _ _ _ D _ _ _ _ _
9.              _ E _ _ _ _ _
                  L
```

Clues

1. When an electric charge moves from your body to a metal object, you feel a ________.
2. An extra supply of electrons.
3. The negatively charged particle of matter.
4. A material that does not allow electrons to move easily.
5. The movement of electrons.
6. A lack of electrons.
7. The positively charged particle of matter.
8. A material that allows electrons to move easily.
9. Having the same number of protons as electrons.

5-4. ELECTRIC CIRCUITS

Until relatively recently, most household chores had to be done by hand. For hundreds of years, people cleaned with brooms and dust mops, heated irons on wood-burning stoves, and lighted their homes by candlelight or, later, gaslight. Today, vacuum cleaners, electric irons, and electric lights have made life easier for most people. These and thousands of other electric appliances work because of electrons flowing through them.

When you finish lesson 4, you will be able to:

- Describe how an electric current can be made to flow in an *electric circuit*.
- Distinguish between *direct current* and *alternating current* in terms of movement of electrons.
- ○ Demonstrate the flow of electric current in an electric circuit.

Switching on a light and turning on a faucet are similar. Opening the faucet lets water flow from a pipe. Turning on an electric light switch permits electrons to flow in a conductor. See Fig. 5-22. Water will not flow in a pipe, however, unless a force pushes it along. That force could be supplied by gravity, causing the water to flow downhill. A pump could supply the energy needed to move the water. Electrons flowing through a conductor also need a force to cause them to move.

About the year 1800, an Italian scientist named Alessandro Volta discovered a way to make electrons flow. Volta found that a combination of two different metals and salt water could move electrons through a conductor. A chemical reaction between the metals and the salt solution caused this movement of elec-

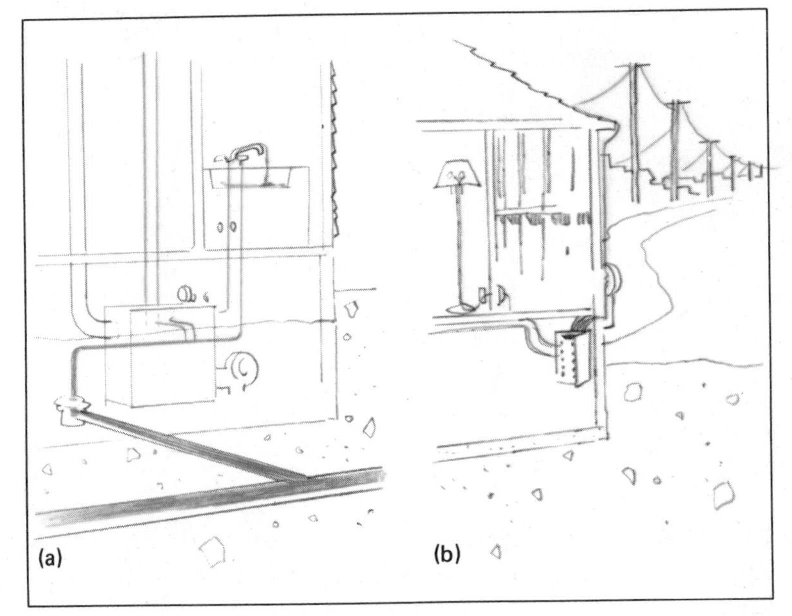

5-22. a. *Turning on a faucet causes water to flow in the pipes.* **b.** *In a similar way, closing a switch causes an electric current to flow in the wires.*

trons. This arrangement of two metals together with a solution causing a flow of electrons is called an *electrical cell*. An automobile battery is made up of several electrical cells. Electrical energy put into the battery is changed into chemical energy and stored in the cells. Later the cells in the battery can change the stored energy back into electricity to produce a flow of electrons.

An ordinary flashlight battery is also a kind of electrical cell. It is often called a *dry cell* because it does not contain a liquid. Instead, a moist chemical mixture takes the place of the liquid.

A dry cell works because of chemical changes taking place inside the cell. These chemical changes provide the energy needed to cause electrons to move through a conductor. Electrons will flow through any conductor that is connected between the part of the cell marked negative (−) and the part marked positive (+). See Fig. 5-23. Electrons flowing in a conductor attached to a dry cell make an **electric circuit.** An *electric circuit* is a path that allows electrons to flow through a conductor. The movement of electrons through a conductor in a circuit is called an electric current.

When you plug an electrical appliance into an electric outlet, you are completing an electric circuit. See Fig. 5-24. The two parts of the plug and the attached

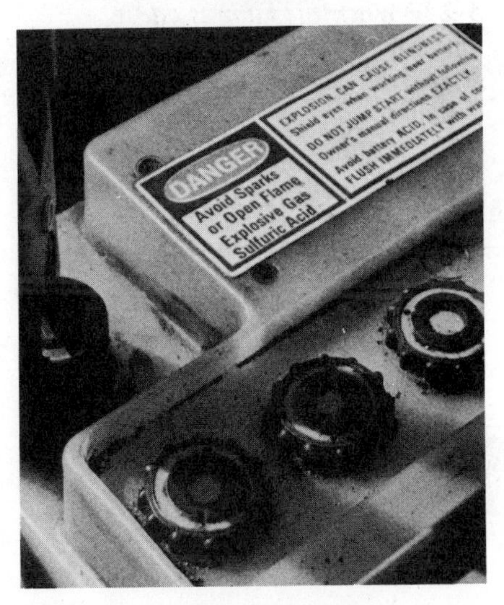

5-23. *The terminals on a dry cell or battery are marked + for positive and − for negative.*

Electric circuit
A complete path that allows electrons to move from a place rich in electrons to a place poor in electrons.

lamp
lamp cord
to source
of current
wall plug

5-24. *When the lamp is on, a continuous electric circuit exists between the power station and the lamp.*

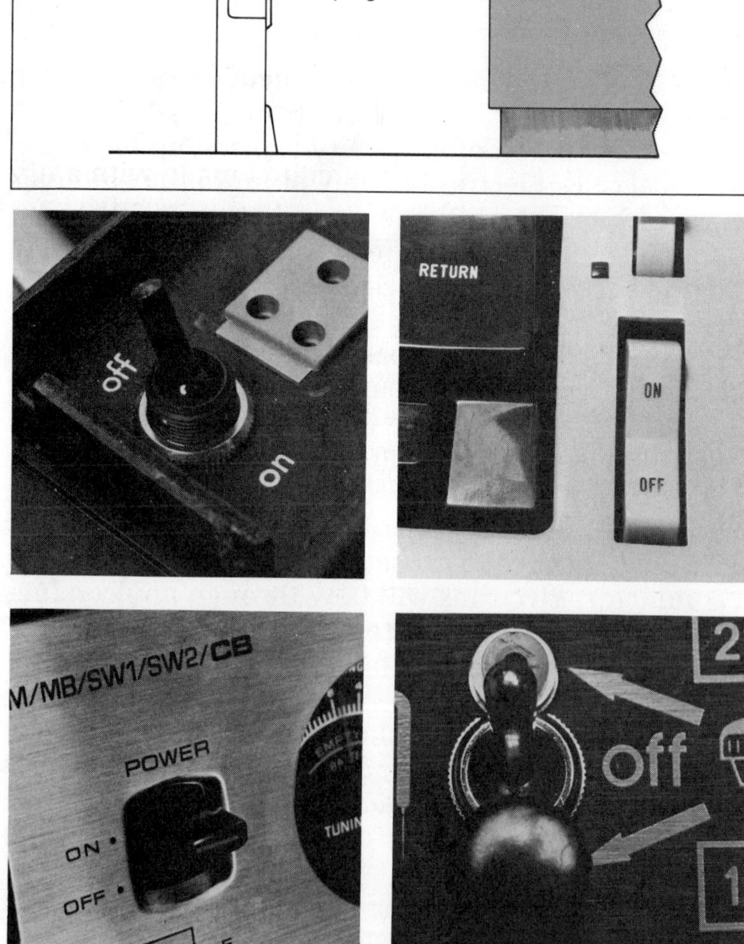

5-25. *The electric circuits controlled by these switches will be complete only when the switches are flipped to "on."*

wires provide a complete circuit for electrons to flow through whatever appliance is plugged into the outlet. Electric circuits in a house are complete only when a switch is turned on. Examples are the circuits that operate lights. See Fig. 5-25. While electrons flowing in a circuit are like water flowing in a pipe in many ways, there is one difference. If you could follow one electron along a wire carrying a current, you would find that the electron itself moved very slowly, only a fraction of a centimeter per second. Yet when you switch on a light, the electric current takes effect almost instantly. You do not have to wait for the electrons to move along the wires. This is because the electrons repel each other. Electrons repel each other because they carry like charges. An electron in a wire repels other electrons in the wire. Electrons all along the wire pass along this movement from one to the next. This effect travels rapidly along the wire. This is what is meant when electrons are said to "flow" along a conductor.

Electrons do not always flow in the same direction in all electric circuits. When a circuit is made with a dry cell, the electrons always move from the negative connection toward the positive connection. This is called **direct current** (DC). In a circuit carrying *direct current*, the electrons always flow in one direction. However, the most commonly used current is not direct current. Electric current supplied by the machines in power stations changes direction many times each second. This is called **alternating** (**awl**-tur-nay-teeng) **current** (AC) because it continuously alternates or changes its direction of motion.

Direct current
An electric current that flows in one direction in an electric circuit.

Alternating current
An electric current that changes direction in an electric circuit.

ACTIVITY

Materials
compass
battery
wire
light bulb

Making an Electric Circuit

This activity will show you a way to tell that electrons are moving in a wire.

A. Obtain the materials listed in the margin.

B. Lay a compass on a tabletop.

1. In what direction does the compass needle point?

C. Connect a battery, light bulb, and wires as shown in Fig. 5-26.

2. Is the compass needle still pointing in the same direction? Be careful not to

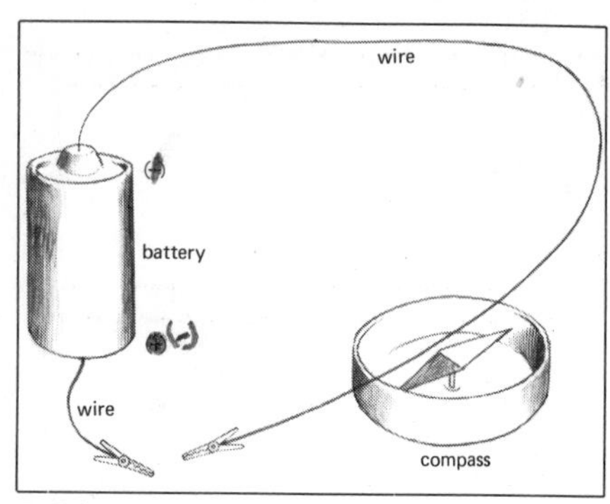

5-26.

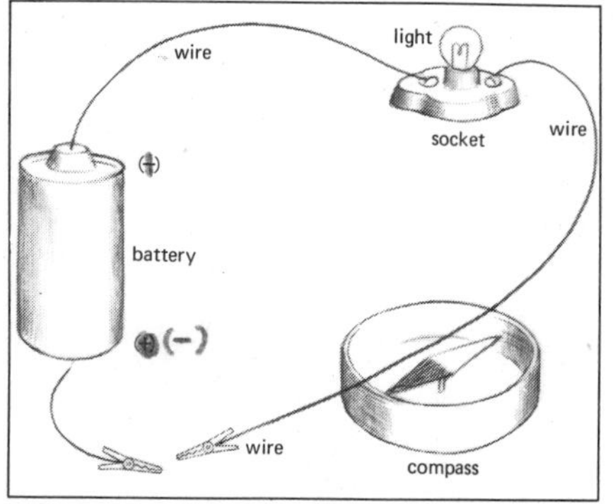

5-27.

bring any metal such as the rings in your notebook near the compass. Metal rings, clips, and watches may change the direction of the compass needle. Be certain that the wire is along the needle.

D. Connect the wires to complete the circuit so that current flows. Observe the light bulb.

3. Does the bulb light? If the bulb does not light, you may not have a complete circuit. Check the connections or check for a dead battery.

E. Now watch the compass needle as you connect the wires to complete the circuit.

4. Does the compass needle move? The compass needle should react to an electric current in the wire by moving.

F. Take the light bulb out of the circuit and connect the wires as shown in Fig. 5-27. Do not join the final wires yet. Be sure the wire over the compass lies in the same direction as the compass needle. Watch the compass needle as you touch the wires to complete the circuit. Do not leave them connected.

5. When the wires were connected, what evidence did you see that an electric current was present? Write a complete sentence.

SUMMARY

Electrons can flow through a conductor only if there is a complete electric circuit. An electric circuit can be completed with a dry cell by connecting the negative and positive parts of the cell with a conductor. The flow of electrons in a circuit can be controlled by the use of switches. Current from a dry cell is called direct current since it travels in one direction. Current used in home appliances is called alternating current since it changes direction rapidly.

QUESTIONS

Use complete sentences to write your answers.

1. Describe what happens to the electrons when you complete a circuit by turning on a wall switch, causing an overhead light to go on.
2. Contrast the flow of electrons in an AC circuit with the flow of electrons in a DC circuit.
3. Refer to Fig. 5-26 to answer the following:
   a. What is necessary in order for a current to flow in the circuit?
   b. When the circuit is completed, do electrons move through it in a clockwise (moving first through the light) or in a counterclockwise (moving first over the compass needle) direction?
   c. When the circuit is completed, what evidence do you see to indicate that there is a current in the wires?

5-5. MEASURING ELECTRICITY

The diver shown in the photos has something in common with a dry cell. Can you guess how the diver and the dry cell are alike? When she is standing on the diving board, the diver has potential energy. What about the dry cell? Why can a dry cell be used to cause a flow of electrons? Lesson 5 will help you to answer these questions.

When you finish lesson 5, you will be able to:

- Describe the relationship of *volts, amperes,* and *ohms*.
- Explain the difference between *series* circuits and *parallel* circuits.
- ○ Demonstrate the effect of changing the voltage and *resistance* on an electric current.

A diver on a diving board has potential energy. She has gained this energy by climbing up to the board. Divers are interested in their potential energy, though they may not use that term, since it determines how hard they will hit the water. The amount of potential energy increases as the height of the board above the water increases. The potential energy of a diver could be found by measuring the height of the board.

A dry cell also has potential energy stored in it. This energy cannot be given off until the dry cell is made part of an electric circuit. Then, the flow of electrons in the circuit will release some of the potential energy. How could you get some idea about the potential energy stored in a dry cell? Some way to measure this energy is needed. Just as the potential energy of a diver is determined by height, the potential energy of electrons from a dry cell is measured in **volts.** A *volt* measures the amount of work done if electrons are moved

Volt
A measure of the amount of work done in moving electrons between two points in an electric circuit.

between two points in an electric circuit. If the flow of electrons is compared to water running down a hill, then voltage is a measure of how high the hill is. An ordinary dry cell, such as the one shown in Fig. 5-28, gives 1.5 volts (V). This would compare to a diver jumping off a low diving board. The combination of dry cells, also shown, gives 6 V. This would be like a diver jumping off a higher board. Consider another example. The single dry cell is similar to water flowing down a low hill. The combination of cells is similar to water flowing down a higher hill. The 6-V combination of cells does four times as much work as the 1.5-V cell. In other words, the 6-V combination *pushes* the electrons harder than the 1.5 V-cell.

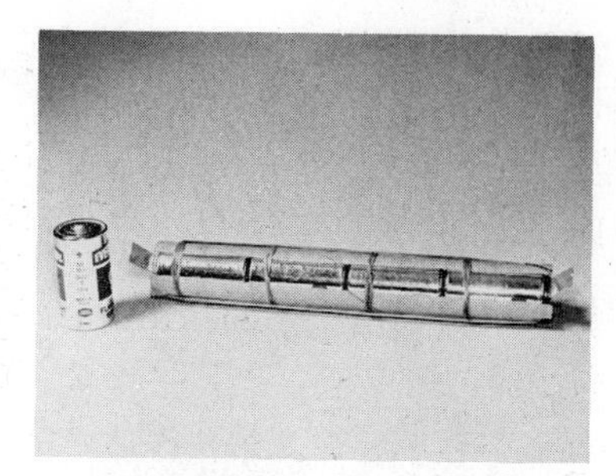

5-28. *Which would supply more volts: the single dry cell or the combination of four dry cells?*

The voltage of an electrical circuit can be measured by an instrument called a *voltmeter*. When a voltmeter is part of an electric circuit, the voltage of the circuit can be read on the dial of the voltmeter. Many automobiles have a voltmeter on the instrument panel. This voltmeter tells the driver if there is enough voltage in the car's electrical system to run the starter, lights, and other parts.

For most electric circuits, we want to know not only the voltage or how hard the electrons are pushed, but also how much current is flowing. To measure the amount of current, we use **amperes** (**am**-pirs). An *ampere* (A) measures the amount of charge moving past a point in a circuit in 1 sec. Ampere is often called "amp" for short. Measurement of voltage and amperage describes the behavior of electric currents. For example, a circuit may have high voltage with low amperage. This would be like a very small but swiftly flowing stream. On the other hand, a circuit with high amperage but low voltage would be like a large but slow-moving river. The amperage of a circuit can be measured by attaching an instrument called an *ammeter* to the circuit.

Ampere
A measure of the amount of current moving past a point in an electric circuit in 1 sec.

If you push a book across the top of a desk, the book will probably slide easily. What will happen if you put a rubber eraser between the book and desk top? Much more force will be needed to slide the book. The ease with which something can be moved changes with different conditions. This is also true for electrons. When electrons move through a material, they meet **resistance** (rih-**zis**-tunts). *Resistance* is the name given to all

Resistance
Any condition that limits the flow of electrons in an electric circuit, for example, a light bulb in a circuit.

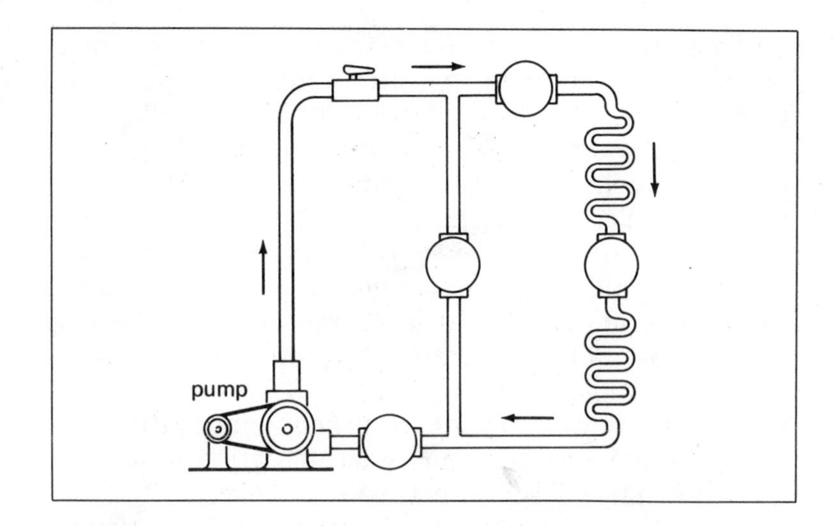

5-29. *Water flowing through pipes is similar to an electric current in a circuit. What part of the pipes would have the highest resistance to the flow of water?*

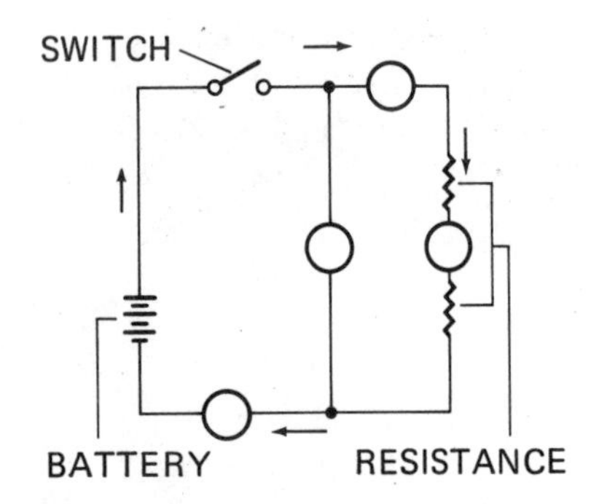

5-30. *If the battery in this circuit provides 6 V and the total resistance is 2 ohms, what is the current in the circuit in amperes?*

Ohm
A measure of the amount of resistance in an electric circuit.

conditions that limit the flow of electrons in an electric circuit. For example, a light bulb adds resistance to an electric circuit.

The amount of current that flows in a particular electric circuit is determined by the voltage. Think of water flowing through a pipe. The amount of water that will pass through the pipe is determined by the force pushing the water. Suppose that the water flows through a narrow place. See Fig. 5-29. Less water could then pass through the pipe. The narrow part of the pipe has the same effect as resistance in an electric circuit. See Fig. 5-30. If electrons flow through a part of the circuit where the resistance is high, then the amount of current flowing through the entire circuit is reduced. Resistance is measured in **ohms** (Ω). A resistance of one *ohm* allows one ampere of current to flow at one volt. The symbol for ohm is the Greek letter omega.

The voltage, amperage, and resistance in an electric circuit are related to each other by a rule known as Ohm's law. This relationship was discovered by a German schoolteacher, Georg Ohm, in the early 1800's. Ohm experimented with electric circuits made up of wires having different amounts of resistance. He discovered a general rule that describes the relationship between voltage, current, and resistance in a circuit. This rule, now known as Ohm's law, can be written as:

$$\text{amperes} = \frac{\text{volts}}{\text{ohms}}$$

5-31. *Many kinds of Christmas tree lights will go out if one bulb in the string burns out. Why is this true?*

For example, an automobile with a 12-V battery has headlights whose resistance is 4 Ω. When the lights are on, the amperage needed is

$$\text{amperage} = \frac{12\ \text{V}}{4\ \Omega} = 3\ \text{A}$$

Most automobile batteries can supply 3 A of current for only a few hours. Thus, a battery can run down if the headlights are left on for several hours without the engine running.

An electric circuit consists of several parts. There must be a source of the electrons to be moved through the circuit. Conductors, usually wires, are needed to connect all the parts. These parts include switches and the appliance to be operated, a light for instance. These items can be connected one after another. A string of lights on a Christmas tree is a good example of such an arrangement. See Fig. 5-31. This arrangement is called a **series circuit.** In a *series circuit*, all parts of an electric circuit are connected one after another. A series circuit can cause some problems. For example, no part of a series circuit can be switched off without turning off everything. If the lights in a house were connected in series, they would have to be all on or all off at once.

Series circuit
A circuit in which all the parts are connected one after the other.

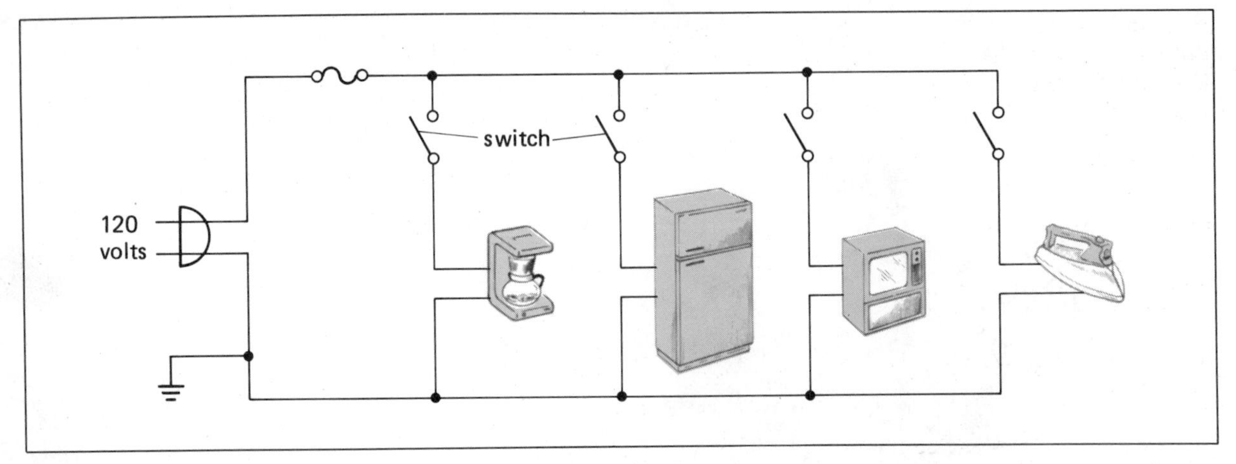

5-32. *Can you see why this type of circuit is called a parallel circuit?*

Parallel circuit
An electric circuit with the various parts in separate branches.

Another way to connect the parts of a circuit is shown in Fig. 5-32. This arrangement is called a **parallel circuit.** In a *parallel circuit,* the various parts are on separate branches. Each branch of a parallel circuit can be switched off without affecting the other branches. The various circuits in a house are arranged in parallel.

ACTIVITY

Materials
compass
2 batteries
2 light bulbs
3 wires

5-33.

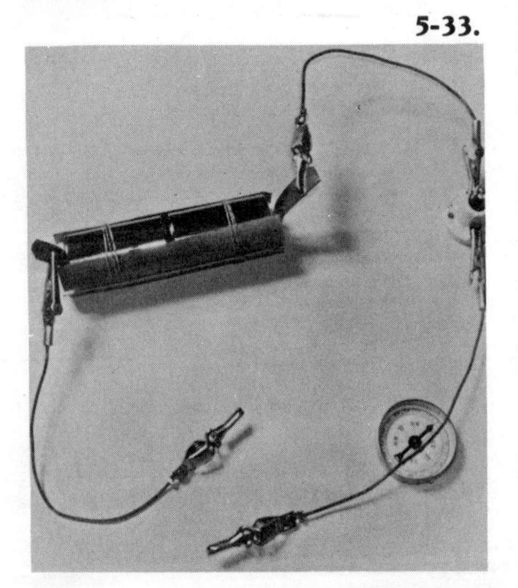

The Amount of Current in a Circuit

In this activity, you will investigate some conditions that affect the flow of current.

A. Obtain the materials listed in the margin.

B. Set up an electric circuit just like the one in Fig. 5-25, lesson 4.

C. Arrange the compass so that the needle points north on the numbered scale. This scale is in degrees.

D. Arrange the wires so that the wire over the compass is in the same direction as the needle. Watch the compass needle as you make the final connection.

E. Read the angle, in degrees, where the compass needle finally comes to rest. Tap the compass to make sure the needle is not stuck.

1. When the current is flowing, how many degrees from north does the compass needle point?

F. Next, increase the number of batteries in the circuit to two. See Fig. 5-33.

2. What effect should increasing the number of batteries have on the current?

G. Watch the compass as you make the final connection. Tap the compass to free the needle.

3. Compared with the first circuit, what evidence do you see that more current is flowing now?

The effect of increasing the resistance in a circuit can also be shown. The brightness of the light bulb in an electric circuit is one way to tell how much current is flowing. The brighter the bulb, the larger the current.

H. Again set up an electric circuit as shown in Fig. 5-33. It is the same circuit used in step F, but the compass is not needed.

I. Look at the light bulb and notice how bright it is.

J. Now add a second bulb to the circuit, in *series* with the first bulb. See Fig. 5-34. Since each bulb has resistance, the resistance in the circuit has been increased and less current should flow.

4. Compared to a single-bulb circuit, what evidence is there that less current is flowing when there is more resistance?

5. In your own words, explain two ways to detect current flowing in an electric circuit.

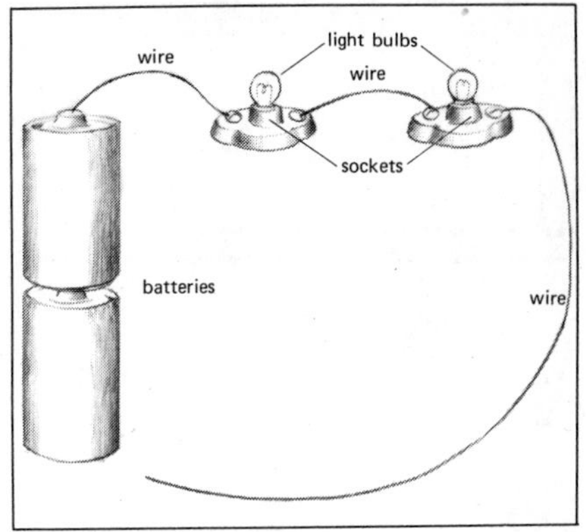

5-34.

SUMMARY

Can you see how a diver on a board is like a dry cell? Both have potential energy. For the dry cell, potential energy is measured in volts. If the energy in a dry cell sends current through a circuit, the amount of current moving is measured in amperes. Whatever resistance the current meets is measured in ohms. Circuits may be arranged so that the parts are connected either in series or in parallel.

QUESTIONS

Unless otherwise indicated, use complete sentences to write your answers.

1. If a TV set has a resistance of 330 Ω, how much current will it require when plugged into a 100-V outlet and turned on?

2. The starter motor on an automobile has a resistance of 0.06 Ω. The battery is 12 V. How much current flows in starting the car?

3. Match the description with the electrical term that it properly describes.

1. ohm
2. ampere
3. volt
4. parallel circuit
5. series circuit
6. resistance

a. The measure of potential energy of a dry cell.
b. The way electric lights are wired in a house.
c. The unit used in measuring the amount of electrons flowing in a circuit.
d. The kind of circuit in which current stops flowing when any part of the circuit is turned off.
e. Any condition that limits the flow of electrons through a circuit.
f. The unit used in measuring resistance.

5-6. ELECTRIC POWER

The photograph was made during a power blackout in New York City in July 1977. The electricity suddenly went off for as long as twenty-four hours in many parts of the metropolitan area. Is your life changed when the electricity goes off? We usually realize how much we depend upon this form of energy only when it goes off. When the power fails, lights go out, the TV is dead, food spoils in the refrigerator, and even most of the clocks stop. A city becomes helpless. Traffic signals stop working, elevators stop, airports close, and computers lose their memories as the flow of electricity stops.

When you finish lesson 6, you will be able to:

- Explain why electricity is such a commonly used form of energy.
- Describe how electric circuits are used to make electricity available in a home.
- Indicate what is measured by an electric meter in a home.
- ○ Make and observe the action of an electric fuse.

Electric energy is used in so many different ways for several reasons. See Fig. 5-35. Most important is the fact that electricity can be changed into almost any other form of energy. Electric motors change electricity into motion. Lamps change electricity into light. The temperature inside buldings is controlled by air conditioners and heaters run by electricity. Your voice can be carried great distances by controlling the behavior of electrons. The telephone is called an *electronic* machine because it works by controlling the movement of electrons. See Fig. 5-36. Other electronic machines, such as TV and radio, make up the lines of

5-35. *People use many different types of electrical appliances everyday. How many electrical appliances shown in this photo can you name?*

communication that connect all parts of the earth. Computers are electronic machines that store and arrange information.

Electric energy also can be sent over long distances through wires. Because of this property, factories can be located far from a source of power. The same wires allow each building to be connected to a source of

5-36. *A telephone line repairer maintains and repairs the cables and wires that form our communication system. This job can involve climbing telephone poles, as shown here, or working underground or underwater. A lineperson must also be able to operate the necessary electrical power equipment.*

energy. Think about how your life would change if those wires were not there to bring electricity to your home.

To provide a building with electricity, at least two wires must be connected to the power lines. The voltage sent out through the main power lines is very high. This high voltage must be reduced to a lower voltage before being sent into homes. Voltage can be lowered by use of a **transformer.** A *transformer* is used to change the voltage of alternating current. The electric company sends out high-voltage alternating current because it is the cheapest way to send large amounts of electric energy over long distances. You may have seen the transformers on power poles. See Fig. 5-37. These transformers change the high voltage into 110 V needed in homes. Some household appliances, such as electric stoves and clothes dryers, need 220 V. To provide this voltage, buildings often have three wires coming from the power lines. When all three wires are connected to a circuit, the voltage is doubled to give 220 V.

The electricity supplied to homes is alternating current. Alternating current changes its direction of flow

5-37. *A transformer lowers the voltage in the power lines for use in homes.*

Transformer
A part of an electric circuit that changes the voltage of an alternating current.

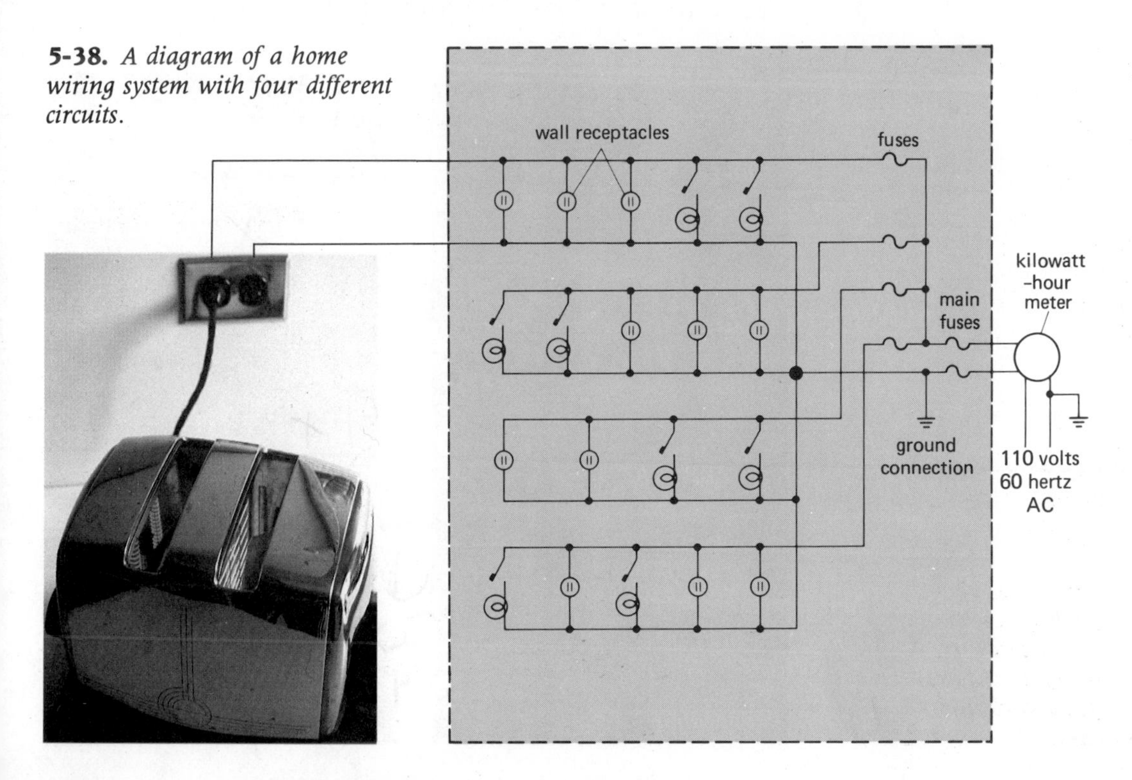

5-38. *A diagram of a home wiring system with four different circuits.*

5-39. *The wire or filament in this bulb becomes hot because of its high resistance. When it gets hot, it begins to glow.*

many times each second. If it is called 60 cycle AC, the current changes its direction 60 times each second.

Wires bringing electricity into a home are connected to parallel circuits in the home. See Fig. 5-38. To use the electric energy, the appliance is connected to one of the circuits. For example, a toaster is plugged into a wall socket. The plug and the cord on the toaster make the toaster a part of the circuit. Electrons flow through special wires in the toaster that have high resistance. When electrons meet resistance, they are robbed of some of their energy. The energy lost by the electrons is changed into heat. The heat toasts the bread.

Many other common appliances, such as electric stoves and irons, work because electric energy is changed into heat energy. When the current flowing through an appliance meets high resistance, the electric energy is changed into heat. A common electric light bulb also works this way. The bulb contains a thin metal wire. Because of its high resistance, this thin wire becomes hot enough to give off a bright light. See Fig. 5-39. Too much current flowing in a house circuit is dangerous. For example, if you attach a toaster, a steam iron, and an electric coffeepot to the same circuit, more than 20 A of current would be needed. The

5-40. *Common types of fuses and circuit breakers used in household circuits.*

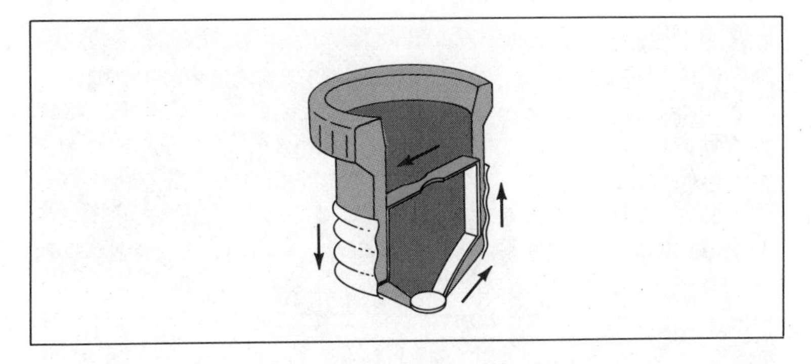

5-41. A type of fuse that must be replaced if the circuit is overloaded. The part shown in color melts to prevent too much current from flowing in the circuit.

wires of the circuit could become overheated and start a fire. To prevent this overheating, each circuit in a house is provided with a **fuse.** A *fuse* is a part of a circuit that prevents too much current from flowing. Some fuses contain a part that melts and breaks the circuit if there is too much current. A more common arrangement is a special switch, called a circuit breaker. The circuit breaker opens when too much current flows. All home circuits must have some kind of overload protection to prevent a serious fire threat. See Figs. 5-40 and 5-41.

Fuse
A part of an electric circuit that prevents too much current from flowing in the circuit.

It is possible to receive dangerous electrical shocks from the house circuit. The danger increases with the amount of current that flows through your body. To prevent contact with dangerously large currents, you should always follow these basic safety rules:

1. Never touch a part of an electrical circuit such as a switch while wet or standing in water. Moisture lowers the resistance of your skin to the flow of current.
2. Do not use electrical appliances while you are also touching a metal object, such as a water pipe, which is connected to the earth. This may allow an electric current to flow through you to the earth.
3. Never come close to the wires on power poles by climbing the poles or nearby trees and buildings. These wires often are at very high voltages. Do not touch in any way wires that have fallen from power poles or buildings.
4. Never put metal objects into electrical outlets or appliances that are plugged in.
5. Never allow the covering of electrical cords to become worn.
6. Never overload an electrical outlet by plugging several appliances into it.

5-42. *The meter measures the use of electricity in kilowatt-hours. The electric company reads the meter each month and charges the customer for each kilowatt-hour used.*

Watt
A unit used to measure the rate with which electric energy is changed into other forms of energy.

Kilowatt-hour
The amount of energy supplied in one hour by one kilowatt of power. It is used to measure how much electric energy is consumed.

You have probably noticed that many electric appliances are marked to show how much electric energy they use. For example, a light bulb might be marked as 60, 75, or 100 watts. A **watt** (W) measures the rate at which electric energy is changed to other forms. *Watts* are useful in measuring how much electric energy is consumed. The amount of electricity used by a light bulb, for example, takes two things into account: (1) how fast the bulb used the electricity by changing it into another form of energy and (2) how long the bulb was turned on. A 100-W light bulb burning for 1 hour could be said to use 100 *watt-hours* of electricity. A watt-hour (W-hr) is a small amount of electric energy. It is more common to use a **kilowatt-hour** (kW-hr) which is a thousand times larger. A *kilowatt-hour* is the amount of energy supplied in 1 hour by 1 kW of power. Homes have a meter attached to their lead-in wires. See Fig. 5-42. The meter measures how much electricity is consumed in kilowatt-hours. The electric company then charges a certain amount for each kilowatt-hour used. Some electric companies charge a lower rate if more electricity is used. Do you think that this practice encourages people to conserve energy?

Most electrical appliances have labels attached, telling how many watts of electricity they use. See Fig. 5-43. Table 5-1 shows the typical wattage of some common appliances. In order to find the cost of operating an appliance, first multiply its wattage times the number of hours it is used. This gives you the number of watt-hours. Then divide this result by 1,000 to find kilowatt-hours. Multiply the number of kilowatt-hours by the amount the electric company charges per kilowatt-hour. For example, if electricity costs $0.06 per kilowatt-hour, a 300-W TV set operating for 5 hrs would cost:

$$300 \text{ W} \times 5 \text{ hrs} = 1{,}500 \text{ W-hr}$$
$$\frac{1{,}500 \text{ W-hr}}{1{,}000 \text{ W/kW}} = 1.5 \text{ kW-hr}$$
$$1.5 \text{ kW-hr} \times \$0.06/\text{kW-hr} = \$0.09$$

TABLE 5-1.
At $0.06 per kilowatt-hour, how much would it cost to run each of these appliances for 2 hr?

Appliance	*Typical Wattage*
color television	300 W
hair dryer	1,000 W (1 kW)
refrigerator	400 W
toaster	750 W
range/oven	3,000 W (3 kW)
vacuum cleaner	500 W

Room Air Conditioner
Capacity: 9500 BTU/hr

(Name of Corporation)
Model(s) XXXXXXX

ENERGYGUIDE

Models with the most efficient energy rating number use less energy and cost less to operate.

Models with 9300 to 9799 BTU's cool about the same space.

6.1

Least efficient model **5.7**

THIS MODEL

Most efficient model **10.2**

Energy Efficiency Rating (EER)

This energy rating is based on U.S. Government standard tests.

How much will this model cost you to run yearly?

Yearly hours of use		**250**	**750**	**1000**	**2000**	**3000**
		Estimated yearly $ cost shown below				
Cost per kilowatt hour	**2¢**	$8	$23	$31	$62	$94
	4¢	$16	$47	$62	$125	$187
	6¢	$23	$70	$94	$187	$281
	8¢	$31	$94	$125	$250	$374
	10¢	$39	$117	$156	$312	$468
	12¢	$47	$140	$187	$374	$562

Ask your salesperson or local utility for the energy rate (cost per kilowatt hour) in your area. Your cost will vary depending on your local energy rate and how you use the product.

Important Removal of this label before consumer purchase is a violation of federal law (42 U.S.C. 6302)

GPO 863 391 (Part No. 20648)

5-43. *All major electrical appliances now carry energy labels such as this one. These labels help consumers estimate the cost of operating the appliance. They can also compare the energy use of competing models.*

Materials

paper
pencil

The Cost of Energy Use

In this lesson, you learned a method for calculating the cost of operating an electrical appliance if you know the wattage and the length of time for which it is used. Review the method if you need to.

In your notebook copy a data table like the following:

Appliance	Watts (W)	Kilowatts (kW)	Hours Used per day (hr)	Kilowatt-hours (kW-hr)	Condition of Cord
television light bulb toaster clock etc.					

Total Energy Used Each Day
__________ kW-hr
Total Cost of Energy Used
$__________

A. Go through your house room by room and list all the electrical appliances you find, as well as the power needed to operate each one. The power is measured in watts. It is printed on a plate connected to each item or on the item itself, like on a light bulb. Unplug any item that you need to move in order to record the wattage. While you are doing this survey, inspect the cord on each appliance. Is the insulation brittle? Is bare wire showing through the insulation? Record this information in your data table. Report any bad cords to your parents.

B. Convert the watts to kilowatts for each appliance and record this number.

C. Make the most careful guess you can as to the number of hours per day, on the average, you think the appliance is used. Record your estimate. If your estimate is in minutes, change it to hours before recording.

D. Multiply the kilowatts by the time in hours to get kilowatt-hours and record this in your table.

E. Find the total energy in kilowatt-hours recorded in your data table. This tells you the total energy used each day.

F. To find the cost of this energy, ask your parents the amount of the total electric bill for an average month and the total number of kilowatt-hours used that month. Divide the cost by the kilowatt-hours used.

G. Calculate the daily cost of electricity by multiplying the cost per kilowatt-hour by your estimate of the number of kilowatt-hours used in your home each day.

H. Change this cost per day into cost per month by multiplying by 30, the number of days in an average month.

I. Compare your result in step G with the cost for an average month.

SUMMARY

How would your life be different without electricity? If you think of all the ways in which your life would change if the electricity went off, you will see why electricity is one of the most commonly used forms of energy. Electricity is useful mainly because it can be changed into other forms of energy so easily. Electricity can also be sent over long distances. Electricity is supplied to a home by several low-voltage circuits that are connected to the power lines. Each circuit must be protected from overloads. Electric energy consumed in a home is measured in kilowatt-hours. If you know the rate charged per kilowatt-hour, you can calculate the cost to run electric appliances.

QUESTIONS

Unless otherwise indicated, use complete sentences to write your answers.

1. Give at least two reasons why electricity is such a commonly used form of energy.
2. Describe what you must do and what happens in the circuit when you use an electric toaster.
3. Why are fuses made a part of electric circuits in a home?
4. What does an electric meter in a home circuit measure?
5. In your own words, list at least five safety rules you should follow in using electricity in your home.
6. How much does it cost to burn a 150-W light for 8 hr if electricity costs $0.05 per kWhr?
7. It takes a dryer 20 min to dry a load of clothes. If the dryer uses 1,500 W, how much would it cost to dry 5 loads if electricity costs $0.06 per kWhr?

Materials
2 batteries
3 light bulbs and sockets
4 wires

5-44.

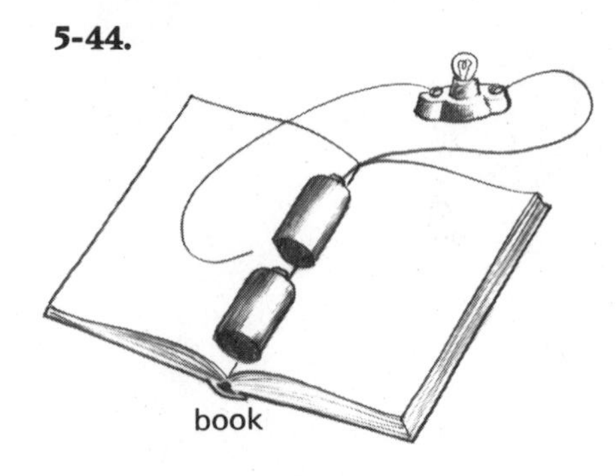

Electric Current

Purpose

Using different combinations of bulbs and batteries, you will use the brightness of the light bulbs to show how much current is flowing in a circuit. The brighter the bulb, the larger the current.

Procedure

A. Set up a circuit using one battery, as shown in Fig. 5-44. If battery holders are available, use them. If not, place the batteries in the crease of a book; this will keep them from rolling and make it easier to line them up when using more than one. When you complete the circuit, notice how bright the light bulb is. (Be sure to check all connections. Is the bulb firmly in the socket?) You will compare this brightness with other cases. Do not leave any circuit connected when you are not using it.

1. What is the evidence that there is a current? Remember, the brightness of the bulb is one indication of how much current is flowing in the circuit. The brighter the bulb, the larger the current.

B. Now connect two batteries in series. Watch the bulb as you make the final connection. Do not leave it connected.

2. When the circuit is completed, is the bulb brighter than with one battery?

3. When two batteries are connected in series, is the amount of current greater than with one battery? If your answer is yes, it is because the second battery is causing more current to flow. The increased brightness of the bulb should show this.

C. Next add a second bulb to the circuit. See Fig. 5-45. Watch the bulb as you make the final connection. In order to compare this setup with the standard brightness of one bulb and one battery, you may need to reconnect the circuit as in Fig. 5-44.

4. With two batteries in series and two bulbs in series in the circuit, how does the brightness compare with one battery and one bulb (brighter, the same, dimmer)?

5. How does the amount of current with two batteries and two bulbs compare with one of each?

Think about the following. One battery cannot light two bulbs in series as brightly as it can light one bulb. Also, two batteries can light two bulbs just as brightly as one battery and one bulb: Before making the circuit, can you predict how bright the bulbs will be when three bulbs are connected to two batteries?

6. What is your prediction?

D. Now wire the circuit and make the final connection to check your prediction.

7. Was your prediction in question 6 correct?

8. What evidence is there to support that your prediction was correct? (If your prediction was wrong, how did you know you were wrong?)

You can see the effect of connecting two bulbs in parallel as follows.

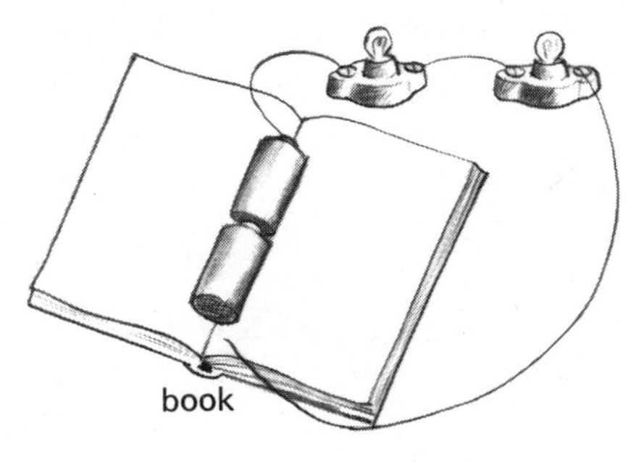

5-45.

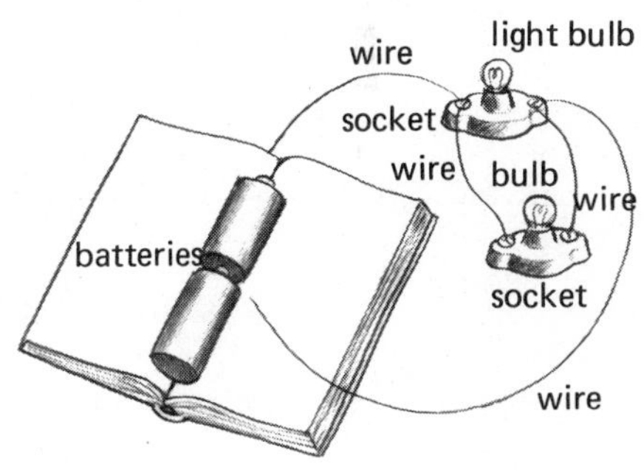

5-46.

E. Set up an electric circuit with one bulb. See Fig. 5-44.

F. Look at the bulb to see how bright it is.

G. Now put a second bulb in the circuit, in parallel with the first. See Fig. 5-46. If the voltage is the same across both bulbs as it was with one bulb, they will be the same brightness as the single bulb.

9. Look back over your answers to the previous questions. When two bulbs were connected in parallel, was the current in each bulb the same as when only one bulb was used? What evidence did you see?

Summary

a. Write a paragraph summarizing the results you obtained in steps A through D. The main topic of this paragraph is to be "the comparison of light bulbs in series circuits using one or two batteries in series."

b. Look over this summary and write a complete sentence giving a general statement relating the current in series circuits to the number of bulbs and batteries used.

VOCABULARY REVIEW

Match the number of the term with the letter of the phrase that best explains it.

1. alternating current
2. electric charge
3. electric circuit
4. electric field
5. electron
6. insulator
7. negative
8. neutral
9. proton
10. scientific model

a. The space in which an electric force exists.
b. Having equal numbers of protons and electrons.
c. A very small, positively charged particle.
d. The charge given to a rubber rod when rubbed with wool.
e. Electrons continuously reversing their direction of flow.
f. Having a given amount of electric energy.
g. An arrangement that allows the flow of electrons.
h. A very small, negatively charged particle.
i. A mental picture of something that cannot be seen directly.
j. A material that does not allow easy movement of electrons.

REVIEW QUESTIONS

Complete each statement by choosing the best word or phrase, or by filling in the blank.

1. An object that has an electric charge will cause another charged object to **a.** be attracted **b.** be repelled **c.** be attracted or repelled **d.** remain unchanged.
2. An object that has the same amount of positive charge as it has negative charge is ________.
3. If the size of the electric charge on two objects were increased, the electric force would **a.** increase **b.** decrease **c.** remain the same.
4. An electric current is the result of **a.** electrons moving from a place where protons are to a place lacking protons **b.** protons moving from a

place where electrons are to a place lacking electrons **c.** electrons moving from a place where there are many of them to a place where there are few **d.** rubbing an insulator with a conductor.

5. Materials that allow electrons to move easily through them are called __________, while materials that do not allow easy movement of electrons are called __________.
6. The current supplied by a dry cell or battery is called direct current because electrons __________.
7. The current supplied to most homes from power stations is called alternating current because the electrons __________.
8. Use the words amperes, ohms, and volts correctly to complete the following sentence: In a circuit, the force with which the electrons are pushed is measured in __________, the amount of current flowing is measured in __________, and the resistance to current flow is measured in __________.
9. The kind of circuit in which a branch may be shut off without affecting the other branches is called a **a.** parallel circuit **b.** series circuit **c.** direct circuit **d.** alternating circuit.
10. The most important reason why electric energy is commonly used is that it **a.** is clean **b.** can be changed into most other forms of energy **c.** is cheap **d.** can be made without causing pollution.

REVIEW EXERCISES

Give complete but brief answers to each of the following exercises.

1. Describe one way in which you could convert the energy of motion into electric energy and become electrically charged.
2. Predict whether the electric force acting between two charged objects would increase or decrease when the following changes are made:
   a. one object is given more charge
   b. some charge is taken from both objects
   c. the objects are moved closer together
   d. the objects are moved farther apart
3. Give an example of how you could demonstrate that a force field surrounds an electrically charged object.
4. If matter is composed of electrons and protons, explain how it can become either positive, negative, or neutral.
5. Select from the list those objects which would usually be classified as insulators: rug, doorknob, cat's fur, penny, plastic bag, kitchen fork, wooden spoon, pencil lead.

6. Use Ohm's law (amperes = volts/ohms) to calculate the missing values for the electric circuits in the following table:

	Amperes	Volts	Ohms
flashlight	0.30	3.0	___
table lamp	___	111	200
pencil sharpener	1.0	___	110
toaster	10.0	110	___
hand mixer	0.90	___	123
cooking burner	6.8	220	___
microwave oven	___	110	20
electric knife	___	110	125

7. A house circuit breaker is rated to carry 15 A. Look at the results in the table you completed in question 6 and list several groups of appliances that would overload the circuit.
8. Describe, in volts and amperes, the current in a circuit in which a large number of electrons are passing through, but with very little push behind them.
9. Explain what happens in a circuit when the resistance is greatly increased.
10. Make a drawing to show how electric appliances are connected in a home circuit. See Fig. 5-35.

EXTENSIONS

1. The units amperes, volts, and ohms were taken from the names of famous scientists. Look up information about these scientists in the library. Write a short biography, including an interesting story about each.
2. Using diagrams, explain the workings of some common household appliance, for example, steam iron, toaster, electric coffeepot.
3. Find out how a generator (or alternator) works and give a report to the class. Include an explanation of how an AC generator differs from a DC generator.

CHAPTER 6

Magnetism

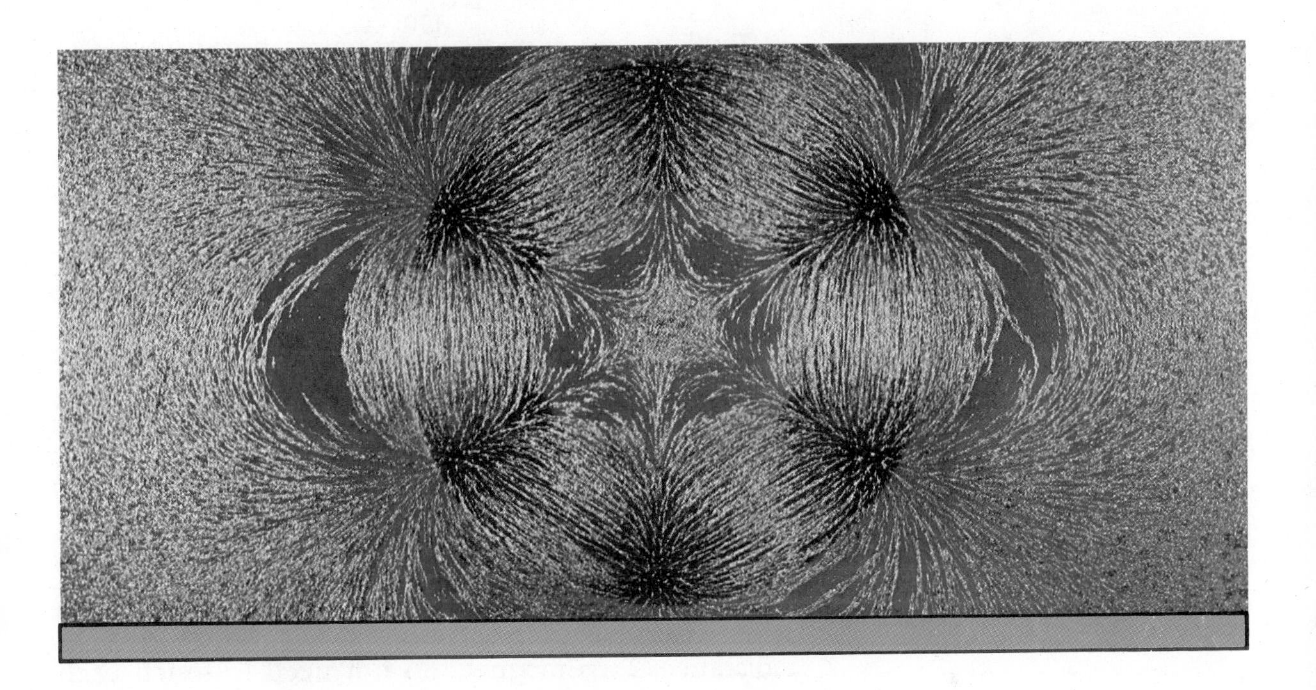

6-1. NATURE OF A MAGNET

What force is acting on you at this moment? Unless you are weightless and floating around the room, you must be experiencing the force of gravity. A second force also acts on you occasionally. This force is the electric force. Walking across a carpet and then touching a doorknob may cause you to experience an electric force in the form of an electrical shock. There is also a third force at work. This force is similar to gravity and the electric force in that it is always present. However, you never experience this force as you experience gravity and sometimes feel an electric force. This third force is magnetism. In this lesson you will explore magnets and magnetic forces.

6-1. *Magnets, like this horseshoe magnet, have two poles.*

When you finish lesson 1, you will be able to:

- Explain how magnets are similar to objects with electric charges.
- Use examples of the action of magnets to explain what *magnetic poles* are.
- Explain how you can locate a *magnetic field.*
- Use two magnets to demonstrate the effect of magnetic poles on each other.

The magnetic force can both attract and repel. In this way, the magnetic force and the electric force are alike. Two magnets will push or pull on each other the most when their ends are brought together. The part of a magnet where the magnetic forces seem to be strongest is called a **magnetic pole.** Every magnet has at least two *poles.* If a magnet is shaped like a bar, the poles are at the ends. Magnets with other shapes may have poles anywhere. See Fig. 6-1. Some magnets have several sets of poles.

Magnetic pole
The part of a magnet where the magnetic forces are strongest.

The magnetic force is similar to the electric force in other ways. For example, both kinds of force work at a distance. Two magnets do not need to touch each other in order to produce attractive or repulsive forces. Also, both the magnetic force and the electric force become weaker with greater distance. As two magnets are brought closer together, the push or pull between them becomes much greater.

6-2. *A bar magnet allowed to swing freely will line up in a north–south direction.*

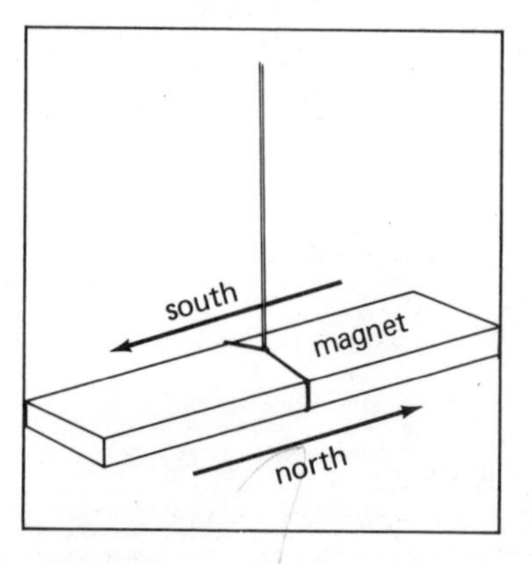

A simple experiment shows another important fact about magnets. See Fig. 6-2. In this experiment, a bar-shaped magnet is held in such a way that it can swing freely. In this case, the magnet will always point in a north–south direction. One pole points north. The other pole then must point south. Because magnets always act in this way, the two poles are given names that reflect this behavior. The pole of a magnet pointing to the north is called its *north pole.* The opposite pole is the *south pole* of the magnet.

Experiments with magnets prove that like poles repel and unlike poles attract. If two magnets are brought together with north pole next to north pole,

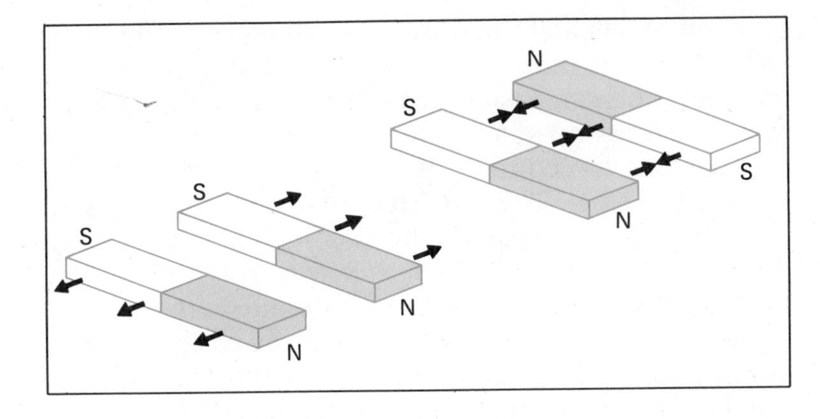

6-3. *Like magnetic poles repel; unlike magnetic poles attract.*

there will be a repulsive force between them. If two magnets are brought together north pole to south pole, there is an attractive force between them. See Fig. 6-3.

What would happen if you cut a bar magnet in half? You might think that the north and south poles would be in separate pieces. The actual result is shown in Fig. 6-4. Each piece becomes a complete magnet with both north and south poles. Each magnet seems to be made of many small magnets. This result leads to a theory about how a piece of metal such as a bar of iron can be made into a magnet.

In an unmagnetized iron bar, the small magnets in the metal are arranged randomly. See Fig. 6-5A. They are like a crowd of people sitting in a park and facing in all directions. When the iron bar is magnetized, each of the small magnets lines up. See Fig. 6-5B. They are more like people sitting in a theater in rows. When the

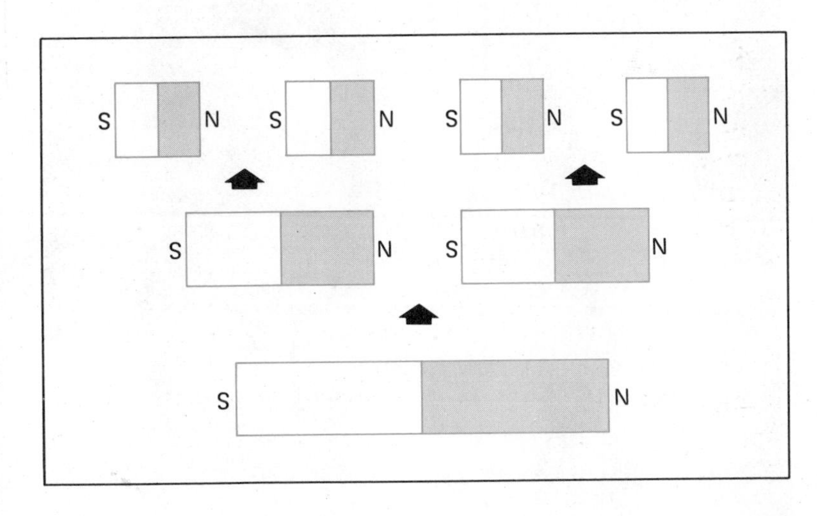

6-4. *When a magnet is cut in half, two magnets are created.*

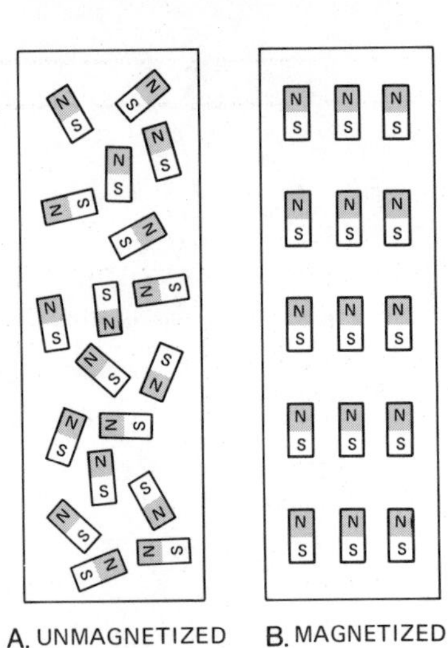

6-5. A. *In an unmagnetized piece of iron, the small magnets in the metal are not lined up.* **B.** *In a magnetized iron bar, the small magnets are arranged so that their poles all face in the same direction.*

small magnets line up, their poles all point in the same direction. This makes the whole iron bar a single large magnet. You can easily turn a steel needle into a magnet. Rub one end of a magnet along the needle, always in the same direction. This rubbing disturbs the particles in the steel. The particles then line up mainly in one direction. This causes the needle to be magnetized. In lesson 3, you will see another way in which a piece of unmagnetized iron can be made into a magnet.

Some metals, such as soft iron, can be easily changed into magnets. The small magnets within the soft iron can be lined up without much difficulty. However, soft iron magnets also easily lose their overall magnetism. For example, hammering on such a magnet can cause the small magnets to lose their orderly arrangement. Magnets that easily lose their magnetism are called *temporary magnets.* Some harder metals, such as steel, are harder to magnetize but tend to keep their magnetism better. A magnet made of a material that tends to keep its magnetism is called a *permanent magnet.* Small permanent magnets are useful for such purposes as making latches to keep cupboard doors closed.

Magnetic field
A region of space around a magnet in which magnetic forces are noticeable.

Since magnets are able to affect each other without touching, they must be surrounded by **magnetic fields.** A *magnetic field* is a region of space around a magnet in which magnetic forces are noticeable. Electrically charged objects, as you remember, are also surrounded by electric fields. A field around a magnet can be seen if small pieces of iron are sprinkled around the magnet. See Fig. 6-6.

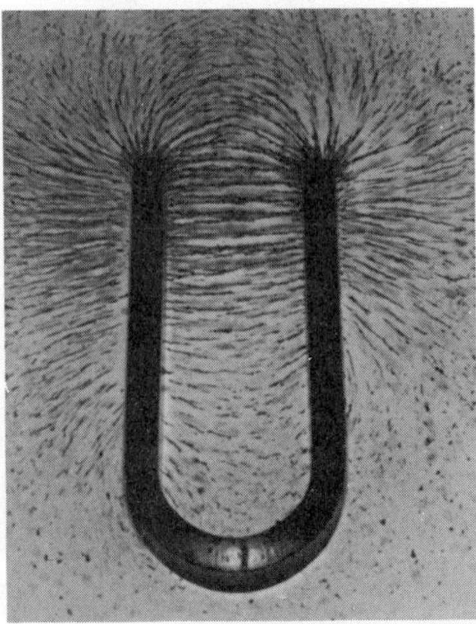

6-6. *The iron filings line up in the field of the magnet. The filings collect mostly around the poles where the magnetic forces are stronger.*

ACTIVITY

Observing the Magnetic Force

Materials
2 bar magnets
various objects (wood, pencil, eraser, paper, glass, plastic, paper clip, aluminum foil, staple, chalk)

Probably the only time you become aware of magnetic forces is when you observe magnets. In this activity, you will use two magnets to make some observations about magnetic force.

A. Obtain the materials listed in the margin.

B. Mark one end of a magnet with a pencil.

C. Line both magnets up on top of a table. See Fig. 6-7.

D. Bring the marked end of the first magnet near one end of the second magnet.

1. Do the ends attract each other or repel each other?

E. Now bring the marked end of the first magnet near the other end of the second magnet.

2. Do the ends of the magnets attract or repel each other?
3. Do both ends of the magnet behave the same or are they different?

F. Cover the marked end of the magnet with a piece of paper and repeat steps D and E.

4. Does the paper have any effect on the attraction or repulsion of the magnets?

G. Copy Table 6-1 in your notebook.

H. Bring the magnet near each of the objects listed in the table. Record the results in the table.

5. What kind of materials were attracted by the magnet? Write complete sentences.

6-7.

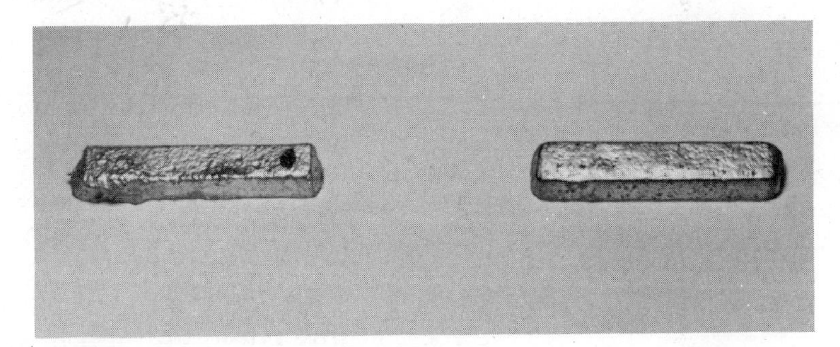

TABLE 6-1

	Attracted	*Repelled*	*No Effect*
1. wood			
2. pencil lead			
3. eraser holder on pencil			
4. eraser			
5. paper			
6. glass			
7. plastic			
8. paperclip			
9. aluminum foil			
10. staple			
11. chalk			

SUMMARY

You can now identify magnetism as a third kind of force. You never feel the effect of magnetism the way you feel gravity and electricity. You can describe ways in which the magnetic force and the electric force are alike. Observations of magnets show that they always have at least two poles. These poles are the north and south poles of the magnet. Like electric charges, magnets are surrounded by a field.

QUESTIONS

Use complete sentences to write your answers.

1. Describe three ways in which magnets are similar to objects with electric charges.
2. Given only a metal bar and a piece of thread, how could you tell if the bar is a magnet?
3. How could you show that a magnet has two poles?
4. What is the difference between a permanent magnet and a temporary magnet?

6-2. EARTH AS A MAGNET

About four hundred years ago, one of the laws of the sea said: "Any sailor caught touching the lodestone, if not punished by death, shall have his hand fastened to the mast with a dagger." The lodestone was a small piece of magnetic iron ore in the crude compass used on ships at that time. The magnetic lodestone helped the sailors to determine the direction they were sailing. The harsh punishment for touching the lodestone is evidence of how sailors depended upon magnetism to find their way. A compass is simply a magnet used to show directions.

When you finish lesson 2, you will be able to:

- Explain how to make a compass.
- Use a compass to show that the earth acts like a magnet.
- Explain why a compass does not always point directly north and south.
- ○ Make a compass from a steel needle.

Any magnet that can turn freely will swing around so that its poles point toward the north and south. For example, a small magnet can be put on a floating cork. See Fig. 6-8. The magnet will then act as a compass. The lodestone used in early compasses is a mineral that is a natural magnet.

According to the laws of magnetic attraction and repulsion, a compass could not work unless the earth acts like a magnet. Imagine a giant magnet inside the earth. See Fig. 6-9. A compass would point north and south because its magnetic poles would be attracted to the opposite poles of the magnet in the earth. Of course, there is not really a giant magnet buried in the

6-8. *A small magnet can be used as a compass.*

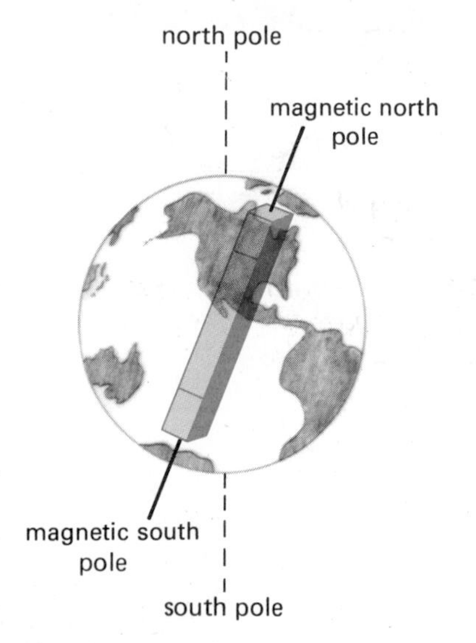

6-9. *The earth acts as if a giant magnet were buried within it.*

earth. Scientists have not yet been able to discover exactly why the earth acts like a magnet. It may be a result of slow movements of very hot materials deep inside the earth. Whatever the cause of the earth's magnetism may be, its effect is the same as that of a huge magnet buried within the earth.

Today, it seems reasonable to use the earth's magnetism to explain the behavior of compasses. However, compasses were used for centuries before this idea was suggested. In 1600, William Gilbert, who was the physician of Queen Elizabeth I of England, first decided that the earth is a magnet. He predicted that the earth would be found to have magnetic poles. When those poles were discovered, the laws of magnetic attraction and repulsion were not completely understood. Thus the magnetic pole of the earth to which the north pole of a compass needle points was incorrectly named the "north" magnetic pole. However, this magnetic pole of the earth must actually be a south pole since it attracts the opposite pole of a compass needle. See Fig. 6-10. Likewise, the earth's "south" magnetic pole is also incorrectly named. It would be too confusing to try to correct the error now. Thus, the earth's north and south magnetic poles are still usually named according to their locations near the imaginary axis marking the true north and south geographic poles.

Notice in Fig. 6-9 that the magnet in the earth is not exactly lined up with the actual north and south geographic poles. The earth's magnetic poles are not in the same place as the true geographic poles. The magnetic pole in the north, for example, is in northeastern Canada about 1,600 km from the north pole. This causes a problem for navigation. Since compasses point toward the earth's magnetic poles, a compass needle does not necessarily show true north and south. This error in a compass is called **magnetic variation.** *Magnetic variation* has to be considered in a compass because the earth's magnetic poles and its true geographic poles are not in the same places. See Fig. 6-11. Magnetic variation does not always cause the same error in a compass. Close to the equator, the error is small. There, the distance to the pole is great, and the magnetic pole and north pole seem closer together. The closer you get to either pole, the greater the amount of error. Magnetic variation must be taken into account by anyone using a compass to find accurate directions.

Magnetic variation
The error in a compass caused by the difference in location of the earth's magnetic and geographic poles.

Like all magnets, the earth is surrounded by a magnetic field. A field around a small magnet can be outlined by small pieces of iron on a piece of paper over the magnet. If there were some way to do this for the earth, you would have a picture of its magnetic field. See Fig. 6-12. Knowledge of the earth's magnetic field helps to explain the beautiful and mysterious northern lights, as you will learn in lesson 3.

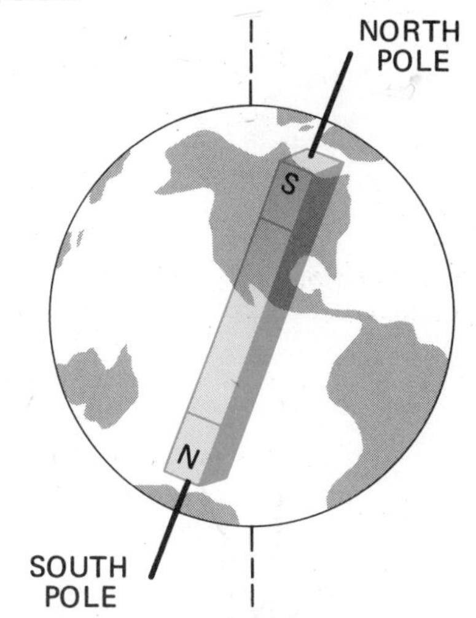

6-10. *The magnetic pole near the north geographic pole is actually a south magnetic pole.*

6-11. *Compass needles do not point directly north and south because the geographic north pole and the magnetic north pole are not in the same place.*

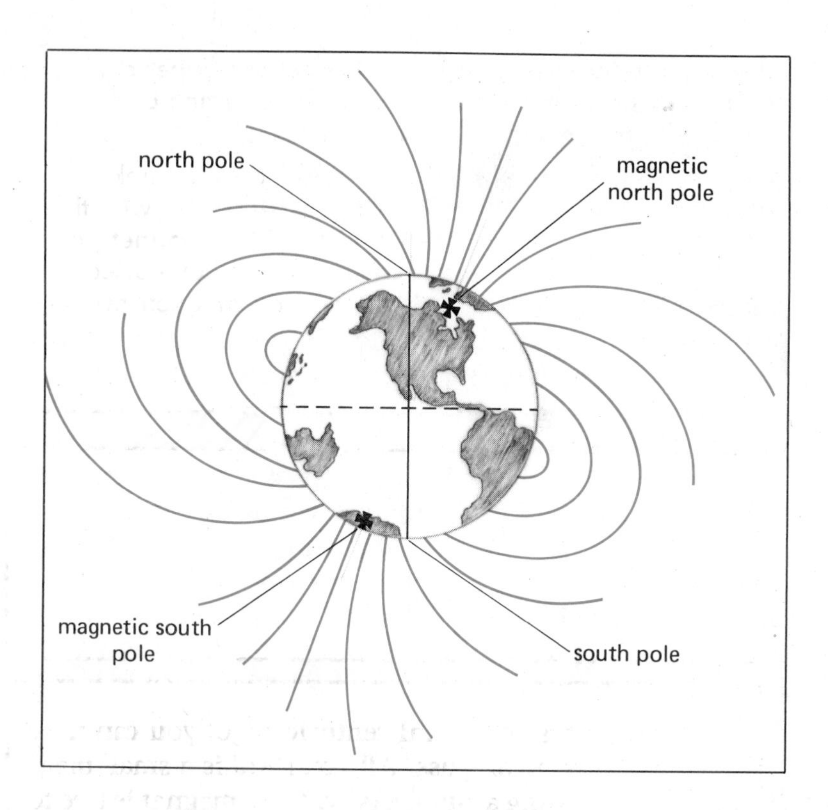

6-12. *The lines shown in color indicate the shape of the earth's magnetic field.*

ACTIVITY

Materials
bar magnet
steel needle
nylon thread, 50 cm
transparent tape
paper clip

6-13.

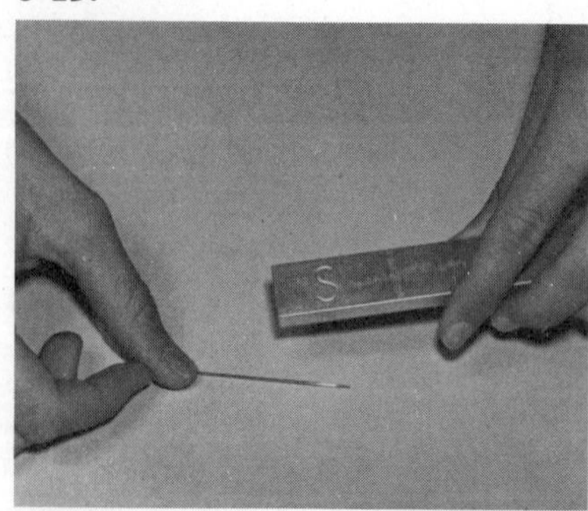

Making a Compass

A. Obtain the materials listed in the margin.

B. Place a needle on the top of your desk.

C. Hold the eye end of the needle down with your finger.

D. Starting near the eye end of the needle, stroke the needle with the south pole of a magnet. Stroke in one direction. When you get to the end of the needle, lift the magnet up and return it to the eye end of the needle. Stroke the needle again. Repeat this 10 or 15 times. See Fig. 6-13.

E. Tie a thread around the center of the needle. Hang the needle by the thread so that it swings freely.

1. In what direction does the needle point?

If the needle does not point in any particular direction, stroke it with the magnet several more times.

2. Which kind of a magnetic pole is the point of the needle?

This hanging needle can be used as a crude compass but it has limitations.

F. Twist the thread in one direction about 20 times

while the needle is hanging from it. Release the thread.

3. Does the needle point in the same direction as before?

G. Bring a paper clip near the hanging needle.

4. Does the paper clip affect the direction of the needle?

5. How do you think rubbing the needle with the north pole of a magnet (see step D) would have affected the needle's magnetism? Try it.

SUMMARY

Like the sailors of several centuries ago, you can now make your own compass. All you need is a small magnet that can swing around easily. If the magnet is free to swing, it will turn until it points north and south. This kind of compass works because the earth acts like a magnet. Like all magnets, the earth has two magnetic poles. However, because these magnetic poles and the true geographic poles are not in exactly the same places, compasses do not point true north or south.

QUESTIONS

Use complete sentences to write your answers.

1. **a.** If you were given a small bar magnet, how would you make it into a compass? **b.** Describe how you could make a compass using a steel needle.
2. Why does a compass needle not point directly north and south?
3. What is the error in compass direction called?

CAREERS IN ELECTRONICS AND HOME HEATING

The careers illustrated on these pages represent two important influences on modern living. The electrical age began in 1831 with the discovery of electromagnetic induction. This discovery was followed by the invention of the electric generator in 1832, the electric motor in 1873, and the electric light bulb in 1879. Today, electrical appliances and electronic devices are essential to our way of life. Equally essential are the machines that cool our homes in summer and heat them in winter. As a result of the energy crisis of recent years, it is now especially important that home heating and air conditioning be as efficient as possible. Both of these fields offer a variety of career opportunities.

Air-Conditioning, Heating, and Refrigeration Technician

Description: Air conditioning, heating, and refrigeration technicians design, manufacture, sell, and service equipment to regulate interior temperatures. Technicians may specialize in a particular area such as heating or in a particular type of activity such as research and development. Technicians often assist engineers and scientists in the design and testing of new equipment.

Requirements: Most employers require specialized technical training although a combination of work experience and education may be a sufficient qualification for a job as a technician. Technical institutes, junior and community colleges, and vocational schools all provide specialized technical training. Training is often available through correspondence schools, apprenticeship programs, and on-the-job training.

For more information:

National Association of Trade and Technical Schools, Accrediting Commission, 2021 L St. NW, Washington, DC 20036

U.S. Department of Education, Washington, DC 20202

Electrician

Description: Electricians install, maintain, and repair lighting systems, generators, transformers, and other electrical equipment. Electricians may work in factories, office buildings, or private homes.

Requirements: Most electricians learn through on-the-job training or apprenticeship programs. Apprenticeship usually lasts 4 years. It includes on-the-job training and classroom instruction in mathematics, electrical and electronic theory, and blueprint reading.

For more information:
International Union of Electrical, Radio, and Machine Workers (AFL-CIO), 375 Murray Hill Parkway, East Rutherford, New Jersey 07073

Electronic Computer Programmer

Description: Because computers cannot think for themselves, computer programmers must write step-by-step instructions (programs) for the machines to follow in solving a problem.

Requirements: Employers of programmers have different needs; thus, training requirements are not universal. Computer programming is taught at public and private vocational schools, colleges, and universities. Many high schools also offer computer courses.

For more information:
American Federation of Information Processing Societies, 210 Summit Ave., Montvale, New Jersey 07645

Extensions

1. Write to your state employment service and ask for information about apprenticeship programs for electricians or home heating technicians.
2. If your school has a computer center, interview the person in charge and find out how the computers are programmed for specific tasks.
3. Look at the photograph on page 172 and Fig. 5-35 on page 186. How many electrical appliances can you identify?

6-3. MAGNETISM AND ELECTRICITY

At certain times of the year, the night sky seen from places near the north or south pole is filled with colored streamers or curtains of light. These silent green and red lights dance over the sky, growing dim or bright as they change shape. After a while, the lights fade away, leaving only a faint glow. What is the explanation of these "northern lights" and "southern lights"?

When you finish lesson 3, you will be able to:

- Explain how magnetism and electricity are related.
- Describe how an electric motor works.
- ○ Demonstrate that a wire carrying an electric current is surrounded by a magnetic field.

Aurora
The northern or southern lights.

The colorful displays of the northern and southern lights are called **auroras** (uh-**rore**-uhz). *Auroras* are most often seen in the far north or far south of the earth. Auroras occur as a result of the shape of the earth's magnetic field. The magnetic field curves around the earth, meeting near the poles. See Fig. 6-14. Experiments have shown that the auroras are caused by electrified particles from the sun. When these electrically charged particles reach the earth, they are guided by the magnetic field toward the poles. The lights are produced as the charged particles fall into the earth's atmosphere.

Any moving object with an electric charge is affected by a magnetic field. There seems to be a connection between magnetism and electricity. The first scientist to discover this connection was Hans Christian Oersted. In 1819, Oersted found that an electric current passing through a wire caused a nearby compass needle to move.

6-14. *The shape of the earth's magnetic field causes electrically charged particles from the sun to be drawn in near the poles.*

When an electric current moves through a wire, that wire becomes surrounded by a magnetic field. The field spreads out over a long wire and can be concentrated by turning the wire into a coil. If a piece of iron is put inside a coil of wire, a very strong **electromagnet** can be made. An *electromagnet* is a temporary magnet made by wrapping a coil of wire around a piece of iron. When an electric current flows through the wire, the iron becomes a magnet. See Fig. 6-15. An electromagnet is different from a permanent magnet in two important ways. First, an electromagnet can be made stronger or weaker by changing the amount of current flowing through the wire coil. Second, an electromagnet can be turned off and on. It is a strong magnet only when the current is on. Superconducting electromagnets can be made at very low temperatures. When very cold (around −273°C), some metals have almost no electrical resistance. This low resistance allows large currents to be sent through these cold conductors. Superconducting magnets operating at very low temperatures are important tools in scientific research.

Electromagnet
A temporary magnet made when an electric current flows through a coil of wire wrapped around a piece of iron.

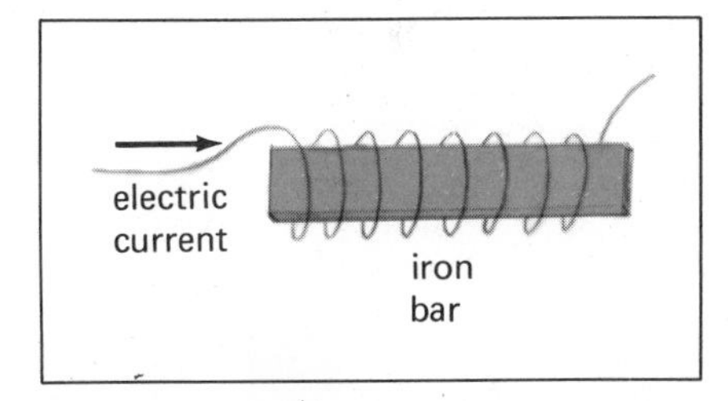

6-15. *A simple electromagnet can be made with an iron bar and a current-carrying wire.*

Electromagnets are different from permanent magnets in one other way. The poles of an electromagnet can be changed. If an electromagnet is connected to a dry cell, a compass can be used to tell which are its north and south poles. See Fig. 6-16. The connections to the dry cell can be switched to make the current reverse direction. The compass will then show that the poles of the electromagnet have also reversed. This

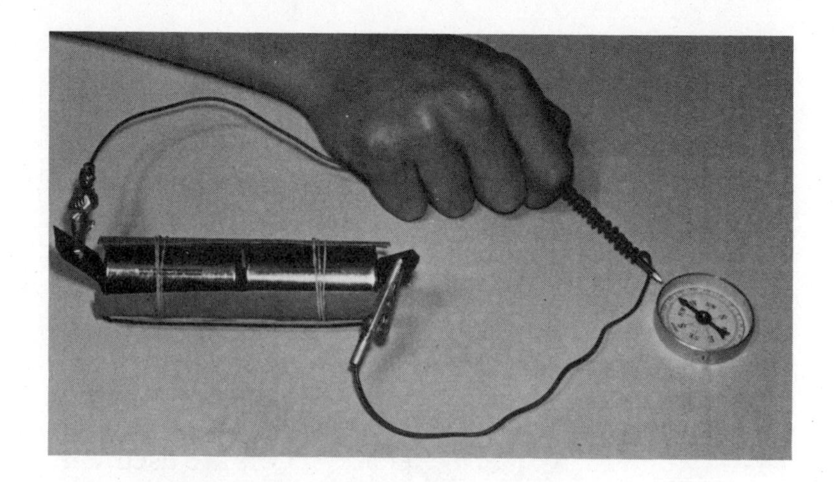

6-16. *A compass can be used to find the north and south poles of an electromagnet.*

reversing of poles in an electromagnet explains the operation of electric motors.

An electric motor is made up of two magnets. One magnet is held in a fixed position on the frame of the motor. See Fig. 6-17. The other magnet is made by sending a current through a loop of wire on a rotating shaft. This causes the wire loop to form a north and south pole. The loop will then turn to line up its poles near the opposite poles of the outer magnet. If the current continued to flow in the same direction through the loop, nothing else would happen. But an automatic switch causes the current in the loop to be reversed. Now the poles of the loop are unlike the poles

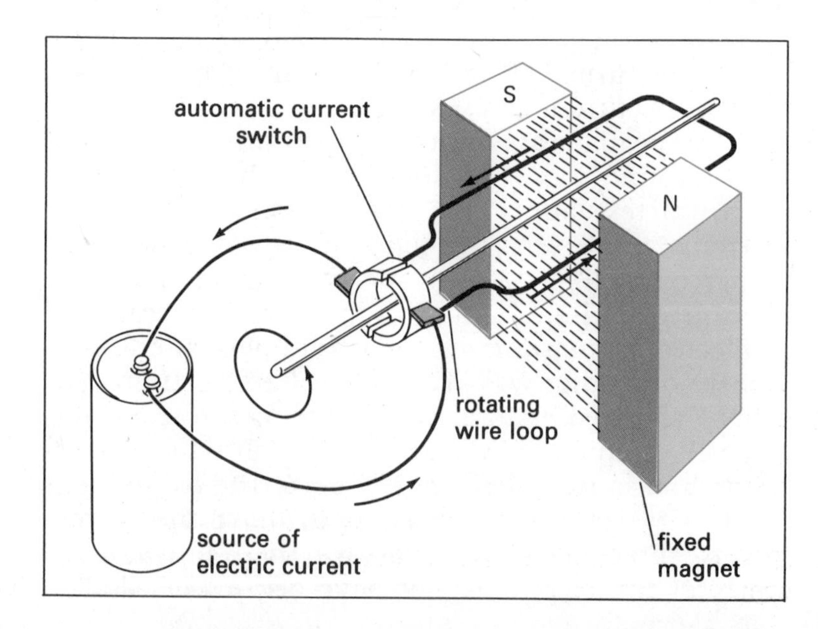

6-17. *A simplified diagram of an electric motor.*

of the outside magnet. The wire loop will be repelled and thus cause the shaft to turn. Again the switch reverses the current in the loop to keep the shaft turning. The motor will keep turning as long as the electric current flows, first one way and then the other through the loop of wire.

Electric motors are made with many wire coils attached to the rotating shaft. The current flows through each coil of wire at the proper moment to keep the shaft turning smoothly. See Fig. 6-18. The outer magnet is usually also an electromagnet with the current always flowing in the same direction through it. Many different kinds of electric motors are used. But no matter how they are made, all electric motors use magnetic forces caused by electric currents.

6-18. *A small electric motor.*

ACTIVITY

A Simple Electromagnet

In this activity, you will test to see if an electric current can produce a magnetic field.

Materials
nail
wire, 1.5 m
2 batteries
light bulb and socket
compass
paper clip
2 wire connectors

A. Obtain the materials listed in the margin.

B. Wrap 1.5 m of wire around a nail about 30 times. See Fig. 6-19.

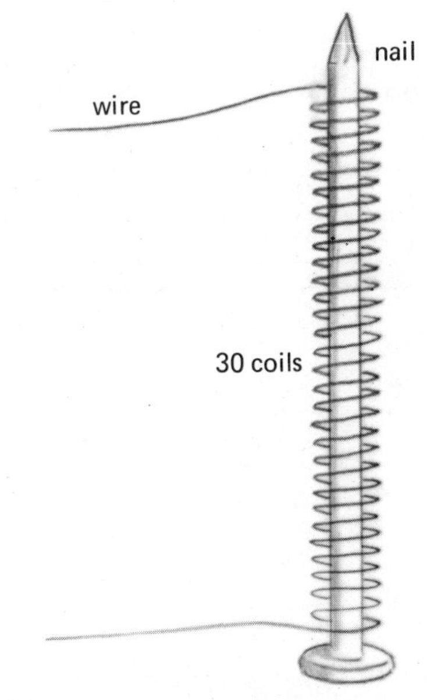

6-19.

C. Set up an electric circuit as shown in Fig. 6-20. Do not make the final connection yet.

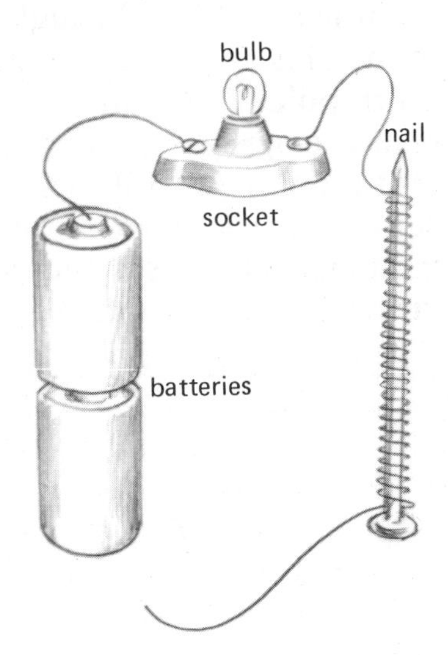

6-20.

D. Bring the head of the nail near a paper clip.

1. Does the nail attract the paper clip?

E. Connect the final wire to complete your circuit.

2. What happens to the light bulb when you complete the circuit?
3. Is there a current flowing in the circuit?

F. While the current is flowing, bring the head of the nail near a paper clip.

4. Does the nail attract the paper clip?

G. Bring the head of the nail near the south pole of a compass.

5. Does the nail attract or repel the south pole?

H. Bring the head of the nail near the north pole of a compass.

6. Does the nail attract or repel the north pole?

7. How do the north and south poles of the compass react to the point of the nail?

SUMMARY

The auroras seen near the earth's poles are caused by electrically charged particles from the sun following the earth's magnetic field. Auroras provide evidence of a relationship between magnetism and electricity. Electromagnets show that electricity can be changed into magnetism. An electric motor operates on this principle.

QUESTIONS

Use complete sentences to write your answers.

1. Give three examples that show how electricity and magnetism are related.
2. Using Fig. 6-17, describe how the electric motor works.
3. Describe how you could show that a wire carrying a current is surrounded by a magnetic field.

6-4. ELECTRO-MAGNETIC INDUCTION

When you plug in a radio or stereo at home, where does the electricity come from? You could find out by following the electric power lines all the way back to the power plant. Inside the power plant you would probably find giant boilers making steam by burning coal or oil. You would also find large electric generators in action. How does this machinery make electricity? You can understand the generation of electricity by knowing the complete relationship between electricity and magnetism. Magnetism can be used to make electricity.

When you finish lesson 4, you will be able to:

- Predict what happens when a wire moves in a magnetic field.
- Define *electromagnetic induction*.
- Explain how an electric generator produces electricity.
- ○ Demonstrate what happens when a wire moves in a magnetic field.

Electricity can produce magnetism. Can magnetism also produce electricity? This question was first answered at about the same time by two different scientists. In England, the discovery was made by Michael Faraday, a former bookbinder's apprentice who became a famous scientist. Faraday found that a current passed through a coil of wire created a magnetic field. That magnetic field caused a current to be produced in a second neighboring wire coil. In America, a mathematics teacher named Joseph Henry made the same discovery. He found that a current was pro-

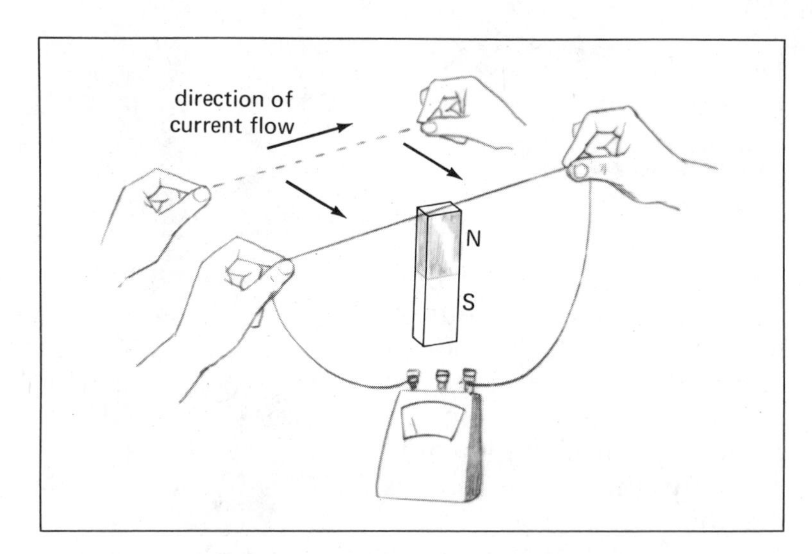

6-21. *An electric current is produced in a wire that is being moved through a magnetic field.*

duced in a coil of wire when a magnet was brought nearby. Both Faraday and Henry had discovered that magnetism can produce electricity when a magnetic field and an electrical conductor move relative to each other. For example, when a wire is moved in a magnetic field and electric current flows in the wire. This generation of current is an example of what is called **electromagnetic induction** (ih-**lek**-troe-mag-**net**-ik in-**duk**-shun). An electric current produced by *electromagnetic induction* is always the result of motion in a magnetic field. When a wire moves past a magnet, a current will flow in the wire. If the wire is held still and the magnet is moved past, a current flows in the wire. The only requirement for producing a current in this way is to cause motion within the magnetic field. Experimenting with a wire moved past a magnet proves that the current flows in a certain direction. See Fig. 6-21. If the wire is moved in the opposite direction, the direction of current flow is reversed.

Electromagnetic induction Production of an electric current by motion in a magnetic field.

A knowledge of electromagnetic induction can be used to build a machine to produce electricity. A loop of wire is put into a magnetic field. See Fig. 6-22. As the wire loop is turned, an electric current is produced. Spinning the loop can produce a current that is fed into wires.

Remember that the direction of current flow changes with the direction of movement of a wire in a magnetic field. When one side of a wire loop moves down in a magnetic field, the current produced will flow in one direction. When the same side moves up

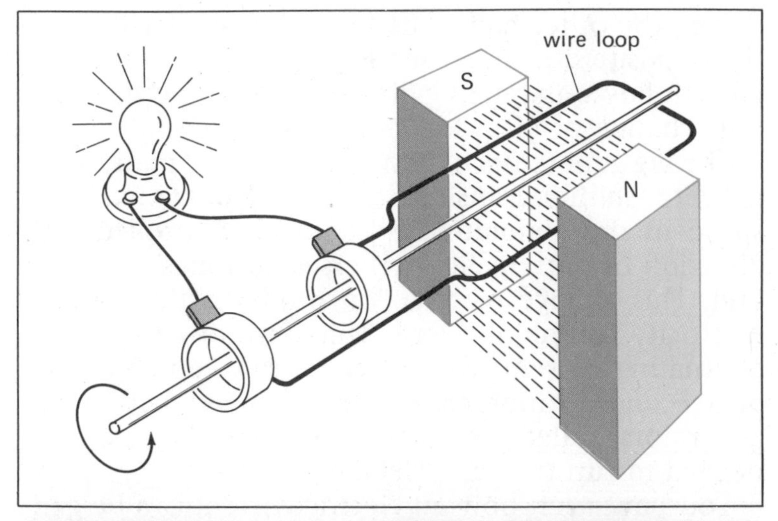

6-22. *A simplified diagram of an electric generator.*

6-23. *(above) A portable electric generator. (left) An alternating current electric generator.*

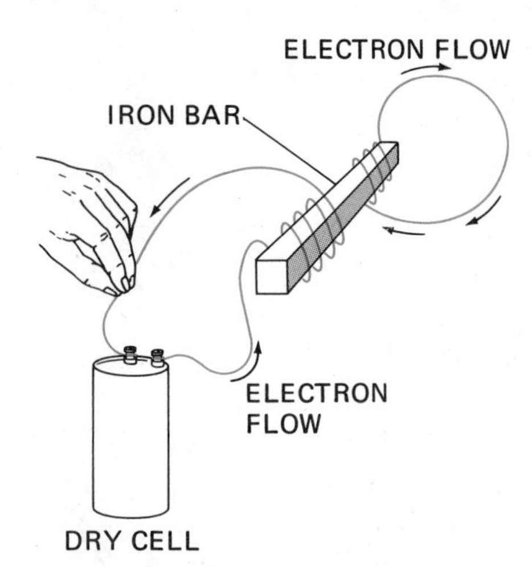

6-24. *Current flowing in a wire coil wrapped around an iron bar can produce a current in a second coil of wire by electromagnetic induction.*

during the other half of its turn, the current flows in the opposite direction. See Fig. 6-23. As a result, the current flows first in one direction and then the other. This change of direction produces *alternating current.*

The big generators in power plants have many loops of wire spinning inside large electromagnets. The speed of the generators is carefully controlled. The direction of current flow reverses 60 times each second. This change of direction produces the form of electricity commonly used. This form of electricity is alternating current with a frequency of 60 Hz (cycles per second). Automobiles are equipped with small generators called *alternators* to supply the energy needed to run the car's electrical system.

You have seen how an electric current can be generated by moving a wire past a magnet. A current may also be produced by electromagnetic induction when a magnet moves past a stationary wire. A third method of electromagnetic induction needs no visible motion.

Suppose two coils of wire are wound around opposite ends of an iron bar. See Fig. 6-24. One of the coils is attached to a source of electric current. When the current begins to flow through the wire coil, the iron bar is magnetized. The second coil of wire then acts as if a magnet were suddenly pushed into it. The effect on the coil of wire is the same as if a magnet had been moved past it. Electromagnetic induction causes an electric current to be produced in the second coil. Each time a current changes in the first coil, a current is produced in the second wire coil.

This principle is the basis of an electrical *transformer.* A transformer consists of two coils of wire wrapped around the same iron core. Transformers are useful because they can change the voltage of alternating current. If the number of turns in the two coils of wire is the same, the voltage produced in the second coil will be the same as that applied to the first coil. If the second coil has *twice* as many turns, its voltage will be *twice* that in the first coil. See Fig. 6-25. Although the voltage is doubled, the amperage is cut in half. If there are fewer turns in the second coil, the output voltage will be less than the input voltage. Transformers can only be used to change the voltage of alternating current. Only alternating current has the start and stop action

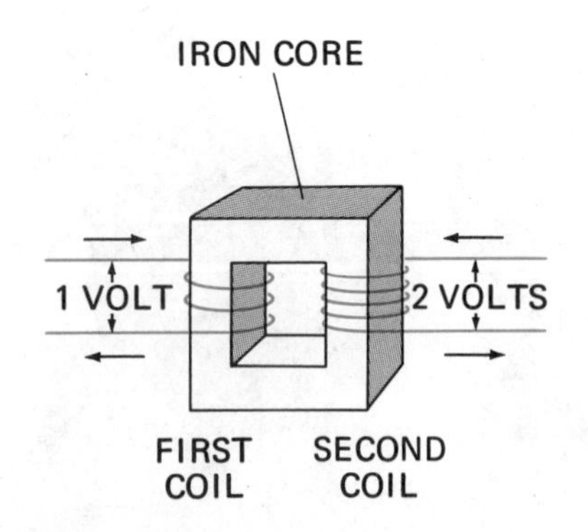

6-25. *A transformer can be used to increase the voltage of an alternating current.*

needed to produce the changing magnetic field required for electromagnetic induction.

Transformers are used to produce the many different voltages needed to operate all kinds of electrical machinery and appliances. The very high voltages needed to send electric currents long distances in power lines are produced by transformers. Other transformers then reduce the voltage again for ordinary use. There is probably such a transformer on a power pole near your house. (See page ~~156~~.) 187

Electromagnetic induction might seem to break the law of conservation of energy. One part of this law says that energy cannot be created. When a wire moves through a magnetic field, the electric energy in the wire seems to come out of nowhere. Remember that the wire must be moving. If the motion stops, the electric current stops. The electric energy really comes from the energy of motion. An electric generator needs a source of energy to cause motion. This source of energy may be heat energy from the burning of fuel. Energy released by falling water can also turn generators. Nuclear energy is becoming more important as a way to produce electricity.

ACTIVITY

Generating an Electric Current

In this activity, you will generate an electric current in a wire by moving a magnetic field past the wire.

A. Obtain the materials listed in the margin.

B. Wrap one end of the wire around a compass so that about five loops go over the top and underneath the compass needle. See Fig. 6-26.

C. At least 15 cm from the compass, wrap the wire into another coil around a pencil. Make about 10 to 15 coils. Then remove the pencil.

D. Join the loose end of the wire to the end of the wire near the compass. This makes a complete circuit. See Fig. 6-26. The coils should be at least 15 cm apart.

The compass will be used to detect any current that flows in the wire.

E. Arrange the compass and its coil so that the compass needle is pointing in the same direction as the wires in the coil around it. Be sure the compass needle is free to move. You may need to tap the compass to be sure the needle is free.

Materials
bar magnet
wire, 1.5 m
pencil
compass

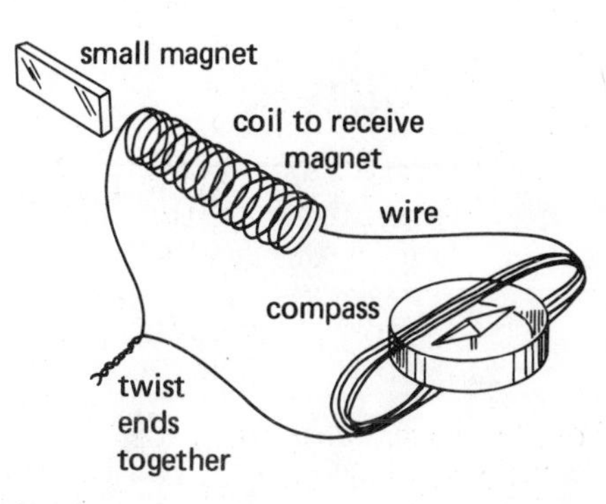

6-26.

F. Insert the magnet in the coil of wire.

G. Watch the compass. Now *quickly* pull the magnet from the coil.

1. Does the needle move? If the coils are 15 cm apart, any movement of the needle is due to an electric current.

2. Explain how you can produce an electric current with a magnet.

SUMMARY

The next time you plug a radio or iron into the wall socket at home, think of the origin of the electricity you are using. Machines are built to generate electricity by electromagnetic induction. Electromagnetic induction involves motion in a magnetic field. An electric generator usually produces alternating current. An electric generator only converts some other form of energy into electric energy.

QUESTIONS

Use complete sentences to write your answers.

1. An electric current is generated in a circuit when a wire is moved through a magnetic field. Predict the effect on this current when **a.** the wire is moved at a faster rate **b.** the wire is moved first in one direction, then back in the reverse direction.

2. Using Fig. 6-22, describe how a generator produces electricity.

3. Describe how you can detect the current in a circuit caused when a magnet is moved in and out of a coil of wire.

4. Who were Michael Faraday and Joseph Henry?

VOCABULARY REVIEW

Match the number of the word with the letter of the phrase that best explains it.

1. aurora
2. electromagnet
3. electromagnetic induction
4. magnetic field
5. magnetic poles
6. magnetic variation

a. The parts of a magnet where magnetic forces are strongest.
b. Formation of an electric current by motion in a magnetic field.
c. The error in a compass caused by a difference in location of the earth's magnetic and geographic poles.
d. A region of space around a magnet in which magnetic forces are noticeable.
e. The northern or southern lights.
f. A temporary magnet made when an electric current flows through a coil of wire.

REVIEW QUESTIONS

Complete each statement by choosing the best word or phrase, or by filling in the blank.

1. A magnet is similar to an object with an electric charge in that **a.** both may attract or repel **b.** their forces act through a distance **c.** their forces increase with a decrease in distance **d.** they are similar in all of the above ways.
2. A magnet will attract or repel another magnet most strongly at its ____________.
3. Small pieces of iron sprinkled around a magnet will make its ____________ become visible.
4. To make a magnet act as a compass it must ____________.
5. When an ____________ moves through a wire, that wire becomes surrounded by a magnetic field.
6. When a wire is moved in a ____________, an electric current flows in the wire.
7. An aurora is produced by the charged particles streaming from the sun and reacting with the earth's **a.** gravitational field **b.** electric field **c.** magnetic field **d.** equator.
8. An electromagnet is **a.** an electric current flowing through a wire **b.** an electric current flowing through a coil of wire wrapped around a piece of iron **c.** an electric wire wrapped around a magnet **d.** a magnet coiled around an electric wire.

9. Electromagnetic induction is the process that produces **a.** an electric current from motion in a magnetic field **b.** a magnetic field from a current in a wire **c.** a permanent magnet from a piece of iron **d.** an electromagnet using electricity in coils of wire and an iron core.
10. An electric generator is a machine that converts **a.** magnetism to electricity **b.** some form of energy into electric energy **c.** the energy of burning coal into electric energy **d.** nuclear energy into electric energy.

REVIEW EXERCISES

Give complete but brief answers to each of the following. Use complete sentences to write your answers.

1. Describe the two ways in which a magnet may behave when brought near another magnet.
2. Explain how you would determine whether a piece of metal is a magnet.
3. How would you demonstrate that the earth has a magnetic field?
4. Describe how you would make a compass.
5. Give two pieces of evidence which show that magnetism and electricity are related.
6. Describe three ways an electromagnet differs from a permanent magnet.
7. Explain how an electric motor works.
8. Predict what would happen in an electric circuit if a magnet is pushed rapidly into a coil of wire in the circuit and then pulled rapidly back out of the coil of wire.
9. Explain how an electrical generator is able to produce electricity.
10. What is a 60-Hz alternating current?

EXTENSIONS

1. The Bermuda Triangle is an area in the ocean where mysterious deaths and disappearances have occurred. Some people believe that magnetic fields are involved. Write a report on the Bermuda Triangle, including possible explanations for the magnetic effects and their effect on ships and planes.
2. Construct a simple electric motor. Consult basic electricity books for a plan. Demonstrate it to the class. How do AC and DC motors differ from each other?
3. Make a dipping needle to measure the inclination of the earth's magnetic field. Demonstrate it to the class. (You will find a description of a dipping needle in an encyclopedia or dictionary.)

CHAPTER 7

Heat

7-1. HEAT ENERGY

Rub your hands together as fast as you can. Do they get warm? Where does this heat come from? This question puzzled scientists for hundreds of years. Their search for an answer led them to some important discoveries.

When you finish lesson 1, you will be able to:

- Explain why heat is considered a form of energy.
- Predict what will happen to particles of matter when heat energy is added.
- ○ Demonstrate what happens to particles of matter when heat energy is added or removed.

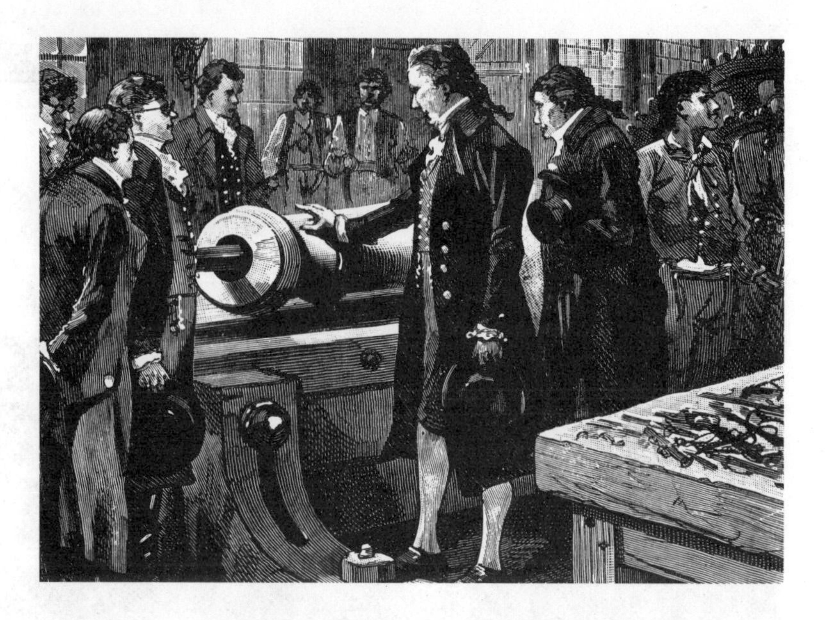

7-1. *Count Rumford proved that heat is a form of energy by studying the heat produced in drilling a cannon.*

What is heat? About two hundred years ago, a scientist would probably have answered this question by saying something like: "Heat is an invisible and weightless fluid. It soaks into an object when the object is heated and drains away upon cooling." According to this theory, ice is changed to water as "heat fluid" is added to it. As more "heat fluid" is added, the water changes to steam. If you burned your hands by sliding too quickly down a rope, the burns were caused by "heat fluid" squeezed out of the rope.

This old theory did explain some observations about heat. For example, heat seems to move from a hot object to a cold object. Heat leaves a warm house in cold weather and must be replaced constantly. This behavior of heat seems to support the idea that heat "flows" from one place to another. The theory of "heat fluid" did not pass the last and most important test of a scientific theory, however. Experiments showed that heat could not be "heat fluid."

One such experiment was done in 1798 by Benjamin Thompson. Thompson was an American who became a government official in Europe where he was called Count Rumford. At one time, he was in charge of a cannon factory. In those days, cannons were made by drilling machines that were run by horses. See Fig. 7-1. Rumford noticed that the cannons became very hot during this process. He set up the following experiment: A cannon was drilled while surrounded by a wooden box containing water. After several hours of drilling, the water began to boil. It continued to boil as long as the drilling went on. To Rumford, the experiment showed that the supply of heat was without limit and therefore could not be a kind of matter contained in the metal of the cannon. Rumford decided that heat was actually a form of energy supplied by the work of the horses. Rumford's experiment, and others that were done later, showed that heat is a form of energy.

If heat is energy, how is this energy contained in a hot object? Modern scientists think of all matter as being made up of tiny particles. One of these particles alone is much too small to be seen. Heat energy causes these particles to move faster. For example, the heat energy caused by Rumford's drilling experiment made the particles in the cannon move rapidly. The more heat that is added, the faster the particles in matter will move. For example, a drop of water contains a huge number of individual water particles. All of these particles are moving. Heating the water will cause the particles to move faster and bump into each other more often. See Fig. 7-2. You cannot watch individual particles to see if heating or cooling changes their speed. But you may have noticed that a colored substance put into water will slowly spread out through the water. For example, carefully put a drop of ink or food coloring into a glass of water without stirring. In time, the motion of the water particles and the particles of the colored substance will cause the color to spread evenly throughout the water in the glass.

7-2. *As water is heated, its particles move more rapidly.*

Because heat is a form of energy, it can do work. Machines can be used to change heat energy into useful mechanical energy. For example, a steam engine uses the heat energy contained in the moving particles that make up the hot steam. The earliest steam engines used steam to push against a metal plate, called a *piston*. The piston moved up and down inside a tube called a *cylinder*. See Fig. 7-3. As the engine worked, steam from a boiler pushed the piston to the top of the cylinder. Cool water was then sprayed into the cylinder. When the cooled steam in the cylinder changed back into water, the piston returned to the bottom of the cylinder and the entire process was repeated. This kind of engine used only a small part of the total heat energy in the steam. Most of the energy was wasted.

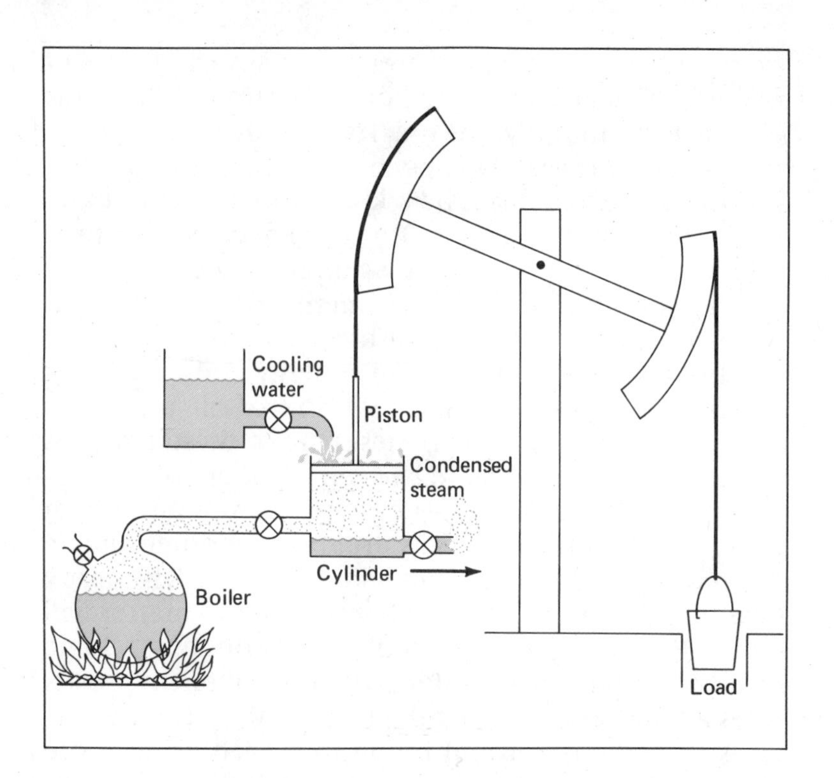

7-3. *A steam engine built in the early 18th century used steam to operate a large "rocking beam." The movement of the beam pumped water from deep mines in England.*

In the late 1700's, James Watt, an instrument maker in Scotland, invented an improved steam engine. Watt developed an engine in which the steam pushed the piston in both directions in the cylinder. See Fig. 7-4. Watt's steam engine was far more powerful than the early wasteful models. It made possible the use of steam power for locomotives and factories. The industrial revolution of the nineteenth century was powered by the heat energy of steam.

Most modern steam engines no longer use the back-and-forth motion of a piston in a cylinder. Instead, the steam pushes against the blades of a *turbine*. A turbine rotates like a high-speed windmill. See Fig. 7-5. Turbines operate smoothly and do not waste much of the energy contained in the steam.

Automobile engines also use heat energy. A burning fuel produces hot gases that move pistons within cylinders. Jet engines also use hot gases produced by burning fuels. However, in a jet engine, the hot gases run a turbine. The action of the turbine turns a compressor that compresses air coming into the engine before it combines with fuel. The engine gets its forward thrust in the same way as a rocket engine. The

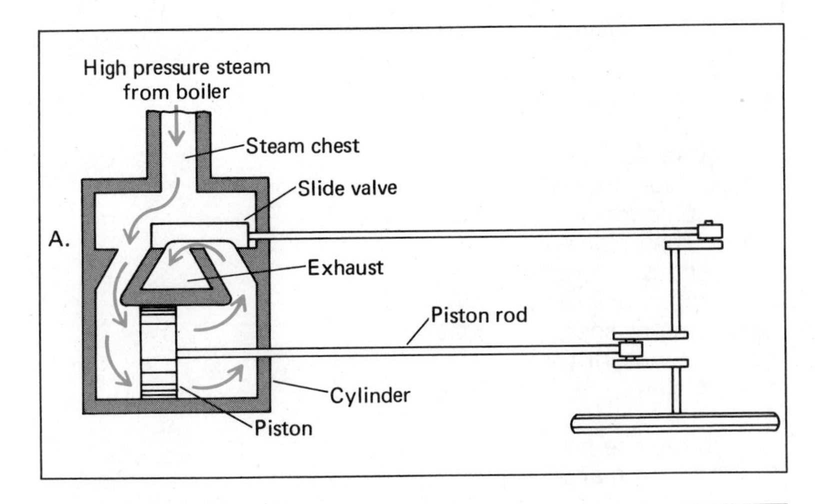

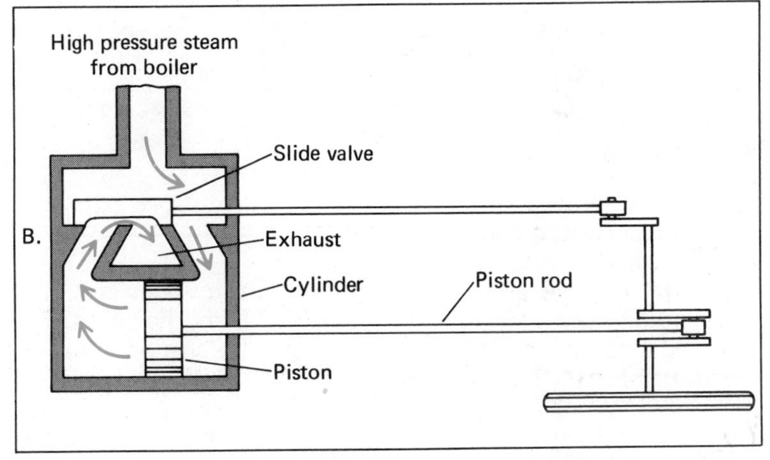

7-4. *The steam engine invented by James Watt used a sliding valve to allow the steam to push a piston back and forth.*

hot gases leaving the rear of the engine at high speed give an equal but opposite forward thrust on the engine according to Newton's third law of motion.

Steam engines and automobile engines are examples of **heat engines.** A *heat engine* is a machine that changes heat energy into mechanical energy. However, there is a problem in using all heat engines. Much of the heat energy in a heat engine is wasted. There is no way to use all of the energy contained in the moving particles of the heated gases that run heat engines. An automobile engine, for example, is only 30 percent efficient. The engine has a radiator to take care of this wasted heat energy. The wasted heat escapes from the engine through the radiator. No one can make a heat engine that will change *all* of the heat energy supplied to it into mechanical energy.

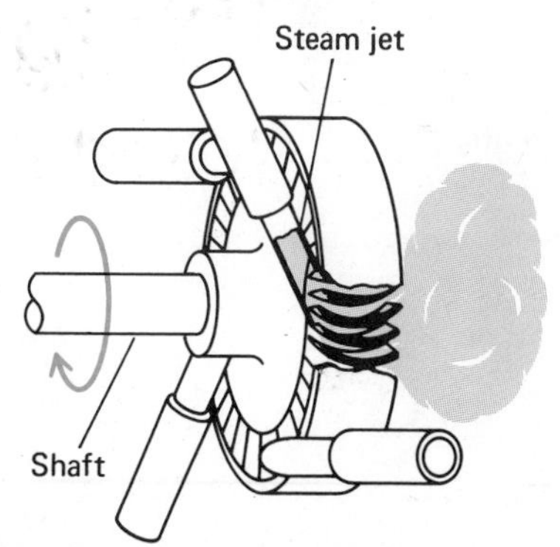

7-5. *In a steam turbine, a jet of steam pushes against the blades of the turbine wheel.*

Heat engine
A machine that changes heat energy into mechanical energy.

ACTIVITY

Materials

2 small glass (or plastic) containers
dark food coloring
ice cube
Bunsen burner
water

Motion of Particles

It is possible to mix a colored substance with a larger amount of water and watch the color spread throughout the water. If the mixture is not stirred, the spreading of the color must be the result of the motion of particles.

A. Obtain the materials listed in the margin.

B. Fill a small clear glass or plastic container about 2/3 full of water. (Use water that is at or near room temperature.)

C. Place one drop of dark food coloring onto the surface of the water. See Fig. 7-6. Observe the coloring and water from the top and sides of the container for 2–3 min.

7-6.

1. Describe, in your own words, the changes you observed after you added the food coloring to the water.
2. In your own words, state a hypothesis that tells what effect you think hot or cold water would have on the rate at which the coloring mixes with the water.

D. Discard the water and food coloring, then fill the container about 2/3 full of water. Add an ice cube to it. After 2–3 min, remove the ice cube.

E. While the water is cooling, fill another container of the same size and shape about 2/3 full of hot water.

F. Place the two containers side by side.

3. In which container (hot or cold) are the water particles moving faster?

G. Wait a minute for the water currents to stop. Now add one drop of food coloring to each container and observe the reactions in each container.

4. In which container (hot or cold) did the mixing appear to take place faster?
5. In your own words, explain your observations in terms of the motion of water particles.

SUMMARY

What happens when you rub your hands together to warm them? Experiments show that heat is a form of energy. When a substance is heated, energy is added to its particles. This causes the particles of matter to move faster. The heat energy is the energy of these moving particles. When you rub your hands, you cause the particles in the outer parts of your skin to move faster. These particles bump other particles until the entire thickness of the skin in your hands is heated. The motion of your hands is changed into heat energy in the skin. Heat engines are machines that change some of the energy of moving particles into motion. Common examples of heat engines are steam engines and automobile engines. Unfortunately, when we try to change heat energy into mechanical energy with heat engines, some of the heat energy is wasted.

QUESTIONS

Use complete sentences to write your answers.

1. What did Count Rumford's cannon-boring experiment show?
2. How is heat stored in matter?
3. Explain what happens in matter as:
   a. heat energy is added to it.
   b. heat energy is removed from it.
4. Describe how you could demonstrate that heat energy added to water causes the water particles to move faster.
5. Briefly describe the difference between two types of steam engines.

7-2. HEAT TRANSFER

A fire in a fireplace is a cheerful source of heat energy. An iron poker put into the fire takes on some of the fire's heat. People sitting near the fire feel the heat. A bird perched on top of the chimney receives some of the heat. Heat can be transferred from the fire to the poker, the people, and the bird. How does heat move from one place to another?

When you finish lesson 2, you will be able to:

- Use examples to show how heat energy can move from one material to another.
- Explain *conduction* and *convection* in terms of the movement of particles in matter.
- Explain how heat is transferred through space by *radiation*.
- ○ Demonstrate the transfer of heat energy by conduction, convection, and radiation.

Conduction
Transfer of heat by direct contact.

A metal poker put directly into a fire is heated by **conduction** (kun-**duk**-shun). Transfer of heat by direct contact is called *conduction*. It is the simplest method of heat transfer. See Fig. 7-7. The metal in the poker, like all other substances, is made up of particles. The rapidly moving particles of the burning wood in the fireplace bump the particles in the poker and make them vibrate faster. The particles in the end of the poker, in turn, hit particles in the cooler part. This motion continues on up the poker until the particles in the handle are also heated. If you touch the handle of a hot poker, you will burn your fingers. The heat is also transferred to your skin by conduction.

7-7. *The heat from the fire causes the particles in the poker to move rapidly. The parts of the poker that are not in the fire are heated by conduction.*

7-8. *Insulation can prevent loss of heat in a home.*

The metal in a poker conducts heat very well. Not all materials conduct heat as well as metals. Wood, for example, is a poor *conductor* of heat. That is why the handles of pokers, pots and pans, and other metal objects that are heated are often made of wood. Nonmetallic solids, liquids, and gases are all poor conductors of heat. They are called *insulators*. Insulators are often used to prevent the loss of heat. Houses and other buildings must be insulated against heat loss. See Fig. 7-8. Without insulation, it becomes very expensive to provide the heat needed to keep a house warm during cold weather. Clothes effectively shield our bodies against loss of heat. Air trapped under clothing acts as an insulator. Many layers of cloth containing air spaces provide better insulation than one heavy layer.

How was the bird on the chimney heated by the fire? In the fireplace, air directly over the flame was heated. This hot air expanded because its particles started moving faster and took up more space. Expansion of the heated air made it lighter than the surrounding air. The warm air moved up the chimney and warmed the bird. See Figs. 7-9 and 7-10. Transfer of heat by the movement of a heated gas or liquid is called **convection** (kun-**vek**-shun). When a gas such as air or a liquid like water is heated unevenly, the heated part rises. This movement of a heated gas or liquid is *convection*. Heat transfer by convection can take place only in gases or liquids.

7-9. *The air above the candles is heated by convection. The warm air then moves against the paddles, causing the whole device to turn.*

Convection
Transfer of heat by movement of a heated gas or liquid.

A fireplace is a poor source of heat for a room because convection carries most of the heat up the chimney. On the other hand, other kinds of room heaters work somewhat effectively by convection. For

7-10. *A hot air balloon makes use of the fact that hot air rises.*

example, steam or hot water is moved through a device that heats some of the air in the room. This warm air rises and moves across to the cooler parts of the room. Cool air sinks and moves toward the heater where it is heated. See Fig. 7-11.

Radiation
Transfer of heat through space by infrared rays.

Convection transfers heat when a part of the heated material (a gas or a liquid) moves. A third method of heat transfer is very different. It is called **radiation** (rade-ee-**ay**-shun). *Radiation* is the transfer of heat through space by infrared rays. Radiation does not cause particles of matter to move. *Infrared radiation* is an electromagnetic wave.

Sources of heat, such as a fire in a fireplace, send out these invisible infrared waves or rays. When these rays reach your skin or any other material, the rays are

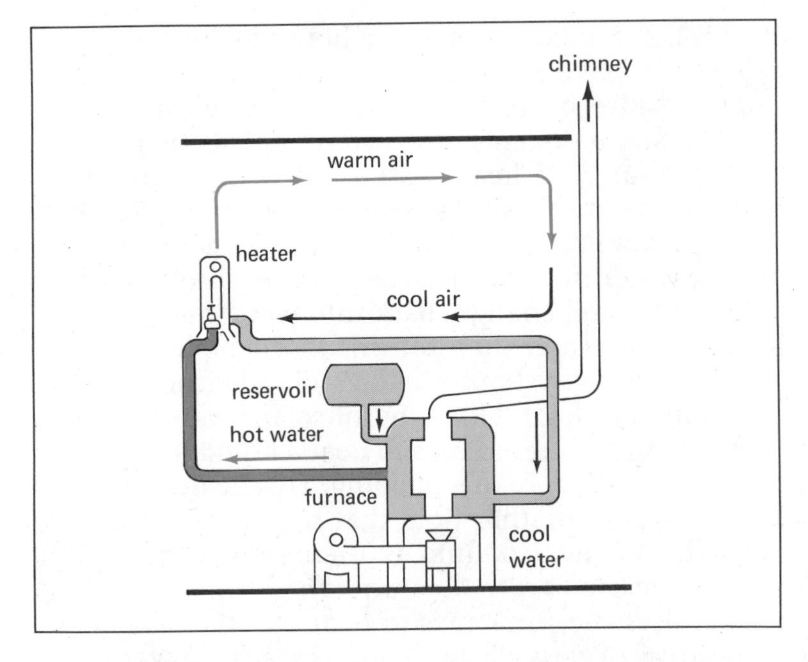

7-11. The system shown in the diagram heats a home by conduction, convection, and to some extent, radiation.

changed back into heat. Infrared rays are responsible for some of the warmth felt by people around a fireplace. Radiation is a very important method of heat transfer. The earth receives heat energy from the sun by radiation. See Fig. 7-12. You feel warm while sitting in the sun because you are receiving heat from the sun by radiation. The sun is 150 million kilometers from the earth. This distance is mostly empty space. Heat could not possibly be transferred by conduction or convection across empty space from the sun to the earth.

Using open fires in fireplaces is not an efficient way to heat a home. Much of the heat is lost because convection causes hot air to go up the chimney. Standing beside the fire only warms the side of you that receives radiation. The other side will still be cool. A great improvement in home heating was made in the 1740's when Benjamin Franklin invented an iron stove that fitted inside a fireplace. The Franklin stove heated room air that came in contact with it. This allowed heat to spread throughout the room by convection. In time, the iron stove itself became the container for the fire and had its own stovepipe chimney.

Many modern houses are heated by burning a fuel such as oil or natural gas. Oil or gas burners are usually located in the basement or a special room of the house.

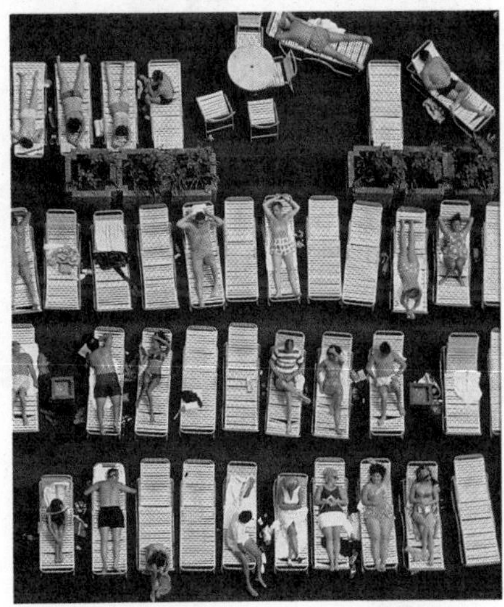

7-12. The earth receives heat energy from the sun by means of infrared radiation.

Heat is distributed through the house by heated water or air. You have already seen how convection can be used to cause hot water to move through a heating system. Some systems also use pumps to help move the hot water. If heated air is used, its convection movements are usually speeded up by a fan. See Fig. 7-13. Steam can also be used. Pressure generated by boiling water in a furnace sends steam into radiators. There the steam changes back into water and gives up its heat. The water then returns to the boiler.

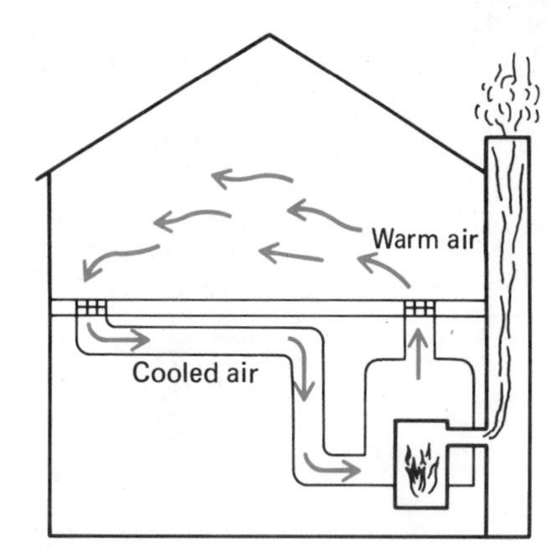

7-13. *The system shown in this diagram uses hot air to heat a home.*

Solar heating systems use the radiation from the sun as a source of heat. One kind of solar heating system uses the sun's infrared rays to heat a liquid in a special collector on the roof of a building. This heated liquid is used to warm another fluid that stores the heat until needed. The heated fluid is then sent through the house. See Fig. 7-14. A second type of solar heating system uses the sun's rays to heat the house directly. Large areas of glass allow the infrared rays to warm the interior of the building. See Fig. 7-15. At night the

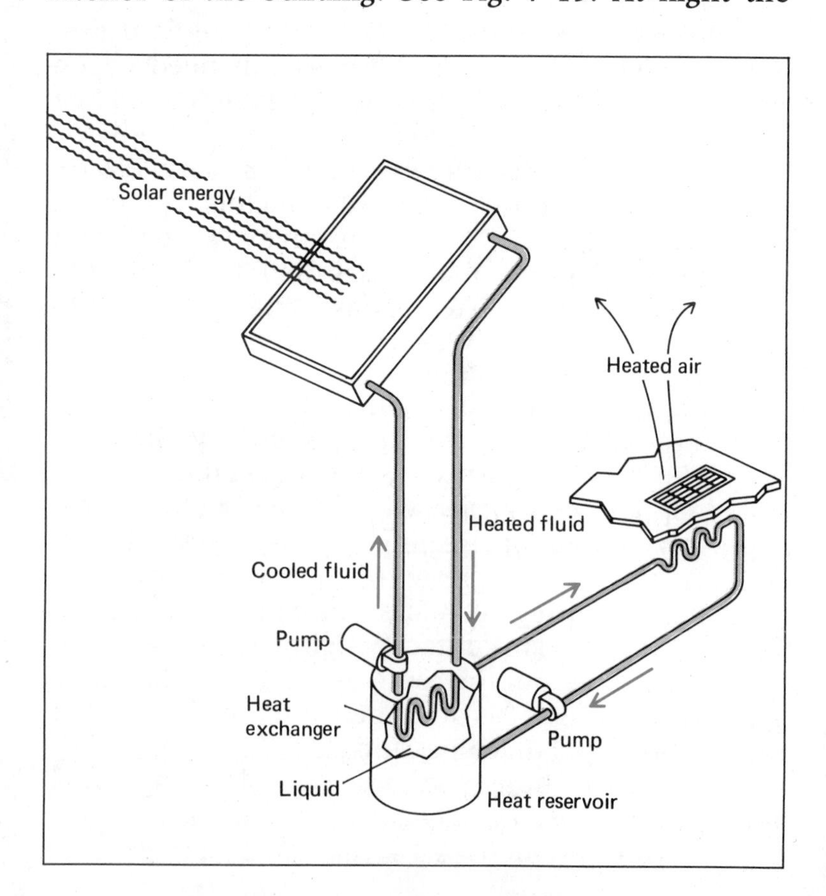

7-14. *The solar heating system shown in this diagram uses the sun's energy to heat a liquid, which then flows through pipes to heat the house.*

7-15. Many new homes are being built with solar heating systems. Older homes can be modified to take advantage of solar energy.

warmed parts of the house give off heat by radiation. Solar heating systems of all kinds usually need a backup system. This backup system supplies additional heat on cloudy days or in very cold weather.

Homes can be heated entirely by electricity. Electric current is sent through special conductors. Compared to copper wire, these conductors have a large amount of electrical resistance that causes them to become hot. Usually these electric heaters are arranged around a room near the floor. Warmed air then rises to flow through the room.

No matter what method is used to heat a home, much energy is wasted if the house is poorly insulated. One method of insulating a building is to fill the spaces around the walls and ceiling with a poor conductor. See Fig. 7-16. Materials used as heat insulators trap air, which is a poor conductor of heat. Sealing the openings around doors and windows also helps to insulate the home by preventing loss of heated air.

7-16. An efficient home heating system includes some type of insulation to reduce heat loss.

You can see if a building is well insulated by using a special camera that records infrared rays. See Fig. 7-17. The colors in the photograph show places where heat is being lost. This type of photograph is called a *thermogram* (from the Greek words *therme* meaning "heat" and *gramma* meaning a "drawing" or "record").

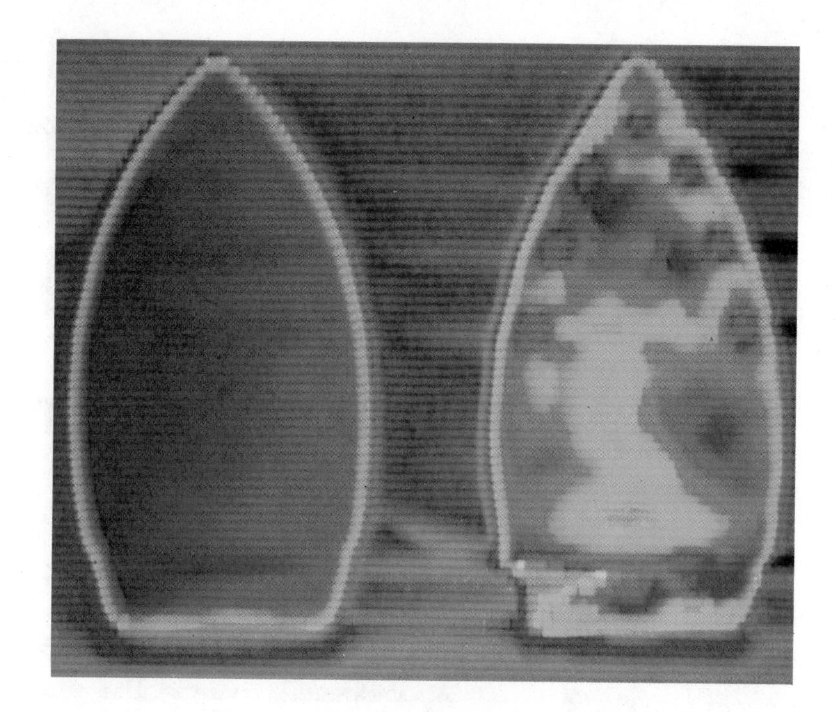

7-17. *A thermogram shows heat energy as a visible picture. The various temperatures appear as different colors or areas of brightness. The hot areas appear as light spots. Thermograms can be used to detect heat loss from a building.*

ACTIVITY

Materials
candle
matches
nail (16d, finishing)
index card
metric ruler

Three Ways to Transfer Heat

Caution: In this activity, be certain you blow out matches before disposing of them. Do not put them in the wastepaper basket. Do not leave hot objects where someone else may pick them up.

A. Obtain the materials listed in the margin.

7-18.

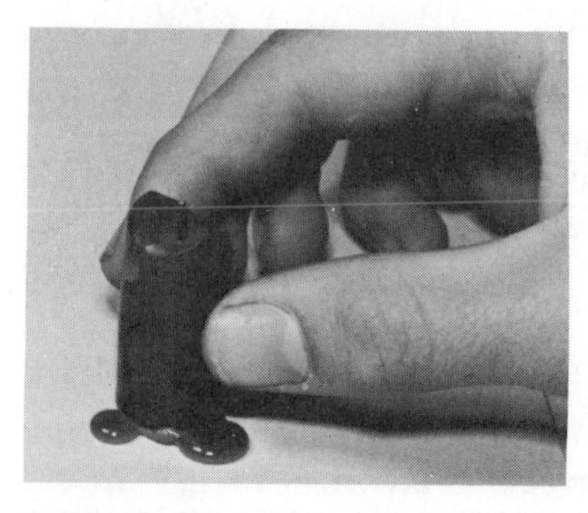

7-19.

B. Light the candle and drip some wax onto the card. Blow out the candle. While the wax is still soft, stand the candle upright in it. See Figs. 7-18 and 7-19.

C. Relight the candle and hold your fingers about 30 cm above the flame. See Fig. 7-20. Move your fingers closer until you can feel the heat from the flame. Be careful not to burn youreIf.

1. How close to the flame must you be in order to feel heat from the flame?

D. Hold your fingers 30 cm to the side of the candle. Slowly move your fingers closer to the candle until you can feel the heat from the flame.

2. How close to the side of the flame must you be in order to feel heat from the flame?

3. Describe the method by which you received heat to the side of the flame.

4. Describe the method by which you received heat above the flame.

5. In either step C or D, would your finger receive heat by *both* radiation and convection? Explain.

E. Hold the nail near its head. Place the point of the nail in the flame for about 15 sec. See Fig. 7-21. Remove the nail from the flame and move it more than 30 cm from the candle flame.

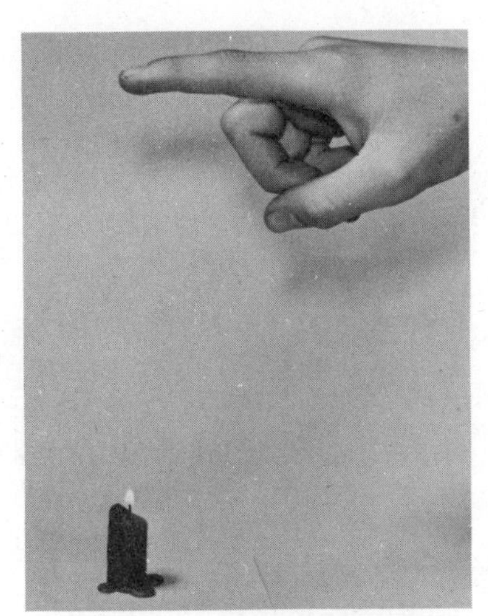

7-20. Caution: *Do not bring your finger too near the flame. You should just be able to feel the heat from the flame.*

6. How long a time is it before you feel any heat at the head end of the nail?

7. Explain why it takes time for the heat to reach the head of the nail. Use complete sentences.

8. Describe the method by which you received heat while holding the nail.

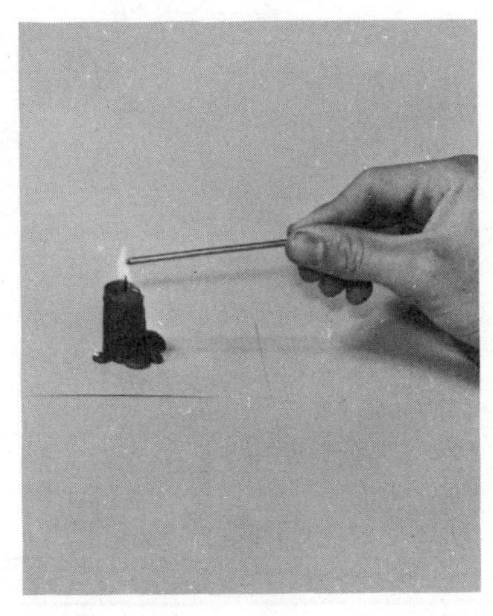

7-21.

SUMMARY

Heat can be transferred in three ways. (1) Direct contact between a source of heat and a cooler substance will speed up the motion of the particles in the cooler material. (2) Gases or liquids become heated unevenly. The heated part moves, carrying with it a supply of heat. (3) Every source of heat gives off invisible rays that heat any material they fall upon. Most common sources of heat, such as the fire in a fireplace, transfer heat in all three ways.

QUESTIONS

Use complete sentences to write your answers.

1. Of the following examples of the transfer of heat energy, which are examples of conduction? Which are examples of convection? Which are examples of radiation?

a. You feel the steam rising from a cup of hot chocolate.

b. You feel the heat from the hot cup handle just before you touch it.

c. You burn your tongue drinking hot chocolate.

d. A spoon in the hot chocolate becomes warm.

e. You are warmed as you stand in front of a fireplace.

f. You are warmed as you stand over a furnace register.

2. Conduction and convection both transfer heat energy through the movement of particles. How do they differ?

3. Given a candle, a match, and a nail, describe how you could demonstrate transfer of heat energy by **a.** conduction **b.** convection **c.** radiation.

4. What type of heating (gas, electric, wood-burning stove, etc.) do you think would be best for your school? Why?

5. What changes in your home would make it easier to keep warm in winter and cool in summer? Be specific. What parts should be insulated, sealed, etc.? If changes were recently made, what were they?

7-3. TEMPERATURE AND HEAT

Which would you rather have accidentally spilled on you—a bucketful of boiling water or a cupful of boiling water? You know by experience that boiling water contains energy in the form of heat. If the hot water is spilled on your skin, the heat energy may cause serious burns. Why would a bucketful of boiling water be more dangerous than a cupful? Both are at the same temperature. There must be more heat energy, however, in the larger amount of water. Temperature alone does not indicate the amount of heat in water. You can continue studying heat energy by finding out how to measure amounts of this form of energy.

When you finish lesson 3, you will be able to:

- Explain the difference between the *Fahrenheit* and *Celsius* temperature scales.
- Distinguish between *temperature* and heat.
- Define a *calorie*.
- ○ Show that the transfer of the same amount of heat does not always produce the same temperature change.

Dip your finger into a glass of warm water and then touch an ice cube. You would probably say that the **temperature** of the water is higher than the *temperature* of the ice cube. Warm water has more heat energy than the same weight of ice. You now know that heat energy is the result of the movement of particles of matter. The water particles in the glass are moving faster than the particles in the ice cube. When you measure temperature, you measure the amount of movement of the particles.

Temperature
A measurement of the movement of particles in matter.

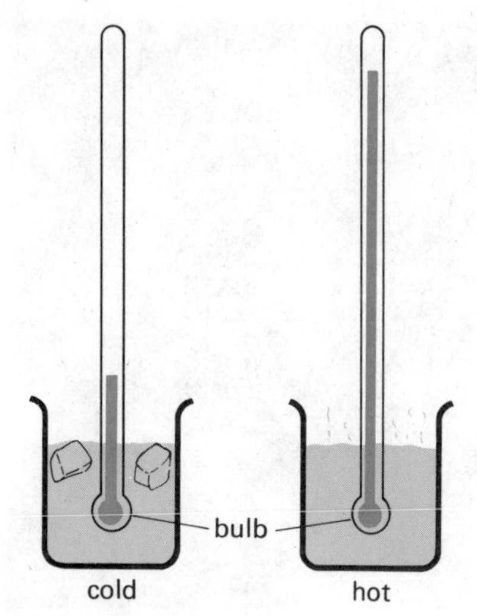

7-22. *The liquid in a thermometer is affected by temperature changes. The length of the column of liquid is a measure of the temperature of the bulb.*

Do you think other people would always agree with your observations of temperature based only on your sense of touch? More accurate measurements of temperature are made with an instrument. Observations made with an instrument are the same for everyone. A *thermometer* is an instrument used for measuring temperature. Most materials expand when heated and shrink when cooled. The most commonly used thermometer is based on this principle. A liquid is sealed in a glass tube. Alcohol is often used and so is the liquid metal, mercury. When a thermometer is heated, the liquid expands and rises in the tube. Cooling causes the liquid to shrink and fall. See Fig. 7-22.

The thermometer tube must have a scale. One kind of temperature scale was invented by a scientist named Fahrenheit in the 1700's. On the Fahrenheit temperature scale, water freezes at 32° and boils at 212°. The Fahrenheit (F) temperature scale was widely used in the past. The temperature scale now used most often all over the world is the **Celsius** (**sel**-see-us) scale. In

Celsius (C)
The name of a commonly used temperature scale. The Celsius scale is generally used in science.

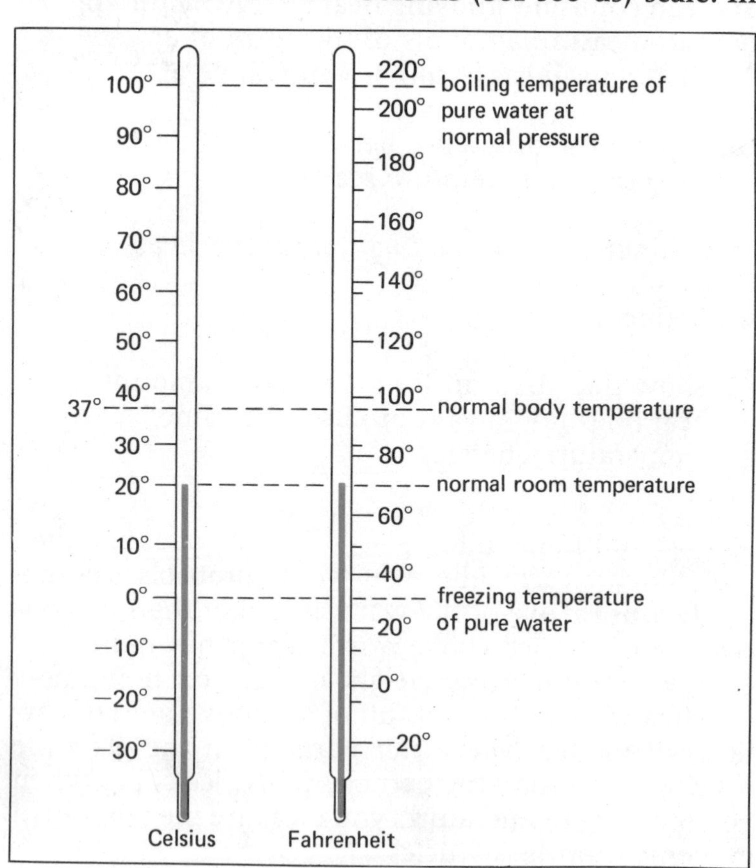

7-23. *A comparison of the Celsius and Fahrenheit temperature scales.*

scientific work, the *Celsius temperature scale* is used most often. The Celsius (C) temperature scale is sometimes also called *centigrade*. On the Celsius scale, water freezes at 0° and boils at 100°. A comparison of the Celsius and Fahrenheit scales is shown in Fig. 7-23.

Not all thermometers are made of a liquid in a glass tube. In some kinds of thermometers, a metal strip bends with changes in temperature. See Fig. 7-24. The amount of bending then moves a needle on a dial to indicate temperature. Another kind of thermometer measures temperature by its effect on the flow of an electric current. See Fig. 7-25. These thermometers are often used to measure body temperature since there is no time spent waiting to read the temperature.

Suppose you had a bucketful of water at a temperature of 25°C. This is about the temperature of water coming from a faucet on a warm day. It would take a large amount of heat to raise the temperature of this water to 100°C. How much heat would be required to change the temperature of a cupful of water from 25°C to 100°C? The cupful of water would need much less heat than a bucketful. The amount of heat in the water

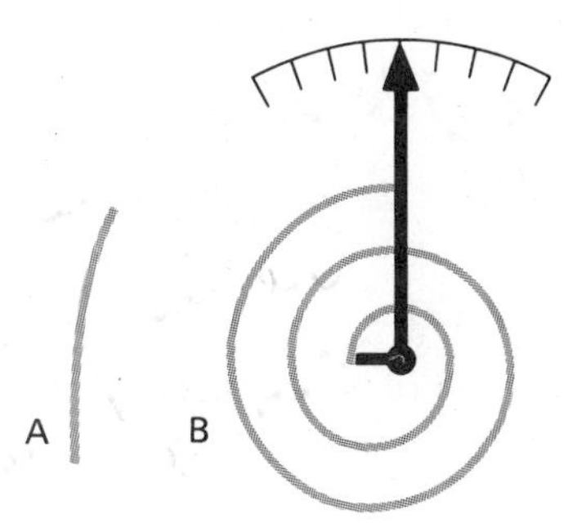

7-24. *A strip made of two different metals will bend with a change in temperature since the two metals respond differently to changes in temperature. The bending of the bimetallic strip can be used to move a needle on a temperature dial.*

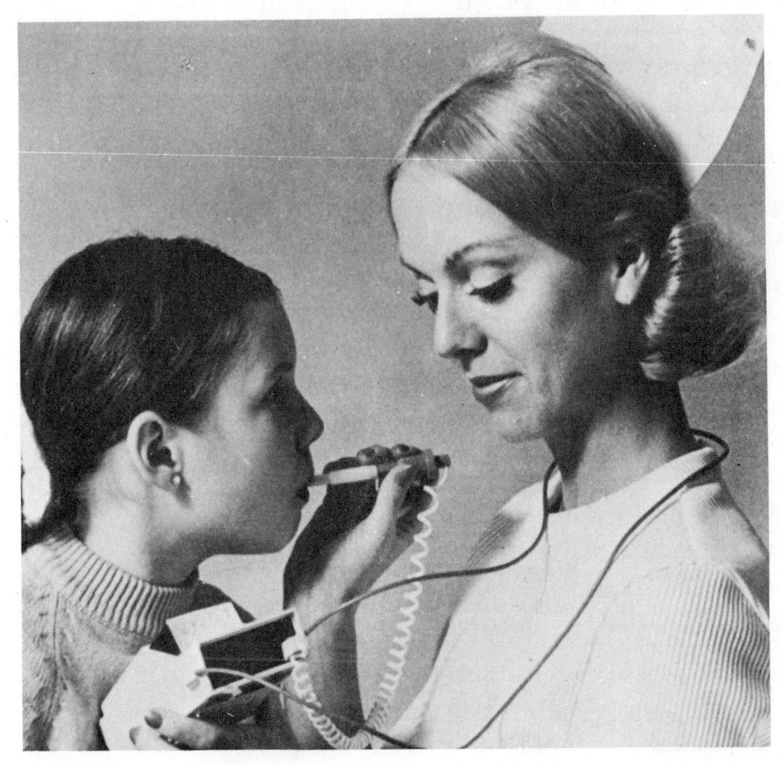

7-25. *Electronic thermometers are often used in hospitals.*

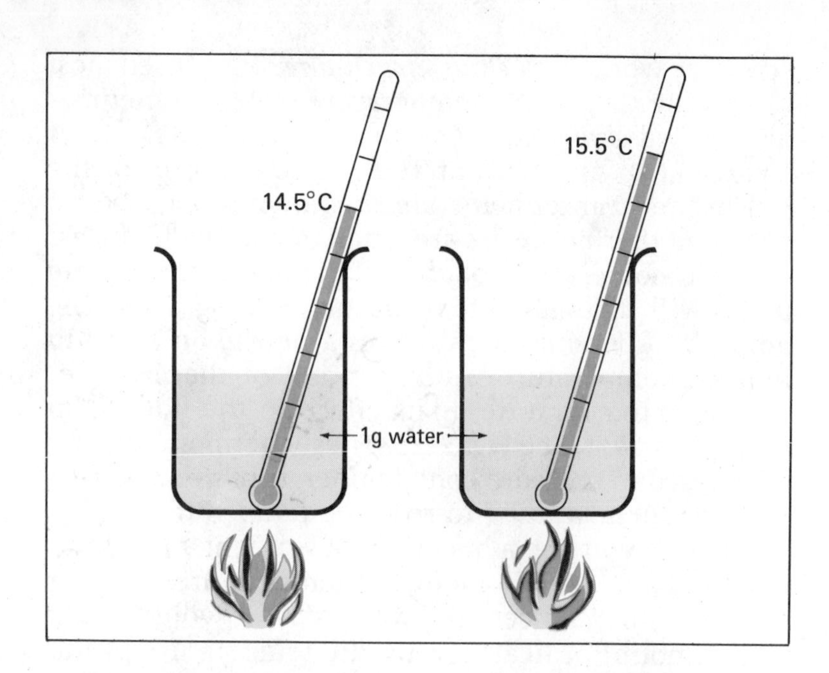

7-26. A calorie is the amount of heat needed to raise the temperature of 1 g of water from 14.5°C to 15.5°C.

depends not only on its temperature but also on the amount of water. For example, a teapot full of boiling water and a cupful of boiling water both have the same temperature, 100°C. The movement of the water particles in both the pot and the cup is the same. However, the pot contains more water than the cup. Therefore, the amount of heat is greater in the pot than in the cup. In order to measure heat, you must include the amount of material heated as well as its temperature change.

calorie
An amount of heat equal to that needed to raise the temperature of 1 gram of water 1°C.

A unit called a **calorie** (**kal**-uh-ree) measures heat in terms of the amount of material and the temperature change. The amount of heat required to raise the temperature of 1 gram (g) of water by 1°C is 1 *calorie* (cal). See Fig. 7-26. For example, suppose that you put 500 g (about 2 cups) of water into an electric coffee maker. The water has a temperature of 20° C. When the coffee maker is started, the water is heated to 100°C. The temperature increase is the difference between 100°C and 20°C, which is 80°C. One calorie of heat is needed to increase the temperature of each gram of water by 1°C. So the amount of heat produced by the coffee maker is:

mass (grams) × temperature change (°C) = heat (cal)

$$M \times T = H$$

$$500 \text{ g} \times 80°\text{C} = 40{,}000 \text{ cal}$$

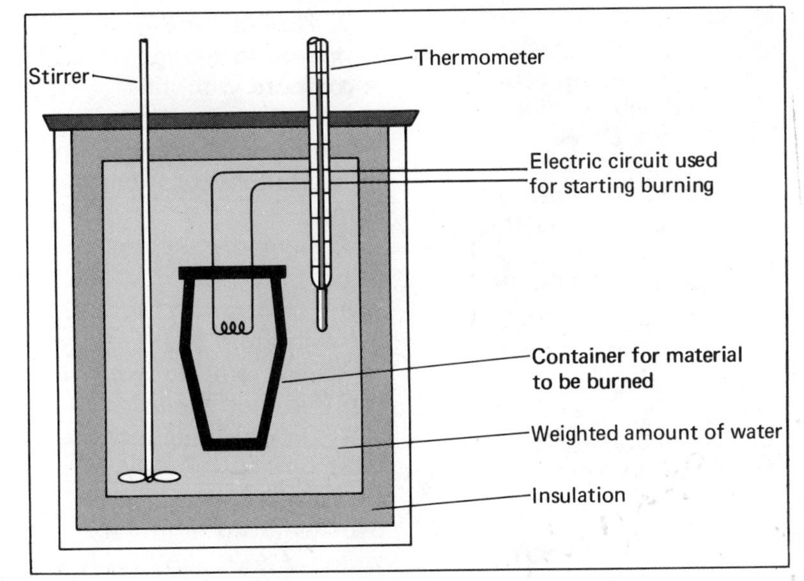

7-27. *A calorimeter is used to measure the amount of heat released when a material is burned.*

You can see that a thermometer tells only a part of the information needed to measure heat. The amount of material heated must also be known. Heat is usually measured by using a special device called a *calorimeter*. A calorimeter holds a known weight of water in an insulated container. The amount of heat used to warm the water is measured in the calorimeter. The resulting temperature change is measured with a thermometer. For example, the heat energy in food can be measured with the kind of calorimeter shown in Fig. 7-27. You are probably familiar with the use of calories to express the energy content of food. The "calorie" commonly used to measure food energy is actually a *kilocalorie*. A kilocalorie is a thousand times larger than a calorie. Sometimes a kilocalorie is called a "Calorie" with a capital C to distinguish it from the ordinary calorie.

ACTIVITY

The Difference between Temperature and Heat

How do the two terms *temperature* and *heat* differ in meaning? In this activity, you can find out.

A. Obtain the materials listed in the margin.

B. Fill one cup about 1/4 full of tap water. Call this cup A.

C. Fill the other cup about 3/4 full of tap water. Call this cup B.

Materials
2 styrofoam cups
2 ice cubes
thermometer (°C)
stirring rod
water

	Cup A	Cup B
Temperature, initial		
Temperature, final		
Change in temperature		

D. Measure the temperature of the water in each cup. Record your observations in a data table like the one shown.

E. Place two pieces of ice of equal size into each cup.

1. If the same amount of heat is needed to melt each piece of ice, in which cup will the temperature change be greater?

F. Use a stirring rod to stir the water in cup A until the ice melts.

G. Read the temperature of the water. Record.

H. Use the stirring rod to stir the water in cup B until the ice melts.

I. Read the temperature of the water. Record.

2. How did the temperature change in cup A compare with the temperature change in cup B? (Remember, the amount of ice in both cups was the same.)
3. Compare the amount of heat removed from the water in cup A to melt the ice with the amount of heat removed from the water in cup B to melt the ice.
4. Do you think that the thermometer correctly measured the amount of heat removed from the water in each cup? Explain your answer. How could you test whether the thermometer is correct?
5. How is the amount of water related to the amount of heat needed to change its temperature?

SUMMARY

A bucketful of boiling water can be dangerous because of the heat energy it contains as well as its high temperature. Thermometers measure temperature in degrees Celsius or degrees Fahrenheit. They are not used to measure heat energy. The amount of heat energy in a material depends upon both the temperature and the amount of the material. Both the amount of material and the change in temperature are used to measure heat energy in calories.

Unless otherwise indicated, use complete sentences to write your answers.

1. Use the clues given to complete the following word puzzle. DO NOT WRITE IN THIS BOOK.

a. F _ _ _ _ _ _ [_] _ _ _

b. _ _ [_] _ S _ _ S

c. _ [_] L _ _ _ _

d. _ E _ _ E _ _ [_] _ _ E

Clues

a. A scale on which ice measures 32°.
b. A scale on which ice measures 0°.
c. Heat required to raise 1 g of water by 1°C.
d. Measurement on a thermometer.

2. Write a definition for the key word found within the box of the word puzzle in question 1.

3. How many calories are needed to raise the temperature of 200 g of water from 20°C to 30°C?

4. To what temperature would 3,000 cal raise 500 g of water at 20°C?

5. A coffee maker produces 9,000 cal of heat energy each minute. How many minutes will it take to raise the temperature of 1,000 g of water from 20°C to 100°C.

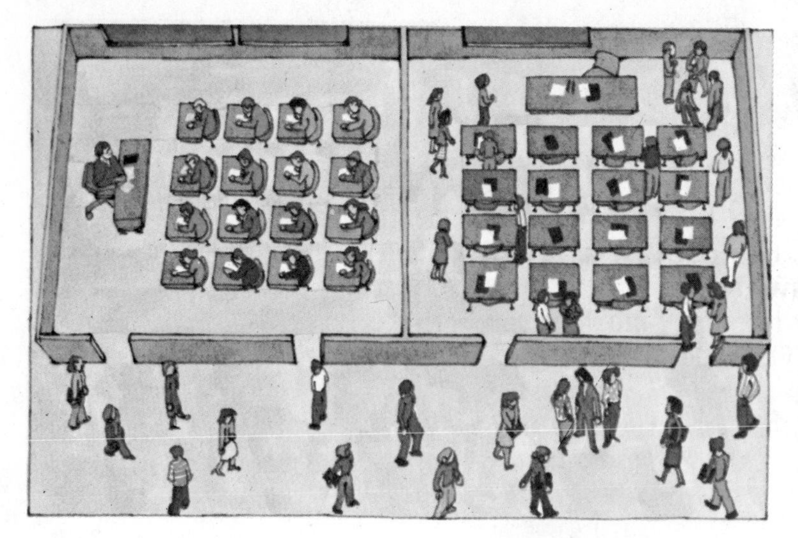

7-4. BEHAVIOR OF GASES

Matter is either a solid, a liquid, or a gas. All matter is made of particles. To understand how these individual particles behave, a comparison might be helpful. Think of yourself and everyone in your class as individual particles. At times the entire class is seated in some kind of orderly arrangement. This situation is somewhat like the behavior of particles in a solid. When the members of the class are out of their seats and moving around the room, they resemble the motion of particles in a liquid. Finally, the class ends and everyone leaves the room, moving apart to fill the space outside in the hall. The movement is similar to the way particles of a gas spread out to fill space. The properties of matter depend upon the behavior of its particles.

When you finish lesson 4, you will be able to:

- Describe how particles move in gases, liquids, and solids
- Use the *kinetic theory* to explain how gases behave.
- Explain why nothing can be colder than *absolute zero*.
- ○ Show how temperature affects the behavior of a gas by making an air thermometer.

Kinetic theory of matter The scientific principle that says that all matter is made of particles whose motion determines whether the matter is solid, liquid, or gas.

The scientific belief that all matter is made of moving particles is called the **kinetic theory of matter.** The *kinetic theory* is one of the most important theories of modern science. By using the kinetic theory, scientists have been able to explain and predict the properties of matter. Each of the three forms of matter is called a *phase*. See Fig. 7-28. Matter in the solid phase is made up of particles that are usually in orderly arrangements. The particles in a solid are usually vibrating back-and-forth but holding their positions beside their neighbors. Thus, under ordinary conditions, a solid like a pencil does not change its shape or the volume it occupies. In the liquid phase, the particles are able to move around each other but remain close together. Thus, liquids may change their shape but still take up a certain volume. Liquid water, for example, can be poured from one glass into another but will occupy the same volume in both glasses. In the gas phase, particles of matter move very fast and spread apart from each other. Gases have neither a definite shape nor volume.

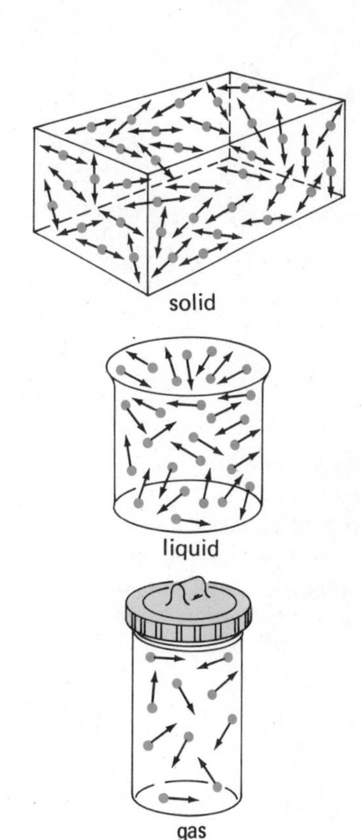

7-28. *The movement of particles in a solid, a liquid, and a gas.*

The kinetic theory can be used to explain the properties of the phases of matter. However, it is particularly useful to explain and predict the behavior of gases. The kinetic theory leads to the following observations about gases.

1. Particles in a gas are moving very fast with large average distances between them.

A gas is mostly empty space. Gas particles move like a swarm of angry bees trapped in a room. Each particle collides with others many times each second. Particles

7-29. *This bicycle pump forces air under pressure into a bicycle tire.*

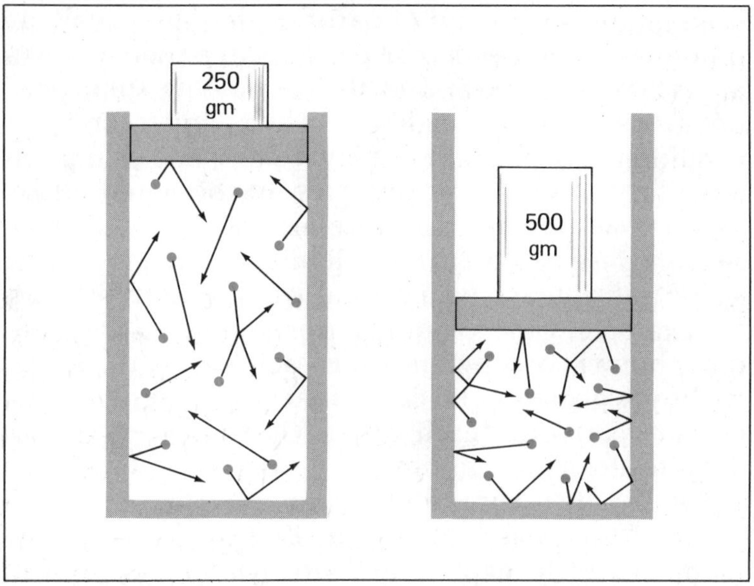

7-30. *Reducing the volume of a gas by one-half will cause the pressure to double. The pressure increases as a result of the crowding together of the gas particles, causing more collisions with the walls of the container.*

are not affected by these collisions. Gas pressure is the result of the rapidly moving particles colliding against their container. Gases have no natural shape but will expand to fill any space available.

2. Gases can be compressed.

Gas particles can be crowded together. When a gas is squeezed into a smaller space, the pressure of the gas goes up. See Fig. 7-29. This rise in pressure is a result of the rapidly moving gas particles hitting the walls of the container more often. See Fig. 7-30. In the same way, a gas will have less pressure if it is allowed to expand to fill a larger space.

3. The temperature of a gas determines how fast its particles move.

If heat is added to a gas, the heat energy causes the gas particles to move faster. See Fig. 7-31. Faster moving particles hit the container holding the gas more often. The more frequent collisions cause higher pressure. If the volume is not changed when the temperature of a gas increases, its pressure also increases.

A gas such as air will expand when its temperature is raised. In the same way, cooling a gas will cause it to shrink as its particles move more slowly. Galileo, one of the greatest scientists of the sixteenth century, invented the *air thermometer*. The air thermometer is simply a quantity of air trapped by water in a tube. If you do the activity at the end of this lesson, you will use modern materials to make something like an air thermometer. You will then be able to study the behavior of the trapped air when it is cooled or warmed.

Can you predict what will happen if a gas is made very cold? If a gas is cooled, its particles move slower and slower. At what temperature would the particles stop moving? Experiments have shown that this temperature is −273°C. At a temperature of 273° below zero Celsius, particles in matter stop moving. The particles do not have the heat energy needed for motion. The temperature at which all motion of particles stops (−273°C) is called **absolute zero.** Since temperature actually measures the amount of particle motion in a substance, there can be no lower temperature than *absolute zero* because particles are not moving at this temperature. Absolute zero is the lowest possible temperature. A temperature scale based on absolute zero is often used in scientific observations. This scale is called the **Kelvin (K) temperature scale.** On the *Kelvin scale,* 0° is equal to absolute zero. A comparison of the Kelvin and Celsius scales is shown in Fig. 7-32.

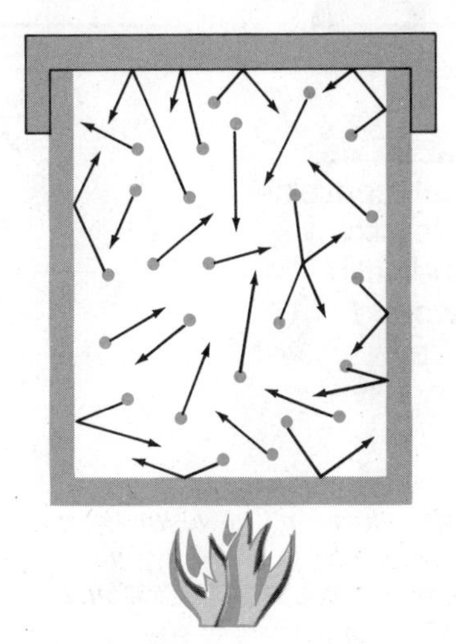

7-31. *Heating a gas in a closed container causes the particles to move faster and collide more often with the walls of the container.*

Absolute zero
The temperature (−273° C) at which particles of matter stop moving.

Kelvin temperature scale
A scale of temperature on which 0° is equal to absolute zero.

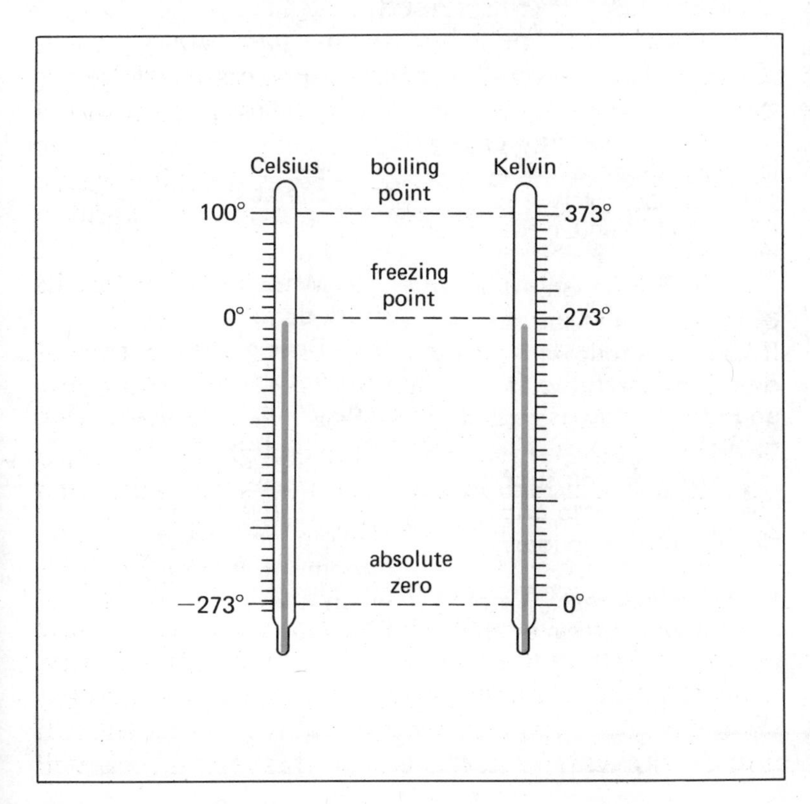

7-32. *A comparison of the Celsius and Kelvin temperature scales.*

ACTIVITY

Materials
small container
soda straw
paper clip
ice cube
water

Making an Air Thermometer

A. Obtain the materials listed in the margin.

B. Fill a small container about half full with water.

C. Place one end of a soda straw in the water, resting it on the bottom. (You may note that the water rises up inside the straw to the level of the water in the container.)

7-33.

D. Bend over the top end of the straw and fasten it with a paper clip. See Fig. 7-33. Squeeze the fold. Closing the top traps some water at the bottom of the straw and some air at the top of the straw. You now have an air thermometer.

E. Raise the straw straight up out of the water, keeping it over the container. Do not squeeze the straw. Hold it near the top under the paper clip.

1. Does the water drip out of the bottom of the straw? If you have followed the directions carefully, the top end of the straw is sealed air tight.

2. Why do you suppose the water tends to drip slowly out the bottom?

F. Make this test. Watch a drop of water form at the bottom of the straw. See Fig. 7-34. Just before the drop falls, touch an ice cube to the side of the straw near the top where you are holding it.

3. What happened? The results you are looking for depend on the top of the air thermometer being air tight, so check for leaks.

G. Remove the ice cube. Hold the straw by cupping your hand around it. Take care not to squeeze it.

4. What will warming the straw do to the trapped air?

5. Does a drop of water start to form again?

6. What result does warming have on the trapped air?

7. On the basis of your observations, explain what happens to the trapped air when it is first cooled and then warmed.

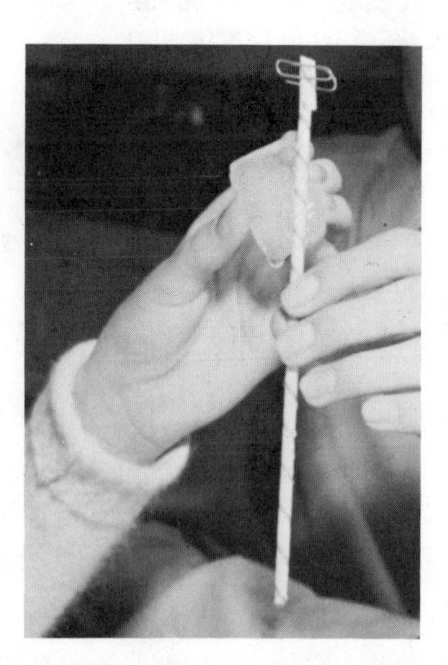

7-34.

SUMMARY

You cannot see separate particles in matter. The kinetic theory describes how particles move in solids, liquids, and gases. The properties of gases can be explained by the rapid motion of their separate particles. Temperature is a measure of how fast particles move. At the lowest possible temperature, there is no motion of particles in matter.

QUESTIONS

Unless otherwise indicated, use complete sentences to write your answers.

1. In which form of matter (solid, liquid, gas) are the individual particles likely to be moving the fastest?
2. Describe a gas in terms of the kinetic theory of matter.
3. Explain why you would not be able to cool anything below absolute zero.
4. Describe how you would make an air thermometer.
5. Normal room temperature is 20°C. Calculate this temperature in degrees Kelvin.
6. Refer to Fig. 7-30. How much mass would be needed to reduce the gas volume to one-fourth the original volume?

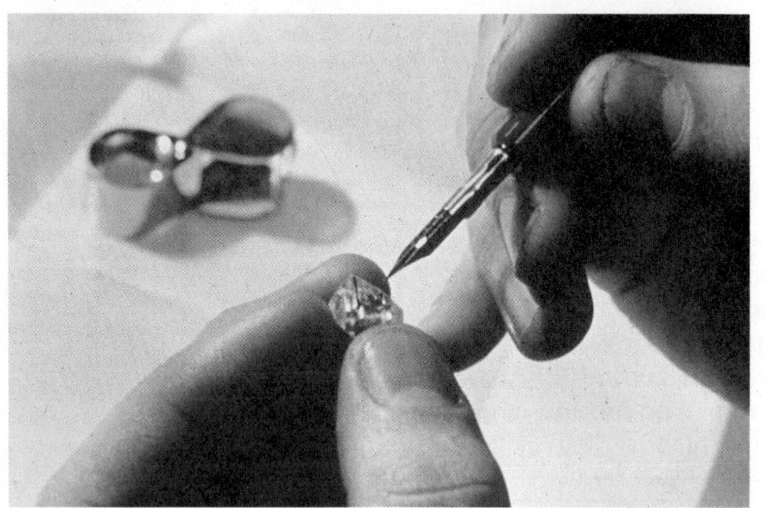

7-5. LIQUIDS AND SOLIDS

By watching a diamond cutter at work, you could learn something about the structure of solids. You might be surprised to learn that the diamond "cutter" actually *breaks* a large diamond into smaller pieces. Diamonds are too hard to be cut. A large diamond to be "cut" is carefully examined. The diamond is then marked with lines. The diamond cutter gently taps on a line. If everything has been done properly, the diamond splits evenly. A mistake may cause a valuable large diamond to shatter into many pieces too small and too irregular to be used for jewelry. How does a diamond cutter know exactly where to tap the diamond? The trained eye of a diamond cutter can see in a diamond a characteristic that is found in most solids: They have natural lines along which they will split. This property of solids is one result of the way particles are arranged.

7-35. *An example of an ice crystal: a snowflake.*

When you finish lesson 5, you will be able to:

- Describe how particles are arranged in most solid materials.
- Explain what happens when a solid melts.
- Explain what happens when a liquid boils.
- ○ Examine crystals of several solids and describe their appearance.

A piece of solid matter cannot change its shape by itself. A diamond, for example, holds its shape unless it is split by a blow. The particles of a solid are not free to move about. They stay in position. The arrangement of the particles forming a solid is usually very orderly. For example, when particles of water freeze to form solid ice, they form a **crystal** of ice. See Fig. 7-35. A *crystal* is a piece of solid matter with a

Crystal
A solid whose orderly arrangement of particles gives it a regular shape.

regular shape. A large piece of ice is made up of many small ice crystals fitted together like the pieces of a jigsaw puzzle. Almost all solids are formed in this way.

When a diamond is split, it is separated along the surfaces that join its separate crystals. In many solids, such as rock candy, the individual crystals are large enough to be seen. See Fig. 7-36. However, in most solids, the crystals are usually too small to be seen with the eye alone. The shape of an individual crystal is determined by the way its particles are arranged. For example, in common table salt each crystal is in the form of a cube. See Fig. 7-37. If you could see the particles in a crystal, each particle would be found in a definite position.

7-36. *Rock candy has large crystals that are clearly visible to the unaided eye.*

In some solids, the particles are not found in an orderly arrangement. These solids are like liquids in that their particles are not arranged in crystal patterns. Glass and some plastics are examples of this kind of non-crystalline solid.

The particles in a solid are always vibrating. If heat is added to a solid, the particles in that solid vibrate faster and faster as the temperature goes up. At some definite temperature, the motion of the particles becomes so great that the particles can no longer hold their orderly arrangement. When this happens, the solid melts and becomes a liquid. See Fig. 7-38. The temperature at which a solid changes into a liquid is called its **melting point.** Each kind of solid made of crystals has a particular temperature at which it melts. *Melting points* of some common substances are given in Table 7-1.

Melting point
The temperature at which a solid becomes a liquid.

7-37. *(left) This is how crystals of ordinary table salt look under a microscope.*

7-38. *(right) When a solid is heated, the particles of the solid begin to lose their orderly arrangement. The solid melts and becomes a liquid.*

7-39. *When a liquid is heated, the particles of the liquid move faster and faster. The liquid boils and changes to a gas. Hot water and steam from within the earth are released in geysers like the one shown here.*

Solids like glass or plastic that are not made of crystals do not have a sharp melting point but soften gradually. The particles in solids like glass are not held in an orderly pattern. A few materials, such as dry ice and moth balls, do not melt but change directly into a gas. The process in which a solid changes into a gas without first becoming a liquid is called *sublimation* (sub-luh-**may**-shun).

The particles that make up solids must be linked together in some way. Otherwise they would not stay in an orderly arrangement. When a solid is warmed up to its melting point, some extra heat energy is needed to unlink the particles. This extra heat energy does not make the molecules vibrate faster and so does not increase their temperature. The heat needed to cause the orderly arrangement of particles in 1 g of a solid to change into a liquid is called its **heat of fusion.** For example, 80 cal of heat is needed to change 1 g of solid ice at 0° C into 1 g of liquid water also at 0° C. The *heat of fusion* of water is 80 cal/g. Melting ice can keep things

Heat of fusion
The amount of heat required to change 1 gram of a solid to a liquid at the same temperature.

cool because it absorbs 80 cal when each gram of ice melts. Ice has a higher heat of fusion than most other substances.

Table 7-1

Melting Points of Some Common Substances	
Substance	*Melting Point (°C)*
iron	1,535
salt	801
lead	327
sugar	186
water	0
mercury	−39

As you have seen, the particles of a liquid move more rapidly than the same particles in a solid. In most liquids at ordinary temperatures, a few particles have enough energy to escape and become a gas. For example, a pan of water left uncovered in a room will slowly evaporate because a few of its particles are constantly leaving the water in the pan and entering the air as water vapor. See Fig. 7-39. If more heat is added to a liquid, the speed of evaporation will increase. Continuing to add heat will finally give all the particles of the liquid enough energy to become a gas. If you measure the temperature when the liquid is changing to a gas, you will have found the **boiling point** of the liquid. The exact *boiling point* depends upon two factors: (1) the amount of heat energy needed to make the particles of the liquid separate to become a gas; (2) the pressure of the air. For example, water boils at 100°C when the air pressure is normal. On a mountaintop where the air pressure is lower, water boils at a temperature below 100°C.

Boiling point
The temperature (at ordinary air pressure) at which the particles of a liquid have enough energy to become a gas.

Extra heat energy is needed to unlink the particles in a solid to form a liquid. In the same way, heat must be supplied to separate the particles in a liquid to form a gas. The amount of heat needed to change 1 g of a liquid into a gas at the same temperature is called **heat of vaporization.** The *heat of vaporization* of water is 540 cal/g. Water evaporating from your skin absorbs its heat of vaporization from your body. This loss of heat makes you feel cool as you dry.

Heat of vaporization
The amount of heat required to change 1 gram of a liquid to a gas at the same temperature.

ACTIVITY

Materials
table salt
plastic sandwich bag
rock salt
small magnifying glass
pencil

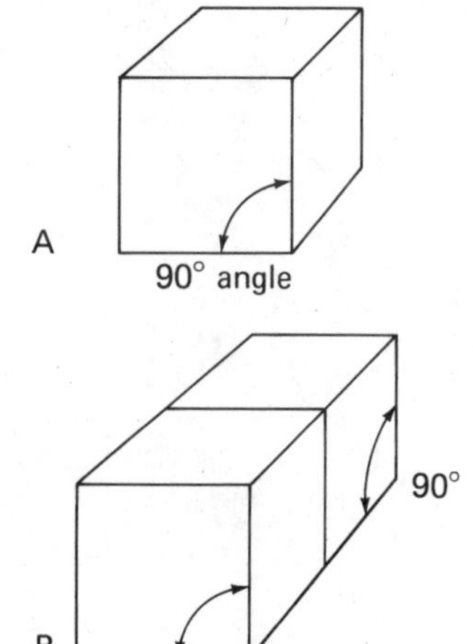

7-40. *The surfaces of a cube meet at 90° angles. Two cubes joined together form a rectangular solid. The surfaces still meet at 90° angles.*

Observing Crystals

Crystals are easily recognized by the beauty of their structure. Their flat surfaces meet, forming definite angles that give the crystals their particular shape. The angles and shape are determined by the kind of particles which make up the solid. For example, you already know the shape of the crystals of common table salt.

A. Obtain the materials listed in the margin.

B. Shake a few crystals of salt into a plastic sandwich bag.

C. Look at the salt crystals with a magnifier.

1. Are most of the crystals cube shaped?
2. Describe any crystals you see that are not cube shaped.
3. Do all the crystals have flat surfaces?

A cube is formed when six square surfaces meet, forming 90° angles. See Fig. 7-40. If you placed two cubes together, you would have a rectangular solid. The surfaces of the rectangular solid would still meet to form 90° angles.

D. Look at the table salt again.

4. Did you find any rectangular crystals?

Another feature of crystals is their ability to break apart, forming pieces with flat surfaces. The broken pieces form the same angles as the original crystal. This can be seen in rock salt. Rock salt is made of the same substance as table salt.

E. Pour the table salt from the sandwich bag back into the container provided.

F. Place a few crystals of rock salt in the plastic bag and look at it with the magnifier.

5. Are there any cube-shaped crystals?
6. Do the crystals have flat surfaces?
7. Do the surfaces meet at 90° angles?

G. Now move several rock salt crystals that are *not* cubes to one side of the plastic bag away from the others.

H. Roll your pencil over these crystals to break them up. You may need to press very hard to do this.

I. Look at these pieces of crystals.

8. Are there any pieces shaped like a cube?
9. Do the pieces have flat surfaces?
10. Do the surfaces meet at 90° angles?
11. In one or two complete sentences, describe your observations of salt crystals.

SUMMARY

Why can something as hard as a diamond be broken with only a gentle tap? The answer is that the particles in most solid materials like diamonds are held together in an orderly pattern. This pattern allows crystals to be separated and causes most solids to have a particular melting point. Increasing the temperature of a liquid will cause it to boil and become a gas. Some solids can change directly into a gas without first becoming a liquid.

QUESTIONS

Unless otherwise indicated, use complete sentences to write your answers.

1. Most solids are formed as ______. Their particles are arranged in an ______ way, giving them a regular shape.
2. Describe what happens to the particles of a solid as it is heated until it melts.
3. What two factors determine the exact boiling point of a liquid?
4. Describe what must happen to water particles in order for the water to become a gas.
5. What are some of the properties you would look for if you had to decide whether a solid was a crystal or not?
6. An ice cube has a mass of about 40 g. How many calories would it take to melt two ice cubes?
7. If an ice cube (40 g) is melted in 250 cm^3 of water, what is the change in temperature in degrees Celsius?
8. How many calories are needed to change 5 g of water to steam?

LABORATORY
Heat of Fusion

Materials
plastic sandwich bag
styrofoam cup
balance
thermometer (-10° to 110°C)
stirring rod
ice, 10 g

Purpose

In order to measure heat energy, you must understand that heat energy is required to raise the temperature of water. You must remove heat energy in order to lower the temperature of water. When you remove a small amount of energy from water, the temperature will decrease until the water reaches 0° C. You will determine the amount of energy that must be removed from each gram of water, at 0° C, to change it to ice at 0° C. This amount of heat is called the heat of fusion. When the water reaches 0° C, much more energy must be removed in order to turn the water into ice without changing its temperature. (This is why ice cubes always freeze a little at a time and not all at once.)

Procedure

A. Obtain the materials listed in the margin. You will need a data table to keep track of the measurements you will make.

B. Copy the following table in your notebook.

1. cup	______	grams
2. cup + water	______	grams
3. water (2 − 1)	______	grams
4. plastic bag	______	grams
5. plastic bag + ice	______	grams
6. ice, weight (5 − 4)	______	grams
7. initial temperature of water	______	°C
8. final temperature of water	______	°C
9. change in temperature of water (8 − 7)	______	°C
10. calories to melt ice (9 × 3)	______	calories
11. heat of fusion in calories/gram (10 ÷ 6)	______	cal/g

C. Leave some space at the bottom of the table for additional data and calculations.

Heat energy is measured in a unit called a calorie. A calorie is the amount of heat energy you must add to 1 g of water to raise its temperature 1° C. Note: $H = (T_2 - T_1)\,M$. In order to find out how many calories have been taken out of a certain amount of water, you must know two facts: **1.** the amount of water (M), and **2.** the change in temperature (T_2-T_1).

1. How many calories does it take to change the temperature of 2 g of water 1°C? 3 g of water 1°C? 1 g of water 2°C?

D. Think carefully about your answers to question 1 and answer the following question.

2. It takes 3 cal to change 3 g of water 1°C. How many calories does it take to change 3 g of water 2°C?

The number of calories is the product of the amount of water in grams and the temperature change in degrees Celsius.

3. How many calories does it take to change the temperature of 110 g of water by 4°C?

The problem you are going to solve is: "How many calories does it take to melt 1 g of ice?" In order to answer this question, you must measure the calories lost by a certain amount of water when a cube of ice melts and divide that number by the number of grams of ice. For instance, if it takes 850 cal to melt 10 g of ice, the answer would be

$$\frac{850 \text{ cal}}{10 \text{ g}} = 85 \text{ cal/g}.$$

E. Weigh the styrofoam cup on a balance. Record this in the data table. See Fig. 7-41.

7-41.

F. Fill the cup about ¾ full of water.

G. Weigh the cup and water together on the balance. Record this in the data table.

H. Measure the temperature of the water in degrees Celsius. Record this as the initial temperature of the water. Keep the thermometer in the water from now on.

I. Weigh the plastic sandwich bag on the balance. Record in your data table.

J. Add a cube of ice to the bag and weigh the bag and ice together. Record.

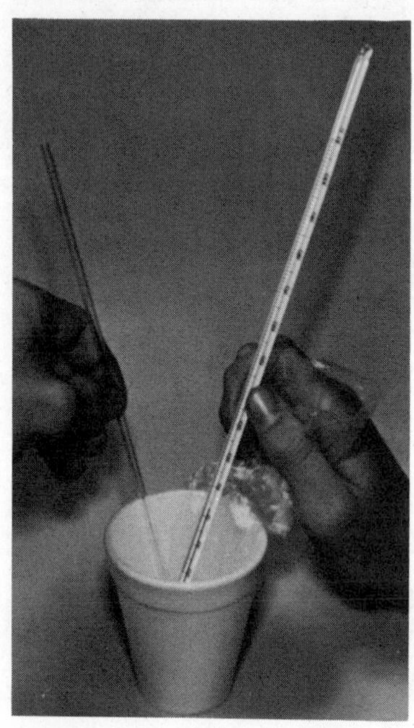

7-42.

K. Subtract to find the amount of ice in the bag. Record in your data table.

L. Put the part of the bag with the ice into the water. Keep the thermometer in the water. See Fig. 7-42. Heat from the water will go into the bag and melt the ice within a few minutes. You should be able to tell if any ice is left by stirring the water with a stirring rod. This will help to quickly average out any cold spots that may have developed.

M. When the ice has all melted, remove the bag from the water *immediately*.

N. Measure the temperature of the water. Record in your data table.

O. To find the number of calories needed to change the temperature, multiply the temperature change (item 9 in your table) by the number of grams of water (item 3 in your table). Record.

P. Find the heat of fusion by dividing the number of calories by the number of grams of ice. Record. The accepted value of the heat of fusion of water is 80 cal/g. If your result differs from 80 cal/g by more than 20 cal/g, you may wish to repeat your measurements. Do the steps quickly and be sure that the ice has all just melted when you remove the bag.

Summary

Write a paragraph describing the procedure to find the heat of fusion of ice. Include the materials that are needed, the measurements taken, the calculations required, and the value you expect to get for the heat of fusion of ice.

VOCABULARY REVIEW

Match the number of the word(s) with the letter of the phrase that best explains it.

1. absolute zero
2. boiling point
3. calories
4. conduction
5. convection
6. crystal
7. heat engine
8. heat of fusion
9. kinetic theory
10. radiation

a. Changes heat energy to mechanical energy.
b. Transfer of heat by infrared rays.
c. Transfer of heat by direct contact.
d. Transfer of heat by movement of a gas or liquid.
e. Heat needed to raise the temperature of 1 g of water 1°C.
f. Matter is made of particles whose motion determines whether the matter is solid, liquid, or gas.
g. The temperature at which particles in matter stop moving.
h. A solid whose orderly arrangement of particles gives it a regular shape.
i. The temperature at which the particles of a liquid have enough energy to become a gas.
j. The amount of energy that must be removed to change 1 g of water into ice at the same temperature.

REVIEW QUESTIONS

Complete each statement by choosing the best word or phrase, or by filling in the blank.

1. Heat is held in matter as **a.** heat fluid added **b.** heat fluid squeezed out **c.** moving particles **d.** crystals.
2. From the results of his cannon drilling experiment, Count Rumford reasoned that **a.** the water boiled as a result of the work done by the horse **b.** heat is a form of energy **c.** the supply of heat was without limit **d.** all of these.
3. The three ways of transferring heat from one object to another are ______, ______, and ______.
4. The transfer of heat by the movement of a gas or liquid is called ______.

5. Temperature is a measure of how hot an object is. To know how much heat is involved in a change in the temperature of water, you would also need to know the ______ of the water.
6. Compared to the Fahrenheit scale, the same temperature change on the Celsius scale would contain **a.** more degrees **b.** the same number of degrees **c.** fewer degrees **d.** these scales are not comparable.
7. The particles of a substance are moving slowest when it is in which form? **a.** gas **b.** liquid **c.** solid **d.** there is no relation between movement and the form of matter.
8. There can be no temperature lower than absolute zero because **a.** no temperature scale goes lower than zero **b.** all particle motion stops at absolute zero **c.** there is too little heat to be detected **d.** particles cannot conduct heat at that temperature.

REVIEW EXERCISES

Give complete but brief answers to each of the following. Use complete sentences to write your answers.

1. Use the formula, $H = (T_2 - T_1)\ M$, where H = heat energy in calories, T = temperature in degrees Celsius, and M = mass in grams, to calculate the heat energy involved in each of the following:
   a. A hot piece of rock is placed in 100 g of water, changing the water temperature from 25°C to 55°C.
   b. An ice cube placed in 200 g of water changed the temperature of the water from 25°C to 15°C as the ice melted.
2. In your own words, describe the difference between temperature and heat.
3. What are the features of the Fahrenheit and Celsius temperature scales that make them different?
4. State what happens to the particles in an object when it is heated.
5. Explain briefly the three ways in which heat is transferred from one material to another.
6. Describe what is meant by the kinetic theory of matter.
7. Discuss the features of the Kelvin temperature scale.
8. State briefly how the motion of particles differs in gases, liquids, and solids.

EXTENSIONS

1. Why has absolute zero temperature (0°K = 273°C) never been reached? Write a report in which you discuss the method used to find that −273°C is an absolute minimum temperature.

the structure of matter

UNIT 3

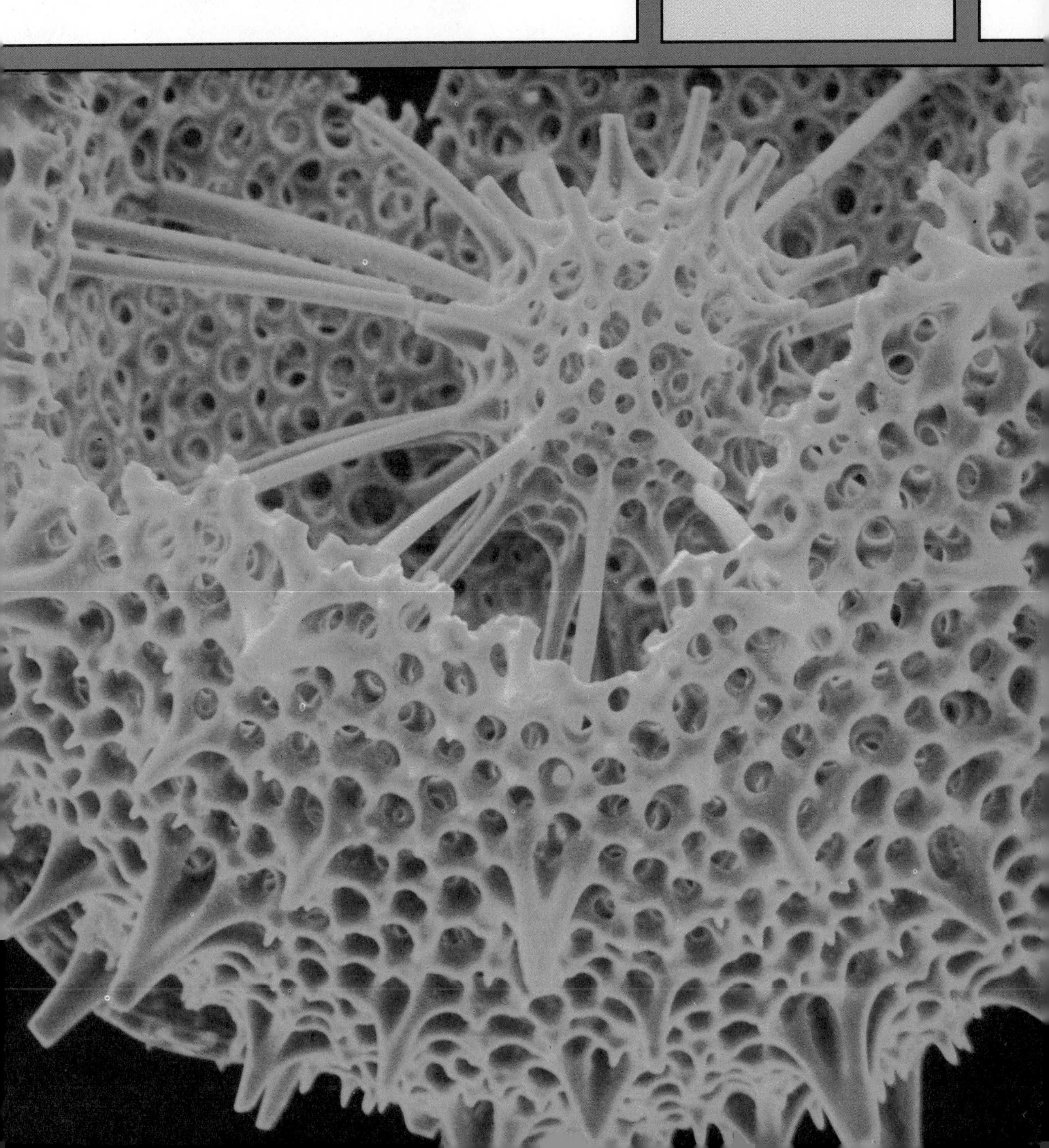

CHAPTER 8

Matter

8-1. MOLECULES

Imagine that you are a skydiver. With your parachute strapped on securely, you jump from an airplane. For a short time you fall freely. Then at a certain height, the parachute opens and you float gently down to the ground.

What might you see on such a jump? At first you would see the whole landscape beneath you. You would see fields, highways, even whole towns. Coming down closer, you would see a smaller area in greater detail. You would be able to see trees, cars on the highway, and individual houses. Finally, on landing you would see grass, soil, and twigs on the ground around you.

In this chapter, you will be a "skydiver" studying matter. You will begin with a view of a whole "landscape" of matter. Then you will close in on the small individual particles that make up the world of matter.

When you finish lesson 1, you will be able to:

- Describe what happens as matter is divided into smaller and smaller parts.
- Explain what is meant by a *molecule*.
- Define and give some examples of a *physical change*.
- ○ Observe examples of changes in matter and decide if they are physical changes.

Matter is defined as anything that takes up space and has weight. Everything you see around you is made of matter. See Fig. 8-1. What is matter made of?

Try dividing a glass of water as many times as possible. First you could pour out half the water. You would still have half a glass of the same water. You could then divide that in half, and so on. What would happen if you could keep dividing the remaining amount of water in half? Imagine that you could keep pouring out more and more water. You would finally get to one tiny particle that would still be water. That small particle of water is called a water **molecule** (**mol**-ih-kyool). A water *molecule* is the smallest particle of water that can be identified as water. If you could divide that particle, you would no longer have water. You would then have two different kinds of matter. This means that a water molecule is made up of separate parts that combine to produce the substance we call water.

Molecule
The smallest particle of a substance, such as water, that can be identified as that substance.

Actually it would be impossible for you to separate out just one water molecule. Water molecules are small. It would take about 60 million water molecules side by side to reach across a penny! If a drop of water were as big as a football field, you would be able to see billions and billions of molecules.

All molecules of water are alike. They all have the same properties.

What would happen if you divided another liquid such as alcohol in the same way as you imagined you

8-1. *Everything that you see and feel in the world around you is made of matter.*

did the water? Eventually you would come to the smallest particle of alcohol that could still be recognized as alcohol. This particle would be a molecule of alcohol. If this particle was divided, you would have two kinds of matter that are not like alcohol.

So alcohol, like water, is made up of separate molecules. Alcohol molecules, however, are different from water molecules. Alcohol burns. Water does not burn. A huge number of different kinds of matter exists in the world. Each has its own kind of molecules. There must be many different kinds of molecules. These different molecules make up the objects we see around us.

Matter is always changing. You know from experience that water is not always a liquid. You can freeze

8-2. *This photo of a stream in winter illustrates the three forms of water: liquid in the stream, solid in the snow and ice, and gas in the vapor.*

pure liquid water to make solid ice. You can also boil water and change it into a gas. See Fig. 8-2. Are the water molecules changed into different kinds of molecules when they are frozen or boiled? Try this experiment at home. Pour a glass of water into an empty ice cube tray. Put the tray in the freezer. After the water freezes, remove the tray and let the ice melt. Does this melted ice look like the water before freezing? Does it taste the same as the water before freezing? Do you think that the individual molecules of water were changed by freezing and melting? This experiment can show you that melted ice has the same properties as the water before it was frozen. Water molecules in liquid water, ice, and steam are all alike. The changes of water from liquid to solid to gas are examples of a **physical change.** After matter such as water has undergone a *physical change*, each of its molecules is the same as it was before the change. The molecules in liquid water, solid ice, and steam are simply arranged differently.

Physical change
A change in matter in which the individual molecules are not changed.

Not all changes in matter are physical changes. Suppose a nail lies on the ground until it becomes rusty. The rust must have been made from the iron in the nail. But does the rust have the same properties as the iron? Careful observation of the rust would show that it is made up of different molecules than iron. Rusting of iron is an example of a **chemical change.** When one kind of molecule is changed into another kind of molecule, a *chemical change* causes a new substance with new properties to be formed.

Chemical change
A change in matter in which one kind of molecule is changed into another kind.

ACTIVITY

Materials
none

8-3.

8-4.

8-5.

8-6.

Physical and Chemical Change

Do you like chocolate? Look at Fig. 8-3. It shows chocolate in its familiar solid form and also in a melted, or liquid, form used for coating candy bars. Chocolate can be melted easily by heating.

1. Does the liquid form of chocolate have the same color as the solid form?
2. Do you think that melted chocolate would taste different from solid chocolate?
3. Do you think that liquid and solid chocolate contain the same kind of molecules?
4. What name is given to a change in matter that leaves the molecules of a substance the same?

Figure 8-4 shows a drop of water and a piece of ice on waxed paper. In Fig. 8-5 you see the drop of water and, now, drops of liquid from the melted ice on the waxed paper.

5. Do both kinds of drops look the same?

If the drops of liquid were tested, they would be found to freeze at the same temperature, boil at the same temperature, and taste the same.

6. With this evidence in mind, do you think the two liquids are made of the same kind of molecules?
7. Is this an example of a physical change?

Figure 8-6 shows ordinary sugar in a bowl and also sugar that has been heated and turned into caramel. Nothing was added but heat.

8. Do sugar and caramel look the same?
9. From your experience, do sugar and caramel taste the same?
10. Do you think sugar and caramel are made of the same kind of molecules?
11. Do you think this is an example of a physical change?
12. In your own words, describe the evidence that made you choose your answer to number 11.

SUMMARY

Taking a closer and closer look at matter, like floating to earth on a parachute, reveals more and more detail. Water consists of small particles called molecules. All matter is made up of particular kinds of molecules. A physical change in matter happens when the molecules are put into a different arrangement but are not changed in any other way. A chemical change causes the molecules to become different kinds of molecules.

QUESTIONS

Use complete sentences to write your answers.

1. Suppose you were standing beside a drop of water and you started to shrink. Describe how the water would look as you got smaller and smaller.
2. Explain what is meant by a molecule.
3. Give two examples of a physical change in matter.
4. When water freezes and then melts, do the molecules change to other kinds of molecules? How could you tell?
5. How could you show that alcohol molecules are different from water molecules?
6. How does a physical change in matter differ from a chemical change? Give an example of each.

8-2. MIXTURES, COMPOUNDS, SOLUTIONS

Have you ever watched a druggist select the materials to fill a doctor's order for a drug or medicine? It takes skill and training to be certain that the materials are put together correctly. A druggist or pharmacist is a person with such skill and training. There are thousands of different drugs and medicines. Pharmacists use many different combinations of materials in preparing these drugs and medicines.

When you finish lesson 2, you will be able to:

- State the characteristics of a *mixture* and give some examples.
- State the characteristics of a *compound* and give some examples.
- Explain the differences between a mixture, a compound, and a *solution*.
- ○ Observe two samples and determine if they are mixtures or compounds.

If you pick up a handful of sand, you will probably be able to see some dark grains among the light ones. Sand is usually made up of more than one kind of material. Since it is a combination of materials, sand must contain more than one kind of molecule. Any matter that contains more than one kind of molecule is called a **mixture.** Sand is a *mixture*. Figure 8-7 shows some common mixtures.

Mixture
Any matter that contains more than one kind of molecule.

The materials in a mixture can be separated by physical means. For example, in a sand mixture you can separate the dark grains from the light ones by physically removing them. Or in mixtures such as sand with

8-7. Many of the things you see around you every day are mixtures.

sugar or salt with pepper, the individual grains of sand, sugar, salt, and pepper can be identified and separated. Each material has its own properties.

Would you be able to separate sugar and water easily? Sugar will dissolve in water. The solid sugar crystals will completely disappear in the water. You can tell that the sugar is mixed with the water because the mixture has a sweet taste. The sugar molecules have left the solid sugar crystals and become mixed with the water molecules. See Fig. 8-8. A mixture formed when one kind of molecule, like sugar, fills the space between another kind of molecule, like water, is called a **solution** (suh-**loo**-shun). *Solutions* are mixtures of separate molecules.

Solution
A mixture formed when one kind of molecule, like sugar, fills the spaces between another kind of molecule, like water.

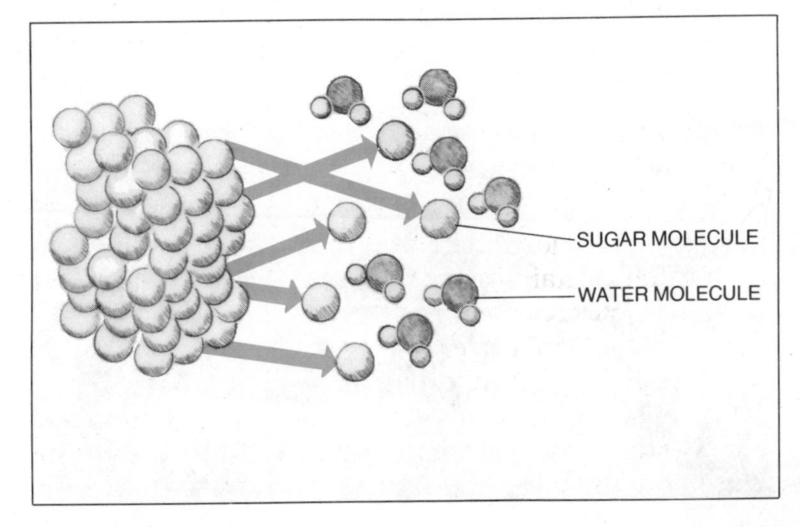

8-8. *When a sugar crystal dissolves in water, the separate molecules of sugar leave the crystal and mix with the water molecules.*

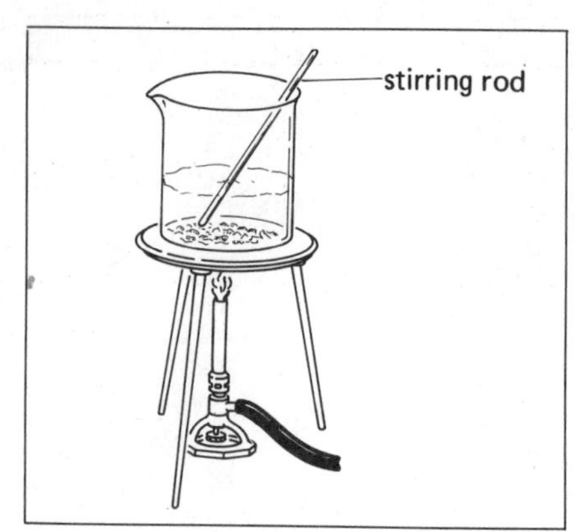

8-9. *If the water in a mixture of salt and water is boiled away, the salt remains.*

Compound
A substance, such as water, that cannot be broken down into simpler parts by a physical change. A compound contains only one kind of molecule.

8-10. *A compound cannot be changed into a simpler material by physical means. For example, if you crush an aspirin tablet, the small particles and powder that remain are still aspirin.*

You cannot separate the different materials in solutions as easily as you can separate grains of salt and pepper. For example, air is a solution of gases. Air contains molecules of oxygen. You cannot see molecules of oxygen in air, nor can you see the air itself. There is no simple way to separate oxygen from the other gases in air. Some solutions seem to be a single kind of matter because the individual kinds of molecules in them cannot be easily separated.

Most mixtures, even solutions, can be broken down by a physical change. Dissolved salt can be removed by heating the mixture. The water can be boiled away. The salt will then be left behind in the form of solid crystals. See Fig. 8-9.

The water used in the above procedure must be pure. It must not have any other material dissolved in it. In this book, when we say water, we generally mean pure water. Pure water remains the same if it is first boiled and then changed from steam back into a liquid. Pure water contains one kind of molecule. This kind of molecule is the same in the solid, liquid, or gas form of water. A substance, such as water, that contains only one kind of molecule is a **compound.**

Compounds cannot be broken down into simpler parts by physical change. Some of the medicines prepared by pharmacists are compounds. For example, aspirin is a compound. Aspirin is made up of only one kind of molecule. If you crushed an aspirin tablet, you would have only smaller particles of aspirin. See Fig. 8-10. It would take more than a physical change like crushing to separate the aspirin molecules it contains. See Fig. 8-11.

8-11. *Cotton candy is a familiar product that is made from sugar, which is a compound. How do you know that sugar is a compound?*

ACTIVITY

Mixtures and Compounds

A. Obtain the materials listed in the margin.

B. Add about 1 teaspoon of sand to about 1 teaspoon of salt and stir them together.

1. Is this combination of salt and sand a mixture or a compound?

C. Place the sand and salt combination in a cup and add 1/2 cup of water. Stir for 2–3 min.

D. Pour the liquid into a beaker. This liquid is a solution containing water and most of the salt. Some salt remains on the sand left in the cup.

2. In your own words, describe the salt–water solution.
3. Describe what must have happened in the formation of this solution.

E. Place two drops of water on a tin can lid. Using a clothespin as a handle, hold the lid above the flame of a candle for several minutes and heat slowly.

4. In your own words, describe what happens.

F. Place two drops of the solution from step D on the tin can lid. Using a clothespin for a handle, hold the lid 10–15 cm above the candle flame for several minutes to heat it very slowly.

5. In your own words, describe what you observe.

G. Wait several minutes for the lid to cool. Add water to the material on the lid.

6. Does the material on the lid go into solution?
7. In your own words, describe how you could separate a mixture of two materials, one of which can be dissolved in water.

Materials
matches
tin can lid
salt
sand
cup
beaker
candle
clothespin

SUMMARY

A pharmacist works with both mixtures and compounds used as drugs and medicines. A mixture always contains more than one kind of molecule. In all mixtures, even solutions, the different molecules can be separated by physical changes. Compounds cannot be broken down by physical change. Compounds contain only one kind of molecule.

QUESTIONS

Use complete sentences to write your answers.

1. Is a combination of sand and sugar a mixture, a compound, or a solution? How could you show that your answer is correct?
2. Is salt in water a mixture, a compound, or a solution? How could you show that your answer is correct?
3. What makes something a mixture? Name at least two mixtures.
4. What makes something a compound? Name at least two compounds.

8-3. ELEMENTS AND ATOMS

Could you change a piece of iron, like a nail, into gold? If you lived about 400 years ago, you might have spent your life trying to do that. In those days, people known as alchemists believed that they could find a way to change ordinary metals into gold. The alchemists studied and tested almost every kind of matter known in their day. Much basic knowledge and many techniques used later by scientists to investigate matter came from the alchemists. In this lesson, you will investigate the breakdown of compounds into simpler forms of matter.

When you finish lesson 3, you will be able to:

- Describe how a chemical change is produced.
- Give examples of compounds being broken down into *elements*.
- Explain the relationship between elements and *atoms*.
- ○ Observe the breakdown of a compound into its elements.

If you dissolve salt in water, you can later recover the same salt. A salt solution is a mixture. The salt and water can be separated without too much difficulty. Can a compound such as water be separated into different materials? The following experiment should answer this question.

Equipment such as that shown in Fig. 8-12 is set up. The equipment consists of a source of electric current such as a battery. The battery is connected to two wires that allow the current to flow through a beaker of water. A small amount of a chemical such as sulfuric acid is added to the water. This is necessary because

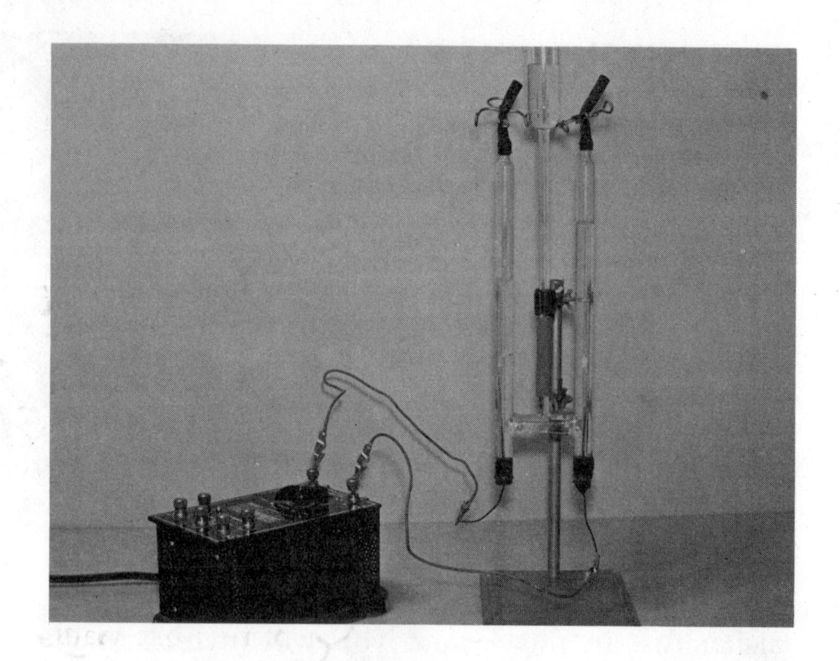

8-12. *When an electric current is passed through water, hydrogen and oxygen are produced.*

pure water does not easily conduct an electric current. When the current flows through the water, bubbles of gas are seen to come from the wires where the current enters and leaves the water. These gases are allowed to bubble up and fill two test tubes. Tests show that one gas is hydrogen. Hydrogen gas burns with an almost invisible flame. The gas in the other test tube is oxygen. Oxygen gas does not burn. But a glowing wooden splint will burst into flame when put inside the test tube of oxygen.

This experiment shows that water can be broken down into two new substances. The hydrogen and oxygen produced are completely different from the water. This breakdown is an example of a chemical change.

Experiments with compounds other than water show that other compounds can also be changed into different materials. Sugar can be broken down into carbon, hydrogen, and oxygen. You may have seen this happen when sugar is burned. The sugar turns into solid black carbon. See Fig. 8-13. Hydrogen and oxygen escape in the form of water vapor.

Investigations have shown that many different compounds can be changed into other substances, such as hydrogen and oxygen. For example, hydrogen peroxide is a compound. It contains more oxygen than the

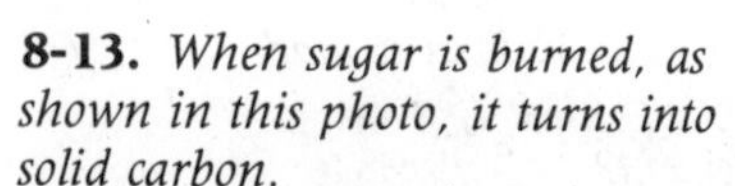

8-13. *When sugar is burned, as shown in this photo, it turns into solid carbon.*

same amount of water. When a given compound is broken down, the compound always yields the same substances in the same amounts. For example, 2 g of water will always provide 11 percent hydrogen and 89 percent oxygen. This means that 20 raindrops weighing 2.0 g could be broken down to give 0.22 g of hydrogen and 1.78 g of oxygen.

$$
\begin{aligned}
\text{hydrogen: } 11\% &= 11/100 = 0.11 \\
\text{oxygen: } 89\% &= 89/100 = 0.89 \\
\text{weight of hydrogen} &= 0.11 \times 2 \text{ g} \\
&= 0.22 \text{ g} \\
\text{weight of oxygen} &= 0.89 \times 2 \text{ g} \\
&= 1.78 \text{ g}
\end{aligned}
$$

A beaker of water weighing 448 g would yield 49.28 g of hydrogen and 398.72 g of oxygen.

The hydrogen and oxygen that come from water cannot be changed further into yet simpler forms of matter. Hydrogen and oxygen are the basic forms of matter in water. Hydrogen and oxygen are examples of **elements.** An *element* is the simplest form of matter. An element cannot be chemically changed into anything else. Up to the present time, more than 100 different elements have been discovered. All matter is made up of these elements. The most common elements found in nature are listed in Table 8-1.

Element
The simplest form of matter.

Atom
The smallest particle of an element.

Water is made up of water molecules. A water molecule is the smallest particle of water possible. What happens if an element such as oxygen is divided in the same way? You would find a small particle that can still be identified as oxygen. The smallest particle of an element is called an **atom.** Each element is made up of a particular kind of *atom*. The element oxygen, for example, is made up of only oxygen atoms. Hydrogen contains nothing but hydrogen atoms; carbon contains only carbon atoms. See Fig. 8-14.

8-14. *Each element is made up of only one kind of atom. Is coal an element?*

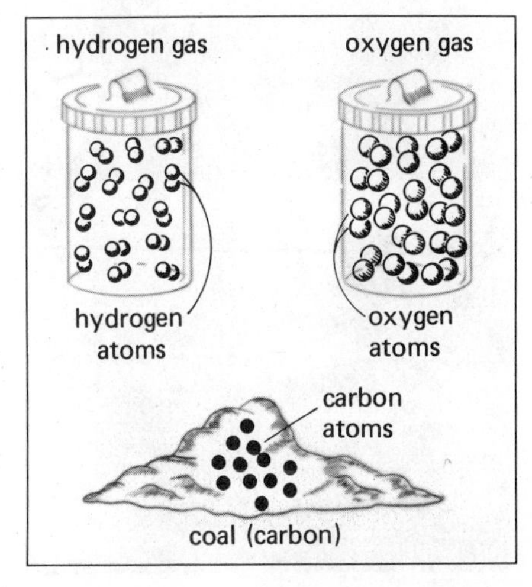

As early as 400 B.C., ancient Greek scientists suggested that all matter was made up of atoms. But the modern atomic theory is based on the work of an English school teacher named John Dalton. In 1808, Dalton first proposed that atoms join together to make up compounds. The atomic theory that developed from Dalton's ideas has become one of the most important scientific theories.

TABLE 8-1
Common Elements Found in the Earth's Crust and in the Human Body.

Most Common Elements in the Earth's Crust		*Most Common Elements in the Human Body*	
Oxygen	49.5%	Oxygen	65.0%
Silicon	25.8%	Carbon	18.0
Aluminum	7.5%	Hydrogen	10.0%
Iron	4.7%	Nitrogen	3.0%
Calcium	3.4%	Calcium	2.0%
Sodium	2.6%	Phosphorus	1.0%
Potassium	2.4%	Potassium	0.3%
Magnesium	1.9%	Sulfur	0.2%
Hydrogen	0.9%	Sodium	0.15%
Titanium	0.6%	Chlorine	0.15%
All others	0.7%	Magnesium	0.019%
		Iron	0.04%
		All others	0.15%

ACTIVITY

Materials
small jar with lid
hydrogen peroxide
manganese dioxide
matches
wooden toothpick
tin can

8-15.

8-16.

Producing Oxygen from Hydrogen Peroxide

In this activity you will break hydrogen peroxide down into two simpler materials, water and oxygen gas. You will study some of the properties of oxygen gas and learn a way to detect its presence.

A. Obtain the materials listed in the margin. Punch a small hole in the center of the lid.

B. Fill a small bottle about 1/3 full of hydrogen peroxide.

C. Look closely at this liquid and answer the following questions.

1. Does hydrogen peroxide have a color? an odor? Does it look like water? If you add the compound manganese dioxide to hydrogen peroxide, the hydrogen peroxide breaks down into water and oxygen gas.

D. Add about 1/2 spoonful of manganese dioxide to the hydrogen peroxide. See Fig. 8-15. Swirl the liquids to mix them.

2. As bubbles form and break, do you see a color or smell an odor?

E. With a match, light one end of a toothpick. Blow out the flame so that the toothpick is just glowing red. Use the glowing end of the toothpick to break one of the larger bubbles of gas in the bottle. See Fig. 8-16.

3. When the bubble breaks, what happens to the toothpick? A glowing piece of wood always breaks into flame and burns rapidly in oxygen. This is the test often used in the laboratory to detect oxygen.

F. Repeat step E several times. Now place the lid tightly on the bottle. Make sure there is a small hole in the lid. You will use this setup again in 5 min.

4. Do you think oxygen itself will burn?

G. Let the bottle producing the oxygen sit for 5 min. Then try to light the gas escaping through the hole in the lid with a match. Swirl the liquid in the bottle.

5. Does the oxygen burn?

H. Bend a toothpick on one end so that it cracks and stays bent. Use the bent end for the glowing splint test for oxygen. See step E.

6. Is oxygen coming out of the hole?

7. In your own words, describe a test that can be used to detect the presence of oxygen.

SUMMARY

Can iron be changed into gold in the same way that water can be changed into hydrogen and oxygen? Unlike the alchemists of centuries ago, today's scientists now know that iron is an element, like hydrogen and oxygen. Iron contains only iron atoms. The iron atoms cannot be made into any other kind of atom by a chemical change. Compounds, such as water, are made up of molecules that can be broken down into simpler elements such as hydrogen and oxygen.

QUESTIONS

Unless otherwise indicated, use complete sentences to write your answers.

1. Describe what is meant by a chemical change and tell how it differs from a physical change.
2. How can water be broken down into the elements from which it is made? What are those elements?
3. How can you test for the presence of oxygen gas?
4. Explain how the terms "element" and "atom" are related.
5. One liter of water has a mass of 1,000 g. What is the mass of hydrogen in 1 L of water? What is the mass of oxygen?

Materials
electrolysis kit
beaker, 250 mL
battery, 6 V (or 1.5 V)
electrolysis solution
water
2 wood splints
matches
2 test tubes

Analysis of Water

Purpose

Is water a compound or an element? You can perform an experiment to find out. Great amounts of energy in the form of heat, light, or electricity are generally required to break down a compound.

Compounds are broken down in order to determine their composition. This is called *analysis* or *decomposition*. You will use electric energy to break down or decompose water. This is called *electrolysis*.

Procedure

A. Obtain the materials listed in the margin.

B. Set up the apparatus shown in Fig. 8-17, or follow the directions in your kit.

C. Fill the glass container about ¾ full with tap water.

D. Connect one electrode in the water to one side of the battery.

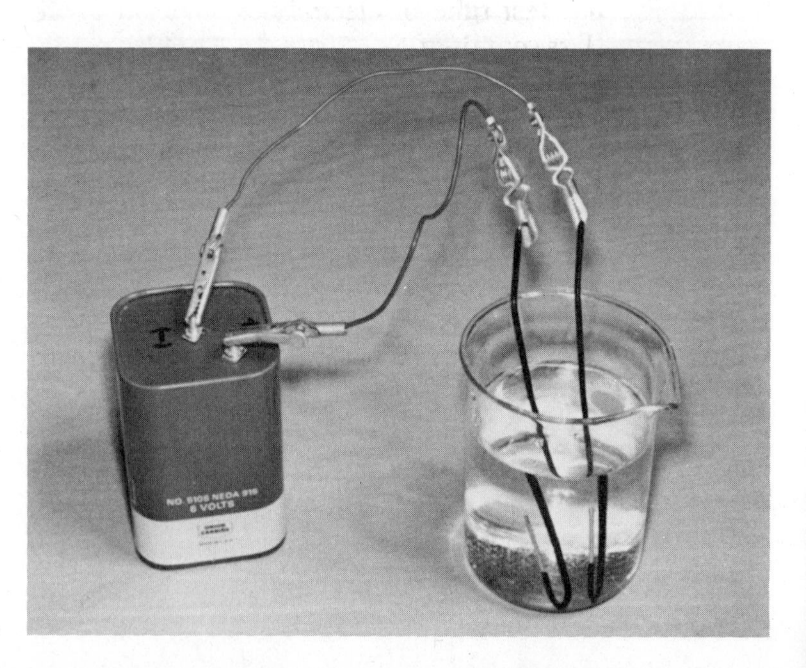

8-17.

E. Connect the other electrode in the water to the other side of the battery. Be sure the electrodes in the water do not touch each other.

F. Watch the electrodes for about 1 min.

1. What is happening around each electrode?

G. If there is no reaction at either electrode, move them closer together. At no time in this experiment should the electrodes be allowed to touch one another.

2. Does the reaction at either electrode appear to be the same?

The reaction that is occurring is a slow one. You would not be able to detect a reaction at all if you were using pure water. Materials dissolved in tap water make the reaction occur fast enough to see. However, the reaction is still not rapid enough for this experiment.

H. Remove the electrodes from the water.

I. Add one full test tube of electrolysis solution to the water in the glass container. See Fig. 8-18. The electrolysis solution makes the water conduct electricity much better than pure water. **Caution:** The electrolysis solution can be dangerous. Immediately wash off any solution that gets on your skin. Report any spill to your teacher.

J. Now fill the two small test tubes to the very top with water from the glass container.

K. Disconnect one end of the battery and set the electrodes back in the water.

L. Carefully place the tubes uspide down in the water. Do not allow air to get in the tubes.

M. Place one of the test tubes over each electrode.

N. Reconnect the battery.

8-18.

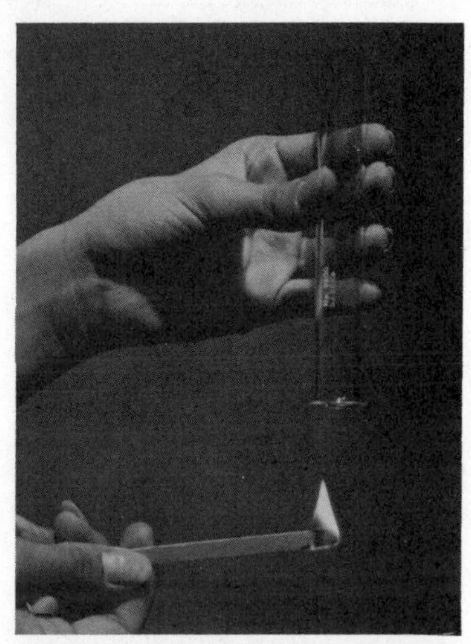

8-19.

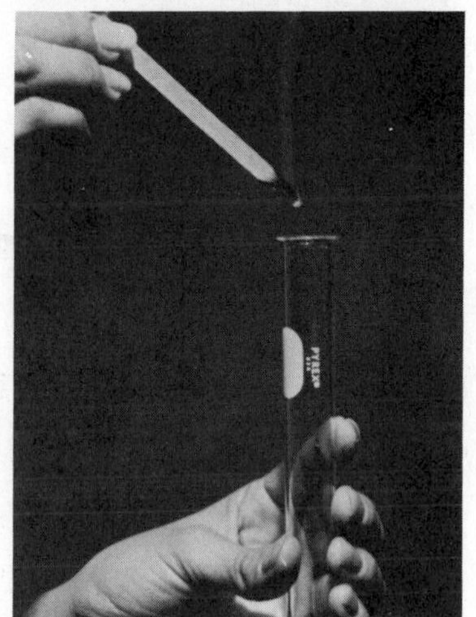

8-20.

O. Make a note of which test tube filled first. When *both* test tubes are full of gas, disconnect the battery. Which tube is over the negative electrodes? Record.

3. Which tube filled first? Can you see any difference in the two gases?

To complete the analysis, tests are usually made to identify the two gases. Hydrogen burns with a colorless flame and makes a popping noise. Oxygen does not burn. However, when a glowing splint is thrust into oxygen, the splint bursts into flame.

P. Copy the following table in your notebook. Record the results from steps Q through S in your table.

Test Tube	*Test Results*	*Gas Identified*
negative electrode positive electrode		

Q. Remove the test tube over the *negative* electrode from the water. Hold the test tube upside down. See Fig. 8-19.

R. Have your partner strike a match and bring the flame to the test tube. Rinse your finger. Record your results.

S. Test the tube over the *positive* electrode in the same way, using a *glowing splint* instead of a flame. Hold the test tube upright. See Fig. 8-20. Record your results and complete the table.

There have been many attempts to break down hydrogen and oxygen into more simple materials. All these tests have failed.

4. Are hydrogen and oxygen compounds or elements?

Summary

Write a paragraph or two summarizing what you did in this laboratory exercise. Include a statement of your results.

VOCABULARY REVIEW

Match the number of the word(s) with the letter of the phrase that best explains it.

1. molecule
2. physical change
3. mixture
4. solution
5. compound
6. chemical change
7. element
8. atom

a. A mixture formed when one kind of molecule fills the spaces between another kind of molecule.
b. Any matter that contains more than one kind of molecule.
c. A change in which a substance such as water is changed into new substances like hydrogen and oxygen.
d. The smallest part of an element.
e. A substance that cannot be broken down into smaller parts by a physical change.
f. The smallest particle of a substance that can be identified as that substance.
g. A change in matter which leaves its molecules the same as they were before the change.
h. The simplest form of matter.

REVIEW QUESTIONS

Choose the letter of the answer that best completes the statement or answers the question.

1. The smallest particle of a substance that can be identified as that substance is a(n) **a.** element **b.** electron **c.** subatomic particle **d.** molecule.

Use the following list to answer questions 2 and 3.

a. heating sugar to form carbon
b. dissolving salt in water
c. freezing water to form ice
d. heating ice to melt it
e. passing an electric current through water to produce two gases

2. Which reactions in the above list are physical reactions?
3. Which reactions in the above list are chemical reactions?
4. In a physical change **a.** new substances are formed **b.** molecules are changed **c.** molecules remain unchanged **d.** only a solid can change into a liquid by heating.

5. Matter that contains more than one kind of molecule is called a(n) **a.** element **b.** compound **c.** mixture **d.** pure substance.
6. A certain substance has only one kind of molecule. This is evidence that this substance is a **a.** compound **b.** mixture **c.** solution **d.** solid.

Use the following list to answer questions 7, 8, and 9.

a. air
b. water
c. raisin bread
d. aspirin
e. sand

7. Which items in the above list are mixtures?
8. Which items in the above list are compounds?
9. Which item in the above list is a solution?
10. Water may be broken down into the elements **a.** carbon and hydrogen **b.** oxygen and hydrogen **c.** salt and oxygen **d.** hydrogen and sugar.

REVIEW EXERCISES

Give complete but brief answers to each of the following. Use complete sentences to write your answers.

1. Describe what happens when you divide matter into smaller and smaller parts.
2. Define the word "molecule."
3. Give two examples of a physical change in matter.
4. Give two examples of a mixture and tell how you know they are mixtures.
5. What is a compound? How does a compound differ from a mixture?
6. How is a solution like a mixture?
7. Describe the arrangement of sugar and water molecules in a sugar–water solution.
8. How does a chemical change differ from a physical change?
9. Give an example of a compound. How can the compound be broken down into elements?
10. Describe how atoms and elements are related.

EXTENSIONS

1. Look up *alchemy* and *alchemists* in the library. What were alchemists trying to do? Did they succeed? Why is alchemy considered the beginning of the science of chemistry? Were any of the elements discovered by alchemists?

CHAPTER 9

Atoms

9-1. ATOMIC PARTICLES

The gold mask shown in the photograph above was buried with an Egyptian king more than three thousand years ago. The second photograph shows a gold article made in modern times. The gold in these two objects was taken from the earth at different times and places. However, if you could compare a single atom from each object, you would find that the atom from the ancient mask cannot be distinguished from that in the modern piece of gold. After many years of work, scientists have discovered why the separate atoms of an element such as gold are identical to one another but different from the atoms of any other element.

When you finish lesson 1, you will be able to:

- Name three kinds of *atomic particles*.

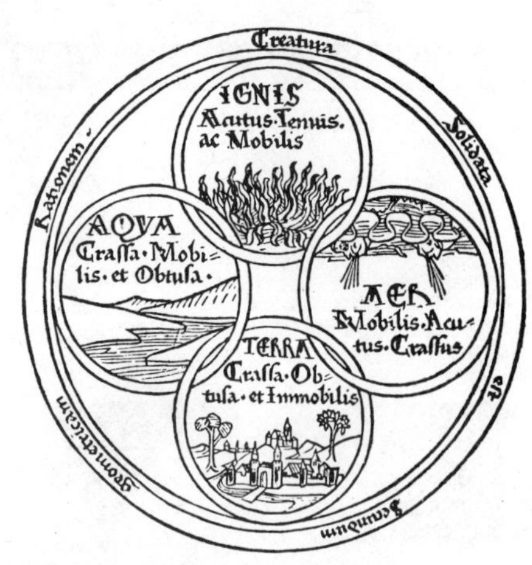

9-1. *In ancient times, scientists believed that matter was made up of four "elements": earth, air, fire, and water.*

- Compare the properties of the three kinds of atomic particles.
- Explain how scientists use mental models.
- Make a model of the contents of a mystery box.

Two thousand years ago, the ancient Greeks believed that all matter was made up of only four materials. These materials were earth, air, fire, and water. See Fig. 9-1. Wood, for example, was said to be made of fire and earth. When wood burned, the fire escaped and the earth remained in the form of ashes. One Greek thinker, Democritus, disagreed. Democritus said that matter was made of small objects that he called "atoms," from the Greek word *atomos* meaning "cannot be divided." But the teachings of Democritus were not accepted. The idea that matter was made up of earth, air, fire, and water was commonly believed for nearly two thousand years.

In 1808, John Dalton was able to show that matter is made up of atoms or atoms combined into molecules. His experiments proved that the atoms in a given element such as hydrogen all weigh the same, but differ in weight from the atoms of other elements. However, Dalton's experiments did not show what atoms are made of. He thought that hydrogen atoms were the smallest bits of matter since they are the lightest atoms.

The first evidence showing that electrically charged particles are hidden within atoms came in 1892. In that year, J. J. Thomson in England discovered the existence of negatively charged particles, which he called *electrons*. When the mass of an electron was later found to be 1/1,836 that of a hydrogen atom, it became clear to scientists that particles smaller than atoms existed.

Since Thomson's discovery, scientists have discovered many kinds of particles that are smaller than atoms. However, only three kinds of particles have been found to form the basic structure of atoms. These basic parts of atoms or **atomic particles** are listed below:

Atomic particles
The basic building blocks of atoms.

1. *Electrons* are extremely light particles. Each electron has a negative electric charge.

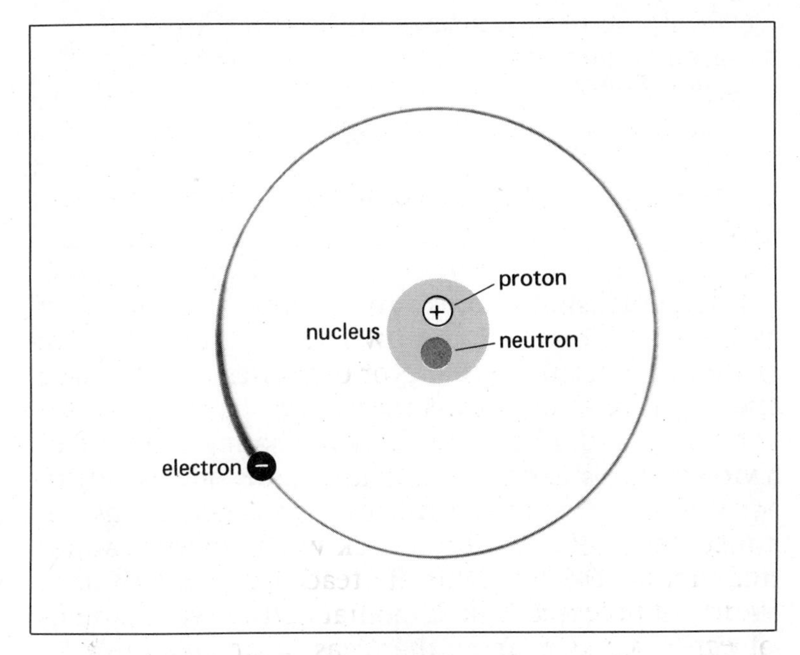

9-2. *An atom consists of a positive nucleus containing protons and neutrons, surrounded by negative electrons.*

2. *Protons* are positively charged particles. The positive electric charge on each proton is the same size as the negative charge on an electron. The charges on an electron and a proton cancel each other.
3. *Neutrons* (**noo**-tronz) are particles that are electrically neutral. A neutron has the same mass as a proton.

These three particles are the same in all kinds of atoms. An electron in a hydrogen atom is exactly the same as an electron in an oxygen atom. All of the different kinds of atoms that exist are made up of the same kinds of *atomic particles*. Atoms of different elements contain different numbers of electrons, protons, and neutrons. See Fig. 9-2.

The three atomic particles differ in electric charge and in mass. All three particles have very small masses. For example, about 6×10^{23} (6 followed by 23 zeros) protons have a mass of about 1 g. Neutrons have nearly the same mass as protons. But electrons have a much smaller mass. These values are so small that scientists use a special unit for the mass of atomic particles. This unit is the **atomic mass unit** (amu). Protons and neutrons have a mass of 1 *atomic mass unit*. Electrons are much lighter. An electron has a mass of 1/1,836 amu.

Atomic mass unit
A unit used to express the masses of atomic particles and atoms;
1 amu = 1.657×10^{-24} g.

9-3. *Models help us to better understand the objects that they represent.*

TABLE 9-1

Atomic Particle	*Mass (in amu)*	*Charge*
Electron	1/1,836	−
Proton	1	+
Neutron	1	0

Table 9-1 lists the mass and the electric charge of each of the three atomic particles. Since atoms are made up of atomic particles, the masses of atoms are also given in atomic mass units.

A typical atom has a diameter of about 2×10^{-8} cm. It would take more than a million atoms side by side to equal the thickness of this page. Individual atoms are much too small to be seen. Therefore scientists cannot study atoms directly. Instead, they make mental models of atoms. The scientific model of the atom is a mental picture of something that cannot be seen directly. See Fig. 9-3. The earliest scientific models of atoms pictured them as tiny solid balls. But the discovery of electrons and other atomic particles showed that the solid-ball model could not be correct. In lesson 2, you will see how a famous experiment led to the development of the modern scientific model for the atom.

ACTIVITY

Materials
sealed box containing several objects
pencil
paper

A Mental Model

The most widely accepted model of the atom is the result of many experiments. Scientists observed the way atoms behaved under different conditions. This activity will give you a chance to form your own scientific model.

One or more objects are hidden in the box you have been given. You must find out as much as you can about these objects without opening the box. After you have made your observations, you will make a model by describing the objects.

A. Obtain the materials listed in the margin.

Box number ____________
Number of objects in the box (your guess). ____________

Object	*Test*	*Observation*	*Model*
1, etc.			

B. Copy the preceding table in your notebook.

C. Spend about 10 minutes trying as many tests as you can, shake the box, turn it upside down, on end, etc. Do not open or otherwise damage the box. Record all your observations.

D. When you have completed your observations, make a sketch of what you think each object in the box looks like.

SUMMARY

Atoms are made up of three kinds of particles. These particles are called electrons, protons, and neutrons. Each particle has a specific electric charge and mass. Scientists use models to describe atoms.

QUESTIONS

Unless otherwise indicated, use complete sentences to write your answers.

1. Copy the following table on a separate sheet of paper. Complete the table by naming three atomic particles and giving their properties.

Particle Name	*Properties of Particle*
a.	
b.	
c.	

2. What is an atomic mass unit and what is its abbreviation?
3. How did the ancient Greek idea of matter differ from today's beliefs?
4. What caused scientists to realize that there were particles smaller than atoms?
5. Why do scientists use models in studying atoms?

9-2. ATOMIC STRUCTURE

Suppose you were given all the separate parts needed to make a watch. You would know that all the various parts fit together. But could you make all the parts into a watch that works? This problem is similar to the one that scientists studying the atom had to solve in the early part of this century. By 1900, it was known that atoms were made up of negatively charged electrons and positively charged protons. However, no one knew how these particles were arranged in an atom. Then, in 1911, a British scientist from New Zealand named Ernest Rutherford reported the results of a brilliant experiment. In lesson 2, you will learn how this experiment showed how an atom is put together.

When you finish lesson 2, you will be able to:

- Describe the atomic *nucleus*.
- Explain how to determine the number of protons and electrons in an atom.
- Use atomic numbers to identify atoms.
- ○ Probe a clay "atom" to determine its structure.

In Rutherford's experiment, a beam of fast-moving, positively charged particles was used. These particles were aimed at a very thin sheet of gold. Gold metal was used as a target because it can be made into sheets that are only a few atoms in thickness. The purpose of the experiment was to see how the paths of the particles would be changed when they hit the gold atoms. The results of the experiment were very surprising. They showed that most of the particles went straight through the gold as if nothing were there. But a few of the particles bounced off the gold atoms as if they were solid.

9-4. If you were to throw pebbles at a wire fence, most of them would go right through it. But some of the pebbles would strike the fence and bounce back. This is similar to what happened in Rutherford's experiment, which led to a model of the atom.

This experiment can be compared to what happens when you throw pebbles at a wire fence. See Fig. 9-4. Most of the pebbles go right through the fence because it is mostly empty space. However, a few pebbles may hit wires and bounce back.

Rutherford's experiment led to a new model for the atom. In Rutherford's model, most of the mass of an atom is found in a very small core at its center. This central core of the atom is called the **nucleus.** The rest of an atom is mostly empty space. It is this model of the atom that is used by modern scientists.

Nucleus
The small central core of an atom where most of the mass of the atom is located. This core is made up of protons and neutrons.

All the protons and neutrons in an atom are found in the *nucleus*. The nucleus has a positive charge because it contains the positive protons. The electrons are found in a cloud surrounding the nucleus. See Fig. 9-5. There is an attractive force between the positive nucleus and the negative electron cloud.

An atomic nucleus is very small. Picture an atom the size of a football field. See Fig. 9-6. The nucleus of that atom would be the size of a flea! Most of the space in an atom is taken up by the electrons around the nucleus. Since electrons are small and light, an atom, like a wire fence, is mostly empty space. Atomic particles shot at an atom, like pebbles thrown at a wire fence, will usually go right through the atom.

All atoms are made up of electrons, protons, and neutrons. Atoms of different elements have different numbers of these particles. Take the number of protons in the nucleus, for instance. Hydrogen atoms have one proton in the nucleus. Oxygen atoms have 8 protons. See Fig. 9-7. Each element can be described by the number of protons in its nucleus. This number is

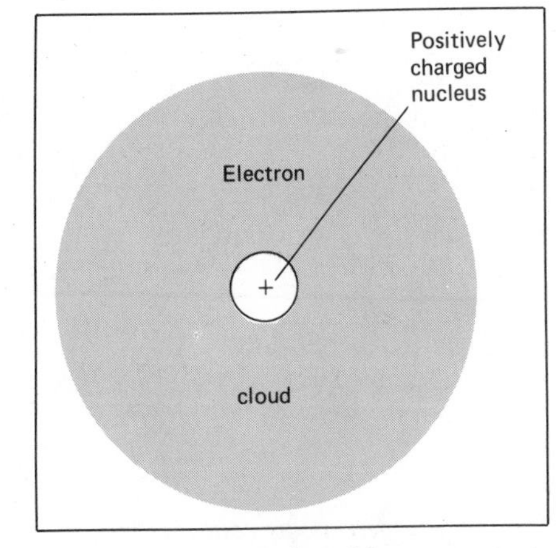

9-5. Electrons are found in a cloud around the nucleus.

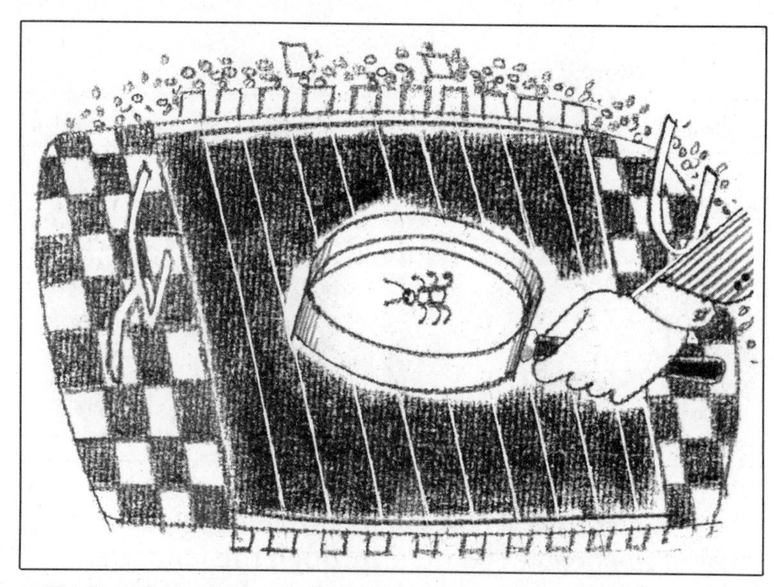

9-6. *If an atom were as big as a football field, the nucleus would be the size of a flea!*

Atomic number
The number of protons in the nucleus of an atom.

called the **atomic number** of the element. Hydrogen has an *atomic number* of 1 because all hydrogen atoms have one proton in the nucleus. Oxygen has an atomic number of 8. Table 9-2 lists the atomic numbers of the first 20 elements.

Atoms normally do not have an electric charge. The positive charges on the protons exactly cancel the negative charges on the electrons.

The number of electrons around the nucleus is the same as the number of protons in the nucleus. Since hydrogen atoms have one proton, they must also have one electron. Oxygen atoms have 8 protons and 8 electrons. Thus the atomic number of an element tells you the number of electrons as well as protons in each atom of that element.

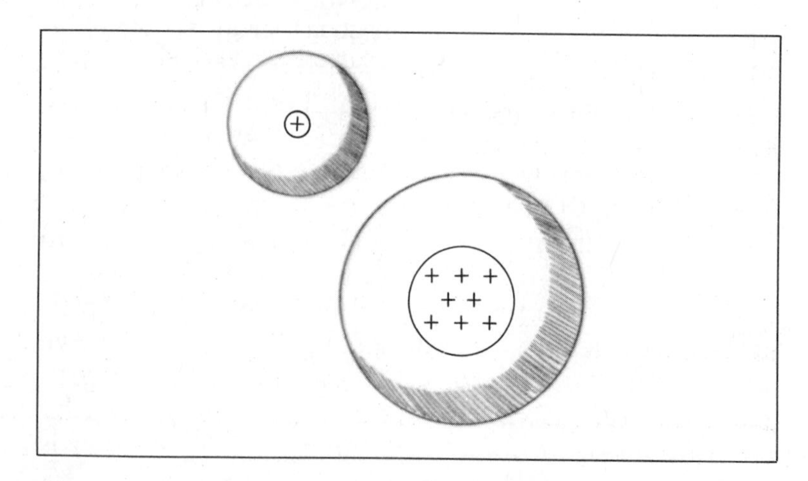

9-7. *Hydrogen atoms contain one proton; oxygen atoms have eight protons.*

TABLE 9-2
Atomic Numbers of the First 20 Elements

Element	Atomic Number
Hydrogen	1
Helium	2
Lithium	3
Beryllium	4
Boron	5
Carbon	6
Nitrogen	7
Oxygen	8
Fluorine	9
Neon	10

Element	Atomic Number
Sodium	11
Magnesium	12
Aluminum	13
Silicon	14
Phosphorus	15
Sulfur	16
Chlorine	17
Argon	18
Potassium	19
Calcium	20

ACTIVITY

The Structure of an Atom

Materials
clay "atom"
paper clip
paper
pencil

A. Obtain the materials listed in the margin.

Scientists shoot atomic particles at atoms in order to study atomic structure. In this activity, you will study the structure of a clay "atom." The clay ball represents the electrons of an atom. In place of atomic particles, you will use a probe made from a paper clip.

B. Use the paper clip probe to find out if the clay ball has a hard object hidden in it. Push the probe into or through the clay only 10 times. Probe with care! See Fig. 9-8.

1. Was your clay "atom" the same all the way through or did it contain a hard object?

C. If there is an object hidden in the clay, use any remaining probes to find out as much as you can about the object.

2. What is the shape of the object?
3. How large is it?
4. Is the object in the middle of the clay ball? If not, where is it located?

D. After making your 10 probes, draw a sketch of your clay "atom" showing its structure. Label this sketch "Model."

E. Now cut open the clay "atom" and see how accurate your model was.

F. Make a sketch of the actual structure of the clay "atom." Label this sketch "Actual."

9-8.

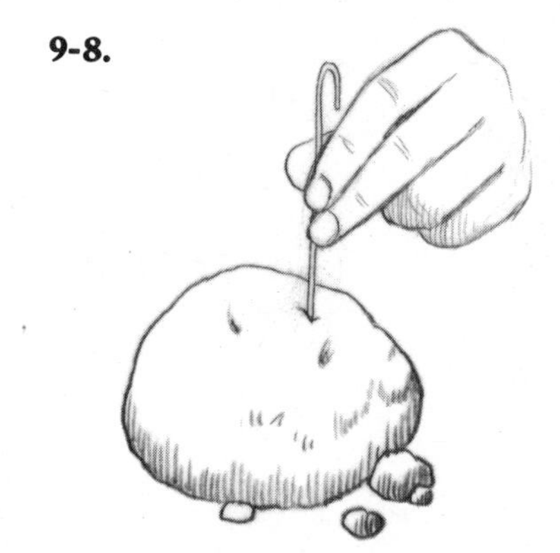

SUMMARY

How do scientists know that an atom is mostly empty space? You can show that a wire fence is mostly empty space by throwing pebbles through it. In a similar way, scientists shot atomic particles at atoms. Most of the particles went right through. Some particles, however, collided with the nucleus. The nucleus is the central core of an atom, and it is made up of protons and neutrons. The atomic number of an element gives the number of protons in an atom of that element. The atomic number also tells you the number of electrons in an atom of an element.

QUESTIONS

Unless otherwise indicated, use complete sentences to write your answers.

1. Describe where electrons, protons, and neutrons are found in atoms.
2. How is an atom like a wire fence?
3. If you know the atomic number of an atom, what particles in the atom will you know the number of also?
4. All atoms contain charged particles. In view of this, how can an atom be neutral?
5. Describe Rutherford's experiment.
6. Copy and complete the following chart.

Element	*Atomic Number*	*Number of Protons*	*Number of Electrons*
Hydrogen			
Helium			
Carbon			
Neon			
Aluminum			

9-3. ELECTRON SHELLS

Suppose that the earth could be squeezed together until the nucleus of every atom was jammed against its neighbor. The entire planet would then be reduced to a globe only about one km in diameter. The earth, like all other forms of matter, is made up of atoms that are filled with empty space. This is the modern scientific model of the atom: atoms consist of electrons whirling around a tiny central nucleus. This lesson deals with the way in which electrons orbit the nuclei of atoms.

When you finish lesson 3, you will be able to:

- Describe the way in which electrons move around an atomic nucleus.
- List the arrangement of the electrons in different kinds of atoms.
- Identify chemical symbols for some common elements.

In some ways, the modern atomic model is like the solar system. The solar system has the sun at its center. An atom has a nucleus at its center. The force of gravity acting between the sun and the planets causes each planet to follow a particular orbit around the sun. Electrical force acting between the negatively charged electrons and the positively charged nucleus causes the electrons to move around the nucleus. But an atom is different from a solar system in one important way. In a solar system, planets can follow many different orbits around the central body. In a million solar systems, there could be a million different arrangements of the planets. This is not true for electrons moving around an atomic nucleus. For example, each atom of gold (atomic number = 79) has 79

9-9. *Niels Bohr (1885–1962) was a young scientist who constructed a model of the atom in 1913.*

electrons around its nucleus. All atoms of gold are alike. All gold atoms would not be alike if the 79 electrons in each atom were arranged in a different way from other gold atoms.

In 1913, Niels Bohr, a Danish scientist who worked in Rutherford's laboratory, provided a theory to explain how electrons move around atomic nuclei. See Fig. 9-9. According to Bohr's theory, electrons moving around a particular kind of atomic nucleus are able to follow only certain orbits. For example, the electrons in two different gold atoms are found to follow orbits that are identical in both atoms. Each electron moves in an orbit that is a definite distance from the nucleus.

This model is a common one used to indicate the number of electrons, protons, and neutrons found in the atoms. Niels Bohr first used this model with circles to represent the paths of the electrons. It is usually called the Bohr model of the atom. Scientists have replaced this model with a modern version based on the mathematics of probability and wave motion. The modern version explains much of the information that exists today concerning the chemical behavior of atoms.

Electron shell
A region around an atomic nucleus in which electrons move.

If you could see an atom, it might look like a fuzzy sphere. See Fig. 9-10. The fuzziness represents a cloud of electrons whirling around the nucleus at high speed. Each electron makes billions of trips around the nucleus in 1 sec. But these electrons do not buzz around the nucleus like a swarm of bees. Each electron must follow an orbit within an **electron shell.** An *electron shell* is a definite orbit that electrons follow as they move around a nucleus. Within a shell, electrons may move in all directions. See Fig. 9-11.

9-10. *Electrons move so fast that they can be thought of as forming a cloud around an atomic nucleus.*

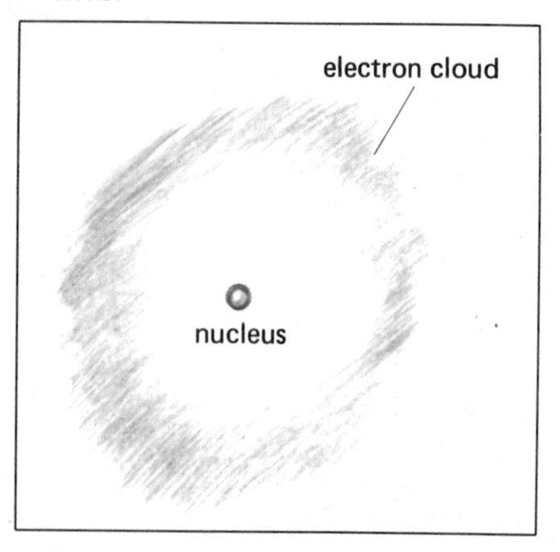

Each electron shell can only hold a certain number of electrons. The one electron in a hydrogen atom moves around the nucleus in the first shell. The two electrons in helium also move in the first shell. Two electrons are the limit for this shell. The next electron shell can only hold 8 electrons. Figure 9-12 shows the electron arrangement of the first 10 elements. This figure shows the electrons in circles where each circle represents an electron shell. Keep in mind that the shells are not flat as shown in the drawings. Table 9-3 shows all the electron shells of the first 20 elements.

There are a total of 7 electron shells. The smallest number of electrons a shell can hold is 2 in the first shell. Eighteen to 32 electrons are found in some of the higher shells.

TABLE 9-3
Electron Shells for the First 20 Elements

Atomic Number	*Element*	*Shells* 1	2	3	4
1	Hydrogen	1			
2	Helium	2			
3	Lithium	2	1		
4	Beryllium	2	2		
5	Boron	2	3		
6	Carbon	2	4		
7	Nitrogen	2	5		
8	Oxygen	2	6		
9	Fluorine	2	7		
10	Neon	2	8		
11	Sodium	2	8	1	
12	Magnesium	2	8	2	
13	Aluminum	2	8	3	
14	Silicon	2	8	4	
15	Phosphorus	2	8	5	
16	Sulfur	2	8	6	
17	Chlorine	2	8	7	
18	Argon	2	8	8	
19	Potassium	2	8	8	1
20	Calcium	2	8	8	2

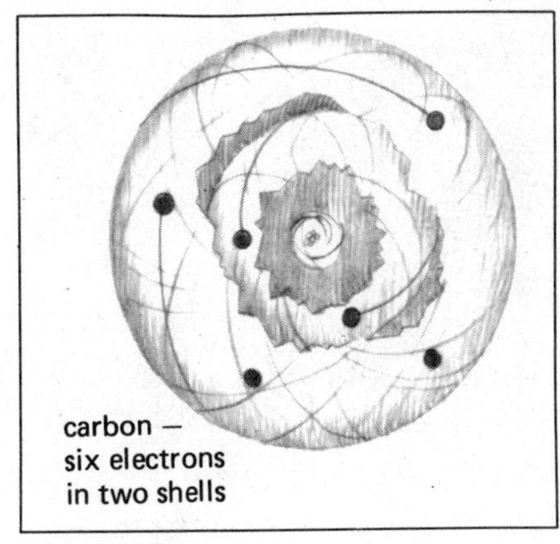

9-11. *Electron shells are made up of electrons moving at a certain average distance from the nucleus.*

The number of electrons in the electron cloud around the nucleus differs in each element. To describe a particular atom you must know its atomic number. The atomic number tells how many protons are in the nucleus and also how many electrons are moving around the nucleus. The electrons fill the shells in which they are able to move. The first shell closest to the nucleus is filled first, then the second shell is filled, and so on until the total number of electrons is used up. For example, a sodium atom with atomic number 11 has 11 electrons. Two of these electrons are found in the first shell, 8 more are in the second shell, and the remaining one is found in the third shell.

Up until now, the name of each chemical element has been written out (hydrogen, helium, and so forth). For many years, chemists had to do the same thing. The chemist Berzelius knew that **symbols** were used

Symbol
One or two letters used to represent an atom of a particular element.

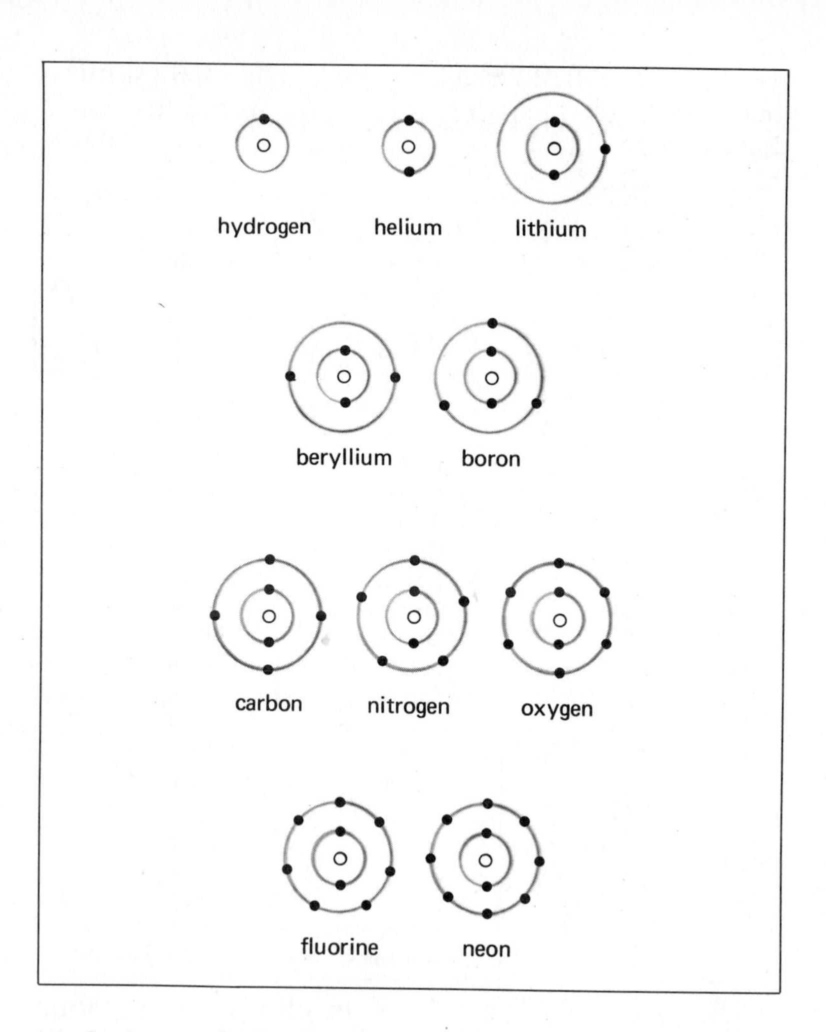

9-12. *The electron arrangements of the first ten elements.*

in algebra and physics. A *symbol* is a form of shorthand. Berzelius suggested that scientists also use symbols for the chemical elements. These symbols consist of one or two letters of the element's name. The symbol for hydrogen is H. The symbol for helium is He. When there are two letters in a symbol, the first letter is capitalized, the second letter is not.

The symbol for mercury is Hg. Does this surprise you? H and g are not part of the word "mercury." The answer is simple. Different languages have different words for mercury. The scientists decided to use the Latin name of the element. "Mercury" in Latin is *hydrargyrum*. Table 9-4 is a list of some common elements and their symbols.

Try to remember the symbols for these elements. Being able to use the symbols of the elements will make your work in science easier.

Table 9-4
Symbols of Some Common Elements*

Name	*Symbol*
Hydrogen	H
Boron	B
Carbon	C
Nitrogen	N
Oxygen	O
Fluorine	F
Phosphorus	P
Sulfur	S
Iodine	I
Helium	He
Lithium	Li
Beryllium	Be
Aluminum	Al
Silicon	Si
Calcium	Ca
Cobalt	Co
Nickel	Ni
Germanium	Ge
Selenium	Se
Bromine	Br
Barium	Ba
Magnesium	Mg
Chlorine	Cl
Zinc	Zn
Sodium *(natrium)*	Na
Potassium *(kalium)*	K
Copper *(cuprum)*	Cu
Gold *(aurum)*	Au
Silver *(argentum)*	Ag
Mercury *(hydrargyrum)*	Hg
Iron *(ferrum)*	Fe
Lead *(plumbum)*	Pb
Tin *(stannum)*	Sn

*The first section lists elements with one-letter symbols. The second lists elements with two-letter symbols. The third section lists elements whose symbols are taken from the Latin name of the element. The Latin name is in parentheses.

ACTIVITY

The Symbols of the Elements

The following puzzle will help you review the symbols and the names of some of the elements mentioned in this lesson. Each clue gives the symbol for an element or an idea of how the element was originally named. Refer to Table 9-4 to find the correct name of the element. Copy the puzzle. DO NOT WRITE IN THIS BOOK.

Materials
paper
pencil

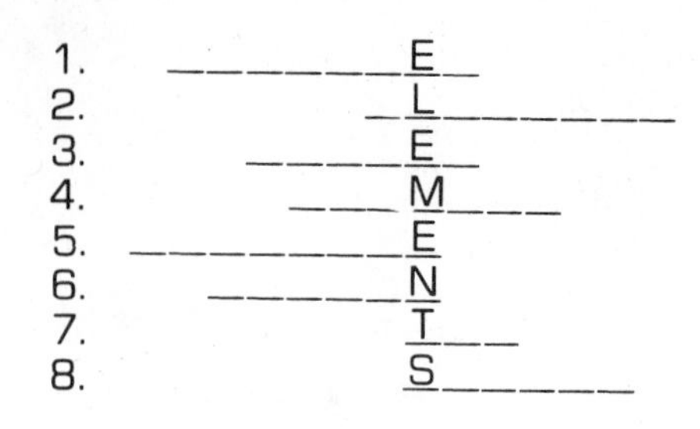

Clues

1. Burns to produce *hydros*, the Greek word for water.
2. Symbol of this metal is Al.
3. A gas found in the air we breathe.
4. Named from the Greek words *bromos*, which means a bad odor.
5. A gas used to purify water, symbol Cl.
6. Symbol is C, the element is life.
7. Another name is *stannum*.
8. Another name for this element is *natrium*.

SUMMARY

An atom can be compared to a miniature solar system. The arrangement of electrons around the nucleus is similar to the way the planets move around the sun. However, unlike the planets, electrons are allowed to follow only certain orbits in electron shells around the nucleus. Each shell can hold only a certain number of electrons. Each kind of atom has its own electron arrangement and can be described by a symbol.

QUESTIONS

Use complete sentences to write your answers.

1. Describe Niels Bohr's model of the atom.

2. Magnesium normally has 12 electrons arranged in 3 shells. What is the number of electrons in its outermost shell?

3. How many electrons can be in the first shell of an atom? How many in the second shell?

4. Name five elements and give their chemical symbols and the number of electrons in each of their shells.

9-4. ATOMIC MASS

An athletic event that takes a great deal of muscle power is the shot put. The athlete tries to throw the shot, which is a heavy iron ball, as far as possible. Suppose that a shot-putter won by using a smaller and lighter shot than the others in the contest. Would this be fair? You know that the person using the shot with the smallest mass would probably win.

Experiments always show that the mass of an object has an effect on how far it can be thrown. Scientists use this simple principle to investigate the mass of atoms. Groups of atoms can be "thrown." Measuring where each kind of atom lands gives a clue about its mass. Discoveries about atomic mass are the subject of this lesson.

When you finish lesson 4, you will be able to:

- Explain how the mass of an atom is determined.
- Find the number of neutrons in a particular atom, using the *mass number* and the atomic number.
- Explain what is meant by an *isotope*.
- ○ Determine the mass number and the number of neutrons for several atoms and draw diagrams of these atoms.

All atoms are made up of electrons, protons, and neutrons. The electrons are found outside the nucleus. The protons and neutrons are found inside the nucleus. Except for the nucleus, an atom is mostly empty space. Most of the mass of an atom, is thus determined by the protons and neutrons in the nucleus. The mass of all the electrons in an atom is so small that it can be ignored.

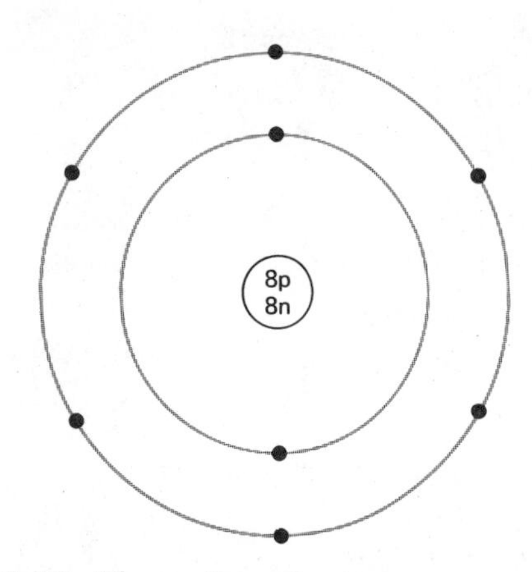

9-13. *The nucleus of an oxygen atom contains 8 protons and 8 neutrons. The diagrams in this lesson represent the Bohr model of the atom with the electrons circling the nucleus.*

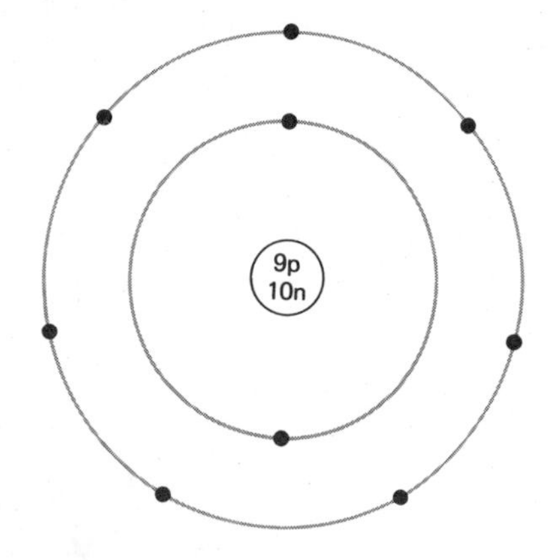

9-14. *This diagram represents a fluorine atom. Can you explain why fluorine has a mass of 19?*

Isotopes
Atoms whose nuclei contain the same number of protons but a different number of neutrons.

Atomic mass
The average of all the masses for the isotopes of a particular element.

The mass of an atom depends on the number of protons and neutrons in the nucleus. Protons and neutrons each have a mass of 1 amu. Oxygen has 8 protons and 8 neutrons in its nucleus. See Fig. 9-13. The mass of oxygen is 16 amu. The mass of an atom, in atomic mass units, is equal to the number of protons and neutrons in its nucleus.

If you know the number of protons and neutrons in a nucleus, you know the mass of the atom. You can also use the mass to find the number of neutrons in the nucleus. For example, you know that the atomic number of oxygen is 8. The atomic number tells you the number of protons. The mass of oxygen is 16 amu. The mass is the sum of the protons and neutrons. You can easily figure out the number of neutrons in an oxygen nucleus: mass (protons + neutrons) − atomic number (protons) = neutrons

16 − 8 = 8

Figure 9-14 shows an atom of fluorine. The atomic number of fluorine is 9 and its mass is 19. Fluorine has 10 neutrons.

Scientists have discovered that not all atoms of an element have the same number of neutrons. A hydrogen atom, for example, may have any one of three arrangements in its nucleus. Most hydrogen atoms have only one proton in the nucleus and no neutrons, as shown in Fig. 9-15. Some hydrogen atoms have one proton and one neutron, as shown in Fig. 9-16. A very few hydrogen atoms have one proton and two neutrons in the nucleus, as shown in Fig. 9-17.

The element known as hydrogen is made up of atoms with all three types of nuclei. Each type produces an atom which is an **isotope** of hydrogen. Hydrogen has three *isotopes*: hydrogen-1, hydrogen-2, and hydrogen-3. Because they have different numbers of neutrons, isotopes have different masses. See Fig. 9-18.

Almost all elements have several isotopes like hydrogen. For example, lithium exists as lithium-6 or lithium-7. The **atomic mass** for a particular element is the average of the masses of all the isotopes of that element. For example, the *atomic mass* of lithium would be given as 6.94 amu. This is the average mass of a mixture of the isotope of lithium with a mass of 6 amu and the isotope with a mass of 7 amu.

The atomic mass given for the mixture of lithium-6 and lithium-7 found in nature is 6.94. This is due to the fact that 93 percent of all lithium atoms have a mass of 7 while only 7 percent have a mass of 6. The atomic mass of the mixture of hydrogen atoms found in nature is given as 1.008. Often, the atomic mass of an element is rounded off to the nearest whole number. This is called the **mass number** of the element. The *mass number* of lithium atoms is 7. The mass number of an atom is always equal to the sum of its protons and neutrons. For lithium, the mass number 7 = 3 protons + 4 neutrons.

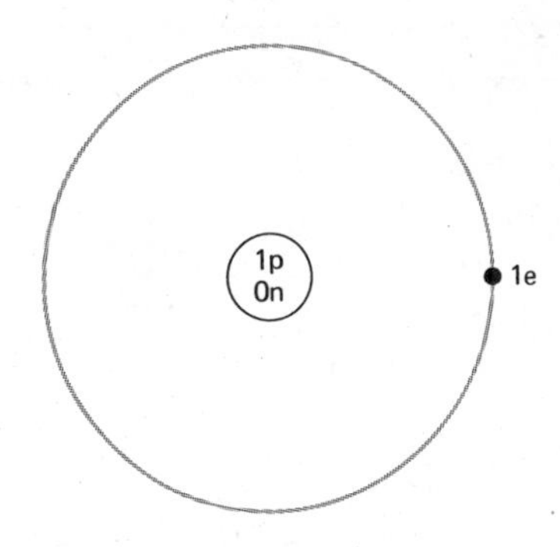

9-15. *An atom of hydrogen-1.*

Mass number
The sum of the protons and neutrons in the nucleus of a particular kind of atom.

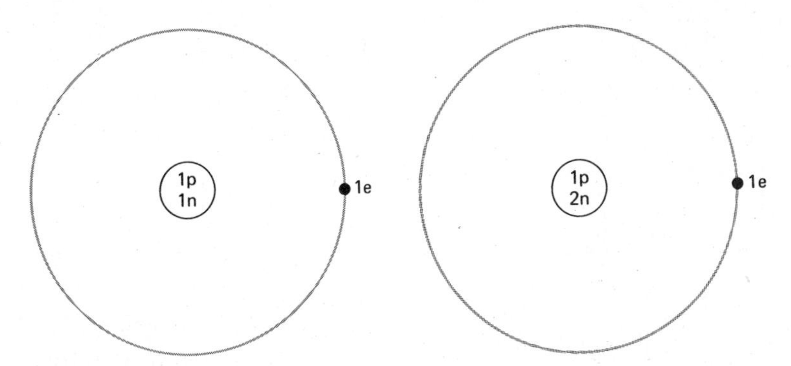

9-16. *(left) An atom of hydrogen-2.*

9-17. *(right) An atom of hydrogen-3.*

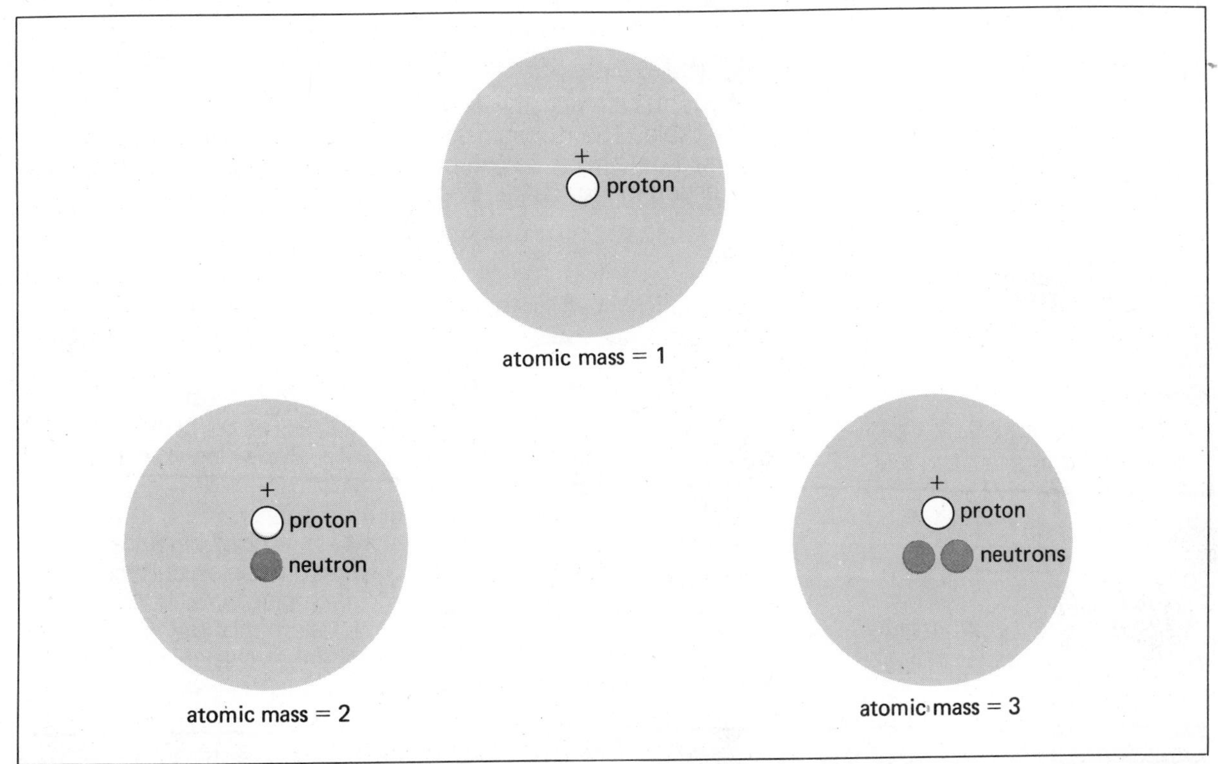

9-18. *The three isotopes of hydrogen.*

ACTIVITY

Materials
paper
pencil
compass

Diagrams of Atoms

A. Obtain the materials listed in the margin.

B. Copy the following table in your notebook. Show the number of protons and neutrons in the nucleus. Show the number of electrons in their proper shells around the nucleus.

Atomic Number	*Element*	*Mass Number*	*Number of Neutrons*
1	hydrogen	1	___
2	helium	___	2
3	lithium	___	4
4	beryllium	9	___
5	boron	11	___
6	carbon	12	___
7	nitrogen	___	7
8	oxygen	___	8
9	fluorine	19	___
10	neon	20	___

C. Based on the information given, fill in the blanks in the table.

D. Draw a diagram of each of the 10 elements listed in the table. The diagrams should be similar to the one of argon in Fig. 9-19. Show

E. Label each diagram with the name of the element and its mass number. For example: hydrogen-1.

F. In the same manner as in step D, draw a diagram of each of the following atoms: magnesium-24, atomic number 12; aluminum-27, atomic number 13; sulfur-32, atomic number 16; chlorine-35, atomic number 17; potassium-39, atomic number 19; calcium-40, atomic number 20. Refer to Table 9-3 for information concerning electron shells.

G. In a brief paragraph, describe how you would make diagrams of the first 10 elements using the Bohr model of the atom.

9-19. *A diagram of an argon-40 atom.*

SUMMARY

What could you find out if you could "throw" atoms? An experiment like this could give information about the mass of the atoms. Whatever mass an atom has comes from protons and neutrons in the nucleus. Knowing that the mass must be the sum of protons and neutrons in a particular atom. The actual number of neutrons in all the atoms of an element may not be the same.

QUESTIONS

Unless otherwise indicated, use complete sentences to write your answers.

1. Which atomic particles account for most of the mass of an atom?
2. The atomic number of oxygen is 8. If its mass number is 16, how many neutrons are in the atom? Show how you got your answer.
3. Explain what is meant by an isotope of an atom.
4. Describe two simple atoms by stating their names and giving the number of each kind of atomic particle present in each of the atoms.
5. Copy and complete the following table.

Element	*Isotope (mass number)*	*Atomic Number*	*Number of Protons*	*Number of Neutrons*	*Number of Electrons*
Lithium	7				
Boron				6	
Carbon	12				
Neon				10	

VOCABULARY REVIEW

Match the number of the word(s) with the letter of the phrase that best explains it.

1. atomic particles
2. atomic mass unit
3. nucleus
4. atomic number
5. electron shell
6. chemical symbol
7. atomic mass
8. isotopes

a. A region around an atomic nucleus in which electrons move.
b. The building blocks of atoms.
c. The central core of an atom containing its protons and neutrons.
d. The number of protons in the nucleus of an atom.
e. One or two letters used to represent an atom of a particular element.
f. Atoms whose nuclei contain the same number of protons but a different number of neutrons.
g. Unit used to express the masses of atomic particles and atoms.
h. The sum expressed in amu of the protons and neutrons in an atom.

REVIEW QUESTIONS

Complete each statement by choosing the best word or phrase, or by filling in the blank.

1. A mental picture of something that cannot be seen describes **a.** an optical illusion **b.** a scientific model **c.** an extrasensory illusion **d.** a physical model.
2. The negative atomic particle is called an __________; the positive atomic particle is called a __________; the neutral atomic particle is called a __________.
3. The two atomic particles found in the nucleus are the __________ and __________.
4. If an atom were the size of a football field, the nucleus of this atom would be about the size of a **a.** baseball **b.** basketball **c.** flea **d.** bird.
5. The normal atom whose atomic number is 5 has __________ electrons and __________ protons.
6. In an atom, the electrons are found **a.** in shells around the nucleus **b.** in shells within the nucleus **c.** in clouds within the nucleus **d.** everywhere in the atom.
7. The number of electrons normally found in the first three electron

shells in atoms is **a.** 1, 2, 3 **b.** 2, 4, 8 **c.** 2, 8, 16 **d.** 2, 8, 8.

8. The chemical symbols for hydrogen, helium, and oxygen are **a.** Hg, H, and O **b.** H, He, and O **c.** H, He, and Ox **d.** Hy, He, and Ox.
9. The atomic number of lithium is 3. The atomic mass of a lithium atom is 7. The number of protons and neutrons must be **a.** 4 and 10 **b.** 3 and 10 **c.** 3 and 4 **d.** 7 and 10.
10. Atoms of the same element can have different atomic masses because they can have different numbers of **a.** neutrons **b.** protons **c.** electrons **d.** protons and electrons.

REVIEW EXERCISES

Give complete but brief answers to each of the following. Use complete sentences to write your answers.

1. Give an example of a scientific model and explain how it is like the real thing.
2. Name three kinds of atomic particles and give the properties of each.
3. Describe Rutherford's experiment and the model of the atom that resulted from his data.
4. Explain what determines the mass of an atom.
5. The atomic number of an element is 5. How many electrons are in the normal atom? Explain your answer.
6. What is an electron shell? How are electrons arranged in the first two shells?
7. In the manner shown in Fig. 9-12 and using information from Table 9-3, draw the electron arrangement for the atoms of elements 11 through 18.
8. Name five of the first 10 elements and give the chemical symbols for each.
9. Describe the atomic structure of the three isotopes of hydrogen.
10. How are mass number and atomic number related to the number of protons, neutrons, and electrons in an atom?

EXTENSIONS

1. Ozone, O_3, is a molecule made of oxygen atoms. Gather information on the importance of ozone in the atmosphere. Include both the good and the bad effects.
2. There are many important isotopes. Carbon-14 and hydrogen-2 (deuterium) are among them. Write a report on at least five important isotopes, including their uses.
3. Write a report on some of the current research being done on subatomic particles.

CHAPTER 10

Kinds of Atoms

10-1. IONIZATION

Many city streets glow at night from the light of electric signs. One commonly seen type of sign is made of thin glass tubes filled with a gas. When an electric current is passed through the gas, colored light is given off. The atoms in the glowing gas produce this light. How can atoms change electric energy into light? You can answer this question with the help of a model of an atom. In this lesson, you will see how an atom can change one form of energy into another form.

When you finish lesson 1, you will be able to:

- Describe what happens inside an atom when it absorbs and then releases energy.
- Explain how an atom can give off energy.

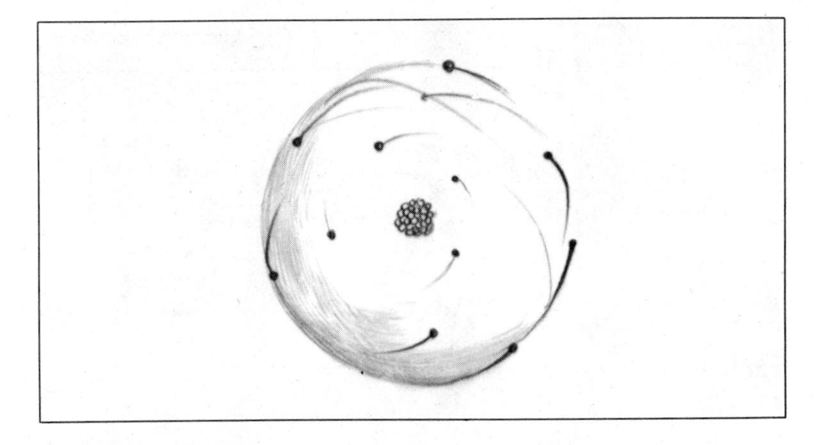

10-1. *An electron in a shell that is far from the nucleus has more energy than an electron close to the nucleus.*

● Distinguish between an atom and an *ion*.

○ Observe *fluorescent* light through a diffraction grating.

Ordinarily, electrons can move endlessly around the nucleus in an atom without gaining or losing energy. What happens if the atom is supplied with energy in some form from outside the atom? An electric current, for example, can add energy to an atom. The extra energy causes an electron in the atom to move farther away from the nucleus. What makes this happen? The attractive force between the positive nucleus and the negative electron has to be overcome. The extra energy does this. It overcomes the attraction between the nucleus and the electron. This allows one or more electrons to jump into a shell farther from the nucleus. It is usually the outer electrons of an atom that make these jumps. Adding energy to an electron,

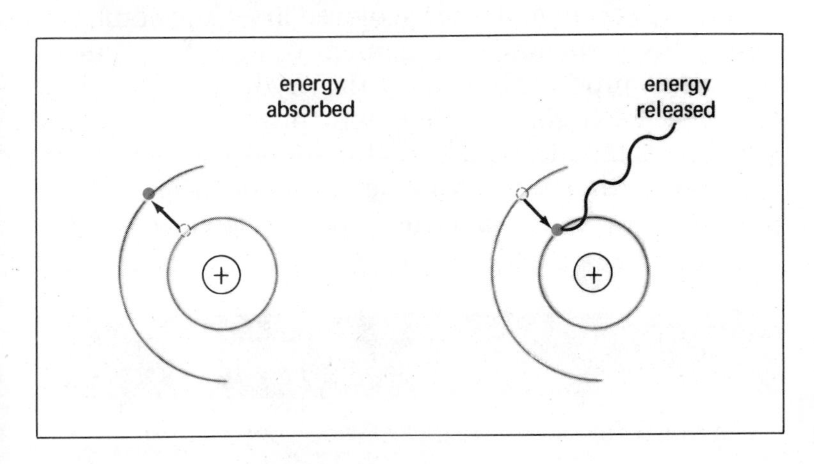

10-2. *When an atom absorbs energy (left), an electron jumps to a higher energy level. Later, the electron falls back to its original position and energy is released.*

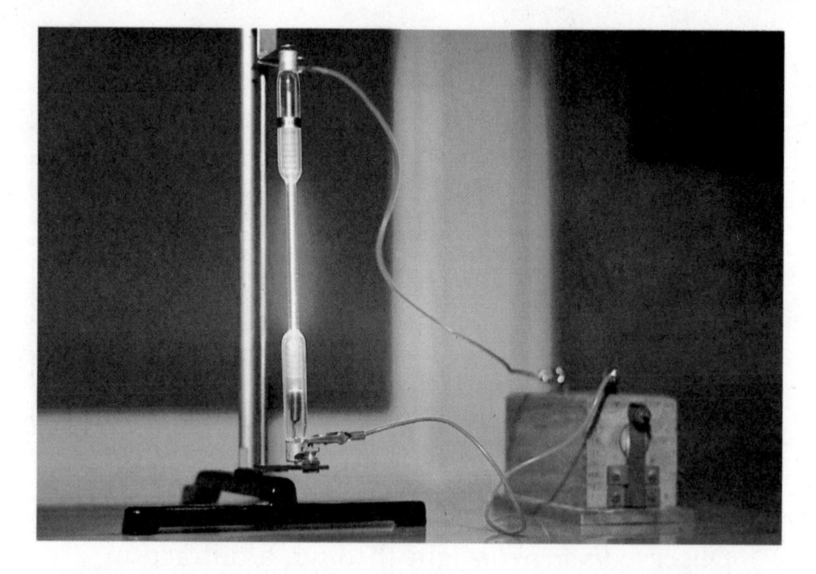

10-3. *In a spectrum tube, atoms of an element give off light of a particular color.*

Unstable electron
An electron that has absorbed energy and moved farther away from the atomic nucleus.

causing it to move away from the nucleus, makes that electron **unstable.** An *unstable* electron will fall back into its original position or another position closer to the nucleus. See Fig. 10-1. This happens because the positive charge on the nucleus always causes electrons to be pulled toward it. When an unstable electron moves back toward the nucleus, its extra energy is released. See Fig. 10-2. This energy is generally given off in the form of light. The energy may also be in the form of other parts of the electromagnetic spectrum, such as infrared or ultraviolet waves. If electricity is passed through a gas inside a glass tube, visible light is given off. See Fig. 10-3. The light is made up of certain frequencies or colors. If the light is passed through a special instrument, each separate color or frequency can be seen as a single bright line. For example, hydrogen produces several violet-colored lines and some red lines. This particular arrangement of light frequencies is always produced by glowing hydrogen gas. The atoms of each chemical element produce a different and characteristic kind of light, called the spectrum, for that element. A particular kind of atom always produces the same spectrum when it gives off light.

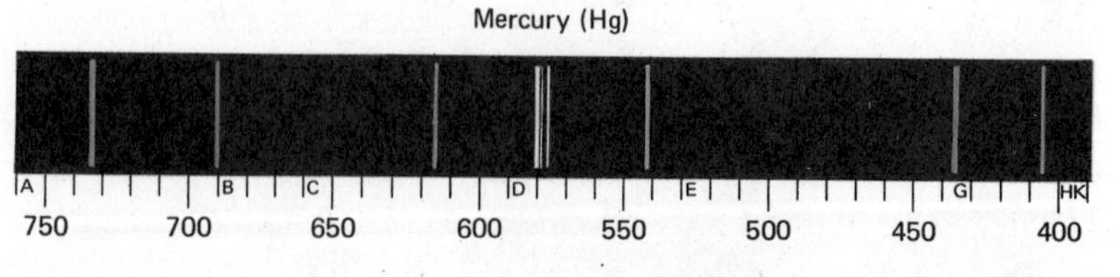

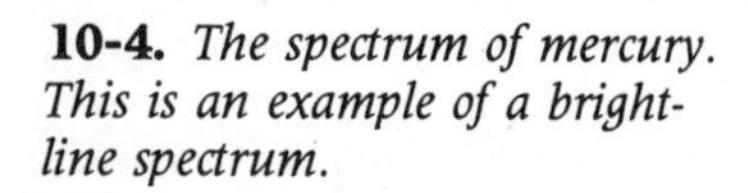

10-4. *The spectrum of mercury. This is an example of a bright-line spectrum.*

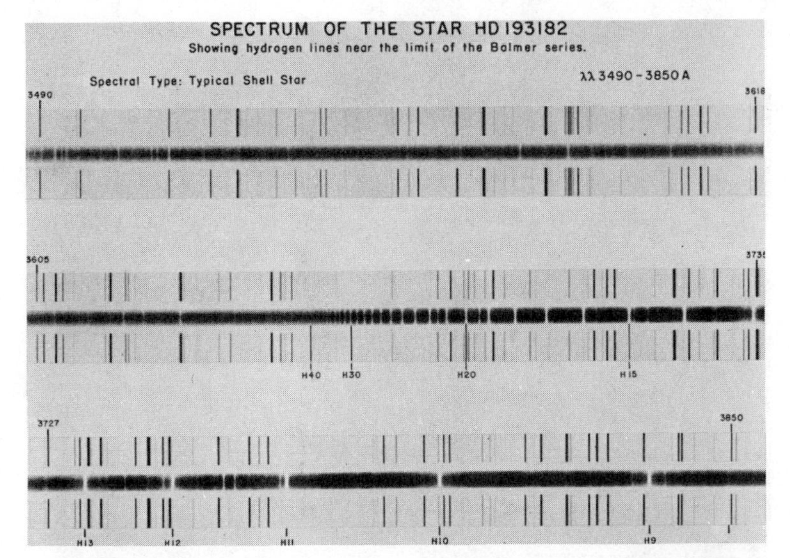

10-5. *Astronomers can learn about the composition of a star by studying the spectrum of that star. This type of spectrum is called a dark-line spectrum.*

See Fig. 10-4. By observing its spectrum, astronomers can tell what kind of atoms are present in a distant star. See Fig. 10-5. Other conditions in the star are also revealed by its total spectrum.

There are several different kinds of spectra. A continuous spectrum is one containing all the colors of the rainbow. A rainbow is the most common example of this spectrum. A bright-line spectrum is seen when the electrons in an atom are excited and moved to higher energy levels. Then, as the electrons move back to their normal energies, light that is characteristic of the particular kind of atom is given off. If light of all colors travels through a cool gas, the gas will absorb the colors that are characteristic of the electron energy levels in the atoms of the gas. In this manner, a dark-line spectrum is created. This is the type of spectrum that astronomers study.

What happens to an atom that has received a very large amount of energy? A very large amount of energy can cause some electrons to move far away from the nucleus. The energy may be enough to cause the negative electrons to overcome completely the attraction of the positive nucleus. One or more electrons may completely escape from the atom. When that happens, the number of electrons will no longer equal the number of protons in the atom. Normally, there are the same number of electrons moving around the nucleus as there are protons in the nucleus. This equal number of positive and negative charges

10-6. *The famous* cyclotron *at the University of California produced this beam of accelerated particles in 1939. The cyclotron was dismantled in 1962 because it was obsolete.*

makes an atom electrically neutral. The charges on the electrons and protons cancel each other.

If the atom loses any electrons, it is left with more protons than electrons. An atom with extra protons is a positively charged **ion.** An *ion* is an atom or molecule that has an electric charge. An atom becomes a positive ion by losing one or more of its electrons. It is also possible for an atom to gain electrons instead of losing them. An atom that gains electrons will have more electrons than protons. The excess electrons will give the atom a negative charge. An atom that gains one or more electrons becomes a negatively charged ion.

Ion
An atom or molecule with an electric charge.

One way to produce ions is to use a machine that produces a beam of moving electrons. See Fig. 10-6. The beam of electrons bombards a group of atoms. The electrons in the beam collide with electrons in the outer shells of the target atoms. These collisions cause electrons to be knocked away from the atom. This loss of electrons changes the neutral atoms into positively charged ions. Ions are also produced in the interiors of stars. The high temperatures that exist there provide enough energy to completely ionize the atoms.

Scientists can measure how hard an electron has to strike an atom in order to change it into an ion. Then they can determine how much energy is needed to remove an electron from the outer shell of an atom.

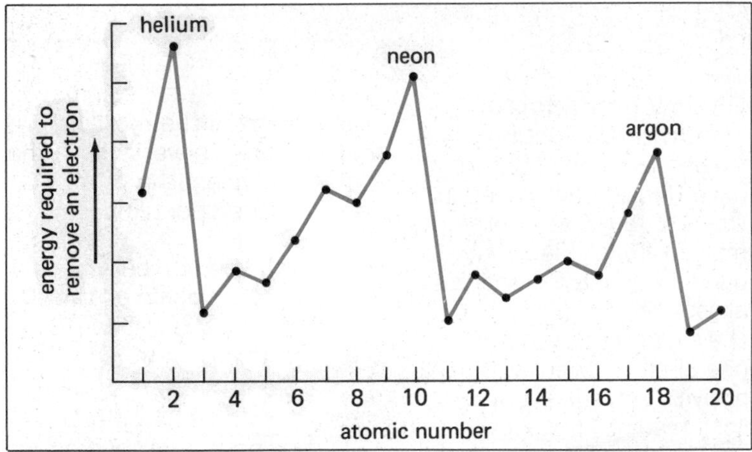

10-7. *This graph shows the energy needed to remove an electron from an atom of the first 20 elements.*

These energies, ranging from hydrogen (atomic number 1) to calcium (atomic number 20), form a pattern. This pattern is shown in the graph in Fig. 10-7. The graph shows that the atoms of the elements helium, neon, and argon require the highest energies to remove electrons.

Helium, neon, and argon atoms have a tighter hold on their electrons than other atoms. Helium (atomic number 2) has a total of 2 electrons. Neon has 10 electrons and argon has 18 electrons. Is there something special about the number of electrons in an atom? Remember how many electrons fill each shell around the nucleus of an atom? The first electron shell is full with 2 electrons. The second shell fills with 8 electrons. Thus helium has all of its electrons in one completely filled shell. Neon, with 10 electrons, has filled the first and second shells (2 + 8). Argon, with 18 electrons, has its first, second, and third shells filled (2 + 8 + 8). See Fig. 10-8.

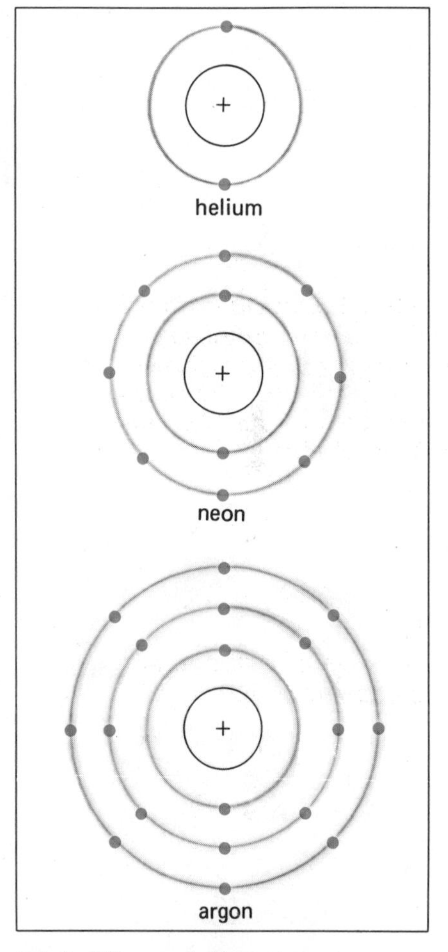

10-8. *Why are helium, neon, and argon called noble gases? How are they different from the other elements?*

Atoms that have completely filled electron shells do not easily lose electrons. The electron arrangement is **stable.** A *stable* electron arrangement means that an atom does not tend to gain or lose electrons since all its electron shells are completely filled. The elements with stable atoms are gases. These gases are called the **noble gases.** Experiments have shown that there are a total of six *noble gases* among all the elements. In addition to helium, neon, and argon, other noble gases are krypton, xenon, and radon. Krypton has 36 electrons, xenon has 54 electrons, and radon has 86 electrons. With these numbers of electrons, the electron shells of krypton, xenon, and radon are completely filled.

Stable electron arrangement
An arrangement in which all of the electron shells are filled.

Noble gases
The six elements whose atoms have completely filled electron shells.

ACTIVITY

Material
diffraction grating

Observing Spectra

The overhead lighting fixtures in many schools have fluorescent tubes in them. The tubes contain mercury gas. Electricity is used to cause the mercury atoms to give off light when their electrons jump from one shell to another. In this activity you will study the light from a fluorescent bulb. You need the diffraction grating you used earlier in this course to break up light into its spectrum. To review this procedure, see chapter 4, page 132.

A. Obtain the material listed in the margin.

B. Look through the diffraction grating at one end of a fluorescent light fixture. Look away from any windows or outside source of light.

1. Do you see a continuous spectrum?

A continuous spectrum is caused by the special coating inside the fluorescent tube. This coating absorbs the light coming from the mercury atoms and converts it to the many wavelengths of a continuous spectrum.

C. With the diffraction grating, look again at the fluorescent tube. Observe the spectrum more carefully.

2. Do any colors appear brighter than the others?

Many people will see a brighter bar or line in the violet or blue region, one in the green region, and one in the yellow region.

D. Look at the color spectrum of mercury shown in Fig. 10-4. This spectrum shows the separate colors given off by mercury gas atoms. When electrons of the mercury atoms absorb energy from the electric current, they move farther from the nucleus to an unstable position.

3. Do the electrons remain in this unstable position?

4. What happens when the electrons move back into their original position?

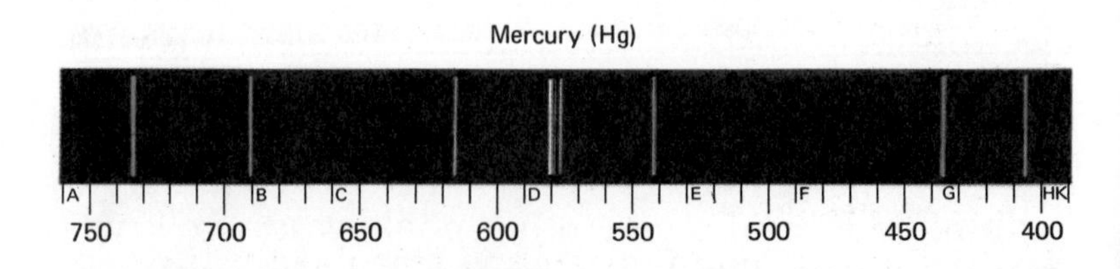

SUMMARY

How can an atom be made to give off light energy? Electric energy must be added to the atom. One or more electrons absorb the energy and jump to new positions. A fraction of a second later, the energy, now in the form of light, is released as the electron falls back toward the nucleus. Electrons can absorb enough energy to escape the atom completely. Changing the number of electrons changes the atom into an electrically charged ion. The noble gases are made up of atoms with a stable electron arrangement.

QUESTIONS

Unless otherwise indicated, use complete sentences to write your answers.

1. The following questions refer to what happens when mercury atoms in a fluorescent tube absorb electric energy.
   a. What do the electrons in the mercury atom do when they absorb electric energy?
   b. When the electrons of the mercury atoms absorb electric energy, they become ______.
   c. What do the electrons do when they give up the energy they absorbed?
   d. The energy that the electrons absorbed is given off as ______.

2. What must happen to make a mercury atom a *positive* ion?

3. What must happen to make a mercury atom a *negative* ion?

4. How are astronomers able to learn about the composition of stars by observing their spectra?

5. In your own words describe what is meant by a noble gas and name three such gases.

10-2. CHEMICAL ACTIVITY

The *Hindenburg* was a famous German passenger airship. In 1937, while attempting to land in New Jersey after crossing the Atlantic, it exploded and crashed in flames. What could have caused this disaster? The *Hindenburg* was filled with hydrogen gas that exploded and burned. Modern airships like the Goodyear blimp are filled with helium. In this lesson, you will see why helium is a safer gas than hydrogen.

When you finish lesson 2, you wil be able to:

- Relate the *chemical activity* of an element to the number of electrons in the outer shell of an atom of that element.
- Explain why certain elements can be grouped together.
- ○ Classify nine elements into groups on the basis of their characteristics.

In order to stay in the air, an airship must be filled with a gas that is lighter than air. The airship then floats in the air in the same way that a piece of wood floats in water. Hydrogen gas is lighter than air. But hydrogen is a dangerous gas. It can burn when mixed with oxygen. A much safer gas for airships is helium. Helium is also much lighter than air. Helium does not burn because it does not combine with oxygen. In fact, helium will not take part in any chemical changes. Helium atoms exist separately. They are never part of a molecule. Why is it that helium does not take part in chemical changes while hydrogen does?

Hydrogen and helium are the only two elements that have only a single electron shell. See Fig. 10-9. This shell can hold only two electrons. Helium already

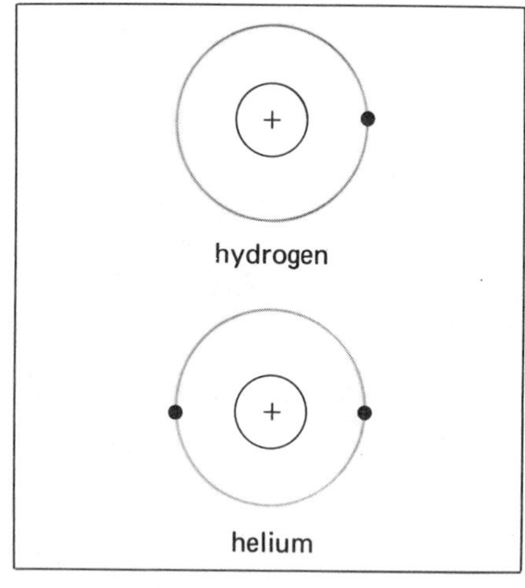

10-9. *Both hydrogen (H) and helium (He) have only one electron shell. What is the difference between H and He?*

has two electrons. Therefore, helium has a completely filled shell. It does not join with other atoms to form molecules. Hydrogen has only one electron. It needs one more electron to fill the shell. Therefore, unlike helium, hydrogen reacts with many other atoms to form molecules. The way an atom reacts with atoms of other elements is called its **chemical activity.** Hydrogen, which reacts readily with other elements, is *chemically active.* Helium, which does not react with other elements, is not chemically active. It is **inert.** All of the noble gases are *inert*.

Chemical activity
Describes the way in which an atom reacts with other kinds of atoms.

Inert
A description of an atom that does not react with other atoms.

Is it possible that the two elements hydrogen and helium behave so differently because of the different number of electrons in their outer shells? If this is true, then atoms with the same number of outer electrons should be alike. In the previous lesson, you learned that atoms with a stable number of electrons do not easily join with other atoms. If an atom has 1, 2, or 3 electrons *less* than a stable number of electrons, it will tend to *add* electrons until a stable number is reached. If an atom has 1, 2, or 3 electrons *more* than a stable number of electrons, it will tend to *lose* electrons until a stable number is reached. For example, think about a lithium atom. The atomic number of lithium is 3. A lithium atom has 3 electrons. There are 2 electrons in the first shell and one electron in the second shell. See Fig. 10-10. Thus lithium has one *more* electron than the stable number of two. Lithium will tend to *lose* that one electron to reach the stable number of two. Table 10-1 lists five other atoms that can also be expected to lose one electron. This group of atoms is called the **alkali (al-kuh-lie) metals.**

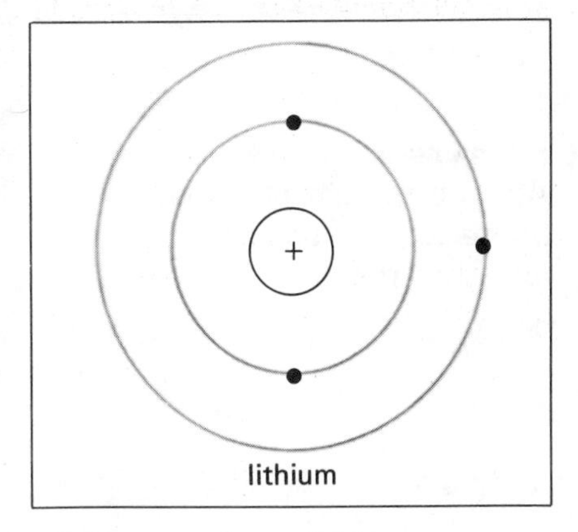

10-10. *How does a lithium atom differ from a helium atom?*

Alkali metals
A group of elements whose atoms all have one electron more than the stable number.

Now think about a fluorine atom. The atomic number of fluorine is 9. Fluorine also has 2 electrons in its

TABLE 10-1
The Alkali Metals

Atoms that lose one electron easily	*Atomic Number*	*Stable electron number after losing one electron*
Lithium	3	2
Sodium	11	10
Potassium	19	18
Rubidium	37	36
Cesium	55	54
Francium	87	86

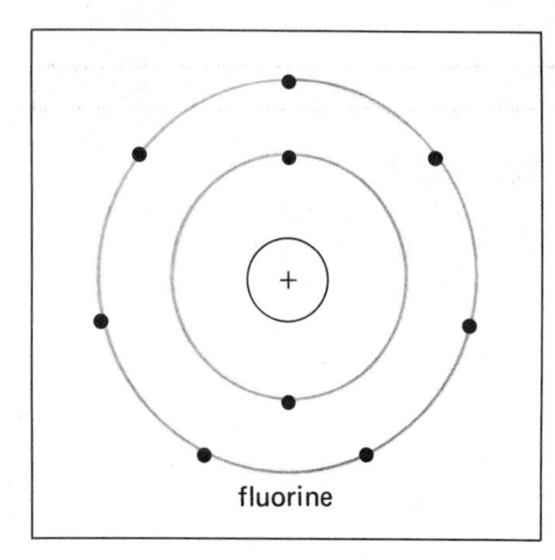

10-11. *How does a fluorine atom differ from a neon atom?*

Halogens
A group of elements whose atoms all have one electron less than the stable number.

TABLE 10-2
The Halogens

Atoms that gain one electron easily	*Atomic Number*	*Stable electron number after gaining one electron*
Fluorine	9	10
Chlorine	17	18
Bromine	35	36
Iodine	53	54
Astatine	85	86

first shell. As shown in Fig. 10-11, fluorine has 7 electrons in its outer shell. Fluorine has one electron *less* than a stable number of 8 in its outer shell. Fluorine must *add* one electron to have a stable electron arrangement. Table 10-2 lists four other elements that will gain one electron to become stable. These elements are called the **halogens** (**hal**-uh-juns). Elements like the *alkali metals* and the *halogens* that lose or gain electrons readily take part in chemical changes.

On the other hand, the noble gases have filled electron shells and do not easily gain or lose electrons. The noble gases do not readily take part in chemical changes. Although the noble gases are not found in great quantities on earth, their lack of chemical activity makes them very useful. The most abundant noble gas is argon. Argon makes up about one percent of the air. Part of the gas used to fill ordinary light bulbs is argon. Its presence inside the bulb helps to slow the rate at which the hot wire filament evaporates, causing the inside of the bulb to blacken. Thus the light bulb stays bright as it is used. Neon is used in advertising signs since it gives off a bright red light when a high-voltage electric current is applied. A mixture of neon and other gases will produce different colors. For example, a mixture of neon and helium will give off a yellow color. As you have already seen, helium is used to fill blimps and balloons. It is also used in certain kinds of welding. The helium gas surrounds the hot metal as the weld is made and prevents gases in the air from reacting with the metal. A mixture of helium and oxygen is often breathed instead of air by deep-sea divers. The reason for this is that the nitrogen contained in ordinary air can cause harm when taken into the lungs under high pressure.

ACTIVITY

Classifying Elements

A. Obtain the materials listed in the margin. Each of the nine cards represents a different element. On the front of the card is the name of the element. Below the element's name is the number of electrons in each shell. For example, see Fig. 10-12.

On the back of each card is a short description of the element. The object of this activity is to classify these nine elements into groups on the basis of their chemical and physical characteristics. You will use Fig. 10-13.

B. Begin with card 1. Read the description of argon on the back of the card.

C. Look at Fig. 10-13 and find the characteristics that describe argon. Follow the arrows.

The last circle in Fig. 10-13 contains the name of the group to which argon belongs.

D. Write the name of this group in your notebook. List argon under this heading.

E. Follow the same procedure for cards 2 through 9.

1. How many groups did you find?
2. What are the names of these groups?
3. Which elements are found in each group?

F. Now look at the front of each card. Beside the name of each element in your lists, write the number of electrons in its outer shell.

4. Describe any pattern you see that relates the groups of elements to the number of electrons in their outer shells.

Materials

9 element cards
paper
pencil

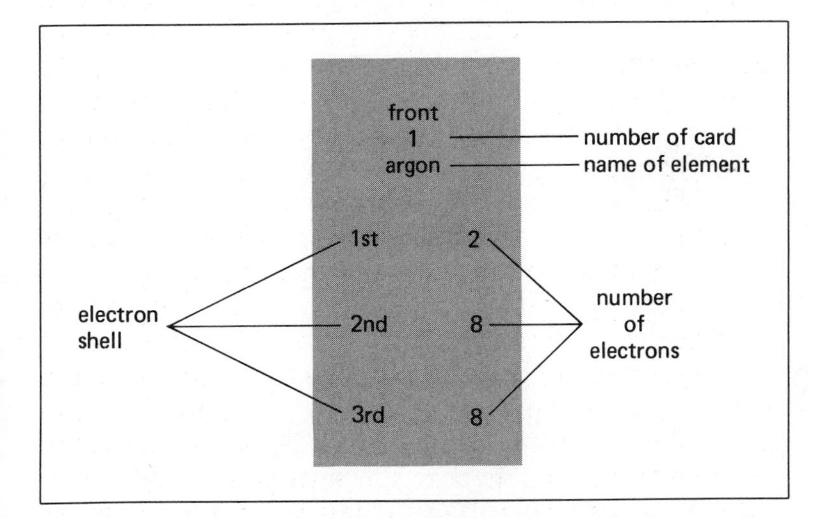

10-12.

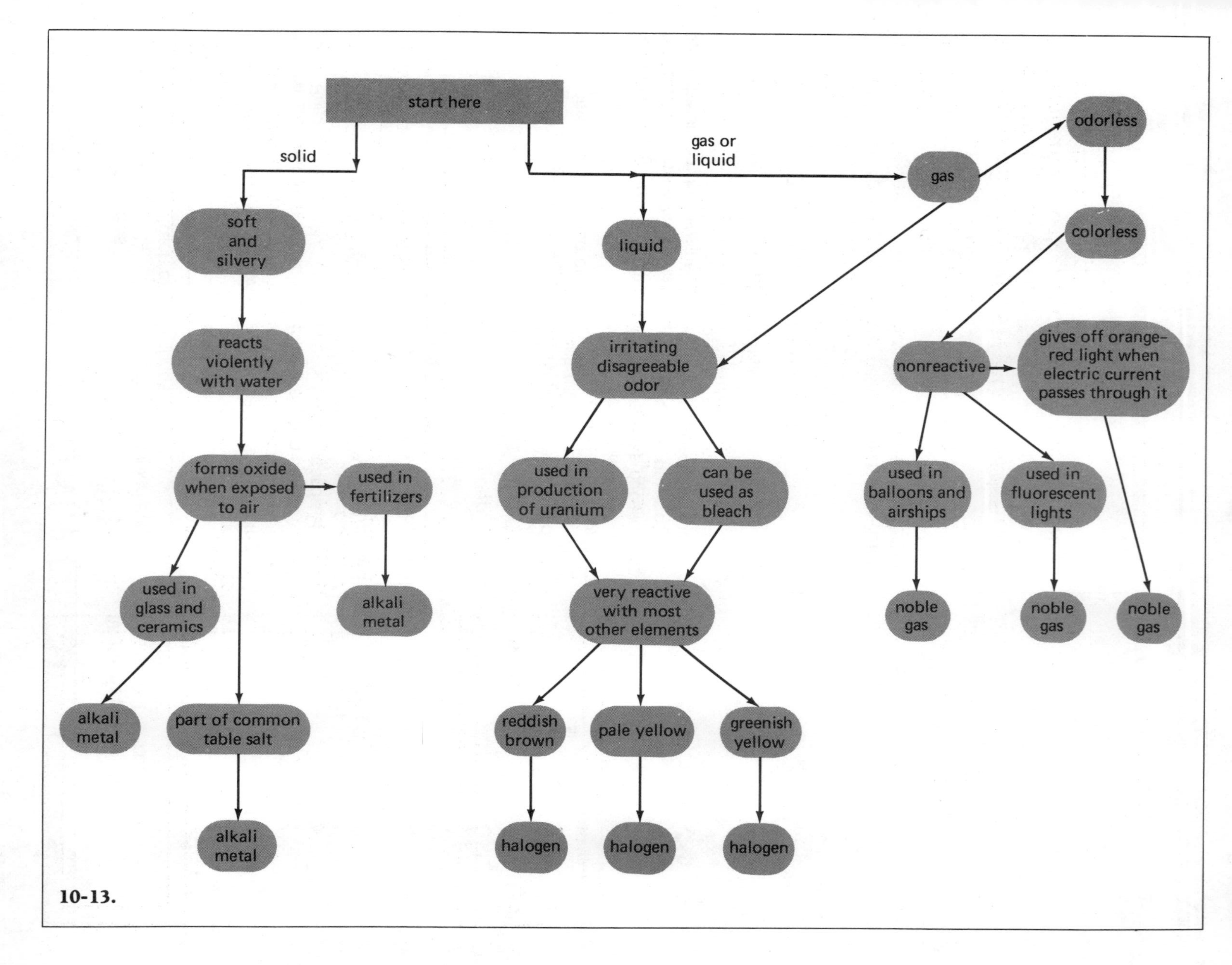
start here
solid
gas or liquid
soft and silvery
reacts violently with water
forms oxide when exposed to air
used in fertilizers
alkali metal
used in glass and ceramics
alkali metal
part of common table salt
alkali metal
liquid
gas
irritating disagreeable odor
used in production of uranium
can be used as bleach
very reactive with most other elements
reddish brown
pale yellow
greenish yellow
halogen
halogen
halogen
odorless
colorless
nonreactive
gives off orange-red light when electric current passes through it
used in balloons and airships
used in fluorescent lights
noble gas
noble gas
noble gas
10-13.

SUMMARY

Airships like the Goodyear blimp are safe because they are filled with helium gas. The atoms of helium are stable. They do not react with different atoms. Atoms that are not stable tend to gain or lose electrons to reach a stable number. Atoms gain or lose electrons by reacting chemically with atoms of other elements. Atoms with the same number of electrons in their outer shells have similar characteristics.

QUESTIONS

Unless otherwise indicated, use complete sentences to write your answers.

Use the following list to answer questions 1–4:

atom A: 7 electrons in outer shell
atom B: 1 electron in outer shell
atom C: 1 electron in outer shell
atom D: 7 electrons in outer shell
atom E: 3 electrons in outer shell
atom F: 8 electrons in outer shell

1. Which of the atoms would be chemically active?
2. Which of the atoms would not be chemically active?
3. What other atom would belong to the same chemical group as atom A?
4. What other atom would belong to the same chemical group as atom B?
5. Name three noble gases and state one use for each of them.

10-3. CHEMICAL FAMILIES

You probably know people who are all members of the same family and look somewhat alike. They share a family resemblance. They share the same name. Sometimes these family members even behave alike. In this lesson, you will learn that atoms also belong to families. The characteristics of an element determine the family to which it belongs.

When you finish lesson 3, you will be able to:

- Describe what is meant by a *chemical family*.
- Explain how you can predict the characteristics of an element by using the *periodic chart*.
- Predict the characteristics of a missing member of two different chemical families.

Scientists have now identified more than 100 different kinds of elements. Additional elements may be discovered in the future. Do you think scientists can easily keep track of such a large number of elements? How would you organize a list of all the different elements? One way to organize a large number of different things is to put the names in alphabetical order. For example, your teacher probably keeps an alphabetical list of all the students in your class.

In the last activity, the cards representing nine different elements were given to you in alphabetical order. You then found a different system for classifying the cards. The nine elements fell into three groups. Each element was placed in a particular group on the basis of its characteristics. Lithium, sodium, and potassium, for example, had similar properties. They were placed in the same group, called the alkali metals. The general name for a group of elements with similar properties is **chemical family.** The alkali metals, halogens, and noble gases are each examples of a *chemical family*.

During the early 1800's, scientists discovered that different elements can belong to chemical families. They listed the elements in order of their increasing atomic masses. When they did this, the scientists discovered that certain physical and chemical properties were repeated at regular intervals. In other words, elements with similar properties occurred *periodically* in the list. This is like the days of the week (Sunday, Monday, Tuesday, and so on), which occur periodically throughout a calendar month.

In 1872, a Russian chemist, Dmitri Mendeleev (duh-**meet**-tree men-d'l-**ay**-uf), tried to arrange the elements in another way. See Fig. 10-14. He arranged the elements in the order of their increasing atomic masses. Instead of listing them one after the other, however, he laid them out as you might deal a deck of cards for a game of solitaire. Figure 10-15 shows what he did.

Look at the three elements in the first vertical column: lithium, sodium, and potassium. These are the three elements that belong to the chemical family of alkali metals. The three elements in the last column are flourine, chlorine, and bromine. These elements belong to the chemical family of halogens.

Chemical family
A group of elements that are alike in their chemical behavior.

10-14. *Dmitri Mendeleev found a better way to arrange the elements in a periodic chart. Using this chart, he was able to predict the properties of elements that had not yet been discovered.*

10-15. *Mendeleev's arrangement of the elements in order of increasing atomic mass.*

lithium atomic mass 7	berylium atomic mass 9	boron atomic mass 11	carbon atomic mass 12	nitrogen atomic mass 14	oxygen atomic mass 16	fluorine atomic mass 19
sodium atomic mass 23	magnesium atomic mass 24	aluminum atomic mass 27	silicon atomic mass 28	phosphorus atomic mass 31	sulfur atomic mass 32	chlorine atomic mass 35
potassium atomic mass 39	calcium atomic mass 40	gallium atomic mass 70	germanium atomic mass 73	arsenic atomic mass 75	selenium atomic mass 79	bromine atomic mass 80

1 H 1.008																	2 He 4.00
3 Li 6.94	4 Be 9.01											5 B 10.8	6 C 12.01	7 N 14.01	8 O 16.00	9 F 19.0	10 Ne 20.12
11 Na 23.0	12 Mg 24.3											13 Al 27.0	14 Si 28.1	15 P 31.0	16 S 32.1	17 Cl 35.5	18 Ar 39.9
19 K 39.1	20 Ca 40.1	21 Sc 45.0	22 Ti 47.9	23 V 50.9	24 Cr 52.0	25 Mn 54.9	26 Fe 55.8	27 Co 58.9	28 Ni 58.7	29 Cu 63.5	30 Zn 65.4	31 Ga 69.7	32 Ge 72.6	33 As 74.9	34 Se 79.0	35 Br 79.9	36 Kr 83.8
37 Rb 85.5	38 Sr 87.6	39 Y 88.9	40 Zr 91.2	41 Nb 92.9	42 Mo 95.9	43 Tc (97)	44 Ru 101.1	45 Rh 102.9	46 Pd 106.4	47 Ag 107.9	48 Cd 112.4	49 In 114.8	50 Sn 118.7	51 Sb 121.8	52 Te 127.6	53 I 126.9	54 Xe 131.3
55 Cs 132.9	56 Ba 137.3	57-71 La-Lu*	72 Hf 178.5	73 Ta 180.9	74 W 183.9	75 Re 186.2	76 Os 190.2	77 Ir 192.2	78 Pt 195.1	79 Au 197.0	80 Hg 200.6	81 Tl 204.4	82 Pb 207.2	83 Bi 209.0	84 Po 209	85 At (210)	86 Rn (222)
87 Fr (223)	88 Ra (226)	89-103 Ac-Lr†	104 Rf (261)	105 Ha (260)	106 (263)	107	108	109	110	111	112	113	114	115	116	117	118

1 — atomic number; H; 1.008 — atomic mass

*Lanthanides	57 La 138.9	58 Ce 140.1	59 Pr 140.9	60 Nd 144.2	61 Pm (147)	62 Sm 150.4	63 Eu 152.0	64 Gd 157.3	65 Tb 158.9	66 Dy 162.5	67 Ho 164.9	68 Er 167.3	69 Tm 168.9	70 Yb 173.0	71 Lu 175.0
†Actinides	89 Ac (227)	90 Th 232.0	91 Pa (231)	92 U 238.0	93 Np (237)	94 Pu (244)	95 Am (243)	96 Cm (248)	97 Bk (247)	98 Cf (251)	99 Es (254)	100 Fm (257)	101 Md (258)	102 No (255)	103 Lr (257)

10-16. *The modern periodic chart of the elements. The numbers shown in the colored boxes represent the atomic numbers of elements that have not yet been discovered.*

Periodic chart
An arrangement of all the elements, showing the chemical families.

Mendeleev discovered that when he arranged all the elements in a table like this, elements in the same chemical family were found in the same vertical column. Elements in each family are found at particular periods or places when you put them in order. Mendeleev's chart is called the **periodic chart** of the elements.

Modern *periodic charts* are slightly different from Mendeleev's original one. A modern periodic chart has the elements arranged in order of their increasing atomic numbers. This causes only small changes from Mendeleev's original arrangement according to atomic mass. In a modern periodic chart, each kind of atom is found in a separate box. See Fig. 10-16. The symbol for the element is given in the middle of the box. Above the symbol is the atomic number of the element. Below is the atomic mass. On some periodic charts, the electron arrangement for an atom of each element is also shown in the box. Other information about the properties of the element may be included within the box.

A map would be helpful if you had to find your way around an unfamiliar city. In the same way, a periodic chart is useful in learning about the chemical elements. Each vertical column of the chart lists elements with similar properties. For example, the noble gases are found in a single column at the right side. Next to the noble gases is the column containing the halogens. The alkali metals are found in a single column at the far left of the chart. All the columns between the right and left sides of the chart list elements with some similarities. The two long rows at the bottom contain elements that would all fit in the third column of the chart. They are placed here to avoid making a very long, single column. Hydrogen is usually put alone at the top of the chart since it resembles both the alkali metals and the halogens. Each horizontal row of the chart is called a *period*. Within each row or period, the properties of the elements generally repeat periodically. Each row begins with an alkali metal element and ends with a noble gas. The elements in the rows between the alkali metals and the noble gases change from metals to nonmetals when moving from left to right across the row. This pattern is repeated periodically within each row. The fourth row contains elements that are unlike any in the above rows. This is the reason for the gap in the top rows on the chart.

Mendeleev discovered a valuable tool for scientific research. Each element has its own position on a periodic chart. If you know the chemical properties of two or three elements, you can predict the properties of a neighboring element. You can even predict the chemical behavior of elements not yet discovered. Mendeleev himself predicted the general properties of several elements that had not yet been discovered. In many cases, when the elements were discovered, they were found to behave almost exactly as Mendeleev had predicted.

ACTIVITY

Predicting the Characteristics of an Element

Materials

cards for: lithium, sodium, potassium, helium, neon, argon, rubidium, and krypton

A. Obtain the materials listed in the margin.

You have three cards describing lithium, sodium, and potassium. These three elements belong to the chemical family called the alkali metals.

B. Look at the back of each card. Reread the description of the element given there.

The fourth member of the alkali metal family is called rubidium. The atomic number of rubidium is 37; its mass is 85.

1. How many electrons would you predict rubidium has in its outer shell?

2. On the basis of the known properties of lithium, sodium, and potassium, list three characteristics that you would expect rubidium to have.

You also have the cards for the noble gases helium, neon, and argon.

C. Reread the description of each element on the back of the card.

The fourth member of the noble gas family is called krypton. The atomic number of krypton is 36; its mass is 84.

3. Predict how many electrons krypton has in its outer shell.

4. On the basis of the known behavior of helium, neon, and argon, predict three characteristics of krypton.

Your teacher will now give you a card for rubidium and one for krypton.

D. Read the descriptions given for rubidium and krypton.

5. Were your predictions in questions 1 and 3 correct?

6. Which properties of rubidium were you able to predict correctly? of krypton?

You should now be able to understand some of the reasons why the periodic table of elements is so useful to scientists.

SUMMARY

Just as you belong to a family of people, elements belong to chemical families. You probably resemble the other members of your family in some ways. The members of a chemical family also have a family resemblance. In a periodic chart of all the elements, the members of a chemical family are found in the same vertical column.

QUESTIONS

Unless otherwise indicated, use complete sentences to write your answers.

An atom of the element beryllium has two electrons in its outer shell. These two electrons are given up rather easily to form an ion having a +2 charge. Beryllium is a solid. It also behaves as a typical metal. Use this information to answer questions 1 and 2.

1. From the following descriptions, choose the one element that would belong to the same chemical family as beryllium.
   a. aluminum: solid, metal, +3 ion, 3 electrons in outer shell
   b. tin: solid, metal, +4 ion, 4 electrons in outer shell
   c. radium: solid, metal, +2 ion, 2 electrons in outer shell
   d. polonium: solid, metal, +2 ion, 6 electrons in outer shell

2. Magnesium belongs to the same chemical family as beryllium.
   a. How many electrons would an atom of magnesium have in its outer shell?
   b. List three other probable characteristics of magnesium.

3. The three main items shown on the periodic table (Fig. 10-16) are **a.** chemical symbol **b.** atomic number **c.** atomic mass. Explain each of these in your own words.

4. What are elements found in the same vertical column of the periodic table called?

Materials
copper penny
zinc plate (approx. 2 cm square)
mixture of ammonium chloride and manganese dioxide (paste), 1/2 tsp
plastic spoon or spatula
flashlight bulb (1 or 2 cell)
2 electrical wire leads
sandpaper or steel wool

Voltaic Cell

Purpose

All metal atoms tend to give up electrons. However, some metals lose electrons more easily than others. In this laboratory, you will use the metals copper and zinc to make a voltaic cell. Copper and zinc are next to each other in the periodic table. Copper has an atomic number of 29 and one electron in its outer shell. Zinc has an atomic number of 30 and 2 electrons in its outer shell. Both copper and zinc tend to lose their outer electrons. Zinc has a greater tendency to lose its electrons than copper. In a voltaic cell, zinc will lose electrons and copper will gain electrons. This transfer of electrons from zinc to copper results in an electric current.

Procedure

A. Obtain the materials listed in the margin. (If you do not have a penny, borrow one from a friend.) Clean the metal with sandpaper, if necessary.

B. Obtain the mixture of ammonium chloride and manganese dioxide from your teacher. This mixture is in the form of a black paste.

C. Coat the piece of zinc with a thin layer of the black paste.

D. Now put the penny down on top of the paste. See Fig. 10-17. Make sure the penny is only in contact with the paste. You have now made a voltaic cell. The penny and the zinc should not touch each other. The paste prevents electrons from flowing directly from the zinc to the copper. Instead, you will use a piece of wire to transfer electrons from zinc to copper.

E. Attach one piece of wire to the zinc and one to the copper penny. You will connect these wires to the light bulb.

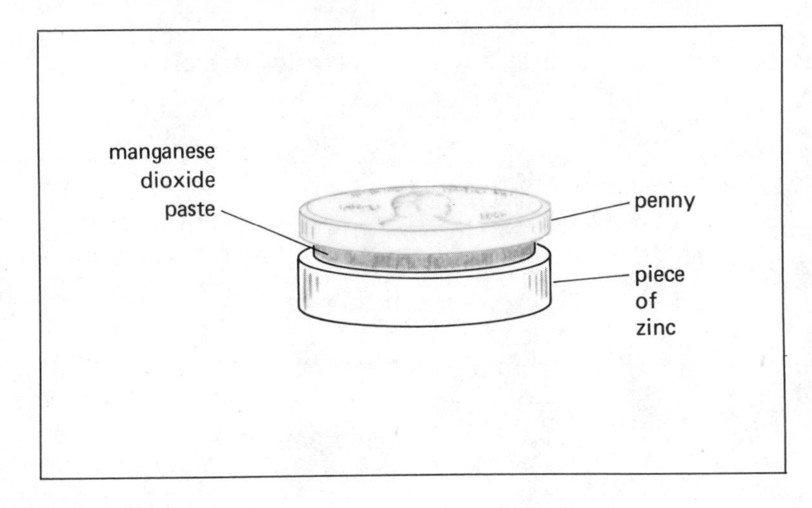

10-17.

1. What function do you think the light bulb will perform in the circuit?

F. Fasten the unattached ends of wire to the two terminals of the flashlight bulb.

2. What happened to your light bulb when the wires were attached?

3. What does this demonstrate?

G. Draw a diagram of your circuit. Include the zinc-copper voltaic cell, the wires, and the light bulb. Label each part. Draw arrows to indicate the direction of flow of the electrons. Indicate what happens to the light bulb when electrons flow through it.

H. Construct the same circuit, using two pieces of copper in place of the copper-zinc combination.

4. Does the bulb light?

5. Do you think two kinds of metal are necessary to make a cell?

Summary

Write a paragraph describing how to set up a zinc in copper voltaic cell. Tell in which direction the current flows and why it flows in that direction. Use a labeled drawing to show the final circuit.

VOCABULARY REVIEW

Match the number of the word(s) with the letter of the phrase that best explains it.

1. inert
2. noble gases
3. chemical activity
4. stable arrangement
5. ion
6. chemical family
7. unstable electron
8. alkali metals
9. halogens
10. periodic chart

a. An atom or molecule that has an electric charge.
b. An electron arrangement of completely filled shells.
c. An electron that has absorbed energy.
d. The way an atom reacts with other atoms.
e. Does not react.
f. A group of elements with similar chemical behavior.
g. The six elements whose atoms have filled electron shells.
h. Group of elements having one less electron than the stable number.
i. Group of elements having one more electron than the stable number.
j. An arrangement of all the elements showing the chemical families.

REVIEW QUESTIONS

Complete each statement by choosing the best word or phrase, or by filling in the blank.

1. When an atom absorbs energy, its **a.** electrons move to a greater distance from the nucleus **b.** protons move faster **c.** electrons move closer to the nucleus **d.** protons move to a greater distance from the nucleus.
2. When an atom releases energy, its **a.** electrons move to a greater distance from the nucleus **b.** protons move faster **c.** electrons move closer to the nucleus **d.** protons move to a greater distance from the nucleus.
3. If an atom gains an extra electron, the atom becomes **a.** a positive ion **b.** a neutral atom **c.** a negative ion **d.** a new element.
4. An atom has a stable electron arrangement if it **a.** requires a great deal of energy to release an electron **b.** has completed electron shells **c.** is a noble gas **d.** all of these are correct.

5. An atom whose electron arrangement is 2, 8, 7 would become stable if it ________.
6. The alkali metals and halogens are chemically active because they **a.** have the same number of electrons **b.** have completed electron shells **c.** easily gain or lose electrons **d.** do not easily gain or lose electrons.
7. The noble gases are grouped together since they **a.** have the same number of electrons in their outer shells **b.** are all gases **c.** are inert **d.** all of these are properties the noble gases have in common.
8. Groups of elements that chemically behave the same **a.** belong to the same chemical family **b.** have the same number of electrons **c.** have the same number of protons **d.** are always inert.
9. The symbols for all the members of the halogen family are ________.
10. The ________ was used by Mendeleev to predict the general properties of several elements not yet discovered at the time.

REVIEW EXERCISES

Give complete but brief answers to each of the following. Use complete sentences to write your answers.

1. Explain how an atom can absorb and then release energy.
2. Describe what happens when an atom becomes an ion.
3. What happens when electricity is passed through a glass tube containing a gas?
4. What is meant by a noble gas?
5. Give an example of an electron arrangement that is stable.
6. A certain atom has 2 electrons in its first shell, 8 electrons in its second shell, and 1 electron in its third shell. Will it gain or lose electrons to have a stable arrangement?
7. How is chemical activity related to electron arrangement?
8. Why are some elements, such as the alkali metals, grouped together?
9. Describe what is meant by a chemical family.
10. How can knowing the position of an element in the periodic chart be of value?

EXTENSIONS

1. Use the library to find more complete descriptions of the members of the alkali metal family. How are they similar? different?
2. Look up characteristics of the halogen family. Describe some uses of halogen compounds.

CHAPTER 11

Atomic Bonds

11-1. CHEMICAL BONDING

Diamond is the hardest substance known. Diamonds can cut through stone or metal as easily as a knife slices bread. A diamond crystal is made only of carbon atoms. Carbon is also found in other forms besides diamond. The hardness of a diamond is a result of the strength with which the individual carbon atoms cling to each other. Compare carbon, in the form of a diamond, with helium. As you know, helium is a very light gas. When the temperature drops to 4.2K, helium changes from a gas into the coldest of liquids. What causes carbon atoms to cling so tightly to their neighbors while helium atoms do not? The purpose of this lesson is to investigate how atoms join together. We will begin with a close look at hydrogen atoms because they are the simplest of all atoms.

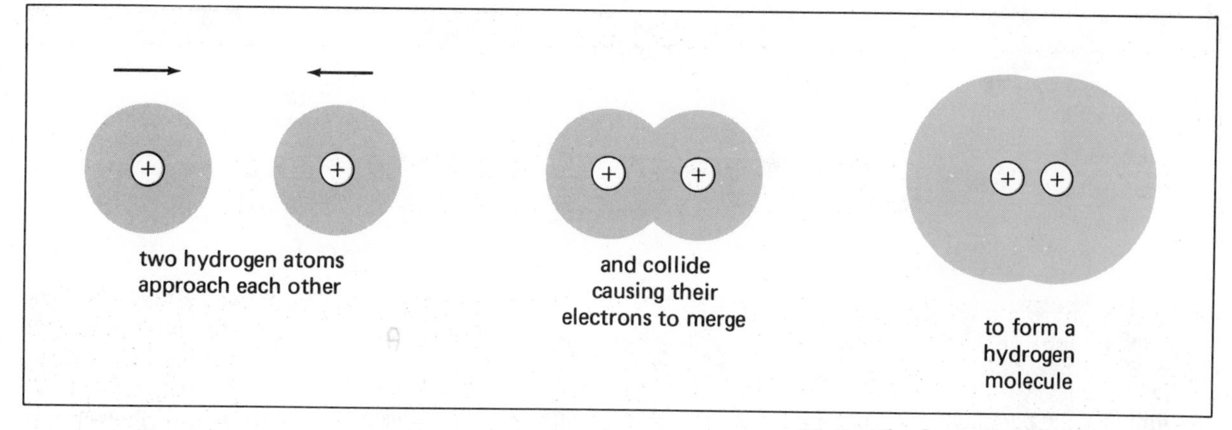

11-1. *The formation of a hydrogen molecule.*

When you finish lesson 1, you will be able to:

- Explain what is meant by a *chemical bond*.
- Use *electron dot models* to explain why no more than two hydrogen atoms can join to form a hydrogen molecule.
- Explain what is meant by *valence*.
- ○ Draw electron dot models for several atoms and simple molecules.

A general rule describes much of the chemical behavior of atoms. The rule says that all atoms are of two general kinds: (1) atoms that tend to gain electrons to become stable; (2) atoms that tend to lose electrons to become stable. A hydrogen atom has only one electron. Two electrons will fill hydrogen's only electron shell. Therefore hydrogen will tend to gain one electron to become stable.

Suppose that two hydrogen atoms come close to each other. Each atom needs one electron. Each atom tends to gain one electron. Neither atom tends to lose one electron to the other. Instead the two hydrogen atoms *share* the two electrons. In doing so, they form a hydrogen molecule. The steps by which they join are shown in Fig. 11-1.

When each of the two hydrogen atoms shares electrons with the other, the two atoms are joined by a **chemical bond.** Atoms are held together by a *chemical*

Chemical bond
A force that joins atoms together.

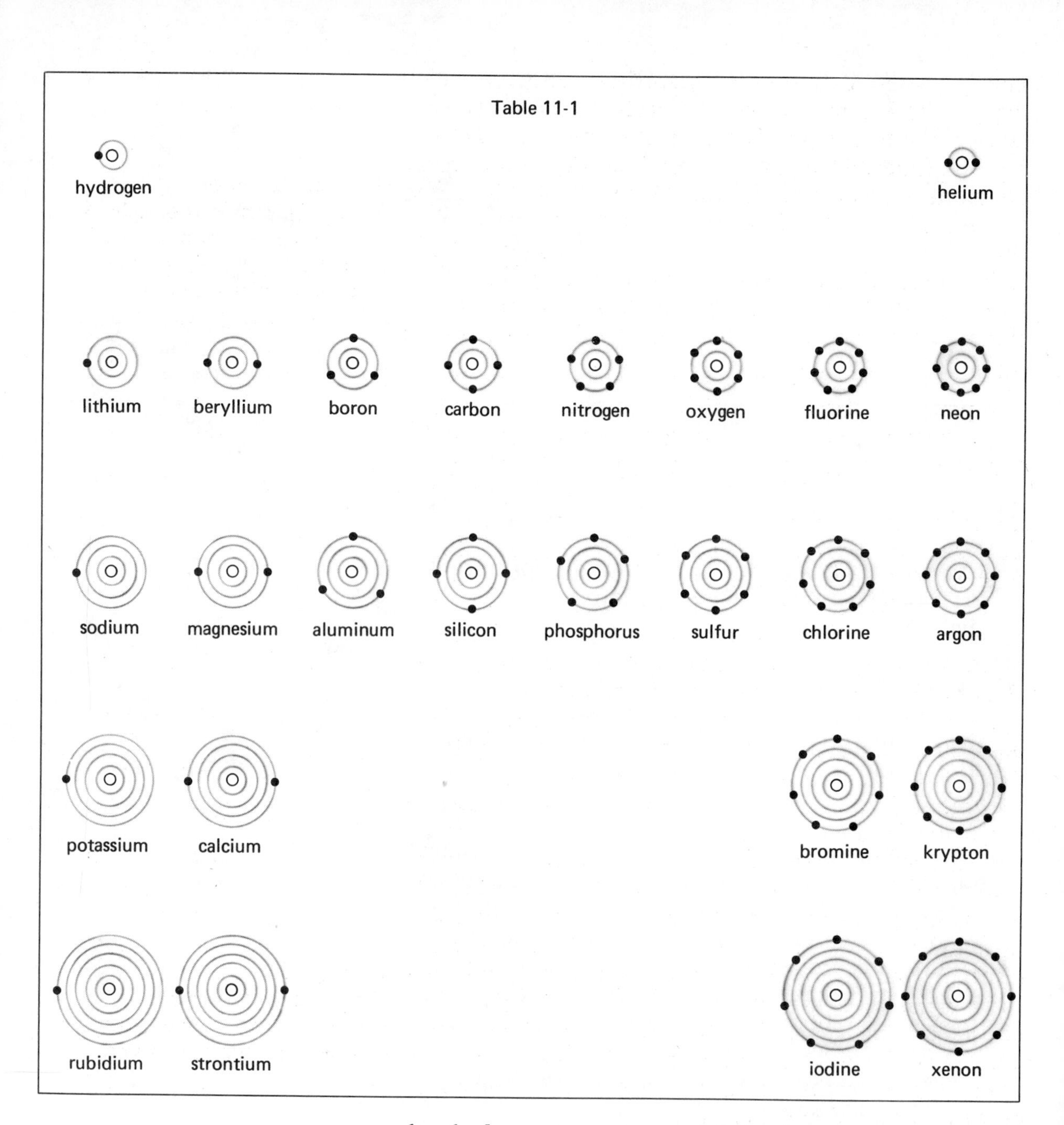

bond when one or more of their electrons are attracted to and move around the nuclei of both atoms. For example, in a hydrogen molecule the electrons from two hydrogen atoms move around the nuclei of both atoms, which remain separate but close together. In lesson 2 of this chapter, you will learn about the different kinds of chemical bonds.

In the 1920's, an American scientist named Gilbert

Lewis suggested a way of picturing atoms to help explain how they formed chemical bonds. In Lewis' system, the symbol for the atom represents the nucleus and the inner electrons. Dots around the symbol represent the outer electrons. This kind of symbol is called the **electron dot model** of the atom. For example, the *electron dot model* of hydrogen is H•. You can then write the electron dot model of a hydrogen molecule as H:H. When there are more than four dots around a symbol, each additional dot pairs with one already there. For example, nitrogen is $\cdot\overset{\cdot\cdot}{\underset{\cdot}{N}}\cdot$, fluorine $:\overset{\cdot\cdot}{\underset{\cdot\cdot}{F}}\cdot$.

Electron dot model
A way of picturing the outer electrons of an atom in which dots representing the outer electrons are placed around the atomic symbol.

A particular molecule is always made up of a certain number of atoms. For example, a hydrogen molecule is always made up of two, and only two, hydrogen atoms. To see the reason for this, look at the electron dot model for hydrogen. Each hydrogen atom has one outer electron, H•. When a hydrogen molecule is formed, two electrons are shared by two hydrogen atoms, H:H. In a molecule, the outer electron shell of each of the two hydrogen atoms is filled. The two atoms in a hydrogen molecule do not gain any additional electrons from other hydrogen atoms.

Your study has shown that the outer electrons of an atom determine much of the atom's chemical behavior. These outer electrons determine how an atom will join other atoms. Table 11-1 shows the outer electron arrangement of a number of atoms. This table is arranged in the same way as the periodic chart to show some chemical families. Knowing the outer electron arrangement of an atom makes it possible to predict how the atom will form chemical bonds. This can be done by using the **valence** (**vay**-luns) of each atom. *Valence* describes the number of electrons an atom gains, loses, or shares to form chemical bonds. An atom that loses electrons is said to have a *positive valence*. An atom that gains electrons from a bond has a *negative valence*. See Table 11-2.

Valence
The number of electrons gained, lost, or shared by an atom when it forms chemical bonds.

Look at the atoms in the second row of Table 11-1. The first atom, lithium, has one outer electron that will be lost. The valence of lithium is +1. This means that a lithium atom will lose one electron to form a bond. Now skip down the second row to oxygen. This atom has six outer electrons and has a valence of −2. An oxygen atom will gain two electrons in a bond.

TABLE 11-2
Valence Numbers of Some Common Elements

Name	*Symbol*	*Valence*
Aluminum	Al	+3
Calcium	Ca	+2
Hydrogen	H	+1
Magnesium	Mg	+2
Potassium	K	+1
Silver	Ag	+1
Sodium	Na	+1
Zinc	Zn	+2
Bromine	Br	−1
Chlorine	Cl	−1
Fluorine	F	−1
Iodine	I	−1
Oxygen	O	−2
Sulfur	S	−2

Valence electrons
Electrons in the outer shell of an atom that take part in a chemical bond.

The number of outer electrons determines the valence of a particular atom. For this reason, the outer electrons in an atom are called **valence electrons.** Lithium has one *valence electron* and oxygen has six. Notice that atoms such as carbon with four valence electrons have a half-filled outer shell. This means that carbon can have valences of both +4 and −4 depending upon the other atoms with which it combines.

Radical
A group of atoms that remain together in a chemical reaction just as if they were a single atom with a single valence.

Experiments show that there are certain groups of atoms that remain together in chemical changes. Such a group of atoms is called a **radical.** One *radical* is made up of one sulfur atom joined to four oxygen atoms. This is called the *sulfate radical*. Since this group of atoms acts chemically like a single atom, it is given a single valence of −2. The names and valences of some other common radicals are given in Table 11-3.

How can valences be used? Perhaps you can already answer this question. Valences can be used to predict how atoms will join together. For example, if you know that the valence of hydrogen is +1 and oxygen is −2, you can predict how a water molecule will be formed. Since hydrogen loses one electron and oxygen gains two, two hydrogen atoms will join with one oxygen atom to make a stable molecule. The following rule is another way of saying this: *The total number of positive and negative valences of the atoms in a simple compound must add up to zero.*

To predict how two kinds of atoms will combine, first write the symbols for the atoms. For example, if hydrogen combines with chlorine, you would write:

$$HCl$$

Then write the correct valences at the upper right of each symbol. The valence of hydrogen is +1 and of chlorine is −1:

$$H^{+1}Cl^{-1}$$

These valences add up to zero: (+1) + (−1) = 0. Therefore one atom of hydrogen will combine with one atom of chlorine.

TABLE 11-3
Valences of Some Common Radicals

Name of Radical	*Valence*
Nitrate (NO_3)	−1
Carbonate (CO_3)	−2
Hydroxide (OH)	−1
Phosphate (PO_4)	−3
Ammonium (NH_4)	+1
Sulfate (SO_4)	−2

To predict how magnesium will combine with chlorine, first write:

$$MgCl$$

The valence of magnesium is +2 and of chlorine is −1.

$$Mg^{+2}Cl^{-1}$$

These valences do not add up to zero. A magnesium atom tends to give up two electrons but each chlorine atom can gain only one electron. Thus *two* atoms of chlorine are needed to accept the electrons lost by a single atom of magnesium. This means that one atom of magnesium will combine with two atoms of chlorine. The compound formed is represented by $MgCl_2$. Used in this way, valences can help you predict how atoms will form chemical compounds.

ACTIVITY

Materials
paper
pencil

Electron Dot Models

A. Obtain the materials listed in the margin.

In this activity, you will practice writing the electron dot models for several atoms and simple molecules. Remember that the electron dot models show the symbol of the element with the outer shell electrons around it. Remember that, when there are more than four dots, each additional dot pairs with one already there.

B. Look at the electron dot models of the atoms shown in Fig. 11-2.

1. On your paper, copy the electron dot models of the atoms shown in Fig. 11-2.

C. Have your electron dot models checked before going on.

The electron dot model of one hydrogen atom and one fluorine atom combined is

H:F: .

2. Now write the electron dot models that show the molecules you would have if you joined the following atoms:

1 hydrogen atom and 1 chlorine atom; 2 hydrogen atoms and 1 oxygen atom; 3 hydrogen atoms and 1 nitrogen atom.

Remember that when a molecule is formed, the single electrons of two different atoms pair up.

D. Have your teacher check your work when you are finished.

11-2.

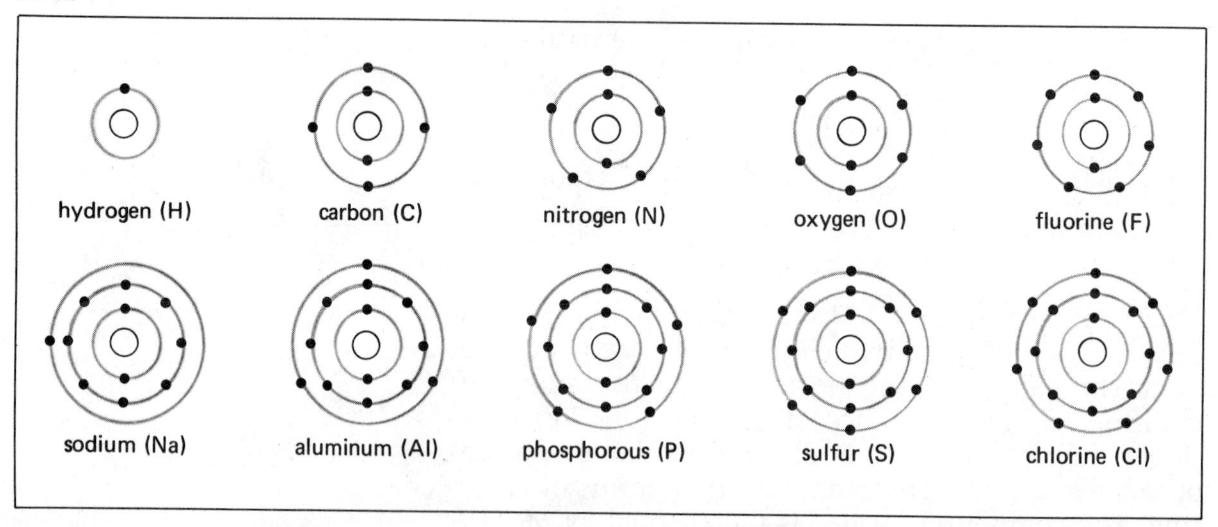

SUMMARY

The two hydrogen atoms that form a hydrogen molecule are joined by a chemical bond. Specific molecules always contain the same number of atoms. That number depends on the number of outer or valence electrons in each atom. The valence electrons of an atom can be shown by an electron dot model for the atom.

QUESTIONS

Unless otherwise indicated, use complete sentences to write your answers.

1. What is meant by a chemical bond?
2. Draw electron dot models for atoms of hydrogen, carbon, oxygen, sodium, and chlorine.
3. Explain why no more than two hydrogen atoms are needed to form a hydrogen molecule.
4. Using an electron dot model, show how hydrogen (H•) and oxygen (:Ö•) form a molecule of water.
5. How many atoms of each element would a molecule of carbon and chlorine contain?
6. Use Table 11-2 to determine how many atoms of fluorine (F) will combine with one atom of aluminum (Al) in a compound.
7. If two hydrogen atoms (H) combine with one sulfate radical (SO_4), what is the valence of the sulfate radical?
8. How many atoms of oxygen (O) will combine with two atoms of aluminum (Al) in a compound?

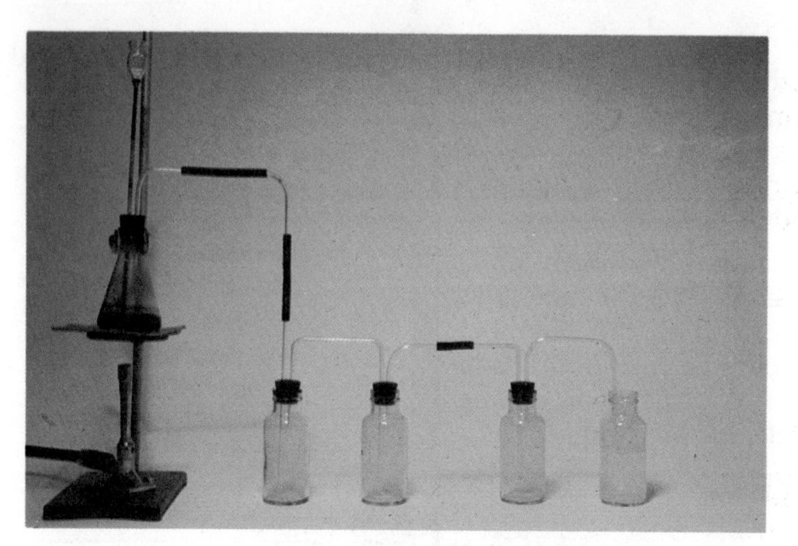

11-2. KINDS OF CHEMICAL BONDS

Pure sodium is a dangerous element. Sodium will burn you if you touch it. Sodium will cause a violent explosion if thrown into water. Sodium is a member of a chemical family called the alkali metals, outstanding for their chemical activity.

Chlorine is also a dangerous substance. Its chemical family, the halogens, is made up of poisonous elements. Chlorine itself is a heavy, greenish-yellow gas commonly used as a bleach and disinfectant.

If sodium metal is put into chlorine gas and heated a little, there is a burst of light and heat as a strong chemical reaction takes place. The only product is sodium chloride, common table salt. Two dangerous elements combine chemically to form a harmless substance, one vital to our health.

Sodium atoms and chlorine atoms join together with chemical bonds. A close look at sodium chloride, however, shows that it results from a different kind of chemical bond from that in a water molecule. Differences in chemical bonds will be explored in this lesson.

When you finish lesson 2, you will be able to:

- ● Explain what is meant by a *covalent bond* between atoms.
- ● Distinguish between covalent and *ionic* bonds.
- ● Explain what is meant by a *chemical formula*.
- ○ Write chemical formulas for some simple molecules.

Chlorine atoms are almost never found alone. Even in pure chlorine gas, each chlorine atom is paired with another chlorine atom. Chlorine atoms are always paired because, like hydrogen, each chlorine atom is missing only one electron to complete its outer shell. Two chlorine atoms will each share one electron with the other to form a chlorine molecule. The following diagram illustrates the formation of a chlorine molecule.

$:\ddot{\underset{\cdot\cdot}{Cl}}\cdot$	+	$:\ddot{\underset{\cdot\cdot}{Cl}}\cdot$	=	$:\ddot{\underset{\cdot\cdot}{Cl}}:\ddot{\underset{\cdot\cdot}{Cl}}:$
chlorine atom		chlorine atom		chlorine molecule

Two hydrogen atoms bond together in the same way as chlorine. This kind of chemical bond is called a **covalent** (koe-**vay**-lunt) **bond.** When atoms like chlorine and hydrogen *share* electrons to fill their outer shells, they form *covalent bonds*.

Covalent bond
A chemical bond formed when atoms share two or more electrons.

Two chlorine atoms form a bond by sharing electrons because neither atom has a tendency to lose electrons. Both atoms tend to gain one electron. Different kinds of atoms may also form covalent bonds between themselves. For example, two hydrogen atoms and an oxygen atom are held together by covalent bonds to make a water molecule. The following diagram illustrates the formation of a water molecule:

$$H\cdot + H\cdot + \cdot\ddot{\underset{\cdot}{O}}: = H:\ddot{\underset{\underset{H}{\cdot\cdot}}{O}}:$$

Now look at a sodium atom. Sodium has 11 electrons. A sodium atom has one outer electron, Na•. Sodium tends to lose one electron to become stable. A chlorine atom could become stable by gaining an electron from a sodium atom. If a sodium atom loses one electron to chlorine, the sodium will also become stable. See Fig. 11-3. When a sodium atom transfers an

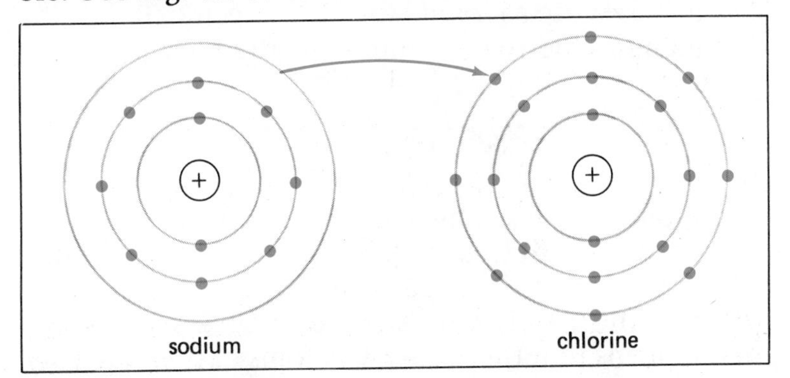

11-3. *Why is an electron transferred from sodium to chlorine, rather than from chlorine to sodium?*

11-4. *Crystals of sodium chloride, which is common table salt, are shown in this photograph. The formation of sodium chloride is the result of an ionic bond between sodium and chlorine.*

Ionic bond
A chemical bond formed when atoms transfer electrons from one to another.

electron to a chlorine atom, the sodium atom becomes a positive ion, Na^+. The chlorine gains an electron and becomes a negative ion, Cl^-. The two oppositely charged ions are then attracted to each other. Bonds that are formed between atoms as a result of a *transfer* of electrons are called **ionic bonds.** Sodium and chlorine form an *ionic bond* to become sodium chloride, which is common table salt.

If you look at salt crystals through a magnifier, you will see many of the salt crystals as small cubes. See Fig. 11-4. Each salt crystal is made up of sodium ions and chloride ions. These ions are held together by their opposite electrical charges. Each positive sodium ion is surrounded by negative chloride ions. Each chloride ion is surrounded by chloride ions. See Fig. 11-15. As a result, the ions form a crystal of sodium chloride (NaCl) with the shape of a cube.

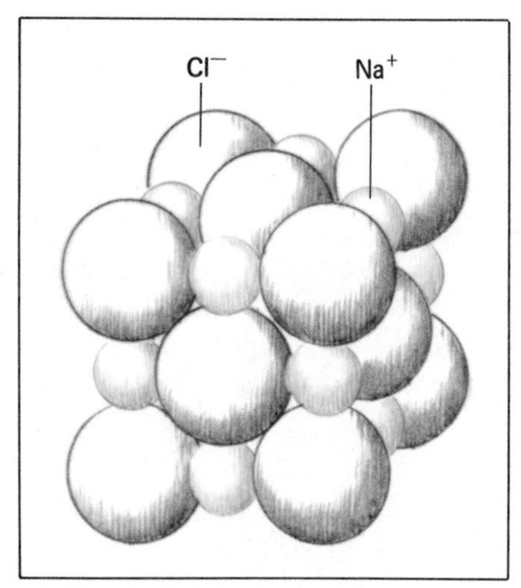

11-5. *A salt crystal is made up of Na^+ ions and Cl^- ions held together in the shape of a cube.*

Once you know the type of bond that is formed, it is possible to write a *formula* for the molecule. Recall the example of chlorine mentioned earlier. Two chlorine atoms form a chlorine molecule:

$$:\ddot{\underset{\cdot\cdot}{Cl}}\cdot + \cdot\ddot{\underset{\cdot\cdot}{Cl}}: = :\ddot{\underset{\cdot\cdot}{Cl}}:\ddot{\underset{\cdot\cdot}{Cl}}:$$

Diatomic molecule
A molecule consisting of only two like atoms, for example, hydrogen gas, H_2.

A molecule consisting of only two atoms of the same kind is called a **diatomic** (die-uh-**tom**-ik) **molecule.** Chlorine gas consists of *diatomic molecules*. The formula for chlorine gas is Cl_2. The numeral 2 in this formula means that there are two atoms in the molecule.

Now look at an example of an ionic bond, sodium chloride:

$Na\cdot$	+	$:\ddot{\underset{\cdot\cdot}{Cl}}\cdot$	=	Na^{+}	+	Cl^{-}
sodium atom		chlorine atom		sodium ion		chloride ion

The formula for sodium chloride is NaCl since there is only one atom of each element in the particle.

Suppose a calcium atom forms an ionic bond with a chlorine atom. A calcium atom has two outer electrons, Ca:. Calcium tends to lose two electrons to become stable. But each chlorine atom can accept only a single electron. Thus two chlorine atoms will combine with one calcium atom:

$$Ca: + \cdot\ddot{\underset{\cdot\cdot}{Cl}}: + \cdot\ddot{\underset{\cdot\cdot}{Cl}}: = Ca^{+2} + 2Cl^{-1}$$

The formula for calcium chloride would then be $CaCl_2$. This formula shows that two atoms of chlorine will join with one atom of calcium.

Valence numbers can also be used to predict correct chemical formulas. For example, to write the correct formula for calcium chloride, first write the valence numbers and the atomic symbols:

$$Ca^{+2}Cl^{-1}$$

If the valences do not add up to zero, write the valence number of each atom as the lower numeral for the opposite symbol. The valence number of calcium is 2; therefore write 2 to the lower right of Cl. Since the valence number of chlorine is 1, write 1 to the lower right of Ca. This criss-cross of valence numbers is shown below:

$$Ca_1{}^{+2} \times Cl_2{}^{-1} \text{ to give the formula } CaCl_2$$

How could valences be used to write the correct formula for aluminum oxide? The valence of aluminum is +3 and oxygen is −2 (see Table 11-2). First write:

$$Al^{+3}O^{-2}$$

Then criss-cross the valence numbers

$$Al_2{}^{+3} \times O_3{}^{-2} \text{ to give the formula } Al_2O_3.$$

Notice that the symbol of the atom with the positive valence is written first in the formula.

ACTIVITY

Materials
paper
pencil

Li •	Ca :	:I• (with dot pairs above and below)	Sr :	:Cl• (with dot pairs above and below)
lithium	calcium	iodine	strontium	chlorine
Na•	Mg :	Rb •	:Br• (with dot pairs above and below)	K •
sodium	magnesium	rubidium	bromine	potassium

11-6.

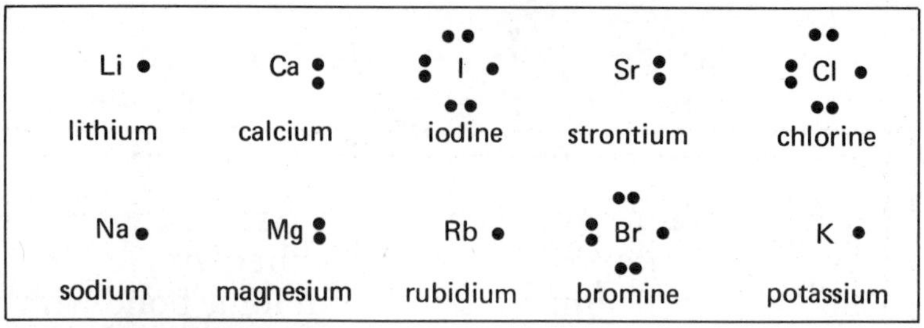

Forming Molecules

A. Look at the electron dot models of the atoms shown in Fig. 11-6.

Notice that bromine has the same number of outer electrons as chlorine. You saw how two chlorine atoms combined to form a chlorine molecule, Cl_2. Bromine atoms combine in the same manner.

:B̤̈r· + ·B̤̈r: = :B̤̈r:B̤̈r:

B. Look again at the electron dot models in Fig. 11-6.

1. What is the name of another atom whose electron dot model has the same number of electrons as chlorine and bromine?

2. On a sheet of paper, draw the electron dot models to show how two of the atoms named in question 1 would combine and share electrons to form a molecule.

You saw how sodium and chlorine atoms combine to form a substance with an ionic bond.

3. List three other atoms with the same number of valence electrons as sodium.

4. Would you predict the atoms you listed in question 3 would also form an ionic bond with chlorine?

As you saw in this lesson:

Ca: + ·C̤̈l: + ·C̤̈l: = $Ca^{+2} + 2Cl^-$

One calcium atom has given up two electrons, one to each of the two chlorine atoms.

C. Select from Fig. 11-6 two other atoms that form an ionic substance with chlorine, as does calcium.

5. Use electron dot models to write equations showing how these atoms react with chlorine.

In Fig. 11-6, you saw how some of the atoms shared electrons to form molecules while others combined to form ionic substances.

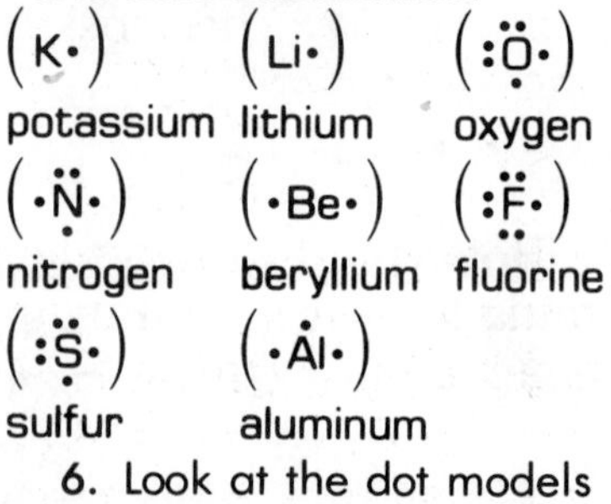

(K·)	(Li·)	(:Ö·)
potassium	lithium	oxygen
(·N̈·)	(·Be·)	(:F̈·)
nitrogen	beryllium	fluorine
(:S̈·)	(·Ȧl·)	
sulfur	aluminum	

6. Look at the dot models above and choose the atoms from the list that most likely would form an *ionic* substance with chlorine.

SUMMARY

Atoms join together by forming either covalent bonds or ionic bonds. That is to say, atoms either share electrons or transfer electrons. When atoms transfer electrons, they become electrically charged ions. You can tell how many atoms make up a molecule by determining the number of electrons that are shared or transferred. Once you know how many atoms are in a molecule, you can write a formula for that molecule.

QUESTIONS

Unless otherwise indicated, use complete sentences to write your answers.

1. Explain what is meant by a covalent bond between atoms.
2. Write the electron dot model for the molecule formed in each of the following:

   a. $\cdot\underset{\cdot\cdot}{\overset{\cdot\cdot}{\mathrm{Cl}}}\!: + \cdot\underset{\cdot\cdot}{\overset{\cdot\cdot}{\mathrm{Cl}}}\!:$

   b. $\mathrm{Na}\cdot + \cdot\underset{\cdot\cdot}{\overset{\cdot\cdot}{\mathrm{Cl}}}\!:$

   c. $\mathrm{K}\cdot + \cdot\underset{\cdot\cdot}{\overset{\cdot\cdot}{\mathrm{F}}}\!:$

   d. $\mathrm{H}\cdot + \cdot\underset{\cdot}{\overset{\cdot\cdot}{\mathrm{O}}}\!:$
3. Write the chemical formula for the molecule formed in each of the cases in question 2.
4. What is the difference between an ionic bond and a covalent bond?
5. Sodium (Na) and chlorine (Cl) combine to form salt. What is the valence of sodium? of chlorine? What is the chemical formula for salt?
6. What is the formula of the most likely chemical compound to form from calcium (Ca) and fluorine (F)?

11-3. CHEMICAL REACTIONS

How would you describe what you see pictured in the photograph? You have probably seen this process, or something like it, many times. You might say simply, "Some pieces of charcoal are burning."

The same process can also be described in a different way often used by scientists.

When you finish lesson 3, you will be able to:

- Give some examples of a *chemical reaction*.
- Explain how a *chemical equation* describes a chemical reaction.
- Use the correct formulas to write and balance a chemical equation

Chemical reaction
A reaction in which a chemical change takes place.

Chemical equation
A description of a chemical reaction using chemical formulas for the substances used and produced.

You have probably seen charcoal burning in a barbecue grill. Did you know that the burning of charcoal is an example of a chemical change? Charcoal is made up of carbon atoms. When charcoal burns, these carbon atoms combine with oxygen. The combination of an element like carbon with oxygen is called *oxidation*. Oxidation is an example of a **chemical reaction.** When charcoal burns, the following *chemical reaction* takes place:

carbon plus oxygen produces carbon dioxide
carbon + oxygen ⟶ carbon dioxide

Another way of writing this reaction is with the formulas for the substances involved:

$$C + O_2 \longrightarrow CO_2$$

A description of a chemical reaction using formulas for the substances involved is called a **chemical equation** (ih-**kway**-zhun). The *chemical equation* for the burning of charcoal says that one atom of carbon reacts

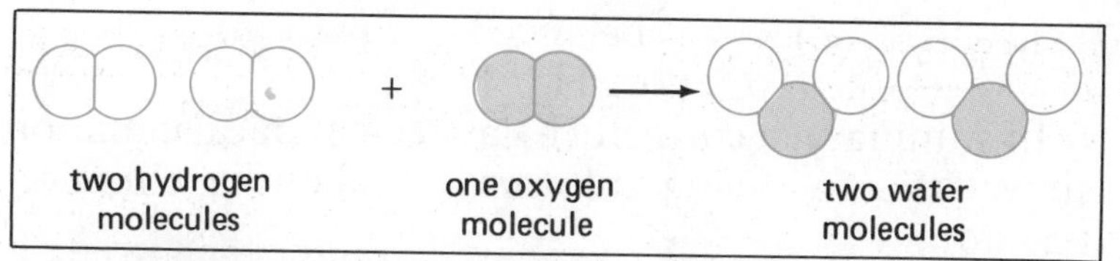

11-7. *A correctly balanced equation.*

with one molecule of oxygen to form one molecule of carbon dioxide.

The following chemical equation shows the formation of water from hydrogen and oxygen:

$$\text{hydrogen} + \text{oxygen} \longrightarrow \text{water}$$

Using formulas, the equation is:

$$H_2 + O_2 \longrightarrow H_2O$$

However, this equation is not complete. Look at the number of oxygen atoms on each side of the equation. There are two oxygen atoms on the left of the arrow but only one oxygen atom on the right. One oxygen atom seems to have disappeared. Scientists have shown that atoms do not disappear during chemical reactions. The *law of conservation of matter* explains what happens during a chemical reaction. This law says that *the same number of atoms exists after a chemical reaction as before the reaction*. To correct the above equation you must *balance* it. Since there are two oxygen atoms on the left of the arrow, there must also be two on the right:

$$H_2 + O_2 \longrightarrow 2H_2O$$

Now the oxygen atoms are balanced but the hydrogen atoms are not. There are four hydrogen atoms on the right. There are only two hydrogen atoms on the left. To correct this, place a 2 in front of H_2:

$$2H_2 + O_2 \longrightarrow 2H_2O$$

This equation is now correctly balanced.

An equation that shows that atoms are not created or destroyed in the reaction is said to be balanced. This is a balanced equation: $2H_2 + O_2 \longrightarrow 2H_2O$. (See Fig. 11-7.)

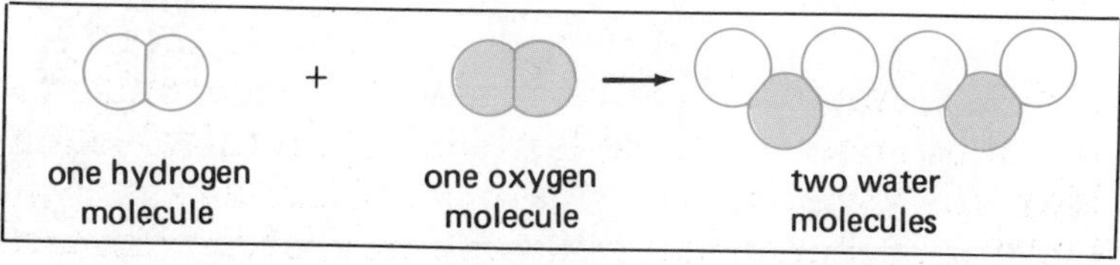

11-8. *An unbalanced equation.*

This is not a balanced equation: $H_2 + O_2 \longrightarrow H_2O$. (See Fig. 11-8.)

In summary, a correctly balanced chemical equation shows the following information about a chemical reaction:

1. The formulas of the molecules that are reacting (reactants) are shown on the left of the arrow (or equal sign).

2. Formulas of molecules that are produced by the reaction (products) are given on the right side of the equation.

3. The total number of each kind of atom used equals the number of those same atoms in the products. No atoms are created or lost during the reaction. See Fig. 11-9.

In some chemical reactions, two substances combine to form a different kind of molecule. You have seen examples of this when carbon reacts with oxygen to form carbon dioxide, or hydrogen combines with oxygen to make water. Another kind of chemical reaction works in the opposite way. A particular molecule can be broken down into different molecules. For example, water can be broken down into hydrogen and oxygen as shown by the following equation: $2H_2O \longrightarrow 2H_2 + O_2$

There are many other kinds of chemical reactions. When a nail rusts, or milk turns sour, or bread is baked, chemical reactions are taking place. Each of these reactions can be described by one or more chemical equations. These equations allow scientists to describe and understand the chemical changes that are taking place around us and inside our bodies.

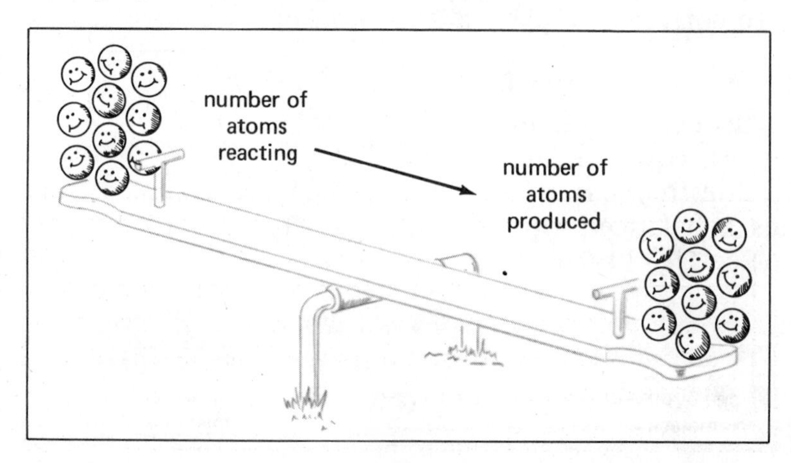

11-9. *The number of atoms reacting must always equal the number of atoms produced.*

ACTIVITY

Materials
small paper cup
water
spoon
plaster of paris (2 tsp)

Temperature Change in a Chemical Reaction

Many chemical reactions result in an increase or a decrease in the temperature of the materials that react. This change in temperature is the result of heat energy being released or used up as the reaction takes place. In this activity, you will observe such a reaction.

A. Obtain the materials listed in the margin.

B. Place two spoonfuls of plaster of paris in a small paper cup. The chemical formula for plaster of paris is $(CaSO_4)_2 \cdot H_2O$. This formula indicates that there is a water molecule (H_2O) attached to the calcium sulfate molecules. $(CaSO_4)_2$, even in the dry form of plaster of paris.

C. Stir in small amounts of water until you have a very thick paste. The plaster of paris will get very hard within a few minutes.

D. As the plaster of paris hardens, feel the cup from time to time.

E. The chemical reaction between plaster of paris and water is described by the following incomplete chemical equation.

1. Copy the following equation and use your observations in step D to fill in the blank. (Hint: A word, not a chemical.)

$(CaSO_4)_2 \cdot H_2O + 3H_2O \rightarrow$
plaster of paris   water

$2CaSO_4 \cdot 2H_2O +$ ______
gypsum

SUMMARY

A chemical equation is a useful form of shorthand describing a chemical reaction. An equation tells you what substances react and what the products of the reaction are. No atoms are created or destroyed in a chemical reaction.

QUESTIONS

Unless otherwise indicated, use complete sentences to write your answers.

1. Give at least two examples of chemical reactions.
2. In your own words, explain the information found in the following chemical equations:
   - **a.** $2H_2 + O_2 \longrightarrow 2H_2O$
   - **b.** $2Na + Cl_2 \longrightarrow 2NaCl$
3. Write the following chemical equations using formulas for the substances. Then balance the equation so that all atoms are accounted for.
   - **a.** hydrogen plus oxygen produces water
   - **b.** carbon plus oxygen produces carbon dioxide
   - **c.** hydrogen plus chlorine produces hydrogen chloride
4. Balance the following unbalanced chemical equations:
   - **a.** $Li + Cl_2 \longrightarrow LiCl$
   - **b.** $Al + O_2 \longrightarrow Al_2O_3$
   - **c.** $H_2O \longrightarrow H_2 + O_2$
   - **d.** $CO + O_2 \longrightarrow CO_2$

11-4. SPEED OF REACTIONS

Imagine a rocket on the pad ready to go. The signal is given to launch; the engine is ignited—but the fuel burns slowly and quietly. This rocket will never get off the pad. Rocket fuel must burn almost instantly or the rocket will not lift off. Some chemical reactions must happen very quickly. Others must go more slowly, such as the cooking of most foods. Your own experiences tell you that there are ways to change the speed of many chemical reactions. What happens when you blow air on a fire? How can you make sugar dissolve faster? Why keep food in the refrigerator or freezer? All of these questions can be answered from common observations that show how to change the speed of reactions.

When you finish lesson 4, you will be able to:

- Explain why crowding molecules together will make them react with each other more often.
- Predict how changes in temperature will affect the speed of a chemical reaction.
- Identify substances that can be used to change the speed of reactions.
- ○ Observe how the speed of a reaction can be changed by the addition of certain substances.

If you want to cause a chemical reaction between two substances, the first thing to do is mix them together. The individual molecules, atoms, or ions in the substances must be made to come together. No new chemical bonds can be made until there is a collision between the particles that will form a new compound. Anything that makes the molecules collide more often will speed up a reaction. Similarly, anything that causes molecules to collide less often will slow a reaction.

Concentration
A description of the number of molecules found in a given space.

One way to make collisions happen more often is to crowd more molecules together. You can speed up the burning of charcoal or coal by blowing air on the fire. This brings in more oxygen molecules to combine with the fuel. The burning reaction speeds up. See Fig. 11-10. This is an example of how **concentration** (kon-sun-**tray**-shun) of the reacting substances affects the speed of the reaction. *Concentration* describes how many molecules are found in a given space. Generally, an increase in the concentration of reacting materials will speed up a reaction and a decrease in concentration will slow a reaction.

When one of the reactants is a solid, the reaction speed can be increased by breaking the solid into smaller pieces. This increases the *surface area* and permits more of the solid to touch the other reacting substances. A piece of wood, for example, can burn dangerously fast if its surface area is greatly increased by being made into sawdust. Explosions can occur in grain elevators, coal mines, and flour mills where the air becomes filled with dust that can burn. A small spark can cause the dust particles to begin burning rapidly with a sudden release of heat causing an explosion. A mixture of sugar and water dissolves faster when stirred because more surface area of the sugar is exposed to water.

Molecules must do more than just touch each other in order to react. They must collide. The force of the collision must be great enough to break the old chemical bonds and allow new ones to form. This means that molecules cannot react unless they have enough energy to collide with a certain force. A particular chemical reaction will not start until its reactants are given the needed energy. The amount of energy needed to start a chemical reaction is called *threshold energy*. The threshold energy needed to start the chemical reactions in a match head is not large. This energy can be supplied by the friction of rubbing the match against a rough surface. On the other hand, the carbon in charcoal has a large threshold energy when it combines with oxygen in the air. The charcoal must be heated to supply the threshold energy needed to cause it to begin burning.

Most cooking of food causes chemical changes in the food. How does a good cook control the speed of these

11-10. *A bellows can be used to add more oxygen to a fire. This additional oxygen causes the fire to burn faster.*

changes? The secret is in the temperature. Chemical reactions generally speed up if the temperature goes up. Lower temperatures usually slow a reaction. The reactions causing food to spoil can be slowed by cooling or freezing the food. In a home refrigerator, the temperature is usually between 10° and 15°C. At these low temperatures, the chemical reactions causing food to spoil are slowed but not completely stopped. Most food will be protected for only a few days. However, in freezers the temperature is held below 0°C. Food can be held at this temperature for several months without spoiling. Some food such as meats can be stored at freezing temperatures for years. If a frozen food is warmed at ordinary temperatures until it thaws, the chemical reactions causing it to spoil will begin again. Once thawed, the food should not be frozen again, since it will not be as fresh as it was before freezing.

As a general rule, the speed of a reaction doubles for every 10°C rise in temperature. This is a result of the greater number of collisions as molecules move faster at the higher temperature.

It is possible to change the speed of a reaction by adding something that remains unchanged when the reaction is finished. For example, you can buy hydrogen peroxide (H_2O_2) at a drugstore to use as an antiseptic. Hydrogen peroxide cannot be kept forever because it slowly breaks down into water and oxygen:

$$2H_2O_2 \longrightarrow O_2 + 2H_2O$$

This reaction is ordinarily slow. However, if a piece of steel wool is dropped into hydrogen peroxide,

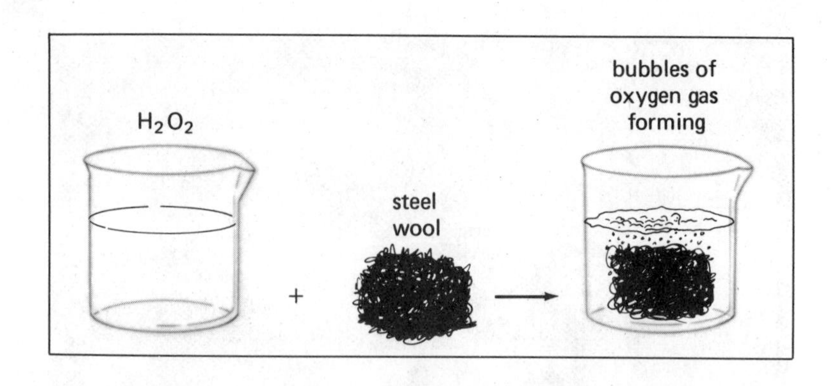

11-11. *Steel wool speeds up the breakdown of hydrogen peroxide (H_2O_2) into water and oxygen. Bubbles of oxygen gas are visible.*

bubbles of oxygen gas can be seen. See Fig. 11-11. The steel wool greatly speeds up the normally slow reaction. When the reaction is over, the steel wool is unchanged. The steel wool acts as a **catalyst** (**kat**-'l-ust). A *catalyst* is a substance that changes the speed of a reaction but is not changed itself by the reaction. Most catalysts cause a reaction to speed up by lowering the amount of threshold energy needed for the reaction to take place. This allows more molecules to collide and react without raising their temperature. Catalysts play an important part in many reactions in living things. Without the right catalyst, many of the chemical reactions needed to carry on life processes could not take place fast enough. Catalysts that control chemical reactions in living things are called *enzymes*. Your body contains thousands of enzymes. Each one controls the speed of a certain chemical reaction. For example, an enzyme called insulin controls the way in which your body uses sugars and starches in your diet. A serious disease called diabetes is caused when a person's body does not produce enough insulin. Most people with severe diabetes can be helped by receiving insulin and carefully controlling sugars and starches in their diets.

Catalyst
A substance that changes the speed of a chemical reaction but remains the same after the reaction.

ACTIVITY

Materials
4 test tubes and holder
hydrogen peroxide
small piece of meat
salt
manganese dioxide

Catalysts in a Chemical Reaction

You can see firsthand how a catalyst speeds up the reaction in which hydrogen peroxide breaks down into water and oxygen gas.

A. Obtain the materials listed in the margin.

B. Copy Table 11-4 in your notebook.

C. Carry out each step in the table. Note that one test tube remains with just hydrogen peroxide in it.

D. Observe the test tubes for about 5 min. Record the rate as *none* (no bubbles), *slow* (just a few bubbles), or *fast* (many bubbles) in your table.

E. Answer the following questions using the data you obtained.

1. Did you see oxygen gas bubble out of the hydrogen peroxide before anything was added?
2. Which substances acted as catalysts?
3. Which substance was the best catalyst?

TABLE 11-4

Steps	*Is a gas given off?*	*Rate gas is given off*
1. Fill 4 test tubes ⅓ full of hydrogen peroxide. 2. Place a small piece of meat in tube 1. 3. Place a pinch of salt in tube 2. 4. Place a pinch of manganese dioxide in tube 3. 5. Leave tube 4 untouched.		

SUMMARY

Some chemical reactions happen very fast, like the burning of rocket fuel. Other reactions take place slowly. The speed at which a reaction takes place can be controlled by changing one or more of these things: (1) concentration, (2) surface area, (3) temperature. The speed of a reaction can also be changed by use of a catalyst.

QUESTIONS

Choose the letter of the answer that best completes the statement or answers the question.

1. Crowding molecules together will make them react with each other more often because the molecules will **a.** move faster **b.** move slower **c.** collide more often **d.** collide less often.
2. Increasing the temperature will usually cause the reaction between the molecules to occur **a.** more often **b.** less often **c.** at about the same rate **d.** first less often, then at about the same rate.
3. A substance that changes the speed of a reaction but remains physically the same after the reaction is called **a.** a fuel **b.** a catalyst **c.** a fossil **d.** a product.
4. Which of the following can be used to control the speed at which a chemical reaction takes place? More than one may be correct. **a.** temperature **b.** surface area **c.** concentration **d.** catalyst.

11-5. ENERGY IN CHEMICAL REACTIONS

An energy shortage is one of the biggest problems the modern world faces. Where does our energy come from? How does an automobile change gasoline into power? How do power plants get the energy needed to generate the electricity for our homes? This lesson will help you answer some of these questions.

When you finish lesson 5, you will be able to:

- Explain how energy is released in chemical reactions.
- Give examples of chemical reactions that require a source of energy.
- Give examples of some *fuels*.
- ○ Determine whether a reaction is *exothermic* or *endothermic*.

Chemical energy may be given off when atoms form bonds with each other. See Fig. 11-12. An atom that does not already have a stable number of electrons is like a car parked on top of a hill. The car can become more stable if it rolls down the hill. An atom can also move to a lower energy condition if it becomes more stable by forming a chemical bond. The energy thus released is chemical energy. See Fig. 11-13.

Chemical energy is usually released in the form of heat. The heat given off in a chemical reaction is often included in a chemical equation. For example, the complete equation for the burning of carbon is:

$$C + O_2 \longrightarrow CO_2 + \text{heat}$$

Chemical reactions that give off heat are called **exothermic** (ek-soe-**thur**-mik) reactions. All burning

Exothermic reaction
A chemical reaction that gives off energy in the form of heat.

11-12. *Fireworks are a spectacular example of the release of chemical energy.*

reactions are *exothermic*. Many other kinds of reactions that do not involve burning are also exothermic. When sodium metal combines with chlorine gas, for example, heat is produced.

Often it is helpful to know how much heat is produced by an exothermic reaction. One way to measure

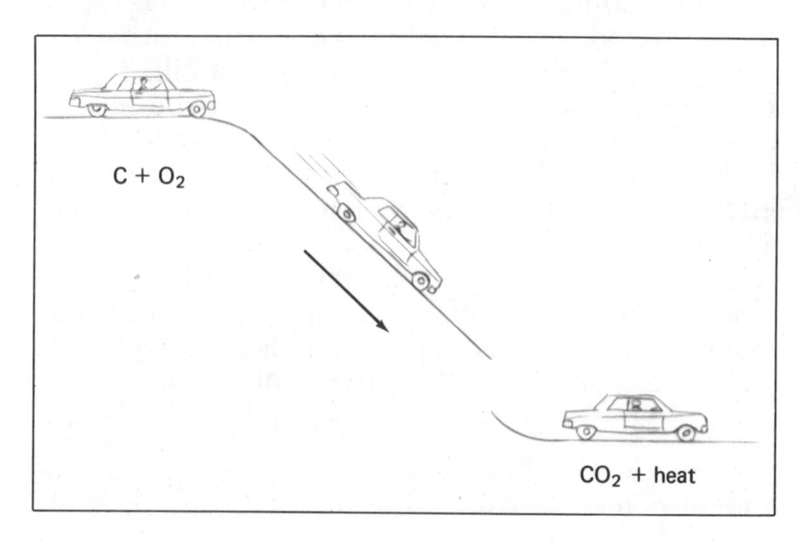

11-13. *A car releases energy as it rolls downhill. Similarly, carbon and oxygen release energy when they react chemically.*

the amount of heat produced is by means of a device called a calorimeter. See Fig. 11-14. For example, by using a calorimeter it can be shown that burning a small lump of carbon weighing 12 g will produce about 94,000 calories.

The energy content of food can also be measured with a calorimeter. A small sample of the food is dried and then combined with oxygen in the calorimeter. The heat released is measured in calories. The number of calories represents the energy released by the food in our bodies. Because most foods produce large amounts of energy, heat content is given in units that are 1,000 times the value of an ordinary calorie. Thus a food Calorie is equal to 1,000 calories or 1 kilocalorie (kcal). Food Calories are also called large Calories and are always spelled with a capital C. See Fig. 11-15.

Not all chemical reactions are exothermic. Some reactions will take place only if energy is supplied; no energy is given off. A chemical reaction that absorbs energy is called **endothermic** (en-duh-**thur**-mik). An example of an *endothermic* reaction is the breaking down of water to form hydrogen and oxygen. Water will only form hydrogen and oxygen if a large amount of heat is supplied or electric energy is passed through the water. An important kind of endothermic reaction takes place in green plants. The plant absorbs energy from sunlight. This energy causes the carbon dioxide

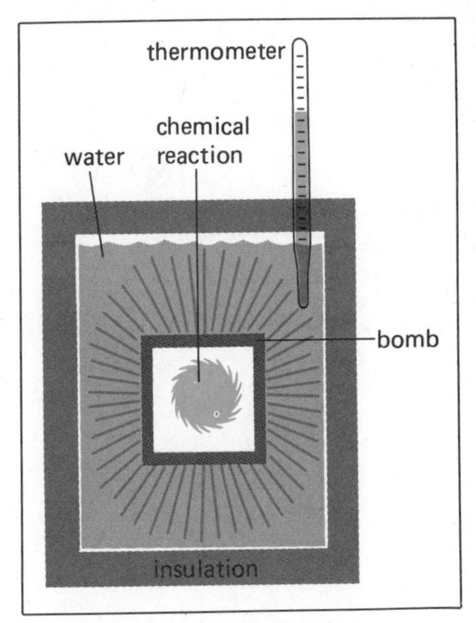

11-14. *The amount of heat given off in a chemical reaction can be measured in a calorimeter. The heat given off by the reaction warms the water. By measuring both the temperature change of the water and its weight, you can calculate the number of calories of heat produced by the reaction. The inner chamber is called a* bomb *because the reactions that take place in it are fast enough to resemble an explosion.*

Endothermic reaction
A chemical reaction that absorbs energy.

11-15. *Different foods contain different numbers of Calories. For example, the number of Calories in a typical balanced meal might be: salad, 43; milk, 68; pudding, 150; baked potato, 93; mixed vegetables, 23; steak, 390.*

11-16. *Coal, the most common solid fuel, consists of the remains of plants and animals that died long ago.*

to react with water to produce glucose. Glucose is a kind of sugar with the formula $C_6H_{12}O_6$. The equation below shows how plants produce glucose and oxygen by an endothermic reaction:

$$6CO_2 + 6H_2O + \text{energy} \longrightarrow C_6H_{12}O_6 + O_2$$

This reaction is part of the process of photosynthesis by which green plants produce food using the energy from the sun.

Exothermic reactions are important to us because they are the source of chemical energy. Any substance used as a source of chemical energy is called a **fuel.** Carbon and some of its compounds are important *fuels*. Carbon fuels are the remains of plants and animals that

Fuel
A substance used as a source of chemical energy.

11-17. *A petroleum refinery. Petroleum, like coal, is a fossil fuel. Petroleum is used to make gasoline, diesel fuel, jet fuel, furnace oils, and other important by-products.*

died millions of years ago. Fuels that come from these ancient plants and animals are called fossil fuels. The fossil fuels are made up of the carbon compounds that were once part of living things. See Fig. 11-16. These compounds have been changed by heat and pressure when they became trapped in the earth. These fossil fuels are found in three forms:

1. *Liquid fuels*. The source of almost all liquid fuel is **petroleum.** *Petroleum* is a mixture of compounds of carbon and hydrogen found deep in the earth. Petroleum can be separated into simpler mixtures like gasoline, kerosene, lubricating oil, petroleum jelly, paraffin, and asphalt. This is done by heating the liquid petroleum (also called *crude oil*). The various liquids

Petroleum
A natural mixture of compounds of carbon and hydrogen.

that make up the petroleum mixture boil at different temperatures. As each kind of liquid boils, its vapor is carried away and allowed to cool and change back to a liquid. These different parts of the petroleum mixture are called *fractions*. The process of separating liquids with different boiling temperatures is called *distillation*. Thus crude oil is separated into its parts by the process called *fractional distillation*. See Fig. 11-17.

2. *Gas fuels*. Natural gas is often found with petroleum. Natural gas also consists of compounds of carbon and hydrogen. The main compound in natural gas is *methane*. Methane has the formula CH_4. *Butane* and *propane* are also carbon–hydrogen compounds obtained from petroleum. They are often used as fuels in places where natural gas is not available. Some gas fuels are artificial. The most common artificial gas is *water gas*. Water gas is a mixture of carbon monoxide and hydrogen. Steam passed over hot carbon forms water gas:

$$\underset{\text{steam}}{H_2O} + \underset{\text{hot carbon}}{C} \longrightarrow \underbrace{\underset{\text{carbon monoxide}}{CO} + \underset{\text{hydrogen}}{H_2}}_{\text{water gas}}$$

Hydrogen gas by itself may become a commonly used fuel in the near future.

3. *Solid fuels*. *Coal* is the most common solid fuel. Coal is made up of the remains of plants and animals that probably died about 250 million years ago. These remains have been changed into a material that is mostly carbon. Coke is a solid fuel made from coal. Coke is made by heating coal without air. This reaction produces almost pure carbon. Most of this coke is used as a fuel by the iron and steel industries. Several useful by-products are produced in the manufacture of coke. These by-products are used to make fertilizers, dyes, perfumes, and medicines.

As the earth's population has grown, our need for energy has also increased. We have already used up much of the earth's supply of carbon fuels. All carbon fuels will probably be gone within a few hundred years. New sources and methods must be found to satisfy the tremendous energy needs of the earth's people. These sources may include solar, geothermal, nuclear, tidal, and wind.

ACTIVITY

Exothermic and Endothermic Reactions

An exothermic reaction releases energy that can usually be detected by a rise in temperature. An endothermic reaction, on the other hand, absorbs energy that can usually be detected by a decrease in temperature. The temperature change determines whether the reactions are endothermic or exothermic.

A. Obtain the materials listed in the margin.

B. Place two spoonfuls of hypo in the glass container. Stir the dry hypo with a stirring rod.

	Hypo	*Washing Powder*
Initial temperature (°C) Final temperature (°C) Change in temperature (°C) Exothermic or endothermic?		

C. Insert a thermometer and record the temperature in a table like the above. See Fig. 11-18.

D. Add one spoonful of water to the hypo and again stir carefully with the rod. Record the temperature. See Fig. 11-19.

1. Is this an exothermic or an endothermic reaction?

E. Rinse the container, the rod, and the thermometer with water and dry them.

F. Place two spoonfuls of washing powder in the container.

G. Stir the washing powder with the rod. Record the temperature of the dry powder in your table.

H. Add one spoonful of water to the washing powder. Stir with the rod and record the temperature.

2. Is this an exothermic or an endothermic reaction?

Materials

glass container
washing soda (sodium carbonate)
hypo (sodium thiosulphate)
water
spoon
thermometer
stirring rod

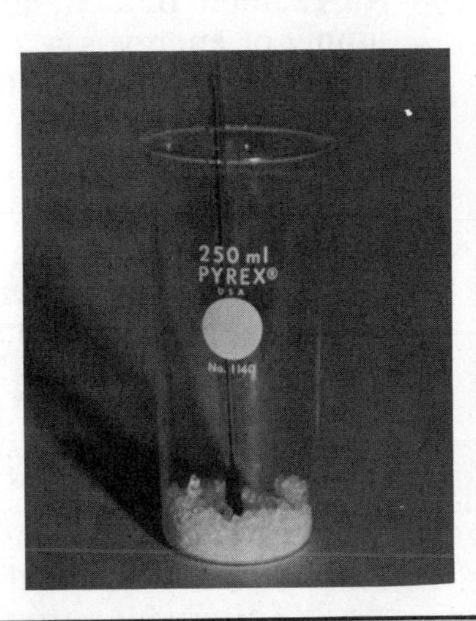

11-18. *(left)*

11-19. *(right)*

SUMMARY

Most of the energy we use every day comes from chemical reactions. Atoms can yield chemical energy when they become more stable by forming bonds with other atoms. Liquid, gaseous, and solid substances, used as fuels, can also yield chemical energy. Almost all fuels are the remains of ancient life. These carbon fuels will be exhausted in the near future.

QUESTIONS

Choose the letter of the answer that best completes the statement or answers the question.

1. Energy is released in a chemical reaction because **a.** the atoms and molecules are at a lower energy condition after the reaction **b.** the atoms and molecules are at a higher energy condition after the reaction **c.** atoms are changed into heat in the reaction **d.** atoms always release heat when they react with each other.
2. An example of a chemical reaction that requires a supply of energy is **a.** combining hydrogen and oxygen to form water **b.** combining carbon and oxygen to form carbon dioxide **c.** breaking down water to form hydrogen and oxygen **d.** combining sodium and chlorine to produce salt.
3. Which of the following chemical reactions is *not* used as a source of energy? **a.** food combining with oxygen **b.** endothermic reactions **c.** exothermic reactions **d.** burning fossil fuels.
4. Any substance used as a source of chemical energy is called **a.** a fuel **b.** petroleum **c.** methane **d.** a fossil.
5. Why does the use of fossil fuels for energy present a problem for the future? What are some possible alternatives to fossil fuels? Use complete sentences to write your answer.

Chemical Reactions

Materials
- 3 test tubes
- medicine dropper
- spoon
- distilled water
- tea (approx. 30 mL)
- lemon juice (1 tsp)
- washing soda (sodium carbonate, 1 tsp)
- epsom salt (magnesium sulfate, 2 tsp)
- baking soda (sodium bicarbonate, 1 tsp)
- vinegar, white (1 tsp)

Purpose

Most chemical reactions cause changes in matter that are easily detected. A change in color, a change of phase (such as a solid forming from a liquid, or a gas forming from a liquid), and a change in temperature are all evidence of chemical reactions. In this laboratory exercise you will carry out three chemical reactions to observe such evidence. In your notebook, copy the data table and record all your observations in it.

Data Table

	Observations
Part I original tea color color of tea after adding lemon juice	
Part II washing soda + epsom salt	
Part III sodium bicarbonate + vinegar	

Procedure

A. Obtain the materials listed in the margin.

PART I. Color Change

B. Fill two test tubes or baby food jars about ¼ full with brewed tea.

C. To one test tube add lemon juice, one drop at a time, until a noticeable color change occurs. Use the other test tube as a guide to tell when the color changes. Record all changes in the data table.

Lemon juice contains citric acid. The reactions between the compounds in tea and citric acid produce compounds with a different color from the original compounds. Natural tea does not have citric acid in it. Most artificial instant teas do contain citric acid. The color change is a good test to tell whether a given tea sample is natural or artificial.

PART II. Formation of a Solid

D. Fill two test tubes or small jars about ⅓ full of distilled water.

E. Dissolve one spoonful of washing soda in one test tube.

F. In the second test tube, dissolve one spoonful of epsom salt. Both samples should be clear and colorless.

G. Now mix the contents of the two test tubes into one container and record your observations.

The chemical equation below describes the reaction in step G.

Na_2CO_3	$+$ $MgSO_4$	$\longrightarrow$	Na_2SO_4 $+$	$MgCO_3$
washing soda	epsom salt		sodium sulfate	magnesium carbonate

Magnesium carbonate does not dissolve well in water. When magnesium carbonate forms, it sticks together in small solid particles. Magnesium carbonate is called a precipitate. When a precipitate forms, a change in phase has taken place as the result of a chemical reaction.

PART III. Formation of a Gas

H. Place one spoonful of sodium bicarbonate in a test tube or small jar. Add one spoonful of vinegar, a little at a time. Record your observations.

When an acid is added to a carbonate or bicarbonate, carbon dioxide gas is formed. The equation for the reaction in step H is as follows:

$NaHCO_3$	$+$ $HC_2H_3O_2$	$\longrightarrow$	$NaC_2H_3O_2$ $+$	H_2O $+$	CO_2
sodium bicarbonate	acetic acid (vinegar)		sodium acetate	water	carbon dioxide

Summary

In your own words, describe the reactions that you observed in this laboratory exercise. Explain your observations.

VOCABULARY REVIEW

Match the number of the word(s) with the letter of the phrase that best explains it.

1. chemical bond
2. covalent bond
3. ionic bond
4. fuel
5. valence
6. chemical reaction
7. chemical equation
8. catalyst
9. exothermic reaction
10. endothermic reaction

a. A substance used as a source of chemical energy.
b. A reaction in which a chemical change takes place.
c. A chemical bond formed when atoms share two or more electrons.
d. The number of electrons gained, lost, or shared by an atom when it forms chemical bonds.
e. A description of a chemical reaction using formulas for the substances used and produced.
f. A force that joins atoms together.
g. A substance that changes the speed of a chemical reaction but remains the same after the reaction.
h. A kind of chemical bond formed when atoms transfer electrons from one to another.
i. A chemical reaction that absorbs energy.
j. A chemical reaction that gives off heat.

REVIEW QUESTIONS

Complete each statement by choosing the best word or phrase, or by filling in the blank.

1. Only two hydrogen atoms join to form a molecule since each hydrogen atom needs ____________ to fill the electron shell of each atom. The electron dot model for the hydrogen molecule (H_2) is ____________.
2. A covalent bond is a bond in which **a.** electrons are shared by more than one atom **b.** electrons are exchanged between atoms **c.** protons are shared by more than one atom **d.** protons are exchanged between atoms.
3. In order to predict how many electrons an atom will gain, lose, or share to form chemical bonds, you must know **a.** the atom's mass

b. the number of neutrons in the atom **c.** the atom's atomic number **d.** all of the above.

4. Magnesium has atomic number 12, atomic mass 24, and has 12 neutrons in its nucleus. The valence of magnesium is ________.
5. Aluminum has a valence of +3 and oxygen has a valence of −2. Using this information, the most likely compound of aluminum and oxygen would be **a.** AlO **b.** Al_3O_2 **c.** Al_2O_3 **d.** Al_2O_2.
6. In order to be correct, the complete chemical equation $H + O \longrightarrow H_2O$ needs the number 2 **a.** after both the H and the O **b.** before the H **c.** before the H_2O **d.** all of the above answers are needed.
7. Crowding atoms together will cause them to react **a.** less often because they collide less often **b.** less often because they collide more often **c.** more often because they collide less often **d.** more often because they collide more often.
8. Which of the following would cause a chemical reaction to speed up? **a.** Raise the temperature and keep the solid in large pieces. **b.** Lower the temperature and keep the solid in large pieces. **c.** Raise the temperature and break the solid into small pieces. **d.** Lower the temperature and break the solid into small pieces.
9. A substance that changes the speed of a chemical reaction but remains the same is called **a.** a product **b.** a catalyst **c.** a reactant **d.** an ionizer.
10. Which of the following is a fuel? **a.** gasoline **b.** methane **c.** coke **d.** all of these are fuels.

REVIEW EXERCISES

Give complete but brief answers to each of the following. Use complete sentences to write your answers.

1. Draw electron dot models for hydrogen, chlorine, oxygen, and water.
2. Use an electron dot model to show how two chlorine atoms join to form a molecule of chlorine.
3. Write the chemical formulas for water, table salt, and carbon dioxide.
4. What is the difference between a covalent bond and an ionic bond?
5. Explain what is meant by the valence of an atom.
6. Explain the following chemical equation:

$$4Al + 3O_2 \longrightarrow 2Al_2O_3$$

7. Explain what happens in a chemical reaction that produces energy.

8. What is the name given to a chemical reaction that occurs when energy is supplied?
9. How does a correctly written chemical equation show that atoms are not lost in a chemical change?
10. What is there about an oxygen atom that allows you to predict how it will form chemical bonds?

EXTENSIONS

1. Advertising suggests that some washing detergents are more effective than others. Plan a fair method of testing the "cleaning power" of detergents you and other students bring from home. How does water temperature affect your results?
2. Diamonds (carbon) and sand (silicon dioxide) are composed of molecules just as are other molecular substances. Find out why the temperatures at which diamonds and sand melt are extremely high. Write a report on the reasons for the high temperatures.
3. What does *biodegradable* mean? Bring several biodegradable products to class. Explain to the class why it is important that these products be made of biodegradable materials.

CAREERS IN CHEMISTRY

Chemistry has had an influence on everything in your life. Clothing, food, fuel for transportation, medicines and drugs, and building materials for schools and homes have all been improved by modern chemistry. New products such as nylon, plastics, synthetic rubber, and penicillin were discovered by chemists. Chemistry is a method of classifying and studying matter. There are many fields of chemistry dealing with different aspects of matter. Today, a wide range of career opportunities are available in chemistry.

Chemist

Description: Most chemists work in research and development. Chemists doing basic research investigate the properties and composition of matter and the laws controlling the combination of elements. This basic research often leads to practical results, such as the production of plastics and synthetic rubber. New products are created or improved in research and development. Some chemists work in production and inspection.

Requirements: Beginning jobs in chemistry usually require at least a bachelor's degree with a major in chemistry or a related discipline. A graduate degree is required for many research and college teaching positions. A beginning chemist should have a wide background in chemistry and good laboratory skills.

For more information:

American Chemical Society, 1155 16th St. NW, Washington, DC 20036

Manufacturing Chemists Association, 1825 Connecticut Ave. NW, Washington, DC 20009

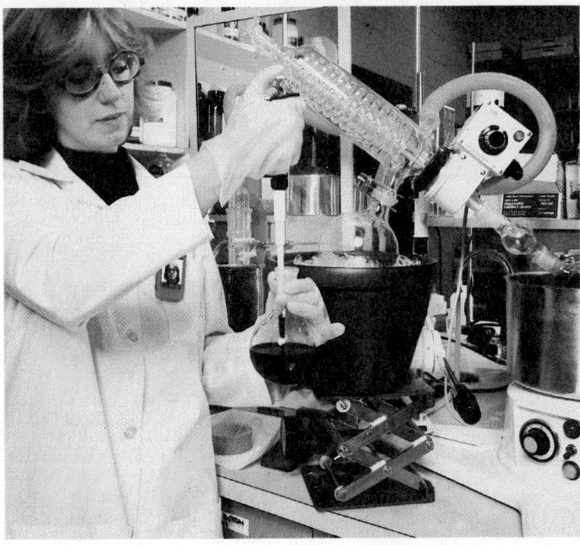

Chemical Technician

Description: Chemists in industry study the properties of chemicals to discover new and improved products and production methods. Technicians conduct tests and record their results. They may perform a variety of routine tests or complicated analyses.

Requirements: Chemical technicians may be graduates of technical institutes, junior colleges, or vocational technical schools. Most chemical technicians begin as trainees or assistants.

For more information:
Manufacturing Chemists' Association, Inc., 1825 Connecticut Ave. NW, Washington, DC 20009

Biochemist

Description: Biochemists study the chemical composition and behavior of living things, including reproduction, growth, and heredity—and particularly, the effects of foods, drugs, and hormones on the body.

Requirements: An advanced degree is usually the minimum requirement for research or teaching positions in biochemistry. A bachelor's degree with a major in biochemistry or chemistry is usually sufficient qualification for jobs as research assistants or technicians.

For more information:
American Society of Biological Chemists, 9650 Rockville Pike, Bethesda, Maryland 20014

Extensions

1. Arrange a visit to a local hospital or clinic. Ask the laboratory technicians to explain several of the chemical tests that they perform.
2. How does the chemical industry contribute to environmental pollution? What are local chemical companies in your area doing to limit the spread of chemical wastes?
3. Choose one or more products that were developed through chemical research, for example, nylon, synthetic rubber or fibers, or drugs such as penicillin. Write a report describing how that particular product was discovered and produced.

CHAPTER 12

Chemical Changes

12-1. METALS AND NONMETALS

The use of metals, such as gold, goes back thousands of years in history. Metals are among our most valuable substances. The secret of the usefulness of metals is in the way their atoms are joined together. In lesson 1, you will learn about this special feature.

When you finish lesson 1, you will be able to:

- List some important properties of all metals.
- Describe what is meant by a *metallic bond*.
- Explain what is meant by a *metalloid*.
- ○ Test several elements to determine if they are metals or nonmetals.

This lesson discusses three important classes of elements: metals, nonmetals, and metalloids.

1. *Metals*. About three-fourths of the elements are metals. All metals are alike in some important ways. They are good conductors of both electricity and heat. They can be formed into different shapes by using a hammer or pressure. See Fig. 12-1. Aluminum, for example, can be rolled into very thin foil without breaking. Most metals can also reflect light. Reflected light makes metals shiny. Some metals, like aluminum and silver, reflect a white, silvery light. Other metals, such as gold and copper, reflect colored light. However, metals often have a coating that hides their brilliance. Iron is often covered with rust and silver can become tarnished. See Fig. 12-2.

12-1. *Jewelry can be made from precious metals such as gold and silver. Like all metals, gold and silver are easily shaped by hammering.*

Similarity of properties suggests a similarity of structure in the atoms of metals. Close examination of these atoms shows that the similarity is in the outer or valence electrons. Metals have only a few valence electrons for forming bonds. Aluminum, for example, has three valence electrons, •Al̇•. In aluminum metals, these electrons are shared equally by neighboring atoms. This sharing of electrons results in a kind of bond. The aluminum atoms are held together by the electrons that are free to move between all neighboring atoms. This kind of bond is called a **metallic bond.** Atoms held together by metallic bonds are like marbles stuck together with honey. The individual atoms can move around each other, but they are still held together. Because of this kind of bonding, metals usually do not break when bent. Metallic bonds allow the atoms to slide past each other without separating. See Fig. 12-3. Thus metals can be hammered or formed into various shapes without shattering like glass.

Metallic bond
A chemical bond formed by electrons that are not tightly held by any particular atom.

Different metals can usually be mixed together to make *alloys*. An alloy is made up of two or more metals. For example, bronze is an alloy of copper and tin. Bronze is harder and lasts longer than either copper or tin. Alloys are usually made by melting the metals together. The metals then stay mixed together when cooled. The kind of mixture depends upon the properties needed in the particular alloy. For example, chromium can be added to iron to prevent rust. Table 12-1 lists a few common alloys and the metals from which they are made.

12-2. Silver can tarnish when exposed to the air.

Corrosion
The eating away of the surface of a metal by chemical action.

TABLE 12-1
The Composition of Some Common Alloys

Alloys	Metals
Brass	Copper, zinc
Solder	Lead, tin
Gold (14 carat)	Gold, copper, silver
Alnico magnets	Aluminum, nickel, cobalt, iron, copper

Bonding in metals also explains their ability to conduct heat and electricity. The electrons that drift between the atoms are free to move. This cloud of moving electrons conducts an electric current through a metal. The same free-moving electrons also conduct heat. As heat is applied to one part of a metal, the free electrons move more rapidly. These moving electrons transmit the heat energy quickly through the metal.

Metal atoms can also easily lose electrons to form other chemical bonds. For example, when exposed to oxygen, iron usually forms rust. See Fig. 12-4. Rusting is an example of **corrosion** (kuh-**roe**-zhun). The eating away of the surface of a metal by chemical action is *corrosion*. A coating, like paint, can protect metal from corrosion. Some metals, such as aluminum and zinc, have a thin protective layer on their surfaces. The protective layer is formed by the metals combining with oxygen.

2. *Nonmetals*. Have you ever tried to describe something and found that the only thing you can say about it is what it is *not*? Unlike metals, nonmetals do not have many properties in common. The fact that chemists refer to them as a group as *non*metals points to one

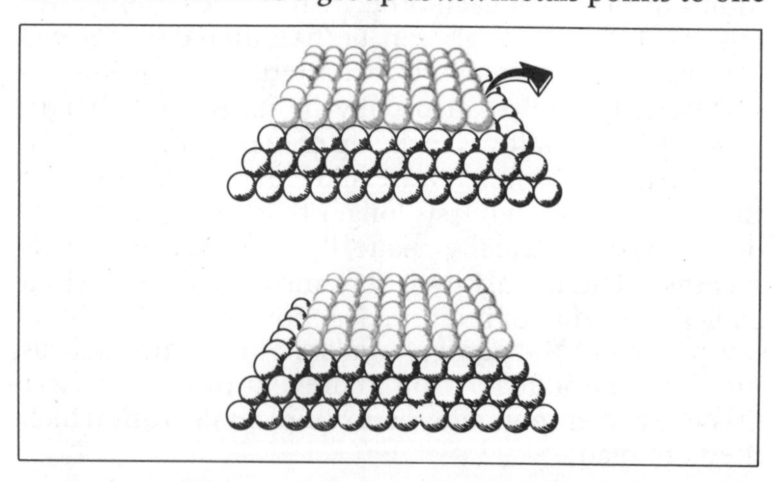

12-3. A pure metal can bend without breaking because the layers of metal atoms are able to slide past each other.

12-4. *Metals such as iron become corroded after being exposed to the air.*

thing they have in common: They are not metals!

Nonmetals are poor conductors of heat and electricity. Most nonmetals are brittle and cannot be easily formed into sheets or other shapes. Some nonmetals, such as carbon and sulfur, are solids at ordinary temperatures. Oxygen, nitrogen, and chlorine, on the other hand, are nonmetals that are gases. Bromine is the only nonmetal that is a liquid at room temperature.

Metals usually look alike. Nonmetals, however, can be colorless like oxygen and nitrogen, or yellow like sulfur. You can see that in studying the nonmetals it is difficult to find many common properties.

Nitrogen and phosphorus are two nonmetallic elements that are very important to living things. Plants require compounds of both of these elements in order to grow. Chemical compounds containing nitrogen and phosphorus are the most important kinds of fertilizers used in farming. If farmers did not use these chemical fertilizers, much less food could be grown on the world's croplands. Nitrogen and phosphorus are also essential to humans. The proteins that are essential in our diet are compounds containing nitrogen. Phosphorus compounds are needed for normal bone growth as well as for other body processes.

Metalloid
An element with some properties of both metals and nonmetals.

3. *Metalloids*. The substances called **metalloids** (**met**-'l-oidz) have some of the characteristics of metals. For example, silicon is a *metalloid* that makes up about 26 percent of the weight of the compounds in the earth's crust. Pure silicon crystals are electrical insulators. However, when small amounts of certain other elements such as arsenic are added to pure silicon, it becomes a *semiconductor*. A semiconductor is a substance that is normally an electrical insulator but can behave as a metal and conduct a current under some conditions. Specially prepared crystals of silicon

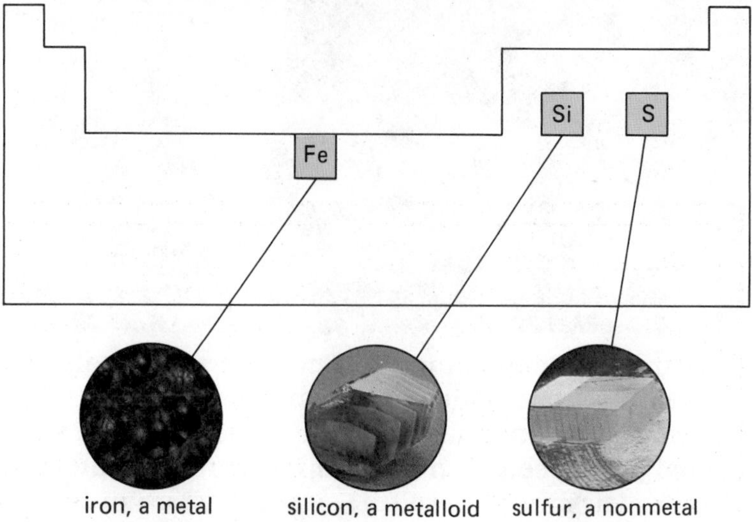

Si

S

Fe

iron, a metal

silicon, a metalloid

sulfur, a nonmetal

12-5. *This diagram shows the positions of iron (a metal), silicon (a metalloid), and sulfur (a nonmetal) on the periodic chart of the elements.*

act as semiconductors in many common electronic devices, such as radios and televisions. Another metalloid, germanium, may also be used as a semiconductor. Other metalloids such as boron, arsenic, antimony, tellurium, and polonium also have some properties of both metals and nonmetals. Figure 12-5 shows the positions of the metals, the nonmetals, and the metalloids in the periodic chart.

ACTIVITY

Materials

carbon
lead
copper
sulfur
hammer
anvil
light bulb
battery
wires

Properties of Metals and Nonmetals

A. Obtain the materials listed in the margin.

B. Copy the following table in your notebook.

C. Test each of the elements listed in the table to determine if the properties are present. Record your observations. (Remember, *all* of these properties must be present in a metal.)

1. Which elements in the table are metals?
2. Which elements are nonmetals?

Element	*Can Be Shaped by Hammering*	*Electric Conductor*	*Luster (shine)*
Carbon Lead Copper Sulfur			

SUMMARY

Metals have many properties that make them valuable to us. These metallic properties are the result of the sharing of valence electrons. These valence electrons form metallic bonds and are also free to form other kinds of chemical bonds. Chemists call elements such as silicon, oxygen, and sulfur nonmetals. Almost the only thing nonmetals have in common is that they are not metals.

QUESTIONS

Unless otherwise indicated, use complete sentences to write your answers.

1. List three important properties of all metals.
2. Describe what is meant by a metallic bond.
3. Name three nonmetals. How are they different from metals?
4. Explain what is meant by a metalloid.
5. Copy and complete the following table:

Element	*Metal or Nonmetal?*	*Practical Use*
Oxygen		
Carbon		
Nitrogen		
Copper		
Lead		
Sulfur		
Silver		

12-2. WATER

Did you know that you are mostly water? On the average, water makes up about 65 percent of the total volume of your body. The loss of only 15 percent of your normal body water can be fatal. We share this need for water with all other living things. When fresh water becomes polluted, plants and fish cannot live in it.

Why is water so important to life? Why must we all work to preserve our supply of clean water?

When you finish lesson 2, you will be able to:

- Explain how the shape of the water molecule determines much of the behavior of water.
- Explain why water pollution is a serious problem.
- ○ Test several water samples for hardness.

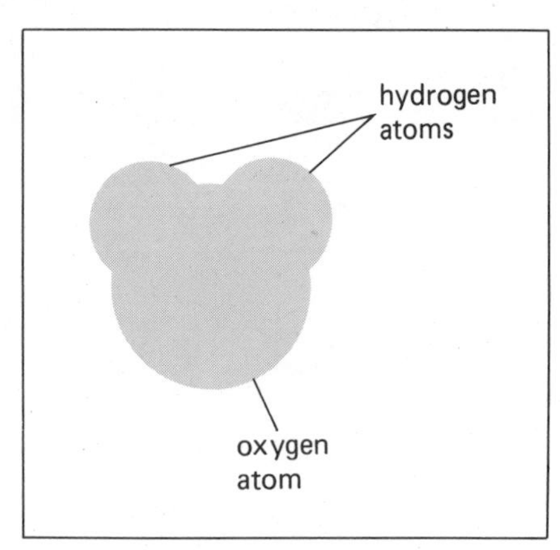

12-6. *The shape of a water molecule is the result of the way in which the atoms are arranged.*

All molecules have a shape. The shape of a molecule is determined by the way atoms making up the molecule are joined. For example, the electron-dot symbol for water is:

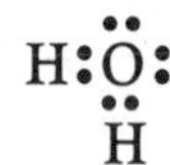

The two hydrogen atoms are joined to the oxygen by covalent bonds. The "bent" shape of the water molecule is shown in Fig. 12-6.

Because of the way electrons are shared in the hydrogen–oxygen bonds, the two hydrogens in a water molecule are slightly positive. The oxygen atom is slightly negative. The fact that the water molecule has positive and negative components makes water a **polar molecule.** Water molecules attract each other. See Fig. 12-7. Each *polar water molecule* can also be attracted to other substances carrying an electric charge. Therefore, water can dissolve many other sub-

Polar molecule
A molecule that carries small electric charges on opposite ends.

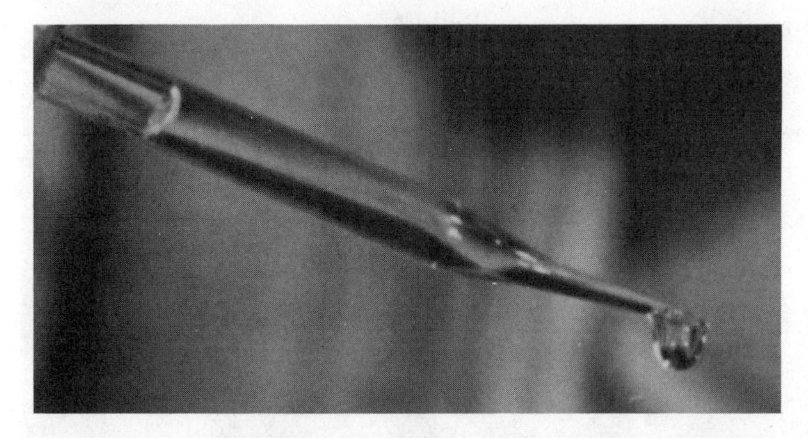

12-7. *Water forms drops because of the attraction of the water molecules for each other.*

stances. For example, common salt is made up of positive sodium ions and negative chloride ions. The ions are held together by the attraction of the opposite charges. If a salt crystal is dropped into water, the polar water molecules are attracted to the sodium and chloride ions. The attraction of the water molecules pulls the ions out of the salt crystal. The ions are then surrounded by the water molecules. See Fig. 12-8. Once that happens, the salt is dissolved. The Na^+ ions and the Cl^- ions are separated and spread throughout the solution. Because water can dissolve so many different things, it is said to be a good **solvent.** A *solvent* is the part of a solution that does the dissolving. The substance that is dissolved is called the **solute.** For example, in a solution made by dissolving salt in water, the *solute* is salt. The solvent is water.

Solvent
The part of a solution that does the dissolving.

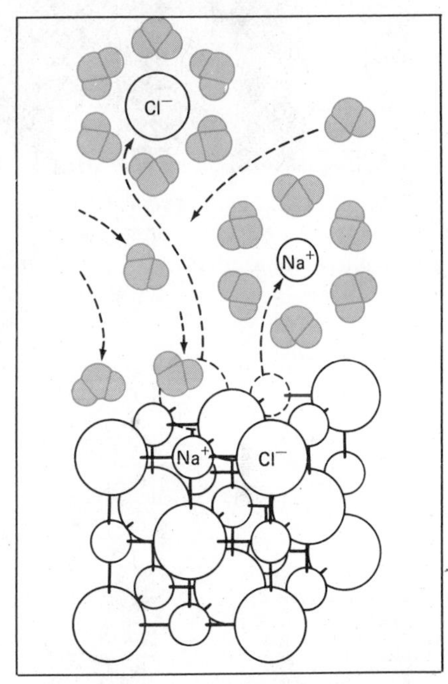

12-8. *Polar water molecules are able to dissolve a salt crystal because the ions in the crystal attract the water molecules.*

The fact that polar water molecules can dissolve many substances makes water important in our lives. The water in your body contains dissolved substances that are necessary to life. For example, your blood contains ions such as Na^+, K^+, and Cl^-. However, this same capacity can cause water to become **polluted.** Water that is chemically *polluted* contains dissolved substances that are harmful.

Most water pollution is caused by human activities. Sewers, for example, carry two kinds of pollution into lakes, streams, and oceans. See Fig. 12-9. The various waste materials in sewage contain harmful bacteria and viruses that can cause disease. In addition, sewage contains chemicals that are harmful. The chemicals called phosphates and nitrates are particularly harmful. These substances encourage the growth of tiny algae and other water plants. When the plants finally

Solute
The part of a solution that is dissolved.

Polluted
A description of water that contains harmful substances.

12-9. *Chemical pollutants can be dissolved in water and spread from one place to another.*

die and decay, the oxygen dissolved in the water is used up. Streams and lakes can usually clean themselves naturally if their water is rich in dissolved oxygen. An overload of sewage can quickly destroy the ability of a body of water to purify itself.

Sewage may also contain chemicals from industrial and agricultural wastes. In time, many of these chemicals decay and become harmless. But others remain unchanged and slowly build up in the water supplies. Over a period of years, the effect of these chemicals can be a threat to the health of anyone drinking the water.

Treatment of sewage to prevent water pollution is carried out in three steps. First, the solid material is allowed to settle. Chlorine gas is added to kill the bacteria and remove odors. A second stage of treatment removes many of the materials dissolved in the sewage water that still remain after the first treatment. This is done by several methods, including mixing the sewage water with air in order to add oxygen. Removal of dissolved chemicals, including phosphates and nitrates, requires a third stage of treatment. This step is difficult and expensive to carry out. However, after this third stage of treatment, the sewage water is often almost completely free of all pollution. Most cities provide only the first two stages of sewage treatment. Only a few places are able to afford the costly third step.

Not all water pollution comes from substances added to the water. *Thermal pollution* is a result of the heating of rivers and lakes. The heat comes from power plants. All types of electric generating plants produce excess heat. This waste heat is carried away by water from streams, lakes, or oceans that is pumped through the plant. The heated water loses some of its dissolved oxygen. This may cause a change in the way plants and animals grow in the water. In some cases, these changes harm the natural environment where the heated water is discharged. Many power plants have arrangements for cooling the water before it is returned to its source.

In some parts of the world, there is another water problem. In these places, the water is *hard*. Hard water is caused by the presence of dissolved minerals such as calcium, magnesium, or iron. The word "hard" in this case means that it is hard to make soapsuds in this water. Washing clothes in hard water is difficult. One way to remove the minerals from hard water is to boil it. This process produces distilled or *soft* water.

Although water is very common, only a small part of the total amount on earth is available for use. Most of the earth's water is in the oceans, in ice at the poles, or deep beneath the surface of the earth. Our need for a steady supply of clean water is filled by less than 1 percent of the earth's water. This limited supply of water must be protected from all forms of pollution.

ACTIVITY

Hard Water

You are going to test water samples for hardness and compare the hardness to that of distilled water.

A. Obtain the materials listed in the margin.

B. Mark the test tube at a point that measures 10 mL. Use a graduated cylinder.

C. Put 10 mL of distilled water in the test tube (up to the mark) and add one drop of soap solution.

D. Shake this mixture and watch for suds to form. If no suds form, add soap solution one drop at a time and shake. Suds must appear and remain for

Materials
test tube
medicine dropper
soap solution
watch with sweep second hand
water samples
graduated cylinder
tape or marking pencil

30 sec. Record the number of drops in a table like the following:

Sample	*Number of Drops*
distilled water tap water ocean water river water etc.	

E. Rinse the test tube with distilled water and add 10 mL of tap water or other water provided by your teacher.

F. Add soap solution one drop at a time until shaking produces suds that remain for 30 sec. Record your observations in the data table.

1. In your own words, explain how you can compare the hardness of two water samples.

SUMMARY

Water is an important part of our lives. Because of the shape of water molecules, water can dissolve many substances—some of them harmful. People are becoming more aware of the problem of water pollution. Efforts are being made to clean up polluted bodies of water. List as many as you can of the ways that water is important in your daily life. Which of these uses of water would be affected if the water supply to your city became polluted?

QUESTIONS

Use complete sentences to write your answers.

1. What property of water molecules makes it possible for water to dissolve many other substances?
2. What is meant by water pollution?
3. Why is water pollution a serious problem?
4. Describe the three steps in sewage treatment.
5. What is hard water? Why is it called "hard"?

12-3. IONS

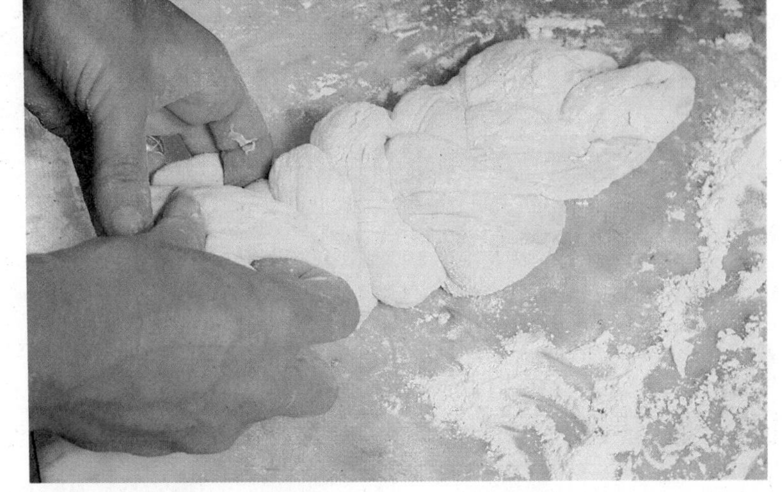

Recipes for breads and pancakes call for baking powder as one of the ingredients. The baking powder you buy at the grocery store is a white powder. If you add a little water to this powder, you will see it change dramatically. As the baking powder dissolves, bubbles of gas appear. This gas is carbon dioxide. Baking powder is used in bread dough because it releases carbon dioxide when in a solution. The carbon dioxide bubbles make the dough expand. The dough is then lighter and the bread will not be heavy.

Baking powder also illustrates an important part of many chemical reactions: water is needed to enable a chemical reaction to take place. Dry baking powder does not release carbon dioxide. In this lesson, you will explore the reasons why most chemical reactions take place between dissolved substances.

When you finish lesson 3, you will be able to:

- ● Explain why some solutions conduct electric currents.
- ● Use examples to show how ions take part in chemical reactions.
- ○ Use a solubility table to predict the identity of a *precipitate*.

You can perform a simple experiment to discover an important difference between dissolved substances. When salt is dissolved in water, it separates into sodium ions (Na^+) and chloride ions (Cl^-). If you test this salt solution, you will find that it conducts an electric current. See Fig. 12-10. You can try the same test on a sugar solution. A sugar solution does not conduct an electric current. Salt is an example of an **electrolyte** (ih-**lek**-truh-lite). When an *electrolyte* is dissolved in water, the resulting solution will conduct an electric current. A salt solution conducts electricity because the solution contains ions. Each ion carries

Electrolyte
A substance that forms a conducting solution when dissolved in water.

12-10. *A diagram of the setup used to demonstrate that a salt solution will conduct an electric current.*

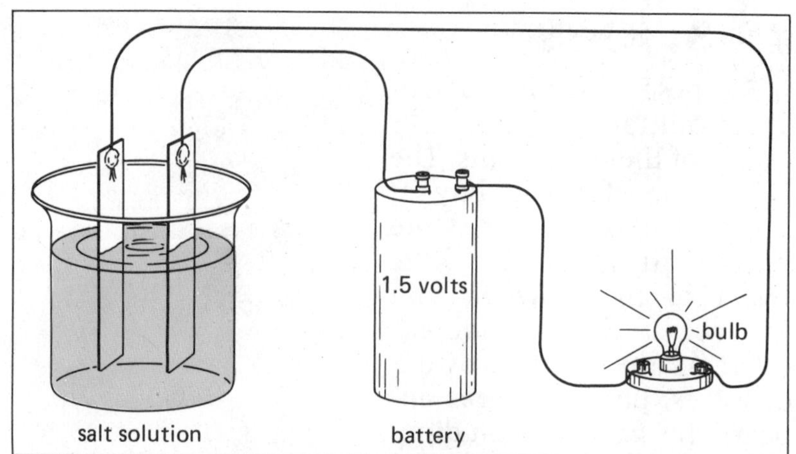

12-11A. *When NaCl dissolves in water, each Na^+ ion is surrounded by the negative ends of the water molecules. Each Cl^- ion is surrounded by the positive ends of the water molecules.*
B. *When sugar dissolves in water, each sugar molecule becomes bonded to a group of water molecules.*

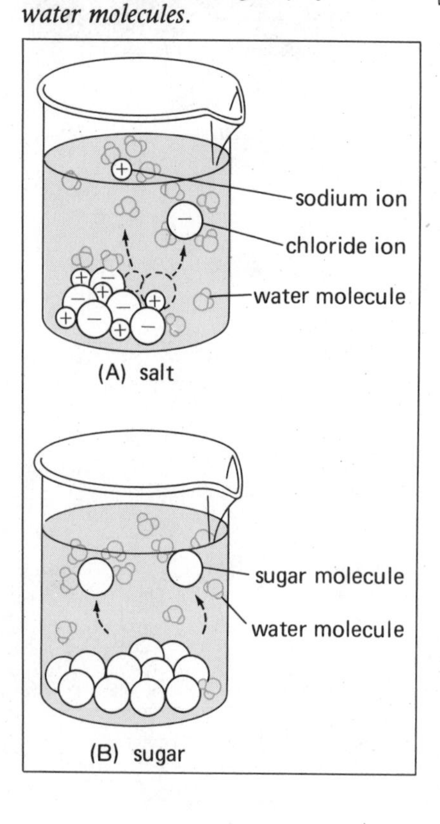

Nonelectrolyte
A substance that forms a nonconducting solution when dissolved in water.

either a positive or a negative electrical charge. The charged ions are free to move around in the solution. When a metal conducts an electric current, moving electrons carry the current. In a solution, moving ions allow the current to flow. See Fig. 12-11.

Some substances do not produce conducting solutions when they dissolve in water. An example is sugar. When sugar dissolves, the sugar molecules separate and spread out in the water. The sugar molecules do not carry electrical charges as ions. There are no electrically charged ions in a sugar solution to cause a current to flow. Substances like sugar are called **nonelectrolytes.** A *nonelectrolyte* does not conduct an electric current.

Many chemical reactions will take place only when the reactants are dissolved in water. For example, the reactants in baking powder will not produce carbon dioxide gas until water is added. Why do many chemical reactions only take place when the reactants are dissolved in water? Study the behavior of the ions released when electrolytes dissolve.

Suppose you mixed solid sodium chloride, NaCl, with solid silver nitrate, $AgNO_3$. Like NaCl, the compound $AgNO_3$ is an electrolyte. When $AgNO_3$ is dissolved in water, it produces two kinds of ions, as shown by the following equation:

$$AgNO_3 \longrightarrow Ag^+ + NO_3^-$$

If solid NaCl is mixed with solid $AgNO_3$, there is no reaction. The ions from both compounds are locked in the solid crystals and cannot move. But if colorless solutions of NaCl and $AgNO_3$ are mixed together, a white solid immediately forms. Such a solid substance that separates from a solution is called a **precipitate** (prih-**sip**-uh-tate). If a *precipitate* forms when two colorless solutions are mixed together, it means that a chemical reaction has taken place. Since the NaCl and $AgNO_3$ solutions contained ions, the chemical reaction forming the precipitate must be caused by the ions.

A mixture of NaCl and $AgNO_3$ solution will contain four kinds of ions as shown below:

$$NaCl + AgNO_3 \longrightarrow Na^+ + Cl^- + Ag^+ + NO_3^-$$

See Fig. 12-12. These ions are all free to move in the solution. As you know, opposite-charged ions are attracted. Therefore, Na^+ will move toward NO_3^- and Ag^+ will be attracted to Cl^-. Sometimes opposite-charged ions can come together to form a substance that remains dissolved in water. This type of substance is said to be **soluble** in water.

Look at Table 12-2, which lists some common ions. By looking at this table, you can tell whether a particular combination of ions makes a soluble compound (S) or forms a precipitate (P). You can also identify the precipitate that forms when NaCl and $AgNO_3$ solutions are mixed. Is the precipitate $NaNO_3$ or AgCl? Which compound is soluble in water?

In some chemical reactions, the ions form a gas when they combine. For example, baking powder produces carbon dioxide gas when dissolved in water. The presence of ions in solutions explains how chemical reactions take place when substances are dissolved.

Precipitate
A solid substance that separates from a solution.

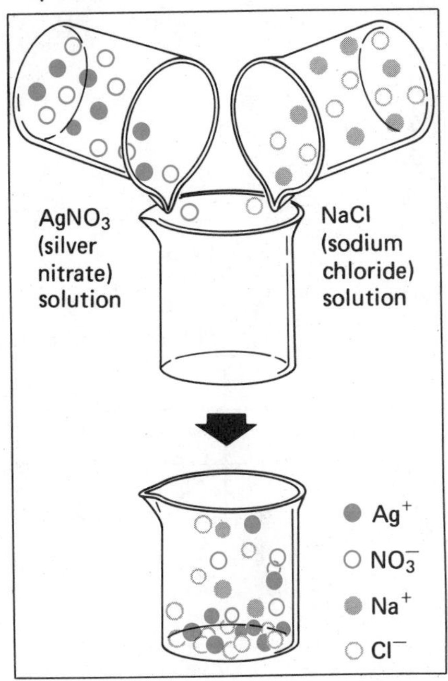

12-12. *A mixture of $AgNO_3$ and NaCl solutions will contain the ions Ag^+, NO_3^-, Na^+, and Cl^-.*

Soluble
A description of a substance that can be dissolved.

TABLE 12-2
Solubilities of Some Ions in Water Solution
(P = Precipitate; S = Soluble)

	nitrate (NO_3^-)	chloride (Cl^-)	acetate ($C_2H_3O_2^-$)	carbonate (CO_3^-)	tetraborate ($B_4O_7^{-2}$)	sulfate (SO_4^{-2})
silver (Ag^+)	S	P	P	P	S	P
sodium (Na^+)	S	S	S	S	S	S
hydrogen (H^+)	S	S	S	—	S	S
ammonium (NH_4^+)	S	S	S	S	S	S
calcium (Ca^{+2})	S	S	S	P	P	P
zinc (Zn^{+2})	S	S	S	P	P	S
magnesium (Mg^{+2})	S	S	S	P	S	S

ACTIVITY

Materials
set of solutions labeled A, B, C, D
unknown solution
medicine dropper
waxed paper

TABLE 12-3			
	A $ZnCl_2$	B Na_2CO_3	C $MgSO_4$
D $Na_2B_4O_7$			
C $MgSO_4$			
B Na_2CO_3			

TABLE 12-4				
	A $ZnCl_2$	B Na_2CO_3	C $MgSO_4$	D $Na_2B_4O_7$
Unknown W, X, Y, or Z				

Solubility

You will now use the solubility table to test all the possible combinations of four ion solutions. You will determine which combination of ions forms a precipitate.

A. Obtain the materials listed in the margin.

B. In your notebook, copy Table 12-3.

The squares in the table represent all the possible combinations of ions in the four solutions.

C. Place a sheet of waxed paper over the table. All the tests will be done on this waxed paper.

D. Using a medicine dropper, place one drop of solution A on each square in column A. (Columns are up and down, rows are across.)

E. Rinse out the medicine dropper with water.

F. Now place one drop of solution B in the squares of column B and one drop of solution C in the squares of column C.

G. Now test the solutions by adding a drop of solution D to each square of row D. Also add a drop of solution C to each square in row C and a drop of solution B to each square of row B.

You now have one drop of two different solutions in each square.

1. How many precipitates formed?
2. Which solutions formed precipitates when mixed with $ZnCl_2$ (solution A)?
3. Which solutions formed precipitates when mixed with Na_2CO_3 (solution B)?
4. Write equations for all the reactions that took place. Circle the precipitates that formed.

H. After removing the waxed paper on your table, place a "P" in the squares in which a precipitate formed.

I. Obtain an unknown solution from your teacher.

J. Copy Table 12-4 in your notebook.

K. Place a clean sheet of waxed paper over this table.

L. Place one drop of your unknown in each of the four squares.

M. Add one drop of solution A, B, C, and D.

N. Compare the precipitates formed in step M with the precipitates formed in Table 12-2.

Your unknown is either solution A, B, C, or D. Therefore, by comparing your two tables, you should be able to identify your unknown.

5. Is your unknown $ZnCl_2$ (solution A), Na_2CO_3 (solution B), $MgSO_4$ (solution C), or $Na_2B_4O_7$ (solution D)?

SUMMARY

Why does baking powder react only when water is added to it? Baking powder is made up of substances that are electrolytes. When these substances are dissolved in water, their ions are free to react. Many chemical reactions take place only between ions in solution.

QUESTIONS

Unless otherwise indicated, use complete sentences to write your answers.

1. Why can some solutions conduct an electric current while others cannot?
2. What happens when silver ions (Ag^+) and chloride ions (Cl^-) are in solution together?
3. Using the solubilities in Table 12-2, determine which of the following are precipitates: $NaNO_3$, $AgCl$, $CaCO_3$.
4. What is an electrolyte?
5. ZnB_4O_7 forms as a precipitate in the following chemical reaction. The remaining atoms become ions. Copy and complete the equation.

   $ZnCl_2 + Na_2B_4O_7 \longrightarrow$ _____ + _____ + _____

6. Using Table 12-2, complete the following chemical equation. (Remember, the nitrate radical NO_3 must be kept as a unit.)

   $AgNO_3 + NaCl \longrightarrow$ _____ + _____ + _____

LABORATORY

Materials
small glass container
3 electrical wires with alligator clips
2 flashlight batteries
flashlight bulb in holder
2 electrodes (flat metal)
paper towels
small plastic spoon
distilled water
samples of: water, table salt (sodium chloride), baking soda, magnesia laxative, battery paste, borax, Epsom salt (magnesium sulfate), household ammonia, sugar, vinegar, others (See your teacher.)

Electrical Conductivity of Solutions

Purpose

In this laboratory, you will test several materials dissolved in water to see if they conduct electricity. Materials that conduct an electric current when dissolved in water are called electrolytes. Materials that do not conduct an electric current when dissolved in water are called nonelectrolytes. The electric circuit you will use is a series circuit. The circuit is put together in such a way that a flashlight bulb will glow when enough current flows through it.

Procedure

A. Obtain the materials listed in the margin. Follow these directions to assemble the equipment.

B. Place the two batteries in the fold of an open book. If a battery holder is available, use it in place of the book.

C. Attach one wire to one of the terminals holding the flashlight bulb. Place the unattached end of this wire near one end of the batteries.

D. Attach a second wire to one of the flat electrodes. Place the unattached end of this wire near the other end of the batteries.

E. Use the third wire to attach the second flat metal electrode to the other terminal of the flashlight bulb holder.

F. While your partner holds the two wire ends tightly to the ends of the batteries, touch the electrodes together. The bulb should glow. If it does not glow, check your connections and try again. Check with your teacher if the bulb will not glow.

G. Copy the table on p. 391 in your notebook and record your results in it. DO NOT WRITE IN THIS BOOK.

H. Fill the glass container about one-half full of water. Place the electrodes in it while your partner holds the other wires to the batteries. Hold the electrodes close to

Material tested	Conductivity: + conducts − does not conduct	Comments
1. water		
2. table salt, dry		
3. salt water		
4. baking soda		
5. magnesia laxative		
6. battery paste		
7. borax		
8. Epsom salt		
9. ammonia		
10. sugar		
11. vinegar		
12. etc.		

each other in the water. Do not allow them to touch. Record the results of this test in your table.

I. Remove the electrodes. Dry them with a paper towel. In a similar way, test a spoonful of dry table salt. Place the electrodes directly in the salt. Record your results.

J. Place the salt in the container of water. Stir to dissolve.

K. Test this solution and record your results.

L. Discard the salt solution and rinse the electrodes and container 2 or 3 times with distilled water.

You are now ready to test the other materials. Use the "Comments" column to record how brightly the bulb glows for positive tests.

M. Repeat steps J, K, and L, making solutions of all the solids to be tested. Rinse the electrodes and container after each test.

N. Liquids are to be tested without adding water. Fill the container one-half full with each liquid and test.

Summary

Summarize the results of your investigation by describing how you tested the materials. Explain the procedure and record your results, listing electrolytes and nonelectrolytes separately.

VOCABULARY REVIEW

Match the number of the word(s) with the letter of the phrase that best explains it.

1. metallic bond
2. corrosion
3. metalloid
4. polar molecule
5. solvent
6. solute
7. polluted
8. electrolyte
9. precipitate
10. soluble

a. A nonmetal that conducts electricity.
b. The part of a solution that does the dissolving.
c. A bond in which electrons are not held tightly by any particular atom.
d. A molecule that carries small charges on opposite ends.
e. A description of water that has harmful substances in it.
f. The part of a solution that is dissolved.
g. The eating away of a surface from chemical action.
h. A solid substance that separates from a solution.
i. A description of a substance that can be dissolved.
j. A substance that forms a conducting solution when dissolved in water.

REVIEW QUESTIONS

Choose the letter of the answer that best completes the statement or answers the question.

1. An element conducts electricity and heat and has a shiny appearance. The element is probably a **a.** nonmetal **b.** liquid **c.** gas **d.** metal.
2. The facts that electrons are shared equally by neighboring atoms and that atoms can slide around each other are true of atoms held together by **a.** ionic bonds **b.** covalent bonds **c.** metallic bonds **d.** nonmetallic bonds.
3. Refer to Fig. 12-5, p. 378. The nonmetallic elements are located in the periodic chart **a.** on the left side **b.** on the right side **c.** in the middle **d.** across the bottom.
4. Much of the behavior of water molecules is the result of their being **a.** polar molecules **b.** bonded ionically **c.** in an uncharged form **d.** spherical in shape.

5. Pollution of water is a serious problem mainly because **a.** most of the earth's water is available for us to pollute **b.** the supply of the earth's water available for our use is limited **c.** it is difficult to pollute water **d.** the harmful materials stay in one place.
6. A substance dissolved in water is called a **a.** polar molecule **b.** solvent **c.** solute **d.** dispenser.
7. Some substances form solutions that can conduct electricity because in solution they **a.** form ions **b.** become all positive particles **c.** become all negative particles **d.** become very stable.
8. When table salt is dissolved in water, **a.** salt molecules form **b.** sodium ions and chloride ions form **c.** salt particles surround water molecules **d.** a precipitate always forms.
9. When Ag^+ and Cl^- ions are in solution together, they **a.** repel each other **b.** prevent electricty from flowing **c.** act as the solvent **d.** form a white precipitate.
10. Using the solubilities in Table 12-2, which of the following will be precipitates? **a.** Ag_2CO_3 **b.** $NaCO_3$ **c.** $CaCl_2$ **d.** NaCl.

REVIEW EXERCISES

Give complete but brief answers to each of the following. Use complete sentences to write your answers.

1. List some ways in which all metals are alike.
2. Look at the periodic chart in Fig. 10-16, p. 326, and describe where the metalloids are located in relation to the metals and nonmetals.
3. What is meant by a metallic bond?
4. If the electron arrangement and shape of the water molecules are as they are shown in the electron-dot symbol in Fig. 12-6 on p. 380, explain why one side of the molecule has a positive charge and the other a negative charge.
5. Describe what happens when a substance such as salt is dissolved in water.
6. Why has water pollution become such a serious problem?
7. Why do some dissolved substances form solutions that can conduct electricity?
8. What is a precipitate?
9. When the following pairs of ions are in solution, which will form a precipitate? Ag^+ and Cl^-, Zn^{+2} and Cl^-, Mg^{+2} and CO_3^{-2}.
10. If the following ions are all in a solution, will a precipitate form? Na^+, $B_4O_7^{-2}$, Zn^{+2}, Cl^{-2}. Give the formula(s) of any precipitate(s) that form.

EXTENSIONS

1. Study the uses of the lesser known metals. You are familiar with the uses of copper, iron, and aluminum. What are some uses of molybdenum, barium, and germanium? Write a report that includes at least six less common metals.
2. Report to the class on how metals are removed from their ores.
3. Write a report on water softeners and how they work. What do they remove from hard water? Do they add anything to the water?
4. Find out how the water you use in your home is treated. Are some types of sewage treatments not done because they are too expensive or take too long?

UNIT 4

the reactions of matter

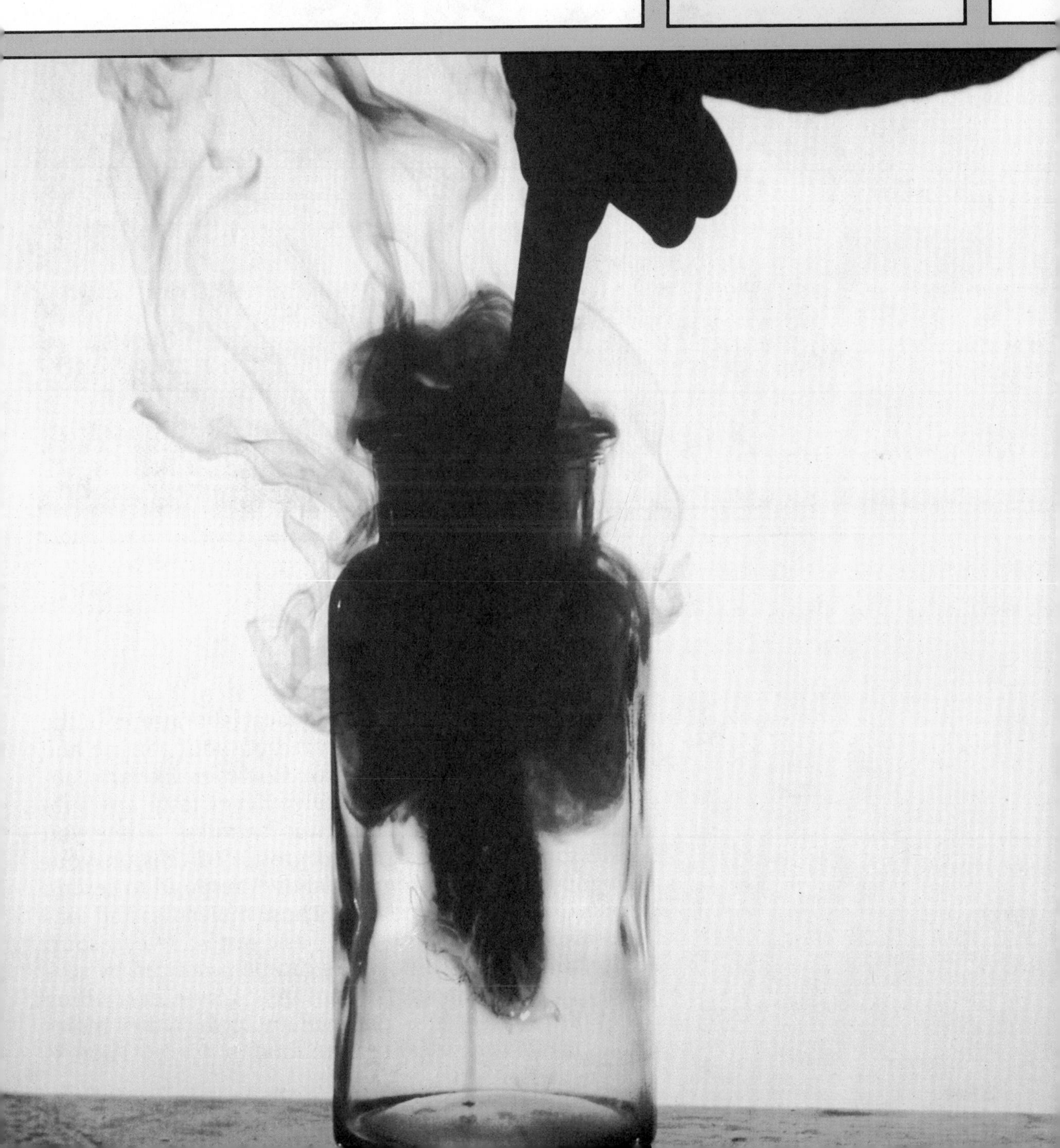

CHAPTER 13

Acids, Bases, and Salts

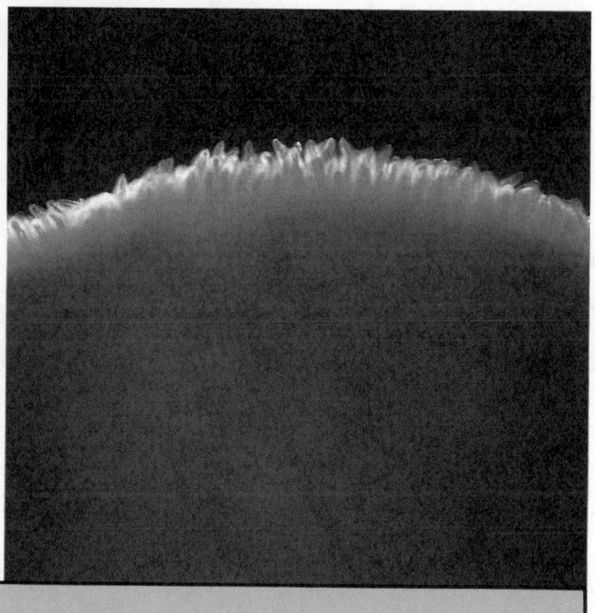

13-1. ACIDS AND THEIR PROPERTIES

You own one of the best chemical laboratories in the world. This laboratory is located on your tongue and determines your sense of taste. However, like any laboratory, it must be used safely. Never taste any substance until you know that it is harmless. When you taste something, bud-shaped groups of cells on your tongue (see the photograph above) respond to certain kinds of molecules and ions. These molecules and ions are responsible for the four basic tastes: sweet, sour, salty, and bitter. Some tastes can be produced by several kinds of molecules or ions. As a result, everything that tastes sweet may not contain sugar. Other molecules can give a sweet taste to food or drinks. There is one taste, however, that is produced by only one kind of ion. In this lesson, you will discover what that taste is and what the ion is that causes it.

When you finish lesson 1, you will be able to:

- Describe what is meant by an *acidic* solution.
- Use examples to show some chemical properties of an acidic solution.
- Name several common substances that form acids.
- ○ Test the acidic properties of two solutions.

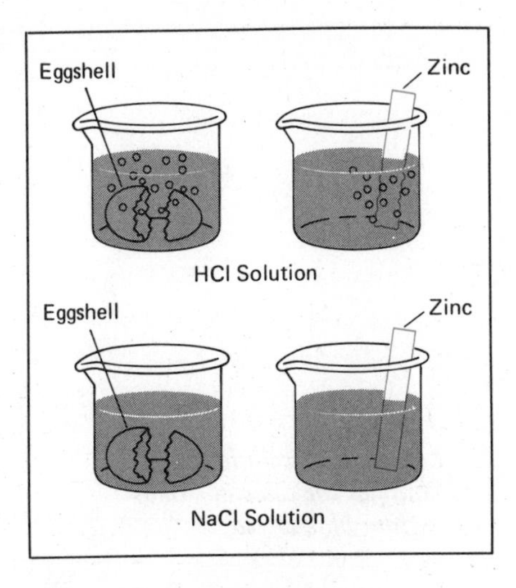

13-1. *The formation of gas bubbles is an indication that eggshell and zinc dissolve in a solution of HCl. No reaction is seen in a solution of NaCl.*

You can begin this lesson by comparing two solutions. One is a salt solution. A salt solution contains sodium ions, Na^+, and chloride ions, Cl^-. The second solution is made by dissolving molecules of the electrolyte hydrogen chloride, HCl, in water. This solution of hydrogen chloride contains hydrogen ions, H^+, and chloride ions, Cl^-. Thus both solutions contain Cl^- ions.

The NaCl and HCl solutions are alike in several ways. Both solutions are colorless, contain Cl^- ions, and can conduct an electric current. But the two solutions are different in some ways. For example, a piece of eggshell dropped into the HCl solution will dissolve. Bubbles of carbon dioxide gas will be given off as the eggshell dissolves. A piece of zinc metal put into the HCl solution will also dissolve. Hydrogen gas will be given off as the zinc dissolves. Neither the eggshell nor the zinc will change when put into the NaCl solution. See Fig. 13-1. If you think about this experiment, you should begin to realize that the H^+ ions in the HCl solution must be the cause of this behavior. Other experiments show that any substance containing H^+ ions can form a solution that dissolves eggshell and zinc.

The presence of H^+ ions in a solution gives that solution special properties. Such a solution is called an **acidic solution.** To make an *acidic solution*, a material is dissolved in water, causing the resulting solution to have more H^+ ions than the plain water. A compound like HCl, which produces H^+ ions when dissolved in

Acidic solution
A solution containing more H^+ ions than pure water.

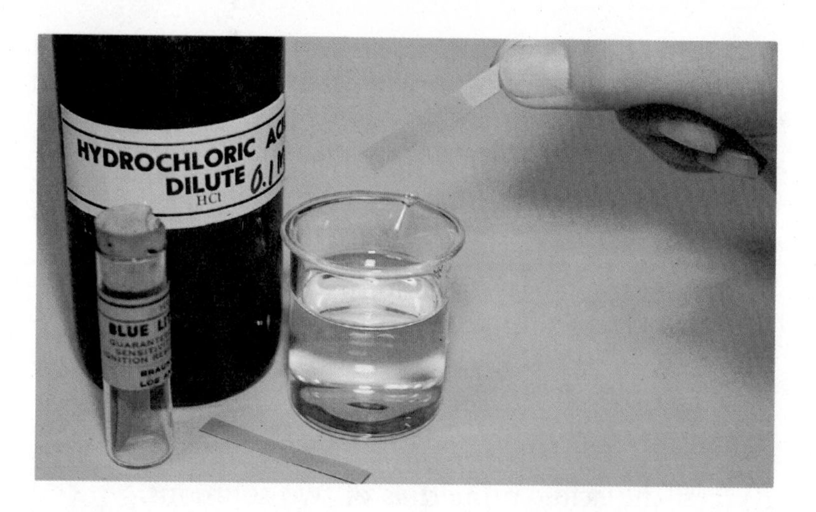

13-2. *An acidic solution such as HCl changes the color of litmus paper from blue to red.*

water, is called an *acid*. A solution of HCl is commonly called *hydrochloric acid*.

You can recognize an acid by the unusual properties caused by the H^+ ion. Acidic solutions always taste sour. Food and drinks containing acids always have a sour taste. It is not safe to taste all kinds of acids. Some acids cause painful chemical burns. Other acids are poisonous.

A safe way to test for acidic solutions is to use certain substances that change color in an acidic solution. Any substance that changes color in an acidic solution is called an **indicator.** One very commonly used *indicator* is *litmus*. An acidic solution causes litmus paper to turn from blue to red. See Fig. 13-2.

Indicator
A substance that changes color when put into an acid solution.

If you test vinegar with litmus paper or some other indicator, you will find that it is an acidic solution. You also know that vinegar has the sour taste common to all acids. Vinegar also is an electrolyte since it forms a conducting solution. However, a vinegar solution is a poor conductor of electricity. This means that vinegar is not able to produce a large number of ions in solution. In other words, vinegar is a weak electrolyte. Acids like vinegar, which are weak electrolytes, are said to be *weak acids*. Other weak acids are lemon juice and vitamin C. Hydrochloric acid, on the other hand, is a strong electrolyte and is also a *strong acid*.

Another strong acid is the one found in automobile batteries. This acid is sulfuric acid, H_2SO_4. In addition to its use in batteries, sulfuric acid is an important chemical used in manufacturing such products as steel and fertilizers. A third strong acid found in chemical laboratories is nitric acid, HNO_3. Nitric acid is also an important acid used in manufacturing. For example, it is used in the production of plastics and explosives.

Sometimes acidic solutions are harmful to our environment. An example is *acid rain*. Rain water can become acidic if certain gases in the air dissolve in the water as it falls through the air. Gases responsible for acid rain are sulfur dioxide and oxides of nitrogen. These gases are released into the air when coal and petroleum fuels are burned. Acid rain can fall in regions where large amounts of these fuels are used. It has occurred in northern Europe, northeastern United States, and Canada. When acid rain falls on soil, it may harm the ability of the soil to support plant growth. Plants exposed to acid rain are easily attacked by disease and pests. Lakes may become so acidic that they can no longer support fish. At the present time, scientists are not certain about the long-range effects of acid rain. However, the possible threat from acid rain requires that careful attention be paid to the release of sulfur and nitrogen oxides into the atmosphere.

ACTIVITY

Acidic Reactions

A. Obtain the materials listed in the margin. CAUTION: Be careful not to spill any acid. CAUTION: Wear goggles for this activity.

B. Fill three test tubes 1/3 full of HCl solution. Label these tubes A1, A2, and A3. Fill three other test tubes 1/3 full of NaCl solution. Label these tubes B1, B2, and B3.

C. Drop a pinch of baking soda into tube A1. Record your observations in a data table similar to the following.

D. Drop a piece of aluminum foil into tube A2. Record the result in your table.

E. Drop a piece of purple cabbage leaf into tube A3. Record the result.

F. Now repeat steps C, D, and E, using tubes B1, B2, and B3.

1. Which solution showed a reaction in all three tests, HCl or NaCl?
2. Both HCl and NaCl solutions contain Cl^- ions. Which ion must have been responsible for the reactions you observed, H^+ or Na^+?

Materials

6 test tubes
goggles
test tube holder
baking soda
aluminum foil
purple cabbage leaf
sodium chloride solution
hydrogen chloride solution

	Reaction with HCl (tubes A1–A3)	*Reaction with NaCl (tubes B1–B3)*
baking soda		
aluminum foil		
purple cabbage leaf		

SUMMARY

A solution containing a large number of H^+ ions is an acid. A person can identify an acidic solution by its sour taste. A safer way to test for an acid is to use an indicator such as litmus. Blue litmus paper turns red in the presence of an acid.

QUESTIONS

Use complete sentences to write your answers.

1. What is meant by an acidic solution?
2. Why are some acids dangerous?
3. Describe a safe test for acids.
4. Name three common substances that are acids.

13-2. BASES AND THEIR PROPERTIES

You may have played a party game that tests your sense of taste. In this game, a person is blindfolded. He or she is then given several different substances to taste. The object is to identify the substance based on its taste. Suppose you are playing this game. You are given some food that has a sour taste. Even if you cannot identify the food, there is one thing you will know immediately. The food must contain an acid.

Just as a sour taste is caused by the H^+ ions in an acid, there is another taste that is caused by a particular ion. In this lesson, you will learn what that taste is and what the ion is that causes it.

When you finish lesson 2, you will be able to:

- Describe what is meant by a *basic solution.*
- Distinguish between an acid and a *base*.
- Name several common basic substances.
- ○ Test some common substances to determine if they are acids or bases.

An acidic solution is one that contains an excess of H^+ ions. These H^+ ions give acidic solutions a sour taste. Have you ever had soapsuds in your mouth by accident? If you have, then you know that soap tastes bitter. This bitter taste is due to the presence of the hydroxide ion, OH^-. Consider the compound sodium hydroxide, NaOH. When NaOH is dissolved in water, it separates into the sodium ion, Na^+, and the hydroxide ion, OH^-. A solution of NaOH in water thus has more OH^- ions than pure water. A compound such as NaOH that increases the number of OH^- ions

Base
A compound that forms a solution with more OH^- ions than pure water.

when it dissolves in water is called a **base.** All *basic solutions* have a bitter taste. It is just as dangerous to taste a *base* as it is to taste an acid. Some basic solutions can cause severe burns and some are poisonous.

Sodium hydroxide is a *strong base*. This means that NaOH is also a strong electrolyte. Bases are able to dissolve grease and some other materials, such as hair. For this reason, the strong base NaOH is used to make drain cleaners. However, you should remember that strong bases are very dangerous to touch. Drain cleaners containing sodium hydroxide or other strong bases must be handled carefully and stored in a safe place away from small children. Never taste household products such as drain cleaners. In addition to its use as a drain cleaner, sodium hydroxide is used in the refining of petroleum and in producing certain plastics.

Many common bases are *weak*. An example of a weak base is magnesium hydroxide, $MgOH_2$. Magnesium hydroxide is used as a medicine in the form of magnesia laxatives. Calcium hydroxide, $CA(OH)_2$, is a weak base often called *lime*. It is used to make plastic and mortar, to soften hard water, and to make soil less acidic.

A very common weak base is household ammonia. It is useful as a cleaner because its basic properties allow it to dissolve greasy dirt. Ammonia solutions are made by dissolving ammonia gas, NH_3, in water. The resulting solution is a base because some of the dissolved ammonia molecules react with water to

13-3. *A basic solution such as ammonia changes the color of litmus paper from red to blue.*

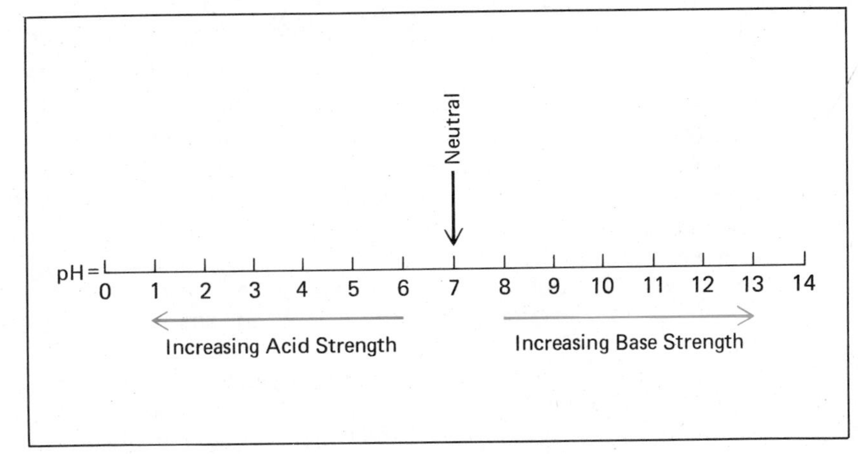

13-4. The pH scale indicates the strength of acidic and basic solutions. Pure water, which is neutral, has a pH of 7.

produce OH^- ions, as shown by the following chemical equation:

$$NH_3 + H_2O \longrightarrow NH_4^+ + OH^-$$

You can test for the presence of a base by using an indicator such as litmus. An acid turns blue litmus paper to red. A base does just the opposite. Red litmus paper turns blue in the presence of a base. See Fig. 13-3.

If a substance does not react with either red or blue litmus paper, then it is neither an acid nor a base. A solution that is neither an acid nor a base is **neutral.** Water, as you can discover by testing, is *neutral*. When the H^+ ions from an acid react with the OH^- ions from a base, water is formed:

$$H^+ + OH^- \longrightarrow H_2O \text{ (or HOH)}$$

This reaction is called **neutralization** (noo-truh-luh-**zay**-shun). A *neutralization* reaction occurs when an acid and a base are mixed, forming water.

Scientists often find it necessary to measure the strength of an acid or a base solution. This is done by using a special scale called *pH*. The pH scale uses numbers to describe how acidic or basic a solution is. The scale goes from 0 to 14. A pH number between 0 and 7 describes an acidic solution. A number between 7 and 14 indicates a basic solution. A pH number of 7 means that the solution is neutral. See Fig. 13-4. A change of one pH number means a change in acidic or basic strength of 10×. That is, a solution of pH 9 is 10 times stronger in basic strength than a solution of pH 8. Table 13-1 shows the pH of some common liquids.

Neutral
A solution or substance that is neither an acid nor a base.

Neutralization
A chemcial reaction that occurs when an acid and a base are mixed.

TABLE 13-1
pH of Some Common Liquids

Liquid	*pH*
digestive juices of stomach	1.6
lemon juice	2.3
vinegar	2.8
soft drink	3.0
orange juice	3.5
milk	6.5
pure water	7.0
human blood	7.4
sea water	8.5
milk of magnesia	10.5

ACTIVITY

Materials

red and blue litmus paper
plastic spoon
following substances in dispensing bottles:
- distilled water
- lemon juice
- household ammonia
- table salt (sodium chloride)
- baking soda (sodium bicarbonate)
- vitamin C (powdered)
- sugar
- vinegar
- others (see teacher)

Classifying Acids and Bases

In this activity, you will test several substances with litmus paper to determine if they are acids or bases. As with acids, wash off any spilled bases thoroughly with water.

A. Obtain the materials listed in the margin.

B. Copy the following data table in your notebook. Use this table to record the results of your tests.

C. To test solids, wet the litmus paper with a drop of water and add a pinch of the solid to it. See Fig. 13-5.

D. Liquid substances, such as vinegar, are already in solution. To test a liquid, simply place one or two drops on the litmus paper.

If the red litmus paper turns blue, place an X in the column headed "Base" next to the substance tested. If the blue litmus paper turns red, place an X in the column headed "Acid." If a substance has no effect on either the red or blue litmus paper, place an X in the "No Reaction" column.

E. Now proceed to test all the substances available. Be sure to record your results in the table.

1. Which substances did you find to be bases?
2. Which substances did you find to be acids?
3. Which substances had no effect on either the blue or red litmus paper?
4. How would you expect the basic solutions to taste?

13-5.

Substance Tested	*Base*	*No Reaction*	*Acid*
distilled water			
lemon juice			
household ammonia			
table salt			
baking soda			
vitamin C (powdered)			
sugar			
vinegar			

SUMMARY

Water is a neutral solution. It has no taste. A solution containing more OH^- ions than plain water is a base. Bases can be identified by their bitter taste. There are safer ways to test for the presence of a base than by tasting. One way is to use an indicator such as litmus. In the presence of a base, red litmus paper turns blue.

QUESTIONS

Use complete sentences to write your answers.

1. Describe what is meant by a base.
2. While it is not a good idea to taste acids and bases in the laboratory, they do have distinct tastes. What are they?
3. Describe a safe test to determine if a substance is a base.
4. **a.** Name two common substances that are bases.
   b. How are they used?

13-3. AN ACID + A BASE = A SALT

Table salt, sodium chloride, is a common chemical compound. The oceans contain a huge amount of it dissolved in their water. Solid sodium chloride is also found as a gray-colored substance called rock salt in large underground deposits. These deposits of rock salt can be mined, as shown in the photograph. Salt is a valuable substance. Two thousand years ago, Roman soldiers were often paid in salt. The Latin word for salt is *sal* and our word "salary" comes from this ancient Roman custom. Salt is still a valuable raw material used to manufacture many chemicals, such as hydrochloric acid and baking soda. Salt is also a vital part of our diet. To the chemist, salt describes not only sodium chloride, it is also the name of a large group of compounds. In this lesson, you will study one way in which different kinds of salt can be made.

When you finish lesson 3, you will be able to:

- Explain how an acid and a base can join to form a salt.
- Explain what is meant by neutralization.
- Predict the results of some neutralization reactions.
- ○ Use an indicator to find the end point in a neutralization reaction.

An acid and a base are chemical opposites. One way to show this is to mix an acid with a base. If the correct amounts are used, the mixture will be neutral. This happens because the H^+ ions from the acid join the OH^- ions from the base. The result is water, as shown by the following equation:

$$H^+ + OH^- \longrightarrow H_2O$$

Heat is also produced in addition to water. However, this equation does not describe the complete reaction between an acid and a base. Consider the equation given below. This equation shows the reaction between sodium hydroxide, NaOH, and hydrochloric acid, HCl.

$$NaOH + HCl \longrightarrow H_2O + NaCl + heat$$

The H^+ ion from HCl and the OH^- ion from NaOH have joined to form water. In addition, the Na^+ ion from NaOH and the Cl^- ion from HCl have formed NaCl. Sodium chloride is, of course, common table salt. This is an example of a neutralization reaction. When an acid and a base are combined, a neutralization reaction always takes place.

When exactly the right amount of an acid reacts with a base, the properties of both the acid and the base are destroyed. This fact allows a chemist to detect the point at which the acid and the base have been neutralized. For example, the indicator *phenolphthalein* (fee-noe-**thal**-een) turns from colorless in acid to pink in base. You can add a few drops of phenolphthalein indicator to the acid you want to neutralize. Then add a base until the solution takes on a slightly pink color. The point at which an indicator shows the first sign of change is called the *end point*.

Suppose you were to perform neutralization reactions between different acids and bases. You would find that one of the products, in addition to water and heat, would always be a compound called a **salt.** Neutralization reactions produce *salts* when the metallic ion from a base joins with the nonmetallic ion from an acid. Another neutralization reaction that shows this is:

Salt
A compound that is made up of the metallic ion from a base and the nonmetallic ion from an acid.

KOH	$+ HNO_3$	$\rightarrow$	$H_2O +$	KNO_3	$+$ heat
potassium hydroxide (base)	nitric acid (acid)		water	potassium nitrate (salt)	

Salts may also be produced by reactions other than neutralization.

13-6. *The oldest method of obtaining salt from sea water is by letting the water dry up in shallow bays or inlets.*

As you learned at the beginning of this lesson, some of our salt is obtained from underground mines. Salt can also be obtained by the evaporation of sea water. See Fig. 13-6. The oceans contain approximately 50 million billion tons of salt. There are 35 g of salt in 1 kg of salt water. See Table 13-2. Salt obtained from sea water is sold as *sea salt*.

TABLE 13-2
Dissolved Substances in Sea Water

Ion	*Percent of Total Dissolved Solids (by weight)*
Chloride (Cl^-)	55.04
Sodium (Na^+)	30.61
Sulfate (SO_4^{-2})	7.68
Magnesium (Mg^{+2})	3.69
Calcium (Ca^{+2})	1.16
Potassium (K^+)	1.10
Bicarbonate (HCO_3^-)	0.41
Bromide (Br^-)	0.19
Borate ($H_2BO_3^-$)	0.07
Strontium (Sr^{+2})	0.04

How would some of your favorite foods taste without salt? Our taste buds record only four tastes: sweet, sour, bitter, and salty. Adding salt to food usually improves its taste. We also need a certain amount of salt in our diets in order to live. Salt must be present in our body fluids in order to maintain the correct balance of water within our bodies. Only small amounts of salt are needed in our diets in order to provide the amount needed in our bodies. Too much salt can cause our bodies to hold too much water. This may cause a rise in blood pressure, which can produce serious health problems. It is a good idea to avoid eating large amounts of salty foods and to try to use the salt shaker less.

ACTIVITY

The End Point in a Neutralization Reaction

A. Obtain the materials listed in the margin.

B. Copy the following table and record your data in it as you perform the tests. CAUTION: Wear goggles.

Substance	*Number of Drops of NaOH*
HCl	

C. Fill a test tube about 1/3 full of HCl solution. See Fig. 13-7.

D. Place one or two drops of phenolphthalein solution in the test tube with the acid. Shake the test tube to mix the indicator with the acid. CAUTION: Use care not to splash the acid. Any acid that gets on your skin should be rinsed off immediately.

1. What color is the acid-indicator solution?
2. What color will the indicator change to at the end point?

E. Using a medicine dropper, add NaOH solution to the test tube one drop at a time. Shake the test tube each time you add a drop of NaOH. Keep a record of the number of drops of NaOH you add. When the light pink color remains, even after shaking the test tube, you have reached the end point. See Fig. 13-8.

3. How many drops of NaOH were required to reach the end point? Record this in the data table.

F. Add several more drops of NaOH to go past the end point.

4. What happened to the color of the solution?

The neutralization of HCl and NaOH results in the formation of water molecules.

5. Write a complete chemical equation for the reaction between HCl and NaOH.

G. In the same manner, neutralize at least two more acidic substances. Record your results in the data table.

Materials

phenolphthalein solution
hydrochloric acid solution
sodium hydroxide solution
medicine dropper
test tube
test tube holder or small beaker
acid samples
goggles

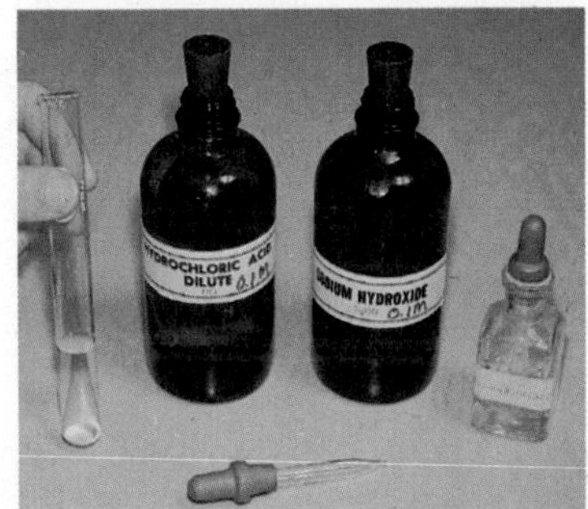

13-7.

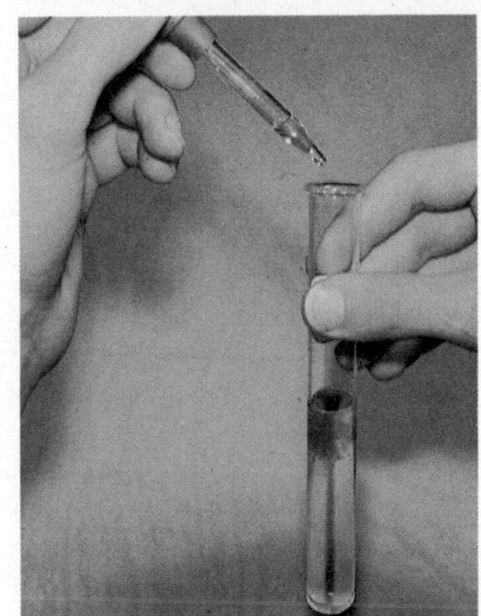

13-8.

SUMMARY

Write a brief description of how you would neutralize an acidic solution. When an acid and a base are combined, a neutralization reaction takes place. The products of neutralization are water, salt, and heat. Many different kinds of salts can be produced when different acids and bases react.

QUESTIONS

Unless otherwise indicated, use complete sentences to write your answers.

1. Write an equation that shows what happens when the base NaOH and the acid HCl are mixed.
2. What is the name of the reaction that occurs when a base and an acid are mixed? What is the product called?
3. What ions are present in HCl, HBr, and HI? What ions are present in KOH, LiOH, and $Mg(OH)_2$?
4. What are the chemical formulas for the products of mixing HCl and NaOH? of mixing HCl and LiOH?

Neutralization of $Ba(OH)_2$ with H_2SO_4

Purpose

Water molecules are always formed when an acid reacts with a hydroxide base. Water is a poor conductor of electricity. The production of another poor conductor during the neutralization process provides a way of detecting the end point using conductivity. In this laboratory, you will neutralize barium hydroxide, $Ba(OH)_2$, with sulfuric acid, H_2SO_4. The equation for this reaction is as follows:

$$\underset{\text{barium hydroxide}}{Ba^{+2}2OH^-} + \underset{\text{sulfuric acid}}{2H^+SO_4^{-2}} \longrightarrow \underset{\text{water}}{2H_2O} + \underset{\text{barium sulfate}}{BaSO_4}$$

Barium sulfate forms as a precipitate; it will not remain dissolved in water. Therefore, when exactly the right amount of H_2SO_4 is added to $Ba(OH)_2$, no ions will remain in the solution. The solution will not conduct electricity. In this laboratory exercise, you will carry out this reaction. You will use the conductivity of electricity to determine the end point of the reaction.

CAUTION: ALWAYS WEAR GOGGLES WHEN WORKING WITH CHEMICALS IN THE LABORATORY.

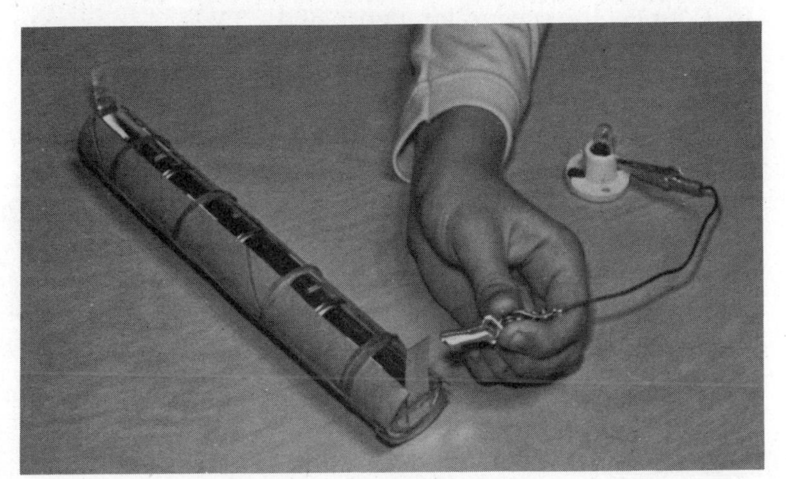

13-9.

Materials

- test tube
- goggles
- test tube holder or small beaker
- 4 batteries
- electric wire connector
- TV antenna lead-in wire
- flashlight bulb in holder
- medicine dropper
- barium hydroxide solution
- sulfuric acid solution

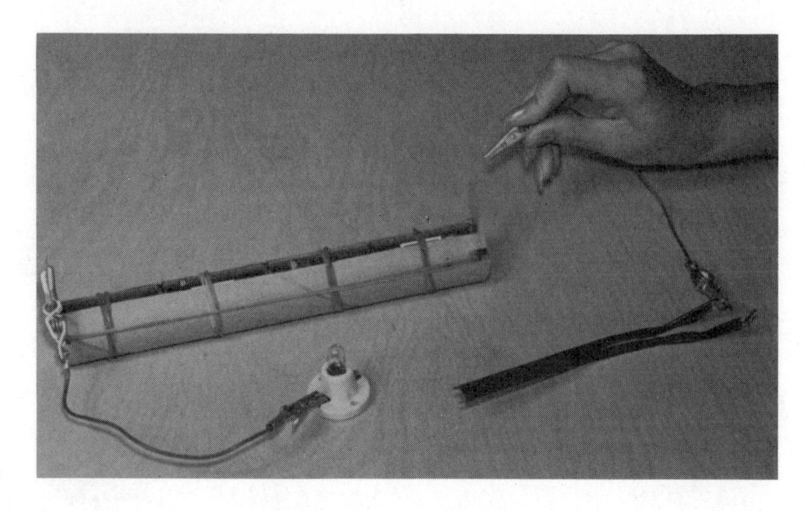

13-10.

Procedure

A. Obtain the materials listed in the margin.

B. Fill a test tube about 1/3 full of $Ba(OH)_2$ solution. Place the test tube in the beaker (or test tube holder).

C. Using the wire connector, attach a flashlight bulb to one end of four batteries in series. See Fig. 13-9.

D. Connect one of the two long wires of the TV lead-in wire to the other end of the batteries. See Fig. 13-10.

13-11.

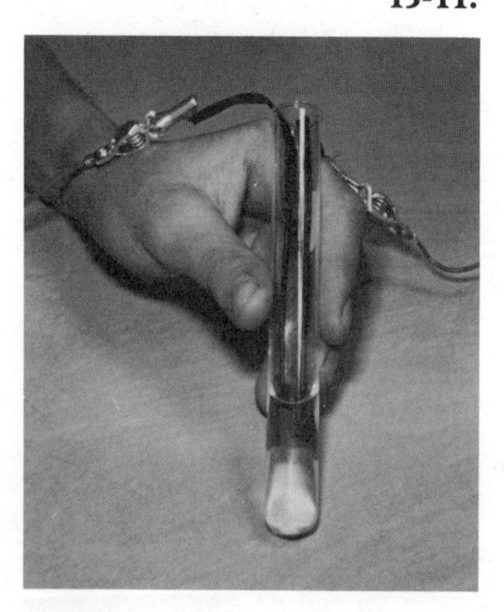

E. Connect the second long wire of the TV lead-in wire to the remaining terminal of the flashlight bulb.

F. Place the end of the TV lead-in wire with the short stubby ends all the way into the $Ba(OH)_2$ in the test tube. See Fig. 13-11. The wires will serve as electrodes. Your setup should look like Fig. 13-12. If all the connections are good, the light bulb should glow dimly.

1. What two ions in the $Ba(OH)_2$ solution cause the bulb to glow?

G. Now add a medicine dropper full of H_2SO_4 to the $Ba(OH)_2$ in the test tube.

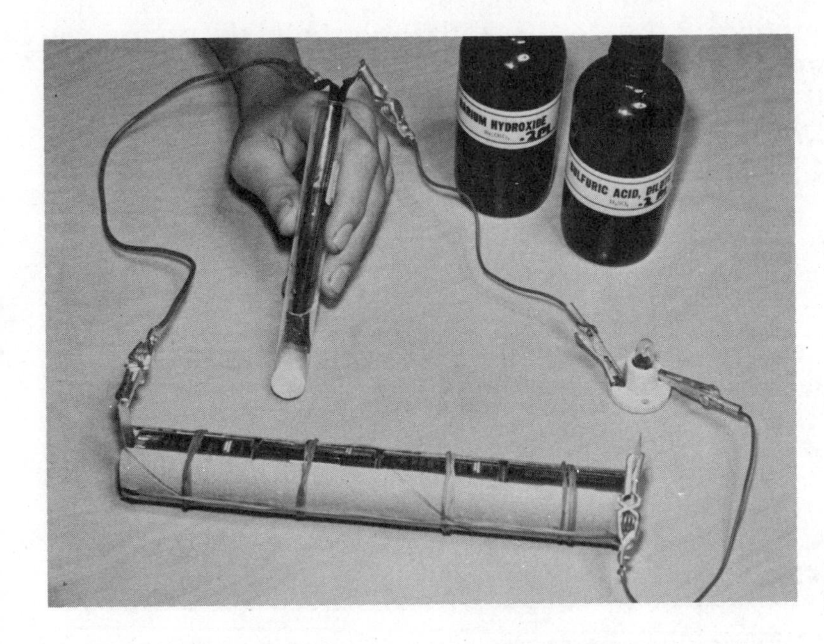

13-12.

2. What is the color of the precipitate that forms?

H. Add two more medicine droppers full of H_2SO_4 to the test tube. Using the electrodes, stir the solution.

I. Now add H_2SO_4, one drop at a time, until the light bulb goes out. Keep a record of the number of drops you add. Stir the solution frequently with the electrodes.

3. How much H_2SO_4 did you add to the test tube before the light went out? Was this amount more than, less than, or the same as the amount of $Ba(OH)_2$ in the test tube?

When the light bulb goes out, you have reached the end point.

4. Why does the light go out at the end point?

5. Would the light go on again if you added too much H_2SO_4? Why?

Summary

In your own words, describe how electrical conductivity can be used to find the end point of an acid-base neutralization reaction.

VOCABULARY REVIEW

Match the number of the word(s) with the letter of the phrase that best explains it.

1. acidic solution
2. indicator
3. base
4. neutral
5. neutralization
6. salt

a. A compound that causes a solution to have more OH^- ions than pure water.
b. A substance that changes color when put into an acid solution.
c. A chemical reaction that occurs when an acid and a base are mixed.
d. A solution containing more H^+ ions than pure water.
e. A compound that is made up of the metallic ion from a base and the nonmetallic ion from an acid.
f. A solution or substance that is neither acidic nor basic.

REVIEW QUESTIONS

Complete each statement by choosing the best word or phrase, or by filling in the blank.

1. An acidic solution **a.** contains more OH^- ions than water **b.** tastes bitter **c.** turns litmus red **d.** all of the above are correct.
2. An HCl solution will **a.** taste sweet and react with baking soda **b.** taste sour and change red litmus paper to blue **c.** taste sour and react with baking soda **d.** taste sweet and change the color of purple cabbage leaves.
3. Which of the following is true? **a.** An acidic solution contains more OH^- ions than water. **b.** An acidic solution contains more H^+ ions than water. **c.** A basic solution contains less OH^- ions than water. **d.** A basic solution contains more H^+ ions than water.
4. Which of the following combinations are both bases? **a.** vitamin C and household ammonia **b.** household ammonia and baking soda **c.** sugar and vinegar **d.** vitamin C and vinegar.
5. A solution that tastes bitter and changes red litmus paper to blue is **a.** a basic solution **b.** an acidic solution **c.** an indicator solution **d.** a neutral solution.
6. When dissolved in water, a white solid turns blue litmus paper red. The solid is a(n) __________.
7. When dissolved in water, a white solid turns red litmus paper blue. This solid is a(n) __________.

8. Choose the steps you would take, in the correct order, to safely identify a white solid as being either an acid or a base. **a.** Place a small amount of the substance on your tongue for a taste test. **b.** Wet red and blue litmus paper with water. **c.** Obtain one piece each of red and blue litmus paper. **d.** Place a small amount of the substance to be tested on the red and blue litmus paper and observe any color changes.
9. Which of the following is a correct chemical equation showing the reaction of an acid with a base?
   a. $Na^+ + Cl^- \rightarrow NaCl + heat$
   b. $KOH + NaOH \rightarrow H_2O + KNaO + heat$
   c. $Ca + ZOH \rightarrow Ca(OH)_2 + heat$
   d. $NaOH + HCl \rightarrow H_2O + NaCl + heat$
10. Which of the following substances will *both* form a salt with HCl? **a.** HBr and HF **b.** HI and KOH **c.** $Ca(OH)_2$ and LiOH **d.** H_2O and $Mg(OH)_2$.

REVIEW EXERCISES

Give complete but brief answers to each of the following. Use complete sentences to write your answers.

1. Describe what is meant by an acidic solution.
2. Give two examples of the chemical actions of an acidic solution.
3. Describe a safe test for an acid.
4. Describe what is meant by a basic solution.
5. Describe a safe test for a base.
6. Which of the following will be bases when in solution: NaCl, NaOH, HCl, HOH, KOH?
7. What is the pH scale? How is it used?
8. When an acid and a base are mixed, what is produced?
9. Write an equation that shows the chemical reaction between an acid and a base.
10. For the acids HCl, HBr, and HF and the bases KOH, $Ca(OH)_2$, and LiOH, write the chemical formulas for all the products possible if each acid was mixed with one base at a time.

EXTENSIONS

1. How many times a day do you drink acid?
   a. Read the labels of soft drink and juice containers and list the acids they contain.
   b. Why is sugar or a sweetener added to soft drinks and juices?
   c. Compare the tastes of unsweetened fruit juice and the same kind of fruit juice with sweeteners added.

2. Several shampoo advertisements claim that their products have a balanced pH, contain no detergent, are nonalkaline, or do not cause tears.
   a. Test several shampoos with red litmus paper.
   b. Test the same shampoos with blue litmus paper.
   c. Report your observations to the class.
   d. Which shampoo do you suggest using?
3. Hydrochloric acid is present in the stomach as an aid to digestion.
   a. What do you think is contained in an antacid used to neutralize excess stomach acid?
   b. Look up and list the ingredients of several antacids sold in drugstores. Do these ingredients agree with your answer to part **a**?
   c. In a reference book, find and then describe the action of HCl in the stomach.
4. The following directions are given on the label of a cleaning solution:
   POISON: External—Flood with water, then wash with vinegar.
   Internal—Drink large amounts of vinegar, lemon juice, grapefruit, or orange juice.
   a. What kind of substance do the vinegar and fruit juices contain?
   b. What kind of substance does the cleaning solution contain?
   c. What process is being carried out by following the directions?
   Another label reads:
   CONTAINS HYDROCHLORIC ACID. First Aid: Drink a teaspoon or more of magnesia, chalk, small pieces of softened soap in milk, or raw egg white.
   d. What conclusion can you draw about magnesia, chalk, softened soap in milk, and raw egg white?
   e. Read the caution labels on several cleaning agents in your home and determine which ones contain acids or bases.
   f. Why is it unlikely that a cleaning agent will contain both an acid and a base?

CHAPTER 14

Organic Chemistry

14-1. CARBON AND ITS COMPOUNDS

The photographs above show cut diamonds and a pencil lead. What do these two substances have in common? In this lesson you will learn how diamonds and pencil lead are related.

When you finish lesson 1, you will be able to:

- Show two ways carbon atoms may form bonds with each other.
- Give examples to show how carbon forms bonds with other atoms.
- Explain what is meant by an *organic compound*.
- Identify some common substances that are compounds containing carbon.

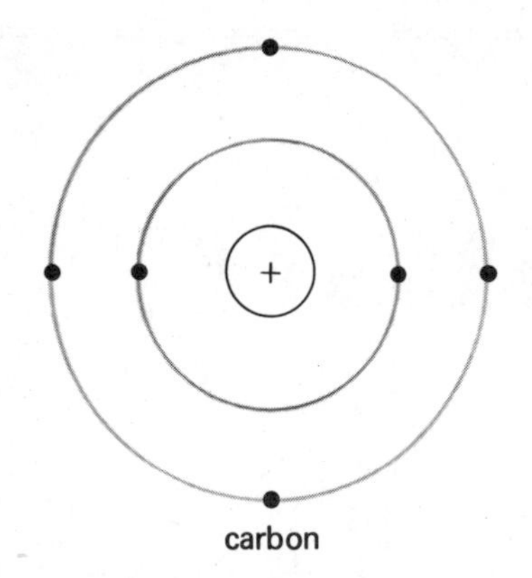

14-1. *Why does carbon most often form bonds by sharing electrons with other atoms?*

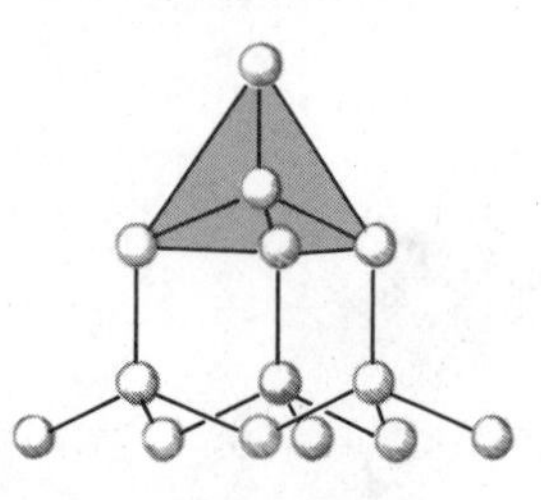

14-2. *In a diamond, each carbon atom is bonded to four others as shown in the diagram.*

Double bond
The bond formed when atoms are joined by sharing two pairs of electrons.

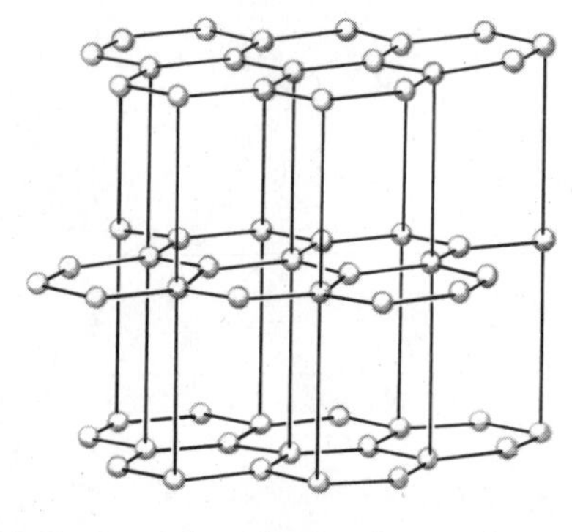

14-3. *In coal and other forms of graphite, carbon atoms are arranged in layers because one of the four bonds on each carbon atom is different from the other three. This arrangement is shown in the diagram.*

Look at the periodic chart in Fig. 10-16, p. 326. Carbon is found in the upper right section of the periodic table. Figure 14-1 shows the electron arrangement of a carbon atom. As you see, a carbon atom has four valence electrons. Carbon forms covalent bonds by sharing these four electrons. For example, one carbon atom may form covalent bonds with four other carbon atoms. Figure 14-2 shows how each carbon atom may link to four others so that each carbon has four equally close neighbors. Carbon atoms bonded together in this way make up a diamond. A diamond crystal is made up of pure carbon. The hardness of diamond results from the strong covalent bonds joining each carbon atom equally to all its neighbors.

Carbon atoms may also join to each other in a different way. Three of the four bonds between the carbons can be strong covalent bonds, as in diamonds. But the fourth bond can be a weaker bond. See Fig. 14-3. This results in a weakness that allows layers of carbon atoms to split off from each other. This form of carbon is called *graphite*. The "lead" in pencils is really graphite. You can write with a pencil because flakes of graphite rub off as you move the pencil over the paper. Thus, both diamond and pencil lead are different forms of pure carbon. The difference is in the way the carbon atoms are bonded to each other.

Carbon also forms bonds with other atoms. When carbon burns, it combines with oxygen:

$$C + O_2 \longrightarrow CO_2$$

$$\cdot\ddot{C}\cdot + :\ddot{O}::\ddot{O}: \longrightarrow :\ddot{O}::C::\ddot{O}:$$

See Fig. 14-4. In carbon dioxide, the carbon and oxygen atoms are joined by two **double bonds.** Four electrons instead of two are shared in a *double bond*. Carbon atoms often form double bonds with other atoms. Carbon and oxygen can also react to form carbon monoxide:

$$2C + O_2 \longrightarrow 2CO$$

Carbon dioxide and carbon monoxide have very different effects on the body. As long as CO_2 is mixed with enough O_2 gas, it is harmless to breathe. Carbon monoxide is poisonous even in small amounts. Carbon monoxide interferes with the substance in the blood that carries O_2. It is the CO produced in automobile

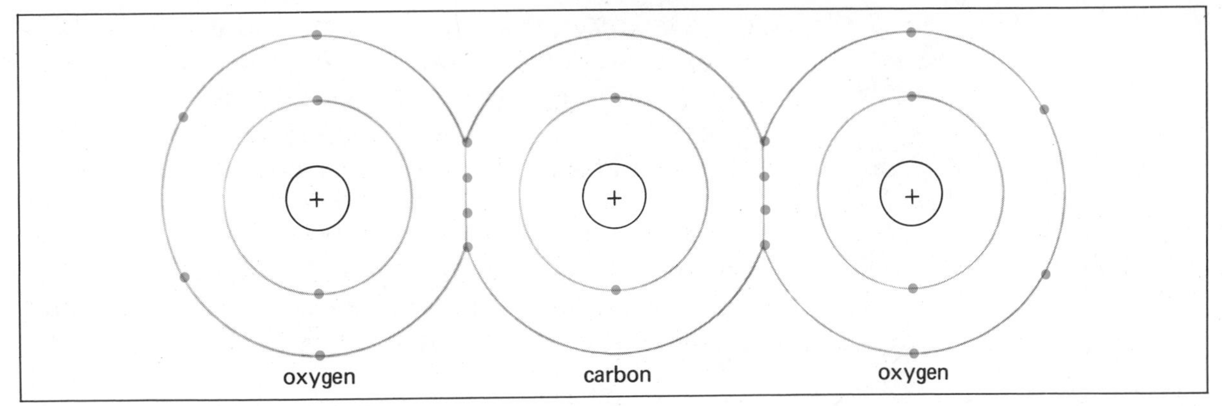

14-4. *The arrangement of bonded electrons in a CO_2 molecule is shown in this diagram.*

exhaust fumes that adds to air pollution. Many cars are now made with special filters to reduce the amount of CO they produce. See Fig. 14-5.

Common fuels such as coal, petroleum, and natural gas contain carbon compounds. When these fuels are burned, large amounts of carbon dioxide are released into the atmosphere. Some scientists think that this build up of CO_2 in the atmosphere may cause the earth's climate to change. This might result from the

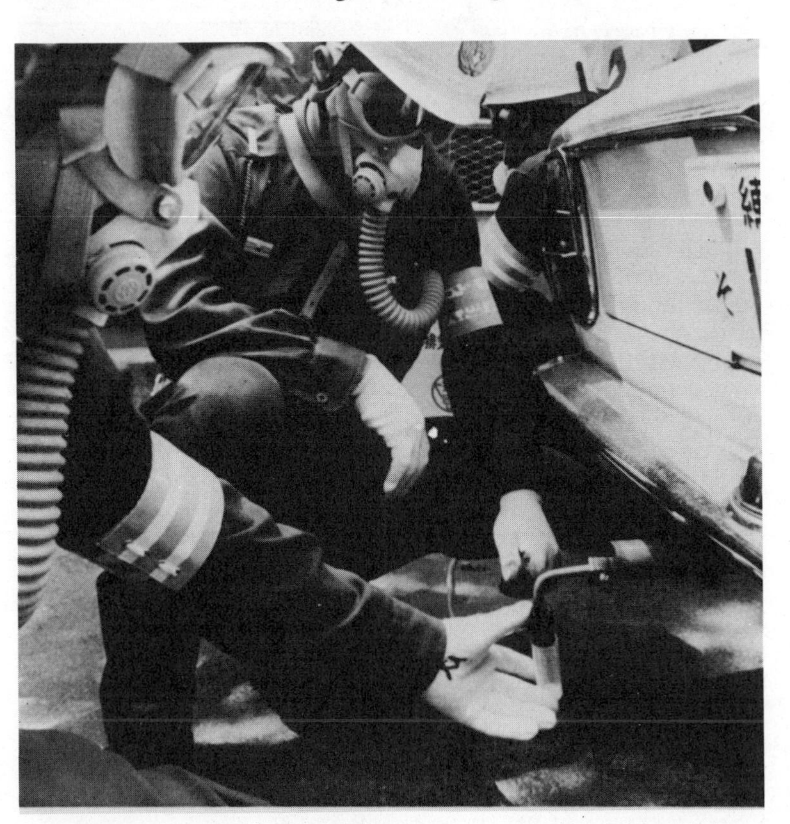

14-5. *Carbon monoxide produced by automobile exhausts is a poisonous gas. In an enclosed space, a person can easily be overcome by CO fumes. Carbon monoxide also adds to air pollution.*

greenhouse effect. Carbon dioxide can trap the sun's heat in the atmosphere in the same way that glass traps heat in a greenhouse. Thus, the release of CO_2 into the atmosphere might cause temperatures all over the earth to increase. No one can be sure how this would change the earth's climate. Some scientists have predicted that if the average temperature of the earth increased by only 5°, this would be enough to cause a large portion of the polar icecaps to melt. The water thus added to the oceans would cover several coastal cities.

Carbon is able to form many other kinds of compounds. Compounds of carbon are found in food, clothing, automobiles, and the paper these words are printed on. Scientists once thought that carbon compounds could only be made by living things. We now know that this idea is false. Chemists can make carbon compounds in the laboratory. **Organic** means "coming from life." The branch of science that deals with carbon and its compounds is known as *organic chemistry*.

Organic chemistry
The chemistry of carbon compounds.

ACTIVITY

Materials
ring stand
tin can lid
index card
candle
samples of: white bread, sugar, salt

Properties of Carbon Compounds

In this activity, you will test three substances: bread, sugar, and salt. All of these substances are important parts of our diets. Bread, sugar, and salt are not all made up of the same type of molecules. Compounds containing carbon will burn or melt when heated. You can classify these compounds on the basis of this property.

A. Obtain the materials listed in the margin.

B. Place a tin can lid on a ring stand.

C. Around the edge of the lid, place small samples of white bread, sugar, and salt.

D. Attach a small candle to an index card by dripping some wax on the card and setting the candle in it.

E. Place the candle directly under the bread sample. The flame should nearly touch the underside of the lid. See Fig. 14-6.

F. Let the candle burn in this position for 3–4 min.

G. After 3–4 min, put the candle under the sugar sample.

1. What happened to the bread?

H. Heat the sugar for 3–4 min.

I. After heating the sugar, put the candle under the salt sample.

2. What happened to the sugar?

J. Heat the salt sample for 3–4 min.

3. What happened to the salt?

You should have observed that two of the samples either burned or melted when heated. One sample did not burn or melt.

4. Examine your samples. Which samples contain carbon? Which does not?

14-6.

SUMMARY

A nylon shirt, a basketball, a wood floor, even a plastic pen—all are made of carbon compounds. Carbon is a special atom. Diamond and graphite are examples of carbon atoms bonded together in two different ways. Carbon dioxide and carbon monoxide are two carbon compounds that have very different properties.

QUESTIONS

Use complete sentences to write your answers.

1. Explain how diamond and graphite are alike and how they are different.
2. Which of the following are most likely to contain carbon or carbon compounds: trees, cotton clothing, milk, meat, wooden pencils?
3. How many valence electrons does carbon have?
4. What is meant by an organic compound?

14-2. HYDROCARBONS

Suppose you are near a swamp or marsh at night and suddenly you see something that almost makes you think a UFO has landed. A faint blue light flickers and then quickly disappears. A little later you see the same blue light in a different place. Is this strange light really caused by visitors from space? Lesson 2 will help you answer this question.

When you finish lesson 2, you will be able to:

- Name some compounds that are *hydrocarbons*.
- Give examples of some practical uses of hydrocarbons.
- ○ Construct models of five different hydrocarbons.

Hydrocarbon
A compound containing only carbon and hydrogen.

Carbon forms a great many organic compounds. One group of organic compounds is the **hydrocarbons.** *Hydrocarbon* molecules contain only carbon and hydrogen atoms bonded together. You have seen how hydrogen atoms combine with other atoms, forming a single bond. Carbon may form one bond

14-7. *(left) Ethane is a component of natural gas. This photo shows a site where natural gas is obtained by drilling.*

14-8. *(right) The propane in these tanks is burned as fuel to heat a home.*

with each of the four electrons in its outer shell for a total of four bonds.

The simplest hydrocarbon molecule is *methane*, CH_4. In a molecule of methane, one carbon atom is bonded to four hydrogen atoms. Methane is sometimes called *marsh gas*. It is formed when plants decay. Sometimes it gives off a blue light that shows up best at night. Natural gas used as a fuel contains mostly methane. A simple way to represent a methane molecule is by a **structural formula.** A *structural formula* is a diagram of a molecule. This diagram has lines to represent the covalent bonds between atoms. The structural formula of methane is:

Structural formula
A diagram of a molecule.

$$\begin{array}{ccc} & H & \\ & | & \\ H- & C & -H \\ & | & \\ & H & \end{array}$$

Ethane is a hydrocarbon molecule containing two carbon atoms. The formula for ethane is C_2H_6. The structural formula for ethane is:

$$\begin{array}{cccc} & H & H & \\ & | & | & \\ H- & C- & C & -H \\ & | & | & \\ & H & H & \end{array}$$

Ethane is also usually a part of natural gas. See Fig. 14-7. The addition of one more carbon atom to an ethane molecule produces *propane*, C_3H_8.

$$\begin{array}{ccccc} & H & H & H & \\ & | & | & | & \\ H- & C- & C- & C & -H \\ & | & | & | & \\ & H & H & H & \end{array}$$

Propane is also found in natural gas and is used as a fuel. See Fig. 14-8. Can carbon atoms keep joining to build up a chain? The next hydrocarbon molecule, *butane*, shows that more carbon atoms can be added. The formula for butane is C_4H_{10}:

$$\begin{array}{cccccc} & H & H & H & H & \\ & | & | & | & | & \\ H- & C- & C- & C- & C & -H \\ & | & | & | & | & \\ & H & H & H & H & \end{array}$$

Do you see any other way in which this molecule

could be put together? A butane molecule might have this structure:

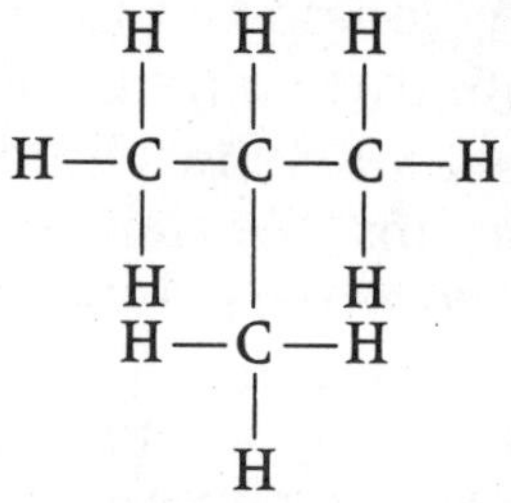

```
        H   H   H
        |   |   |
    H — C — C — C — H
        |   |   |
        H   |   H
        H — C — H
            |
            H
```

Isomers
Two or more molecules with the same formulas but different arrangements of their atoms.

The formula for this molecule is also C_4H_{10}. It is still a molecule of butane. This molecule is an **isomer (ie-suh-mer)** of butane. *Isomers* contain the same atoms arranged differently. This isomer of butane is called isobutane. Some organic molecules have many isomers.

Table 14-1 shows the arrangement of the atoms in the first five hydrocarbon molecules.

The hydrocarbons you have already seen and read about contain carbon atoms joined to each other by

TABLE 14-1

Name	*Formula*	*Two Dimensional*	*Three Dimensional*
a. methane	CH_4		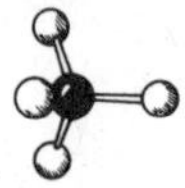
b. ethane	C_2H_6	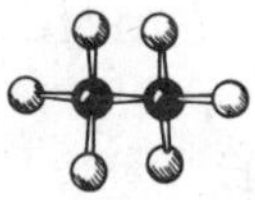	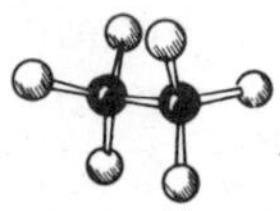
c. propane	C_3H_8	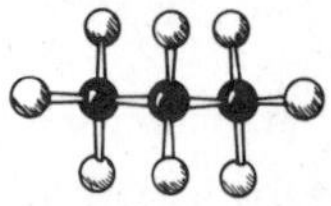	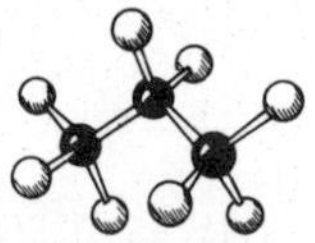
d. butane	C_4H_{10}	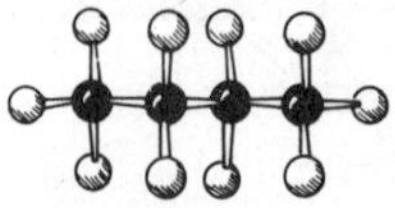	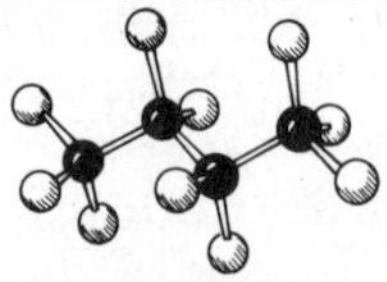
e. pentane	C_5H_{12}	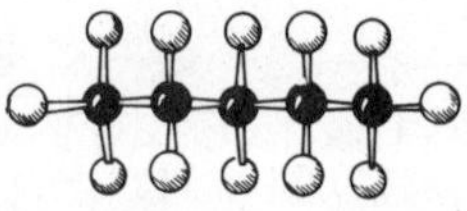	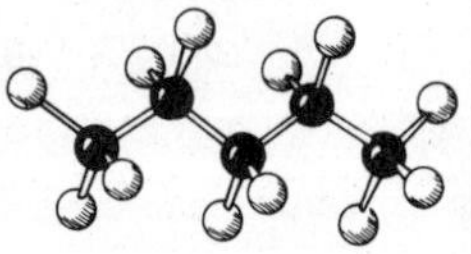

single bonds. Organic compounds containing carbon atoms joined by single bonds only are said to be *saturated*. Many organic molecules contain carbon atoms joined by double or triple bonds. These compounds are called *unsaturated*. An example of an unsaturated hydrocarbon is shown by the following structural formula:

$$\begin{array}{ccc} H & & H \\ & \diagdown \quad \diagup & \\ & C{=}C & \\ & \diagup \quad \diagdown & \\ H & & H \end{array}$$

This molecule is called *ethylene*, C_2H_4. Ethylene is one of the most useful of the hydrocarbons. When C_2H_4 combines with itself, it forms a long chainlike molecule. This molecule is called *polyethylene* and is an example of a plastic. Almost all plastics are made by joining together organic molecules, such as ethylene.

Carbon atoms can also form triple bonds. In a triple bond, three pairs of electrons are shared by two atoms. The simplest hydrocarbon molecule containing a triple bond is *acetylene*, C_2H_2:

$$H-C\equiv C-H$$

Acetylene burns with a very hot flame when mixed with oxygen. Welding torches burn a mixture of acetylene and oxygen to produce a flame hot enough to melt most metals. See Fig. 14-9.

14-9. *This photo shows the use of an acetylene torch in industrial welding.*

You have seen that carbon atoms can form long chains or branching molecules. Six carbon atoms can even arrange themselves in a ring to form *benzene*, C_6H_6:

```
        H
        |
        C
      //  \
  H—C      C—H
    |      ||
  H—C      C—H
      \\  /
        C
        |
        H
```

Benzene is an example of an *organic solvent*. Organic solvents are able to dissolve other organic substances. Most of the dirt that collects on clothing is organic. Therefore, benzene is used in dry cleaning. As you can see, an almost unlimited number of compounds with many different properties can be made using only carbon and hydrogen.

ACTIVITY

Materials
modeling clay
toothpicks

Models of Hydrocarbons

Molecular models of five hydrocarbons are shown in Table 14-1. You will use this table to construct three-dimensional models of several hydrocarbon molecules.

A. Obtain the materials listed in the margin.

B. Look at the three-dimensional views of the hydrocarbon molecules shown in Table 14-1. Use the modeling clay and toothpicks to construct models of the five hydrocarbons: methane, ethane, propane, butane, and pentane.

1. Which of these hydrocarbons has an isomer?
2. What is the name of this isomer?
3. Write the structural formula for this isomer.

C. Use your materials to construct a three-dimensional model of the isomer named in question 2.

SUMMARY

Faint blue lights sometimes seen around swamps are more likely to be caused by methane than visitors from space. Methane is produced when plants decay. Sometimes methane gives off a blue light when it mixes with oxygen in the air. Marsh gas, or methane, is the simplest of the fascinating, large group of molecules called hydrocarbons. Carbon and hydrogen can bond together in different ways to produce a large variety of hydrocarbons.

QUESTIONS

Unless otherwise indicated, use complete sentences to write your answers.

1. Name the hydrocarbon molecules that have one, two, three, and four carbon atoms in them.
2. Draw the structural formulas for ethane, propane, butane, and pentane.
3. What is meant by an isomer of a molecule?
4. Name two hydrocarbons and state how they can be used.

14-3. SUBSTITUTED HYDROCARBONS

The manager is getting desperate. The game is in the bottom of the last inning and the team is one run behind with two out and one player on base. The next scheduled batter is the pitcher! The manager decides to send in a pinch hitter. Maybe a substitute can win the game! In this lesson you will learn how organic molecules can also have substitutions.

When you finish lesson 3, you will be able to:

- Give examples to show how organic molecules can be thought of as substituted hydrocarbons.
- Give an example of a reaction in which two organic molecules combine.
- Name some organic compounds and give examples of how they are used.
- ○ Produce a substituted hydrocarbon from an alcohol.

Baseball and other sports have players who substitute for the regular players in special situations. A similar situation occurs in organic chemistry. There are

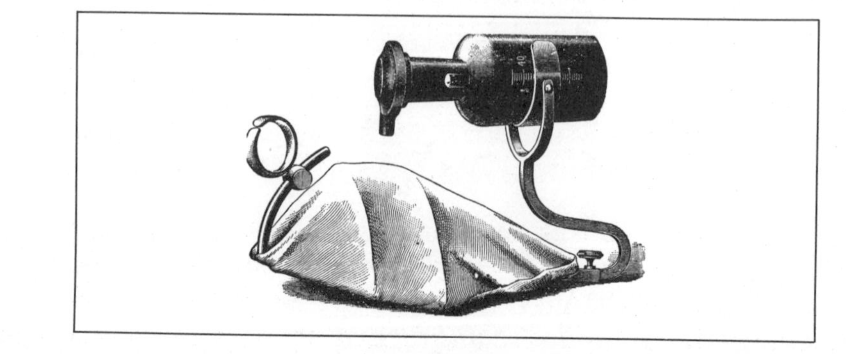

14-10. *Chloroform was often used as an anesthetic in surgery. This illustration shows the cloth mask that was put over the patient's face. Chloroform was slowly dripped from the bottle onto the mask during the operation.*

a huge variety of hydrocarbon molecules. These molecules are built around a skeleton of carbon atoms. Hydrocarbons like methane contain only hydrogen and carbon atoms:

```
     H
     |
  H—C—H
     |
     H
```

Could another atom substitute for a hydrogen atom? Suppose three of the hydrogen atoms in methane were replaced by three chlorine atoms:

```
     Cl
     |
  H—C—Cl
     |
     Cl
```

The molecule now has the formula $CHCl_3$ and is called *chloroform.* The presence of chlorine atoms makes chloroform a totally different compound from methane. Before better drugs were available, chloroform was used to put people to sleep during surgery. See Fig. 14-10. Other halogen atoms can also be substituted for some of the hydrogen atoms in methane. The gas once used as a propellant in spray cans contains both chlorine and fluorine:

```
      H
      |
  Cl—C—F
      |
      Cl
```

This compound is called *freon.* Freon is also used in the cooling coils of refrigerators.

Radicals or groups of atoms as well as single atoms can also be substituted for hydrogens. The OH radical is called the *hydroxide* radical. The hydroxide radical can be substituted for one hydrogen atom in methane:

```
     H
     |
  H—C—OH
     |
     H
```

The formula for this compound is CH_3OH. Because it is built on a methane skeleton, the compound is called *methanol.* Methanol is an **alcohol.** Any organic molecule with OH groups replacing hydrogens is an *alcohol.*

Alcohol
A kind of organic molecule having at least one OH group.

14-11. *Methanol is one of the ingredients in the paint these women are using.*

Methanol is known as *wood alcohol* and is poisonous. See Fig. 14-11. Methanol is not the alcohol found in alcoholic beverages. The alcohol in wine, beer, and alcoholic beverages is *ethanol,* so named because it is built on the basic ethane structure (C_2H_6). The formula for ethanol is C_2H_5OH:

$$\begin{array}{ccccccc} & & H & & H & & \\ & & | & & | & & \\ H & - & C & - & C & - & OH \\ & & | & & | & & \\ & & H & & H & & \end{array}$$

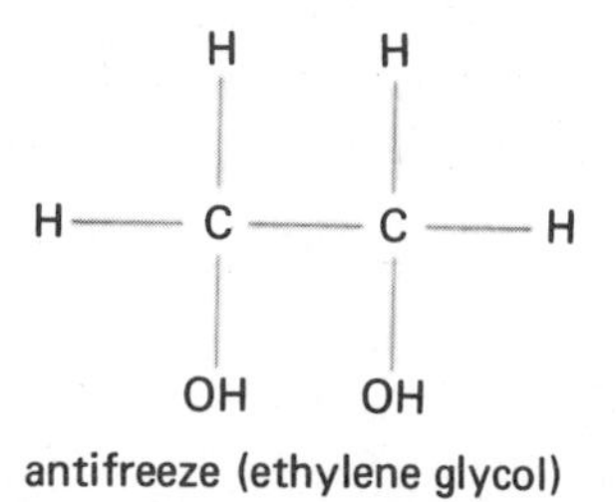

Some alcohols have more than one OH group. See Fig. 14-12.

Another group that can be substituted for hydrogen is the COOH group. When the COOH group is substituted in a methane molecule, the result is *acetic acid:*

$$\begin{array}{ccccccc} & & H & & O & & \\ & & | & & \| & & \\ H & - & C & - & C & - & OH \\ & & | & & & & \\ & & H & & & & \end{array}$$

14-12. *Ethylene glycol, an alcohol, is used as an antifreeze in automobiles.*

You are familiar with acetic acid in the form of vinegar. Acetic acid gives vinegar its sour taste. See Fig. 14-13.

14-13. *Acetic acid gives vinegar its sour taste.*

Acetic acid is an example of an **organic acid.** An organic molecule with the COOH group in place of one or more hydrogen atoms is an *organic acid*. Another example of an organic acid is *formic acid.* Formic acid is produced by ants and was first discovered in ant bites. The citric acid found in fruits such as lemons, limes, and oranges is also an organic acid.

Organic acid
A kind of organic molecule having at least one COOH group.

Organic acids that are important to life are called **amino acids.** See Fig. 14-4. *Amino acids* contain both the COOH group and the *amine* group, NH_2:

$$\begin{array}{ccccccc} & & H & & H & & O \\ & & | & & | & & \| \\ H & - & N & - & C & - & C - OH \\ & & & & | & & \\ & & & & H & & \end{array}$$

Amino acid
A kind of organic molecule having both NH^2 and COOH groups.

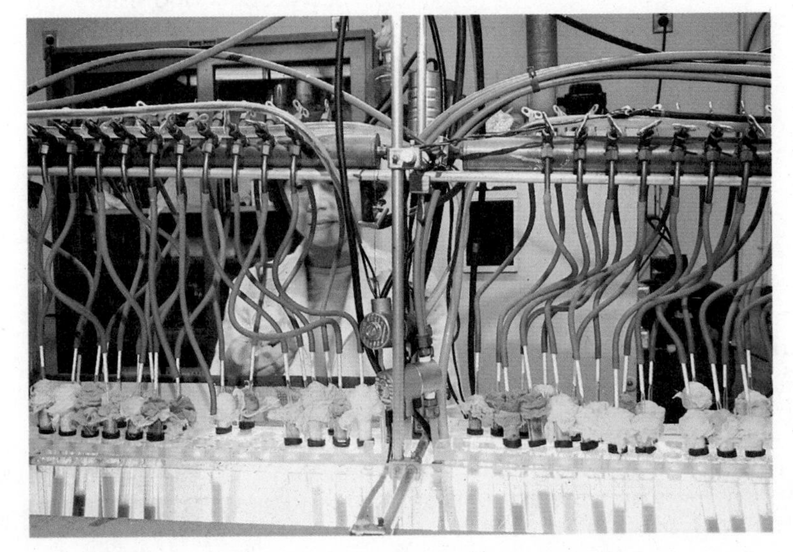

14-14. *Amino acids are the basic building blocks of life. Many amino acids have been made in the laboratory.*

An organic acid can combine with an alcohol. This reaction is similar to the neutralization reaction between an acid and a base:

$$\begin{array}{c} \text{H}\;\;\text{O} \\ |\;\;\;\;\| \\ \text{H}-\text{C}-\text{C}-\text{O}\boxed{\text{H}} \\ | \\ \text{H} \\ \text{acid} \end{array} + \begin{array}{c} \text{H} \\ | \\ \boxed{\text{HO}}-\text{C}-\text{H} \\ | \\ \text{H} \\ \text{alcohol} \end{array} \longrightarrow \begin{array}{c} \text{H}\;\;\text{O}\;\;\;\;\;\;\;\;\text{H} \\ |\;\;\;\;\|\;\;\;\;\;\;\;\;\;| \\ \text{H}-\text{C}-\text{C}-\text{O}-\text{C}-\text{H} \\ |\;\;\;\;\;\;\;\;\;\;\;\;\;\;\;\;| \\ \text{H}\;\;\;\;\;\;\;\;\;\;\;\;\;\;\;\text{H} \\ \text{ester} \end{array} + \begin{array}{c} H_2O \\ \text{water} \end{array}$$

Ester
A kind of molecule formed by combining an organic acid and an alcohol.

The products of this reaction are water and a molecule called an **ester.** The presence of natural *esters* in fresh fruits like strawberries gives the fruits their sweet odor.

ACTIVITY

Materials
copy of crossword puzzle

Substituted Hydrocarbons

We have introduced the names of many organic compounds. All of these compounds can be thought of as substituted hydrocarbons. The substitution of various atoms in these compounds makes them useful in many ways. This activity will help you become familiar with some of these compounds.

A. Complete the puzzle in Fig. 14-15, using the clues given below. DO NOT WRITE IN THIS BOOK.

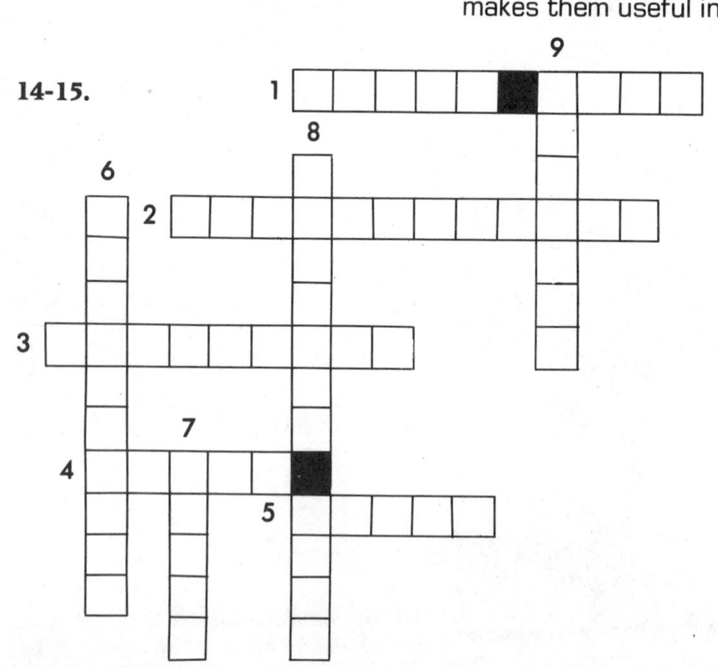

14-15.

Clues

Across

1. Plays an important part in living things.
2. All organic molecules are related to the ____.
3. Alcohols and esters are examples of organic ____.
4. Compound used in spray cans.
5. The NH_2 group is called ____.

Down

6. Its formula is $CHCl_3$.
7. Product when an organic acid reacts with an alcohol.
8. Contains the COOH group.
9. Methanol and ethanol are examples of this kind of compound.

SUMMARY

In organic compounds, as in baseball games, substitutions are important. Organic molecules can be made from hydrocarbons by substituting other atoms or groups of atoms for hydrogen atoms. These types of compounds have different properties and can be thought of as substituted hydrocarbons.

QUESTIONS

Use complete sentences to write your answers.

1. Why is chloroform called a substituted hydrocarbon?
2. Name three organic compounds and state how they are used.
3. Give an example of a reaction in which two organic molecules combine.
4. How does freon differ from methane?
5. Name some fruits or vegetables that contain organic acids.

14-4. MOLECULES NECESSARY FOR LIFE

What did you eat for breakfast today? Perhaps you had fruit juice, milk, eggs, bacon, buttered toast, or cereal. All of these foods contain compounds that are necessary for good health. This lesson will tell you something about a few of these compounds.

When you finish lesson 4, you will be able to:

- Name three types of organic compounds and use examples to explain why they are important to living things.
- Explain how living things get energy from *carbohydrates*.
- Describe the makeup of *proteins* and explain why a huge number of proteins exist.
- ○ Test the solubility of a sugar, a starch, an amino acid, a protein, and a fat.

14-16. *The sugars glucose and fructose are found in fruits and honey.*

We cannot possibly describe all the compounds necessary for life in a book this size. There are simply too many of them. You are already familiar with some of these compounds. Sugars, starches, fats, proteins, and vitamins are parts of your daily diet. In this lesson, you will learn about the three most important types of molecules needed for life: carbohydrates, fats, and proteins.

1. *Carbohydrates*. **Carbohydrates** (kahr-boe-**hie**-drates) are compounds of carbon, hydrogen, and oxygen. The simplest *carbohydrate* is called *glucose*. The formula for glucose, $C_6H_{12}O_6$, is:

Carbohydrates Compounds of carbon, hydrogen, and oxygen whose molecules contain two atoms of hydrogen for every one atom of oxygen.

```
        CH2OH
          |
    H   C----O   H
    | /  |     \ |
    C   OH   H   C
    | \ |     | /|
   OH  C------C  OH
        |      |
        H     OH
```

Glucose is found in many fruits. Glucose is an example of a *simple sugar*. Another simple sugar is *fructose*. Fructose is found in some fruits and in honey. See Fig. 14-16. Glucose and fructose are both ring-shaped molecules. By joining two of these simple-sugar rings together, a double-ring molecule is formed:

```
      CH2OH
        |
  H   C----O   H     CH2OH     O                 H
  | /  |     \ |       |   /      \              |
  C   OH   H   C       C   H                  OH C
  | \ |    | / L--O--J  \ |                    | /|
 OH  C-----C              C--------------------C  CH2OH
      |     |             |                    |
      H    OH            OH                    H
```

This double-ring molecule is called a *double sugar*. The double sugar formed from glucose and fructose is *sucrose*. The formula for sucrose is $C_{12}H_{22}O_{11}$. You use sucrose every day as common table sugar. Sucrose is also known as cane sugar since we obtain most of our supply from the sugar cane plant.

There are many other simple and double sugars. When many simple-sugar rings join together, the product is the carbohydrate called *starch*. Each starch molecule contains several hundred to thousands of glucose glucose

14-17. *When many glucose molecules join together, they form a starch molecule.*

glucose rings joined together. See Fig. 14-17. Plants use starch to store food. You eat starch in bread, cereal, potatoes, and rice. See Fig. 14-18.

Carbohydrates are the body's main source of energy. The starch and double sugars in a bowl of cereal are broken down in your body into simple sugars. The simple sugars then combine with oxygen to release energy:

$$C_6H_{12}O_6 + 6O_2 \longrightarrow 6CO_2 + 6H_2O + \text{energy}$$

Fats
Esters formed by combining a particular kind of organic acid and alcohol.

2. *Fats.* An organic acid and an alcohol can combine to form an ester. **Fats** are a particular kind of ester. A *fat* molecule is formed from an organic acid with a long chain of carbon atoms. Figure 14-19 shows three long-chain organic acids joining with an alcohol to produce

14-18. *Bread and cereal are common forms of starch in your diet.*

a fat. You are familiar with such fats as butter, margarine, fatty meats, and vegetable oils. See Fig. 14-20. Fats store energy. If you eat too much bread and butter, ice cream, candy, or french fries, you will quickly become aware of your own fat layer! An efficient diet is based on the fact that both carbohydrates and fats can be broken down to release energy. If you cut down on the carbohydrates you eat, the fat stored in your body will be used for energy instead of the missing carbohydrates.

3. *Proteins.* **Proteins** (**proe**-teenz) make up your skin, hair, and nails. The substance that carries oxygen in your blood is a *protein.* Proteins regulate many of your essential body processes. You get proteins in your diet in egg whites, meat, fish, and milk. See Fig. 14-21.

Protein molecules are formed by many amino acid molecules joined together. See Fig. 14-22. Protein molecules perform many different functions in your body. In addition, the proteins in your body are different from the proteins found in any other living

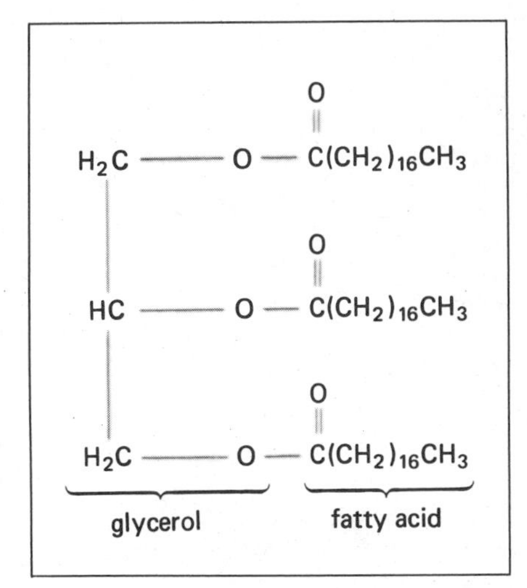

14-19. *A fat molecule is made when three long acid molecules join to a molecule of glycerol, which is an alcohol.*

Proteins
Very large molecules formed when many amino acids join together.

14-20. *Why is it a good idea to cut down on the amount of fats in the food you eat?*

14-21. *Some sources of protein necessary for a balanced diet include fish, meat, and eggs.*

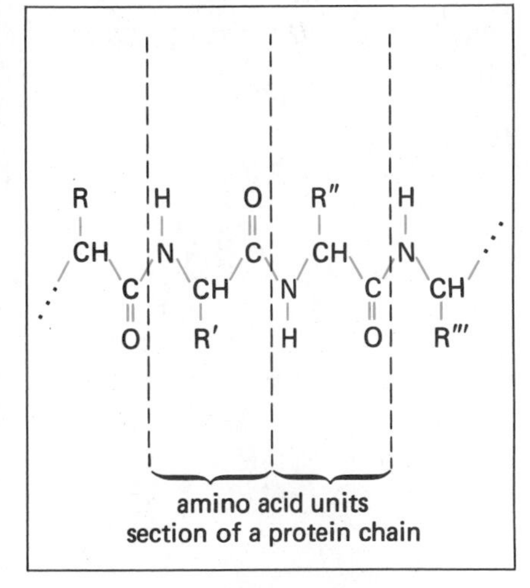

14-22. *Several amino acid molecules join together to form a protein molecule.*

person! All proteins are made up of 23 amino acids. How can only 23 amino acids form so many different proteins?

There is a simple mathematical way to determine which combinations are possible. You need to know the length of the chain and the number of amino acids involved. Suppose you want to arrange two different amino acids in a chain three units long. The number of possible arrangements would be represented by 2^3. This number means that 2 is multiplied by itself 3 times: $2 \times 2 \times 2 = 8$. Eight arrangements are possible.

Most proteins are hundreds of amino acids long. Try to think of the different ways in which 23 amino acids could be arranged in a chain made up of hundreds of acids. If only 10 amino acids were included in a chain 100 units long, the number of possible combinations would be 10, 000, 000, 000, 000, 000, 000, 000, 000, 000,000,000,000,000,000,000,000,000,000,000,000,000, 000, 000, 000, 000, 000, 000, 000, 000, 000, 000, 000, 000, 000.

ACTIVITY

Materials
samples of: sugar, starch, MSG, gelatin, vegetable oil
5 test tubes
water
test tube holder
beaker
Bunsen burner
ring stand
goggles

14-23.

Solubility of Organic Compounds

Organic compounds must be dissolved in water before they can be used by your body. Some organic compounds are more soluble than others. In this activity, you will test the solubility of five organic compounds.

A. Obtain the materials listed in the margin.

B. Copy the following table in your notebook.

C. Label five clean test tubes A, B, C, D, and E. Place the test tubes in order in a test tube rack.

D. Add a small amount of each substance to the test tubes as shown in the table.

E. Add water to each test tube until it is about 1/4 full.

Test tube	*Organic compound*	*Dissolves in cold water*	*Dissolves in hot water*
A	glucose (sugar)		
B	starch		
C	MSG (amino acid)		
D	gelatin (protein)		
E	vegetable oil (fat)		

F. Using a wrist action, shake each test tube well. See Fig. 14-23. Write "yes" in the third column of the table if the compound dissolved in cold water; write "no" if it did not.

G. Fill a beaker 1/2 full of water and place it on a ringstand over a Bunsen burner.
CAUTION: Wear goggles when heating beaker.

H. Put the test tubes containing the compounds that did not dissolve into the beaker. See Fig. 14-24. Heat the beaker over the burner until the water comes to a boil. Then turn off the Bunsen burner.

I. Leave the test tubes in the beaker for 4–5 min. Remove the test tubes one at a time and shake them. Record your observations.

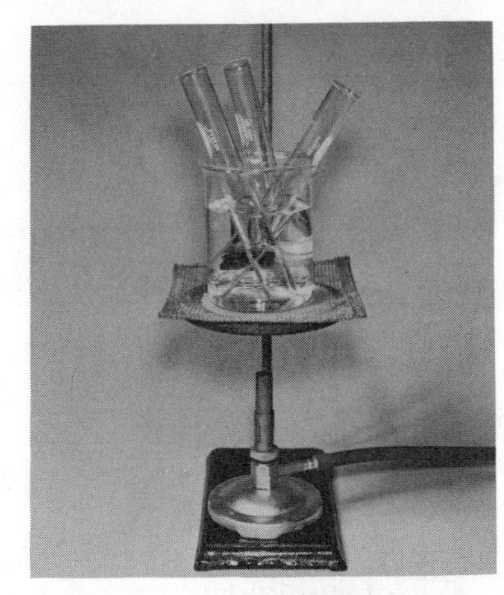

14-24.

SUMMARY

The food you will eat today contains some of the three basic molecules necessary for life: carbohydrates, fats, and proteins. To maintain a healthy body and get the energy you need, you should eat a proper balance of all three at every meal.

QUESTIONS

Use complete sentences to write your answers.

1. Name three types of organic compounds that are important to living things.

2. How are carbohydrates used in your body?

3. What are proteins made of?

4. In addition to three alanines and three glycines, how many combinations of alanine and glycine can be made in three-unit chains? Show these combinations, using the letters A and G.

LABORATORY

Materials

samples of: sugar, starch, gelatin, vegetable oil
4 test tubes
test tube holder
water
beaker
ring stand
goggles
Bunsen burner
Benedict's reagent
iodine solution
biuret reagent
brown paper

Chemical Tests for Organic Compounds

Purpose

You obtain the carbohydrates, proteins, and fats your body needs to stay healthy in the food you eat. However, all foods do not contain all of the necessary organic compounds. How can you find out which foods contain which compounds? In this laboratory, you will perform the chemical tests used to identify sugars, starches, and proteins, as well as a simple test for fats.

Procedure

A. Obtain the materials listed in the margin.

B. Label three test tubes A, B, and C. Copy the following table in your notebook and use it to record your observations.

Test tube	*Compound*	*Color change*
A	glucose (sugar)	
B	starch	
C	gelatin (protein)	

CAUTION: ALWAYS WEAR GOGGLES WHEN PERFORMING CHEMICAL TESTS OR HEATING ANY SUBSTANCE ON A BUNSEN BURNER.

C. Simple sugar test: Place a small amount of glucose in test tube A and add water until it is about 1/4 full. Add an equal amount of Benedict's reagent and heat the test tube in a beaker of boiling water for about 5 min. See Fig. 14-25. Record the color change.

14-25.

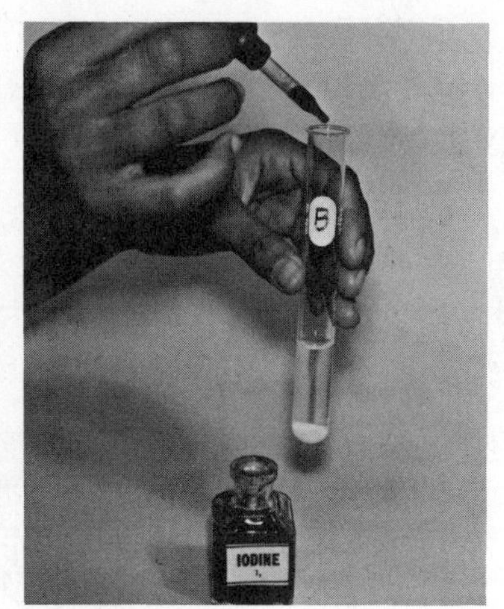

14-26.

D. Starch test: Place a small amount of starch in test tube B and add water until it is about $\frac{1}{4}$ full. Heat the test tube in a beaker of boiling water for about 5 min. Remove, allow the tube to cool, and then add 4 to 5 drops of iodine solution. See Fig. 14-26. Record the color change.

E. Protein test: Place a small amount of gelatin in test tube C and add water until it is about $\frac{1}{4}$ full. Heat the tube in a beaker of boiling water for 5 min. Remove, allow the tube to cool, and then add an equal amount of biuret reagent. Shake the tube to mix the contents. Record the color change.

F. Test for fats: Place a piece of brown wrapping paper on a flat surface. Draw a line down the middle, dividing the paper into two parts. Label the left-hand side "Water." Label the right-hand side "Fat." Now place a drop of water on the left side of the paper and a drop of vegetable oil on the right side.

1. Describe the appearance of the two drops.

Summary

In your own words, describe the chemical tests for sugars, starches, and proteins, and the simple test for fats.

VOCABULARY REVIEW

Match the number of the word(s) with the letter of the phrase that best explains it.

1. organic chemistry
2. hydrocarbon
3. structural formula
4. isomers
5. alcohol
6. organic acid
7. amino acid
8. carbohydrates
9. fat
10. protein
11. double bond

a. Two atoms sharing two pairs of electrons.
b. Two or more molecules with the same formulas but different arrangements of their atoms.
c. An organic molecule having both NH_2 and COOH groups.
d. A diagram of a molecule.
e. An ester formed by combining a particular kind of organic acid and an alcohol.
f. Compounds of carbon, hydrogen, and oxygen whose molecules contain two atoms of hydrogen for every one of oxygen.
g. A kind of organic molecule having at least one OH group.
h. A compound containing only carbon and hydrogen.
i. A kind of organic molecule that has at least one COOH group.
j. A very large molecule formed when many amino acids join together.
k. The chemistry of carbon compounds.

REVIEW QUESTIONS

Complete each statement by choosing the best word or phrase, or by filling in the blank.

1. Look at Fig. 14-4, p. 419, and determine which of the following choices is *not* correct: **a.** Carbon is a nonmetal. **b.** Carbon has a total of 6 protons and 6 electrons. **c.** Carbon has 6 electrons in its outermost shell. **d.** Carbon has 4 valence electrons.
2. Which of the following groups all contain organic compounds?

a. bread, meat, milk, water **b.** meat, milk, water, wood **c.** bread, milk, water, wood **d.** bread, meat, milk, wood.

3. Which of the following groups contain only hydrocarbon molecules? **a.** C_2H_6, C_3H_8, C_5H_{12} **b.** CO_2, C_2H_6, C_3H_8 **c.** C_3H_8, C_5H_{12}, CCl_4 **d.** CCl_4, C_2H_6, C_3H_8.
4. One use of the hydrocarbon propane is as ___________.
5. Ethylene can be made to combine with itself to form the plastic called ___________.
6. When the OH group is substituted for a hydrogen atom on any hydrocarbon molecule, the result is an **a.** acid **b.** amino acid **c.** ester **d.** alcohol.
7. When three chlorine atoms are substituted for three hydrogen atoms on the methane molecule, the substance formed is known as **a.** freon **b.** chloroform **c.** hydroform **d.** iodoform.
8. When an organic acid molecule and an alcohol molecule combine, the result is an **a.** ester **b.** amino acid **c.** organic alcohol **d.** alcohol acid.
9. Amino acids are the building blocks of **a.** carbohydrates **b.** fats **c.** alcohols **d.** proteins.
10. Energy is obtained by living things when carbohydrates combine with **a.** sugar **b.** oxygen **c.** water **d.** hydrogen.

REVIEW EXERCISES

Give complete but brief answers to each of the following. Unless otherwise indicated, use complete sentences to write your answers.

1. Refer to Fig. 14-19, p. 437, to answer the following questions: **a.** What is used to turn hard cider into vinegar? **b.** After the vinegar is aged 30–90 days, what is the next step? **c.** What process is carried out just before bottling the vinegar?
2. Name four common organic compounds.
3. Give the names and chemical formulas for four hydrocarbons.
4. Write the formulas and draw the structural formulas for the three simplest hydrocarbon molecules.
5. State two ways in which hydrocarbons are used.
6. State how some organic molecules can be related to hydrocarbon molecules.
7. Name two kinds of substitute hydrocarbon molecules and the radicals that can be used to identify them.
8. Describe two kinds of molecules that are part of living matter.
9. How are proteins used in living things?
10. What is one way that energy is obtained by living things?

EXTENSIONS

1. Examine the list of ingredients of some substances found at home and write down the names of as many organic compounds as you can find.
2. Look up the chemical composition of aspirin and report on how it acts in the body as a pain reliever.
3. **a.** What is an octane rating for gasoline and why is it important?
   b. Describe the differences between regular and premium quality gasolines. What are the advantages of each?
   c. Why are new cars required to use non-leaded gasoline?
4.

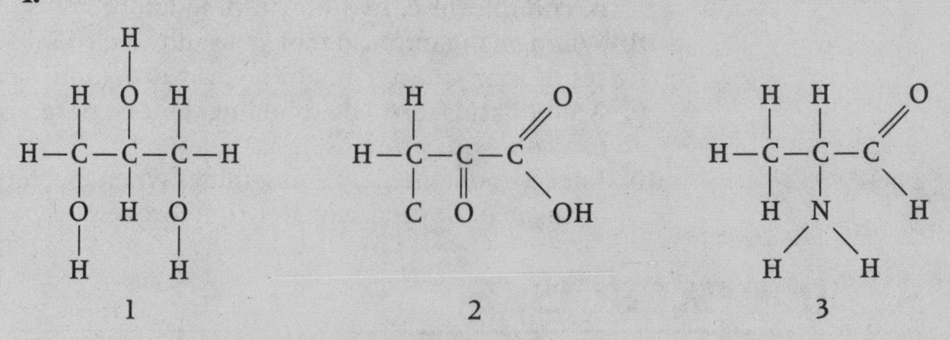

   a. Which of these models represents pyruvic acid? Explain.
   b. Which represents glycerine, an alcohol? Explain.
   c. Which represents an amino acid? Explain.
5. Compare the labels on several kinds of mouthwash.
   a. How many contain an alcohol?
   b. What are the percentages of alcohol (if given)?
   c. Determine the relative strengths of the mouthwashes based on the alcohol content of each solution.

CHAPTER 15

The Nucleus

15-1. RADIOACTIVITY

In 1898, Marie Curie discovered the element radium. She found that radium and other elements like uranium and polonium give off invisible rays. These rays come from the nuclei of atoms. The blue glow in this photograph is caused by the rays from atomic nuclei. In this lesson, you will learn why some atoms produce these invisible but dangerous rays.

When you finish lesson 1, you will be able to:

- Explain why some atoms are *radioactive*.
- Name and describe three types of radioactivity.
- Find a pattern in the makeup of nuclei that makes them stable or unstable.

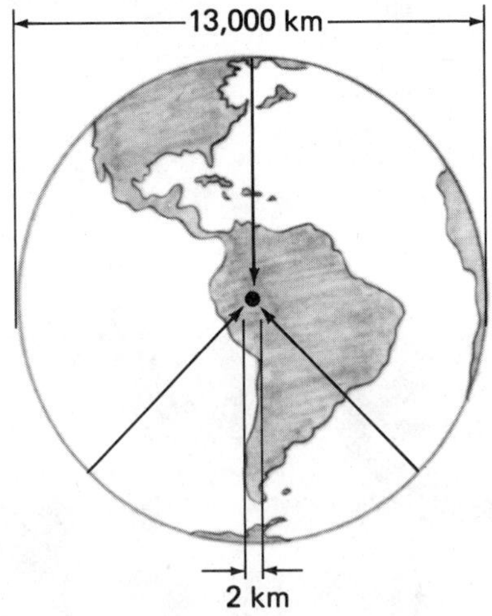

15-1. *If an atom were as big as the earth, which has a diameter of 13,000 km, the nucleus would be about 2 km in diameter.*

Try to picture the nucleus of an atom. Suppose, for example, that the nucleus of a typical atom were the size of the period at the end of this sentence. The electrons belonging to that nucleus would then take up the space of your whole classroom! See Fig. 15-1.

The tiny nucleus contains all of the atom's protons and neutrons. The protons are crowded together in a small amount of space. These positive-charged protons repel each other. It takes a powerful force to hold the protons together in the nucleus. This force is supplied by the **binding energy** of the nucleus. *Binding energy* holds the atomic nucleus together.

Binding energy
The energy that holds the nucleus of an atom together. It is also the energy required to break the nucleus apart.

The differences between atoms are a result of the number of protons and neutrons in their nuclei. For example, a typical carbon atom has 6 protons and 6 neutrons; most oxygen atoms have 8 protons and 8 neutrons.

Each kind of nucleus requires a different amount of binding energy to hold it together. Some nuclei have a small supply of binding energy. The protons and neutrons in these nuclei are not tightly bound. Such nuclei are unstable. Just as an atom with an unstable electron arrangement may change, an unstable nucleus may change to become more stable. An atom whose nucleus changes to become more stable is **radioactive** (rade-ee-oe-**ak**-tiv). *Radioactivity* was discovered by the French scientist Henri Becquerel in 1896. Becquerel observed that substances containing uranium gave off rays that could pass through paper and affect pho-

Radioactive
An atom is radioactive when its nucleus changes by means of decay in order to become more stable.

tographic film. Marie Sklodowska Curie and her husband Pierre became interested in the strange new rays discovered by Becquerel. See Fig. 15-2. After several years of work, they found two previously unknown *radioactive* elements that they named radium and polonium. Other discoveries of radioactive elements followed the pioneering work of Becquerel and the Curies. It is now known that these radioactive elements are made up of atoms whose nuclei are not stable.

An unstable atomic nucleus can become more stable in many ways. For example, rays or particles are thrown out of the nucleus, causing radioactivity. Any change in a nucleus that causes radioactivity is called *decay*. The following are the most common examples of radioactive decay.

1. *Alpha decay.* **Alpha decay** takes place when a nucleus shoots out a particle made up of two protons and two neutrons. This particle is called an *alpha particle*. Alpha particles are thrown out of a nucleus at high speed, but they slow down as they collide with the surrounding atoms. An alpha particle is the same as the nucleus of a helium atom ($^{4}_{2}He$). Each alpha particle can pick up two electrons to become a helium atom.

A radium atom can decay by throwing off an alpha particle. See Fig. 15-3. The atom that remains has two less protons and two less neutrons than radium. This atom is now an atom of the element radon. These radioactive changes can be represented by equations similar to chemical equations. For example, the alpha

15-2. *Marie Curie was born in Poland in 1867. She and her husband Pierre shared the Nobel Prize in 1903 with Henri Becquerel. Curie continued her research alone after Pierre's death and received a second Nobel Prize in 1911. She died in 1934.*

Alpha decay
A radioactive change that takes place when a nucleus shoots out an alpha particle made up of two protons and two neutrons.

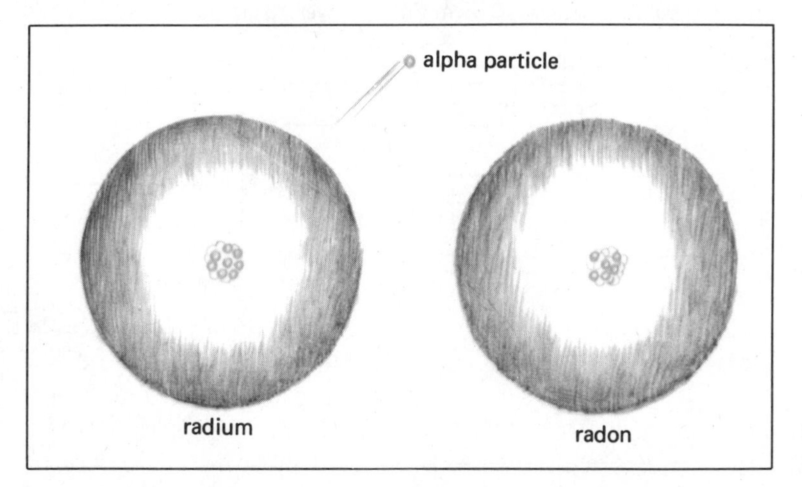

15-3. *A radium atom changes into a radon atom by means of radioactive alpha decay.*

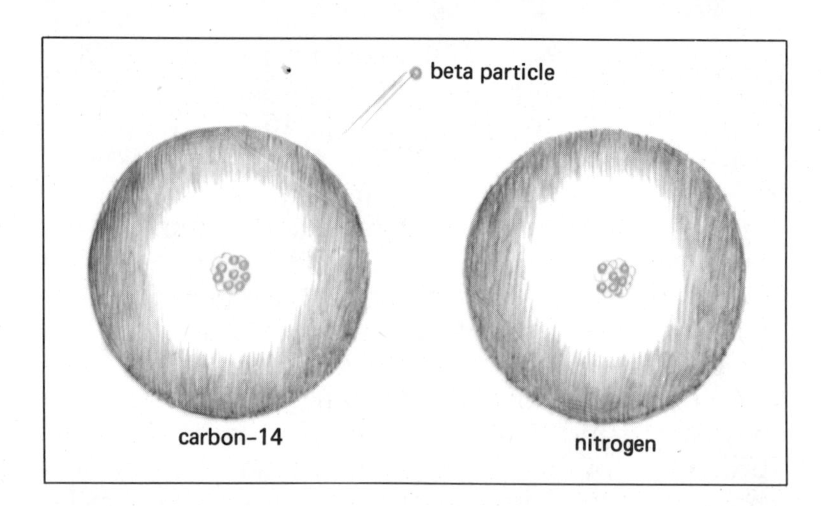

15-4. *An atom of radioactive carbon-14 changes into an atom of nitrogen by means of radioactive beta decay.*

decay of radium is shown by $^{226}_{88}\text{Ra} \longrightarrow {}^{222}_{86}\text{Rn} + {}^{4}_{2}\text{He}$. The atomic nuclei are represented by the chemical symbol with the atomic number at the lower left and the mass number at the upper left.

Beta decay
A radioactive change that causes an electron (beta particle) to shoot out of a nucleus.

Gamma rays
Rays similar to X rays that cause a particular type of radioactive decay.

2. *Beta decay*. **Beta decay** causes an electron to be shot out of the nucleus at high speed. This rapidly moving electron is called a *beta particle*. How did that electron get into the nucleus? A beta particle is formed when a neutron inside the nucleus changes into a proton. A neutron becomes a proton when it loses an electron. This electron leaves the nucleus as a beta particle. When a nucleus gives off a beta particle, it is left with an extra proton. This extra proton changes the atom into an atom of a different element. For example, a

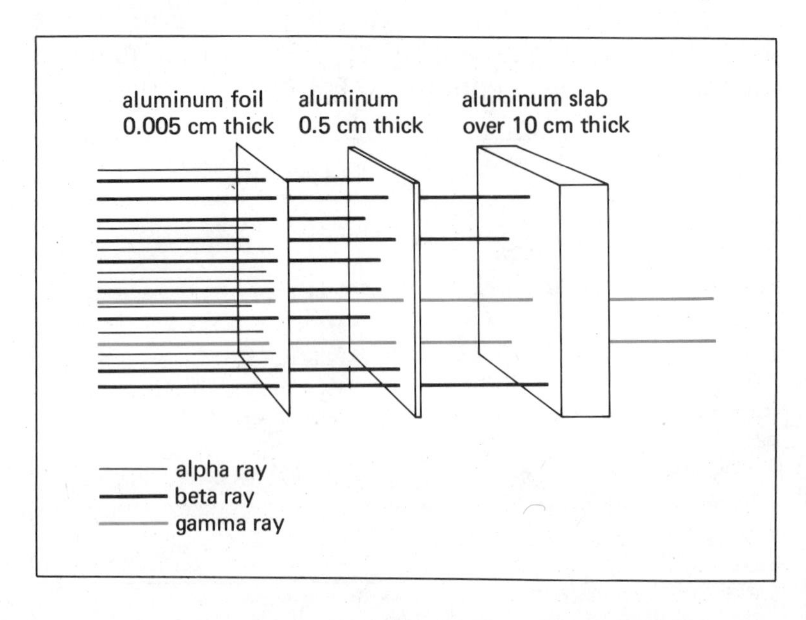

15-5. *By studying this diagram, can you tell which form of radiation is the most dangerous?*

radioactive form of carbon, called carbon-14, changes into nitrogen by beta decay. See Fig. 15-4. Beta decay of carbon-14 is represented by

$$^{14}_{6}\mathrm{C} \longrightarrow {}^{14}_{7}\mathrm{N} + {}^{0}_{-1}e.$$

3. *Gamma decay*. Not all radioactive elements emit particles. When alpha or beta decay takes place, **gamma rays** are also usually given off. *Gamma rays* are a kind of electromagnetic energy similar to X rays. Unlike alpha and beta decay, emission of gamma rays does not cause an atom to change into another kind of atom. Gamma rays are not atomic particles like alpha and beta particles. Gamma rays can penetrate most materials easily. Figure 15-5 shows that it is easier to shield against alpha and beta particles than against gamma rays.

Special instruments must be used to detect radioactivity. One of these instruments is called a *Geiger counter*. One type of Geiger counter is shown in Fig. 15-6. Prospectors use Geiger counters to detect the radioactivity given off by uranium deposits.

15-6. *This photograph shows a common type of Geiger counter. The Geiger counter was invented by Hans Geiger, a German physicist.*

ACTIVITY

Nuclear Stability

Materials
paper
pencil

A. Figure 15-7 represents the makeup of several atomic nuclei. The nuclei are grouped according to their stability.

B. Study the numbers of protons and neutrons in each group.

1. What pattern of odd or even numbers do you see in the numbers of protons and neutrons of stable atoms?
2. What pattern in the numbers of protons and neutrons seems to give the least stability?

There are many unstable nuclei in nature. In addition, about 280 different stable nuclei are now known to exist. Table 15-1 shows the relation between the number of protons and neutrons in each of these stable nuclei.

3. What pattern of protons and neutrons is most likely to give a stable nucleus?
4. What pattern of protons and neutrons is least likely to give a stable nucleus?
5. Are these patterns the same as the ones you found in step B? (Compare answers 3 and 4 with answers 1 and 2.)

There are many conditions that affect the stability of a nucleus. The pattern of protons and neutrons in the nucleus is just one of these conditions.

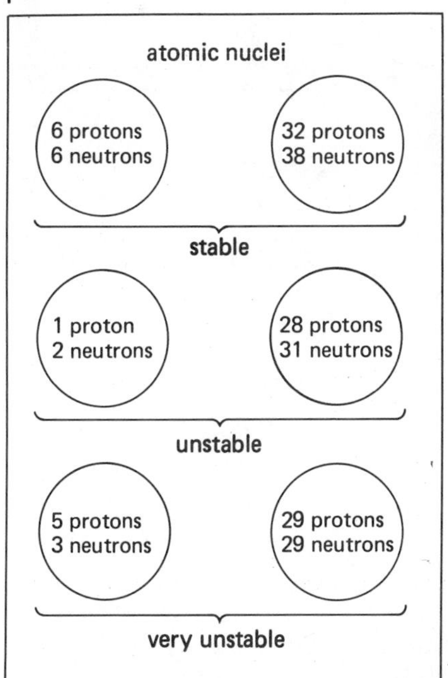

15-7.

TABLE 15-1

Makeup of nucleus	*Number of stable atoms existing*	*Makeup of nucleus*	*Number of stable atoms existing*
even number of protons even number of neutrons	171	odd number of protons even number of neutrons	50
even number of protons odd number of neutrons	55	odd number of protons odd number of neutrons	4

SUMMARY

Every nucleus requires a certain amount of binding energy to hold it together. A nucleus with too little binding energy is unstable. An unstable nucleus will change to become more stable. Radioactivity is the result of decay in an atomic nucleus. This decay can produce alpha particles, beta particles, and gamma rays.

QUESTIONS

Use complete sentences to write your answers.

1. Name and describe three types of radioactive decay.
2. Explain why some atoms are radioactive.
3. How does the nucleus of an atom compare in size to the whole atom?
4. What pattern of protons and neutrons is most likely to give a stable nucleus?
5. Describe the contributions of the following scientists to the study of radioactivity: Henri Becquerel, Marie and Pierre Curie.

15-2. USING RADIOACTIVITY

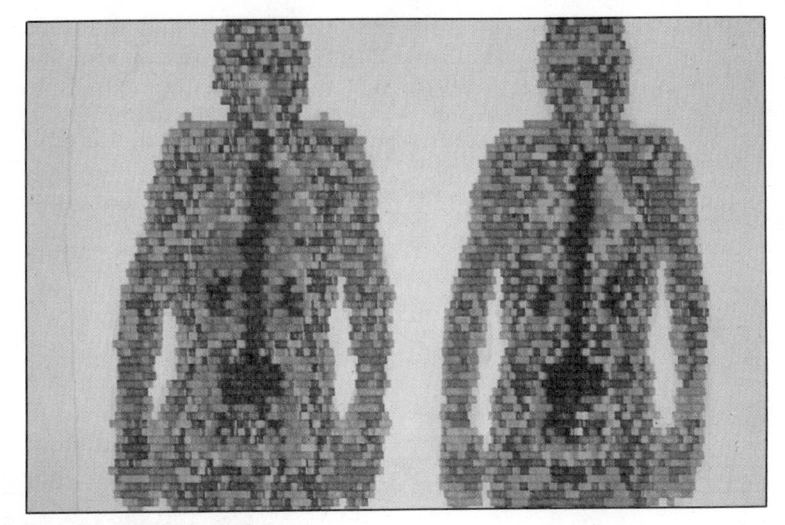

Doctors are able to use radioactive substances in making a diagnosis. They give a patient a solution containing radioactive iodine. The iodine collects in the small thyroid gland near the Adam's apple. The thyroid gland takes iodine from the blood and makes it into another substance needed by the body. By measuring how fast radioactive iodine is taken in by the gland, a doctor can tell if the gland is doing its job properly. A special instrument is used to detect the radioactivity. If the patient's thyroid gland is working too slowly, the doctor can prescribe a medicine to correct the problem. The doctor's diagnosis is an example of how a radioactive tracer element can be followed through chemical reactions.

When you finish lesson 2, you will be able to:

- Explain what is meant by *half-life*.
- Use examples to show how a radioactive *tracer* is used.
- ○ Determine the half-life of a radioactive element.

Many kinds of atoms have unstable nuclei and are naturally radioactive. Radium, for example, is a naturally radioactive element. Compared to most other kinds of atomic nuclei, a radium nucleus has a large mass. It contains 83 protons and 138 neutrons. Most of the naturally radioactive atoms, like radium, have a large number of protons and neutrons. Elements with an atomic number less than that of lead (82) are generally not radioactive. However, there are radioactive forms of some of the lighter elements.

An example of such a radioactive atom is one form of carbon. It has 6 protons and 8 neutrons in its

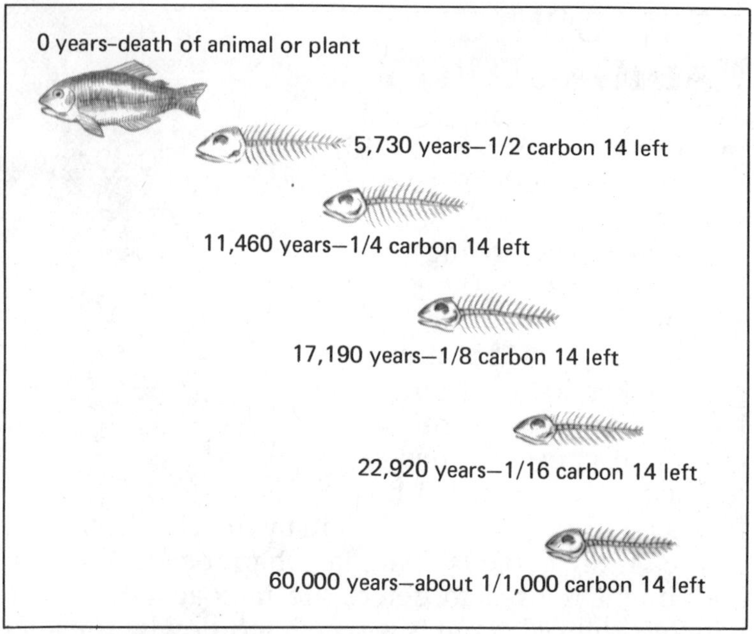

0 years–death of animal or plant

5,730 years—1/2 carbon 14 left

11,460 years—1/4 carbon 14 left

17,190 years—1/8 carbon 14 left

22,920 years—1/16 carbon 14 left

60,000 years—about 1/1,000 carbon 14 left

15-8. *The ages of plant and animal remains can be determined by means of carbon-14 dating.*

nucleus. This isotope is called carbon-14. The common form of carbon, carbon-12, has 6 protons and 6 neutrons. Carbon-14 takes part in life processes, just like ordinary carbon-12. All living things, including you, contain a few atoms of carbon-14. The amount of radioactive carbon-14 in living things remains the same as long as they are alive. After a living thing dies, the amount of radioactivity caused by carbon-14 slowly decreases. The unstable carbon atoms begin to change into more stable atoms. A piece of wood from an old house, for example, is less radioactive than wood from a freshly cut tree.

Experiments have shown that carbon-14 becomes stable at a certain rate. Half of a given number of carbon-14 atoms will become stable after 5,720 years. Carbon-14 has a **half-life** of 5,720 years. The *half-life* is the time it takes for half of a certain number of radioactive atoms to change. Because they know the half-life of carbon-14, scientists can determine the age of plant and animal remains. See Fig. 15-8. The amount of radioactivity of an element decreases by one half after each half-life.

Half-life
The length of time necessary for one half of a given number of atoms to change.

Each kind of radioactive atom has its own particular half-life. Some radioactive elements have a very long half-life. For example, uranium has a half-life of about 4.5 billion years. Other radioactive atoms may have

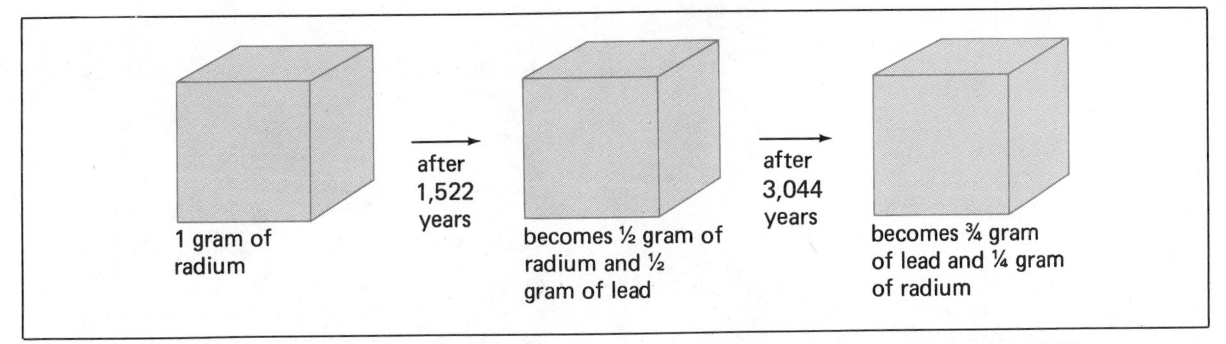

15-9. *During each half-life, one half of the remaining radium atoms changes into lead.*

short half-lives, lasting only several days, minutes, or even fractions of a second. Radioactive iodine has a half-life of 8 days. At the end of each half-life, half of the radioactive atoms will have changed into some stable kind of atom. For example, radium, with a half-life of 1,522 years, changes into lead. If you started with a block of radium, half of the block would be lead after 1,522 years. After two half-lives (3,044 years) had passed, only one fourth of the original radium would remain. See Fig. 15-9.

Scientists can follow radioactive atoms through chemical reactions by using instruments that detect radioactivity. Radioactive atoms that can be followed through the steps of a chemical reaction are called **tracers.** Agricultural scientists use *tracers* to study plant growth. Tracers can help show the effect of fertilizers and other chemicals on crops. Scientists hope to use this information to help solve the problem of feeding the world's increasing population. Knowledge of the way plants use the chemicals in fertilizers can help to produce more crops. See Fig. 15-10.

Tracers
Radioactive substances that are used to follow atoms through chemical reactions.

Large amounts of radioactivity are harmful to living things. When the various radioactive particles and rays pass through living cells, some of the molecules within the cells can be changed. Alpha and beta particles, for example, may collide with molecules in a cell and break some of the bonds holding the molecules together. The cell may be damaged enough to cause it to die. The loss of large numbers of cells by exposure to radioactivity can seriously injure or kill a living thing. Radioactivity can also cause *mutations*. Mutations cause offspring to be produced that are different from the parents. If parents are exposed to too much radioactivity, mutations can produce deformed children.

15-10. *Agricultural scientists use radioactive tracers to study plant growth. However, too much radioactivity can also be harmful to growing plants, as shown in this photo.*

However, radioactivity can also be used to help prevent diseases. For example, a radioactive isotope of iodine, called I-131, can be used as a tracer to measure the working of the thyroid gland in the neck. A person with thyroid problems is given a small amount of I-131. Since the thyroid gland takes up iodine, the I-131 is quickly concentrated there. Measuring the level of I-131 in the thyroid gland helps to show how well the gland is working. The paragraph at the beginning of this lesson describes the measurement of radioactive iodine in a person's body. Iodine-131 has a very short half-life and quickly changes into harmless substances in the body.

Radioactivity can also be used to treat diseases. Some kinds of cancer are treated with radioactivity. See Fig. 15-11. The radioactivity destroys only the cancer without harming any other part of the patient's body. There is still no cure for cancer. Radioactivity is just one weapon in the fight against this disease.

15-11. *A cancer patient is shown here being treated with radioactivity.*

ACTIVITY

Half-Life

Each radioactive element has its own half-life. This half-life is different from the half-lives of other radioactive elements. The half-life of an element is easy to determine. You need to know the number of radioactive atoms present originally. You then determine the number of atoms present after several time intervals. In this activity, you will see how this is done.

Materials
graph paper
pencil

A. Assume that you start with 32 atoms. The number of atoms originally present can be determined by taking the sample weight.

B. Figure 15-12 shows how the number of atoms changes every 5 min for a 20-min period. The number of atoms left after a given time is determined by measuring the radioactivity of the sample.

C. Copy Table 15-2 in your notebook.

D. Look at Fig. 15-12 and count the number of atoms remaining after each 5-min interval. Record the number in your table.

1. After how many minutes were ½ of the original 32 atoms left?
2. What is the half-life of the radioactive atom represented in Fig. 15-12?
3. How many atoms were left after two half-lives?

E. In your notebook, draw and label the axes of a graph, as shown in Fig. 15-13.

F. Plot the data in your table on this graph.

G. Without using a ruler, draw a smooth line joining the five data points on your graph.

4. Is this line straight or curved?

When data for the half-lives of different elements are plotted, the graphs always have the same shape. These graphs show that all radioactive atoms change in the same general way.

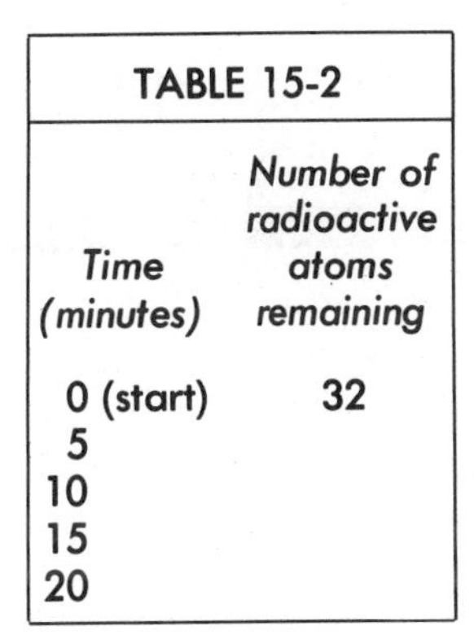

TABLE 15-2	
Time (minutes)	*Number of radioactive atoms remaining*
0 (start)	32
5	
10	
15	
20	

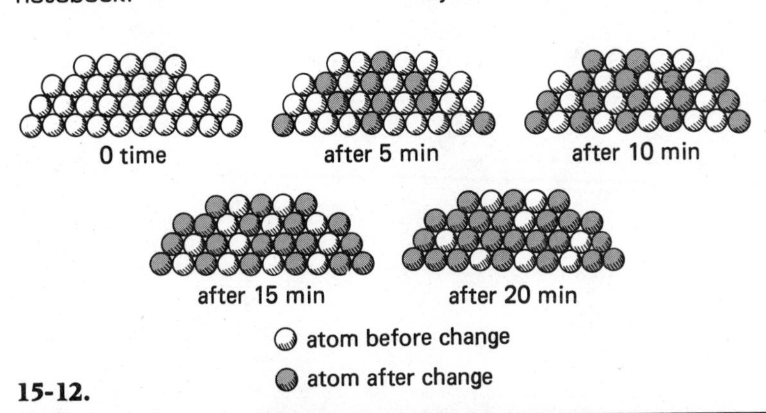

15-12.

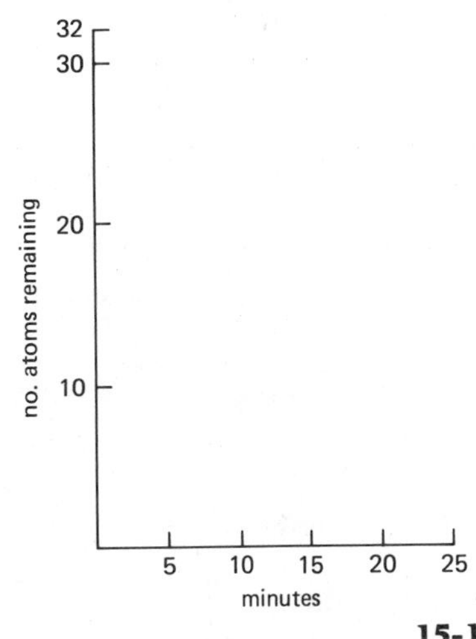

15-13.

SUMMARY

All radioactive elements have a particular half-life. One half of the original number of radioactive atoms will become stable after one half-life. Tracers, such as radioactive iodine, can be used to detect diseases. Radioactivity can also help fight diseases, such as cancer.

QUESTIONS

Use complete sentences to write your answers.

1. Explain what is meant by the half-life of a radioactive element.
2. What is the half-life of carbon-14? of radium?
3. How are radioactive tracers used?
4. Name two radioactive elements and give their atomic masses.
5. Thorium-234 has a half-life of 24 days. If 1,000,000,000 atoms of this element are checked 48 days later, how many will *not* have changed due to radioactive decay?

15-3. NUCLEAR ENERGY

Just before dawn on July 16, 1943, the world's first atomic bomb exploded in a lonely part of the New Mexican desert. For the first time, human beings had released part of the huge supply of energy trapped inside atoms. Before that time, scientists had only known that a small portion of this atomic energy was being released as radioactivity. The explosion of the atomic bomb proved that the energy in the nucleus could be released in other ways. In this lesson, you will see what happens to an atomic nucleus that causes it to release energy.

When you finish lesson 3, you will be able to:

- Give an example to show how energy can be released by splitting a large atomic nucleus.
- Explain how a large number of nuclei can be made to split in a short time.
- Distinguish between *fission* and *fusion*.
- Use photographs to determine which of several situations will result in a *chain reaction*.

Scientists can study atomic nuclei by shooting particles at them. For example, suppose that an alpha particle given off by a radioactive substance collides with the nucleus of another atom. The alpha particle can then combine with the target nucleus and cause it to change. Collision of an alpha particle with a nitrogen nucleus produces an oxygen nucleus. The bombardment of nitrogen gas with alpha particles produces oxygen nuclei according to the following nuclear equation: ${}^{14}_{7}\mathrm{N} + {}^{4}_{2}\mathrm{He} \longrightarrow {}^{17}_{8}\mathrm{O} + {}^{1}_{1}\mathrm{H}$. However, it is difficult to make a positive-charged alpha particle collide with a nucleus. Both the nucleus and the alpha particle carry positive charges. They repel each other.

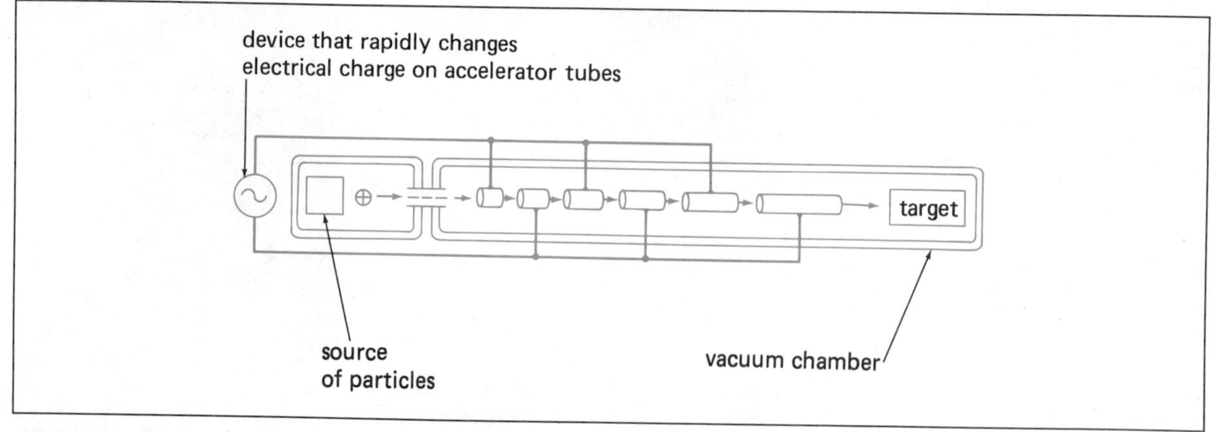

15-14. *In this type of accelerator, charged particles are drawn into a series of tubes. By rapidly changing the electrical charges on the tubes, the particles are first attracted and then repelled. Each time the charge on the tube changes, the particles are speeded up. Because the particles move in a straight line, this kind of machine is called a linear accelerator.*

To overcome this repulsion, machines can be used to speed up the particles. The particles then smash into the target atoms with enough energy to cause a nuclear change. Machines that are used to speed up atomic particles are called *accelerators.* One kind of accelerator is shown in Figs. 15-14 and 15-15. This accelerator is called a *linear accelerator.* Scientists use accelerators to produce beams of different kinds of particles for their

15-15. *A linear accelerator at Stanford, California.*

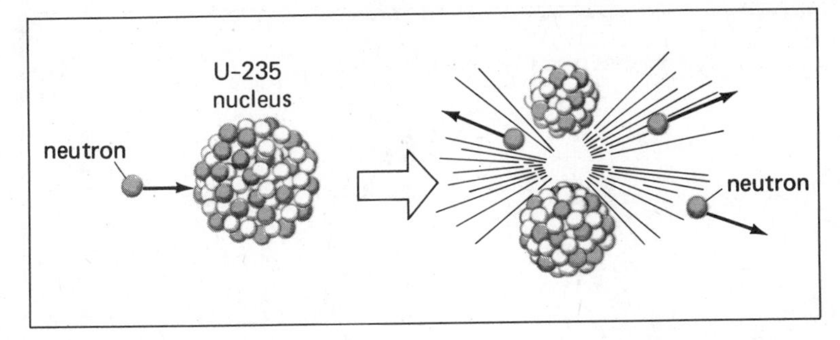

15-16. *This diagram shows the fission of an atom of uranium-235. When a neutron strikes the U-235 nucleus, the nucleus splits into two parts. Two additional neutrons and energy are produced.*

study of the different nuclear changes that result when the particles hit various targets.

A particle commonly used in nuclear experiments is the neutron. Since they carry no charge, neutrons can easily enter the target nucleus. One of the first people to use neutrons in this way was Enrico Fermi, an Italian scientist. In 1938, Fermi was part of a scientific team that made a surprising discovery. They found that a particular type of uranium nucleus changed in an unusual way when it was hit by a neutron.

Figure 15-16 shows what happens when a neutron enters a uranium atom. This uranium atom has a mass of 235 and is called U-235. The common form of uranium has a mass of 238 and is called U-238. In nature, U-235 atoms are found thinly scattered among U-238 atoms. The U-235 nucleus becomes so unstable after being struck by a neutron that it splits in two. The splitting of an unstable nucleus into two smaller nuclei is called **fission.**

Fission
A nuclear reaction in which a large unstable nucleus splits into smaller nuclei.

When *fission* occurs, a small amount of energy is released. The amount of energy released by the splitting of a single U-235 nucleus is not large. The huge amount of energy released in an atomic bomb is the result of the fission of *many* U-235 nuclei. The fission of a single U-235 nucleus produces neutrons. These neutrons strike other U-235 nuclei, causing them to undergo fission. More neutrons are produced. If there are enough U-235 atoms present, the result is a **chain reaction.** See Fig. 15-17.

Chain reaction
A reaction in which each step involves the same reaction as in the previous step.

Uranium-238 is not fissionable. Naturally occurring uranium must be "enriched" with U-235 before it can be used to construct an atomic bomb. An atomic bomb consists of a certain amount of pure U-235, or a similar fissionable material. Neutrons act as the trigger to set off a *chain reaction*. An uncontrolled chain reaction results in an explosion.

triggering neutron from outside

15-17. *This diagram shows the creation of a chain reaction from the fission of a U-235 nucleus.*

In addition to fission, there is another type of energy-producing nuclear reaction. This type of reaction occurs when two atomic nuclei come close to each other. Normally, when two atoms approach each other, their negative electrons and positive nuclei repel each other. However, if the temperature is raised to several millions of degrees, the two atoms move extremely fast. At this temperature, the atoms will collide violently. The collision between these atoms will bring their nuclei close together. The two nuclei then join to form a single, larger nucleus. This type of reac-

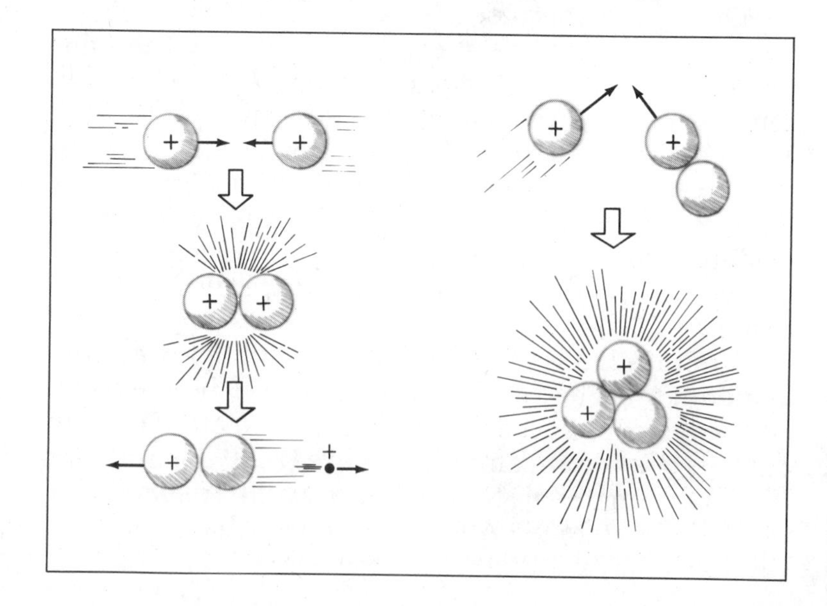

15-18. *Two kinds of fusion reactions are shown in this diagram. On the left, two hydrogen nuclei fuse to form a heavy isotope of hydrogen. On the right, a hydrogen nucleus fuses with a heavy hydrogen nucleus to form a light isotope of helium.*

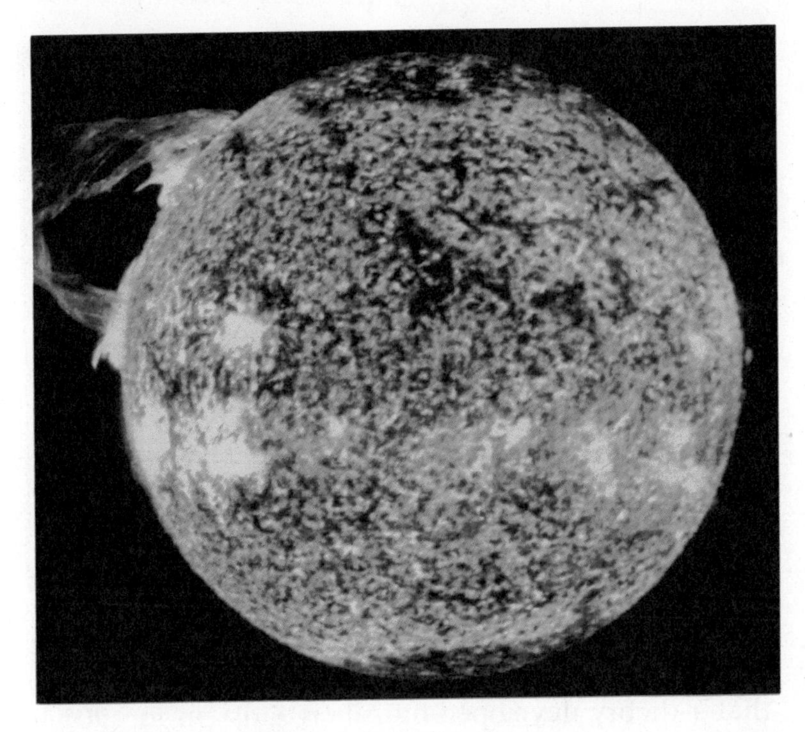

15-19. *The heat and light of the sun are caused by nuclear fusion reactions.*

tion is called a **fusion** reaction. Several kinds of *fusion* reactions are shown in Fig 15-18.

Fusion
A nuclear reaction in which two small nuclei join to form a larger nucleus.

Because very high temperatures are needed, fusion reactions happen only in special places. The insides of the sun and other stars are places where the temperatures are high enough for fusion reactions to take place. See Fig. 15-19. The temperature near the center of the sun is probably about 15 million degrees Kelvin. Such high temperatures can be found on earth only in an atomic explosion. The first atomic bomb exploded in New Mexico was a fission bomb. The high temperatures caused by fission reactions can be used to make a fusion bomb.

Making nuclear fusion take place slowly in a controlled way is much more difficult than causing an uncontrolled explosion. The heat given off by fusion reactions creates very high temperatures. Any materials used to contain the reaction would be vaporized by the great amount of heat released. One solution to this problem is to contain the fusion reaction in a magnetic field or so-called *magnetic bottle*. Another approach uses powerful laser beams to create the high temperatures needed. See Fig. 15-20. Both of these ideas are now being tested. Scientists are working very

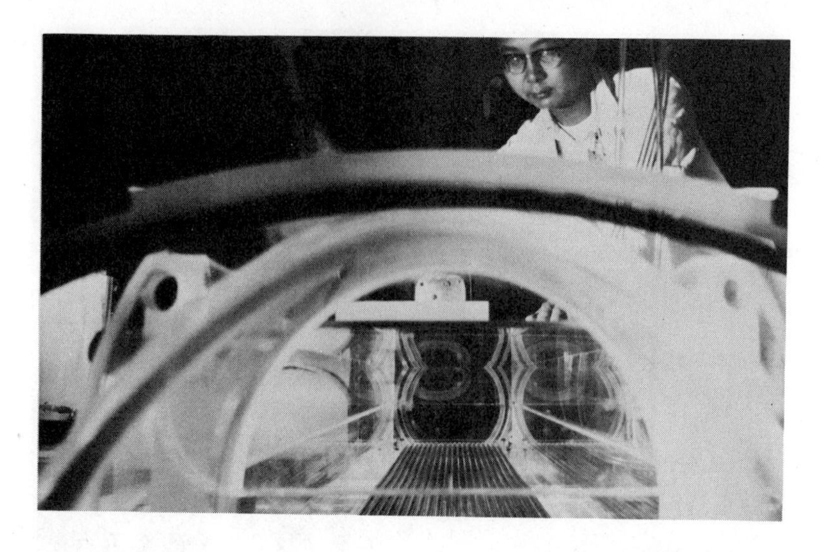

15-20. *Scientists are studying ways to control nuclear fusion as a source of energy. In this machine, laser beams are focused on a tiny fuel pellet, causing it to undergo fusion and release energy.*

hard to find a way to use nuclear fusion as a source of energy. We already use controlled fission as an energy source, as you will see in the next lesson.

The release of energy by nuclear reactions shows that a theory developed by Albert Einstein in 1905 is correct. Einstein's theory stated that matter and energy were different forms of the same thing. This idea is expressed in the famous equation $E = mc^2$. This equation shows that matter or mass (m) can be converted into energy (E). In the equation, c is the velocity of light. The velocity of light is very large: 3×10^8 m/sec. Thus the equation says that a small amount of matter can be changed into a large amount of energy. For example, suppose 1 kg of common sand could be completely converted into energy. This amount of energy could meet the electric power needs of the entire United States for about two weeks. With this energy, you could also drive an automobile around the earth

15-21. *New elements that do not exist in nature have been created in the laboratory through nuclear reactions.*

about 400,000 times. In nuclear reactions, small amounts of matter disappear. This small loss of mass produces the huge amounts of energy released by fission and fusion reactions.

One result of scientific experiments with nuclear reactions has been the discovery of ways to make artificial atoms. The largest natural atom is uranium, with atomic number 92. Heavier atoms, up to atomic number 105 and beyond, have been made by different kinds of nuclear reactions. Only very tiny amounts of most of these artificial elements have been made. Some of these elements have been found to be useful and all of them have added to our knowledge of atoms. See Fig. 15-21.

ACTIVITY

Chain Reactions

Materials
paper pencil

A. Look at Fig. 15-22.

1. Is this an example of a chain reaction?

B. Look at the four photographs in Fig. 15-23.

2. Which of the four photos best represents the fission chain reactions of U-235?
3. In which photo would the arrangement not result in a chain reaction?

Since U-235, as found in nature, is scattered thinly through U-238, the fission reactions that occur do not cause a chain reaction.

4. Which photo in Fig. 15-23 would represent U-235 as it is found in nature?

15-22.

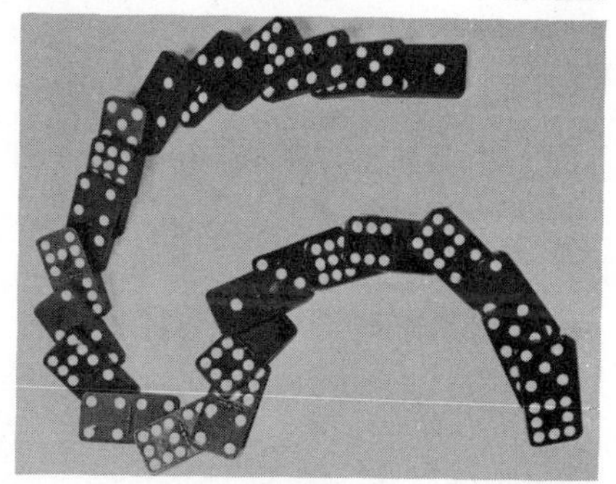

15-23.

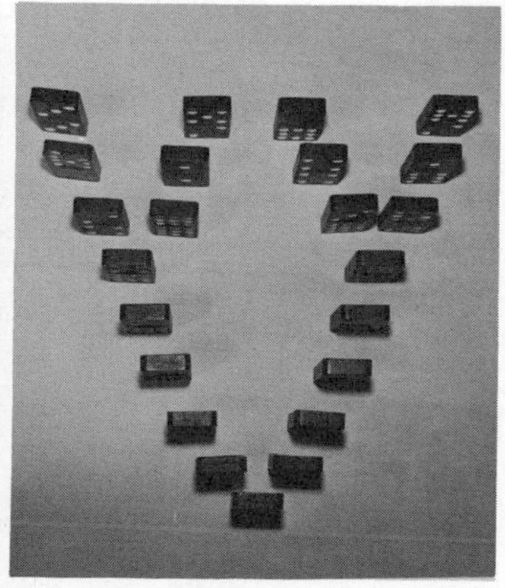

SUMMARY

A tremendous supply of energy is packed into the nucleus of an atom. Some of that energy can be released by two different kinds of nuclear reactions. A fission reaction takes place when large nuclei split into smaller nuclei. A fusion reaction takes place when two small nuclei join to become a larger nucleus.

QUESTIONS

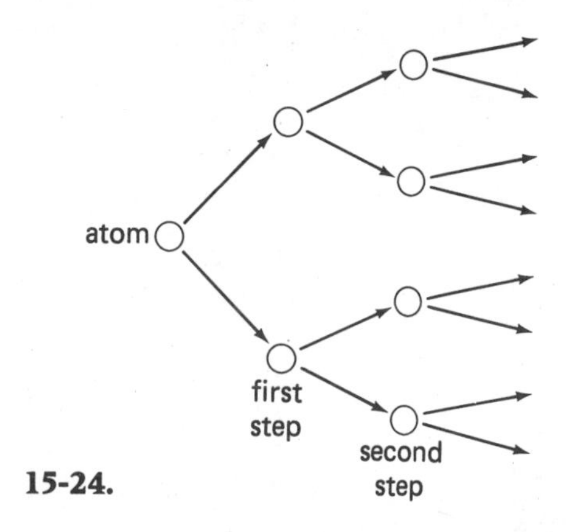

15-24.

Use complete sentences to write your answers.

1. Why does splitting a large nucleus result in a release of energy?
2. How can a large number of nuclei be made to split in a short time?
3. What is meant by nuclear fission?
4. What is meant by nuclear fusion?
5. Suppose that the fission of each atom in a chain reaction causes two additional atoms to undergo fission. See Fig. 15-24. In the twenty-first step of such a reaction, the number of atoms undergoing fission is the product of 2 multiplied by itself 20 times. How many atoms will undergo fission in the twenty-first step?

15-4. CONTROLLED NUCLEAR REACTIONS

Many homes today contain electric appliances, such as can openers, garbage disposal units, toothbrushes, toasters, mixers, and microwave ovens. Consumption of all this energy is rapidly using up our natural resources. Almost all of our energy now comes from carbon fuels, such as petroleum, coal, and natural gas. The supply of these materials is limited. We must develop new sources of energy. One of these new energy sources is the atomic nucleus. In this lesson, you will see how nuclear energy may be used to replace carbon fuels.

When you finish lesson 4, you will be able to:

- Describe how a nuclear *reactor* can be used to control a nuclear fission reaction.

- Explain some of the problems arising from the use of nuclear fission to produce energy.

- Describe two types of nuclear reactors that may be used as energy sources.

- ○ Use examples to explain the difference between a controlled and an uncontrolled chain reaction.

Not all nuclear chain reactions produce an explosion. Fission reactions can be controlled. A controlled fission reaction must be made to take place slowly and at a steady speed. These controlled fission reactions can be used to produce energy. See Fig. 15-25.

An uncontrolled nuclear reaction is like an uncontrolled population of rabbits. Suppose you start with one pair of rabbits. These rabbits mate and, in a short time, produce offspring. The young rabbits soon grow

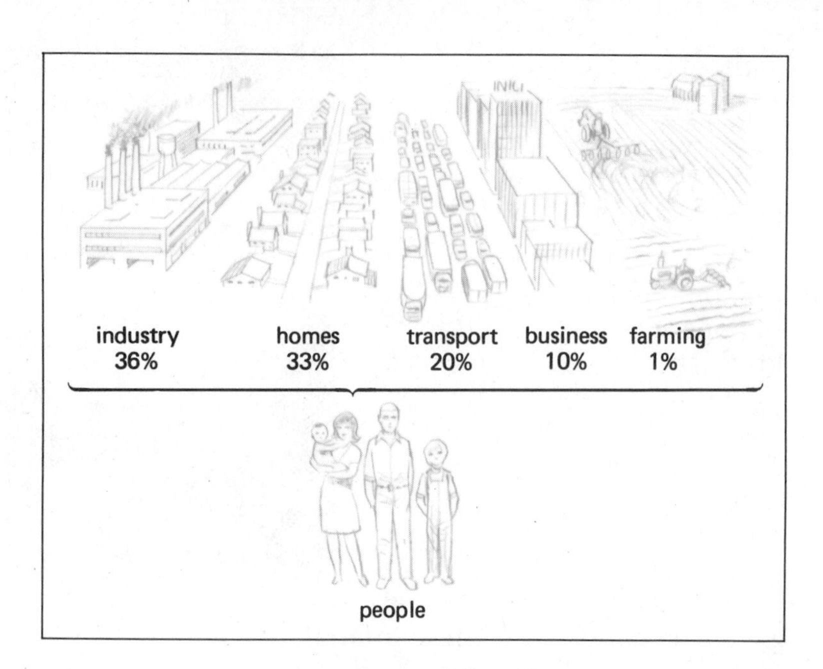

15-25. *This chart shows the breakdown of energy consumption by people.*

up and produce their own offspring. These rabbits in turn produce more rabbits, and so on. Before you know it, you have a population explosion of rabbits!

You can control your rabbit population by removing all but one pair from each new family of rabbits. A nuclear reaction can be controlled in a similar way. Some of the neutrons needed to keep the reaction going must be removed.

Nuclear reactor
A machine used to carry on a controlled nuclear chain reaction.

A controlled nuclear chain reaction takes place in a **nuclear reactor.** Material containing uranium atoms is used as fuel in *nuclear reactors.* See Fig. 15-26. In the reactor, the unstable uranium atoms are bombarded with neutrons. The unstable atoms then split. The fission reaction produces heat and more neutrons. These neutrons hit more uranium atoms. A chain reaction is started. The speed of the chain reaction is controlled by rods made of a substance, like boron, that absorbs neutrons. When the rods are pushed part way into the reactor, neutrons are absorbed. There are fewer neutrons to strike uranium atoms. The reaction slows down. The reaction speeds up again as the rods are pulled part way out of the reactor.

A working nuclear reactor releases heat. This heat can be used just like the heat produced by burning carbon fuels. The heat produced by a nuclear reactor can be used to make steam. The steam can drive generators to make electricity. See Fig. 15-27. The fission of 1 g of uranium produces as much heat as the burning of 3 tons of coal. A piece of uranium the size of a golf ball could run a nuclear-powered car for 100 years. You can see that nuclear power plants could produce large amounts of energy.

Unfortunately, producing energy from nuclear fission also creates some serious problems. All nuclear reactors produce radioactive substances. Since radioactivity is harmful, no radioactive materials can be allowed to escape from the reactor. The reactor must be surrounded by a heavy lead and concrete shield.

Many nuclear power plants are already in operation. These plants are producing another serious problem that has to be faced. In the same way that a kitchen produces garbage, an operating nuclear power plant produces waste materials. These wastes are radioactive and will be dangerous for hundreds of years. Most

15-26. *These tiny nuclear fuel pellets can provide more energy than a much larger amount of coal.*

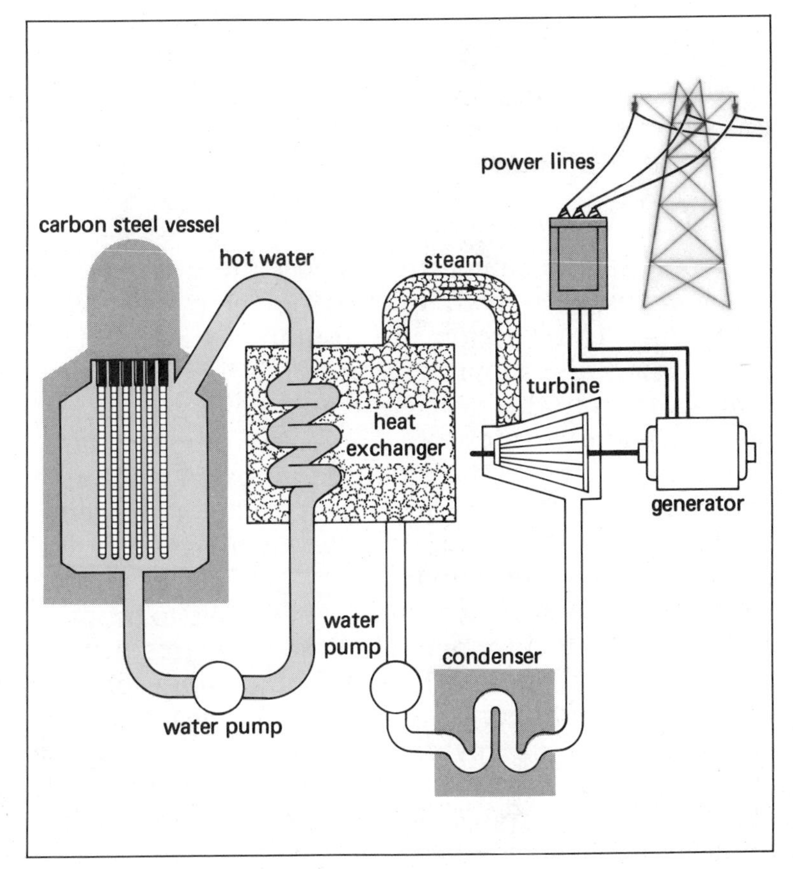

15-27. *This diagram shows how a nuclear reactor can be used to generate electricity.*

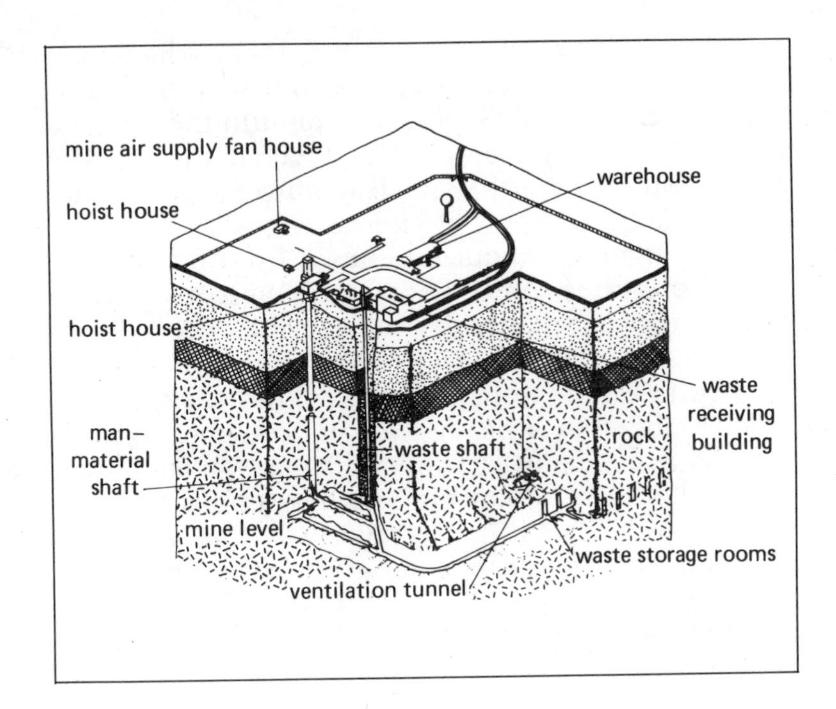

15-28. *One possible solution to the problem of radioactive waste disposal might be to store the wastes deep underground in old salt mines. This type of storage facility is shown in the diagram.*

radioactive wastes are now sealed in underground tanks. Some wastes are sunk in deep ocean water. If more nuclear power plants are built, more radioactive waste will be produced. We must find a safe and inexpensive way to get rid of this nuclear garbage. See Fig. 15-28.

In addition to radioactive waste, there is another problem involved with nuclear power plants. Most plants now in operation use uranium as a fuel. Like carbon fuels, the supply of uranium is limited. A possible solution to this problem would be to use nuclear *fusion reactors.* Fusion reactors also produce large amounts of energy. The fuel used in fusion reactors is hydrogen. Unlike uranium, a great deal of hydrogen is available on earth. Another great advantage of fusion reactors is that they produce less radioactive wastes. At present, scientists have not yet been able to produce and control the extremely high temperatures needed to cause fusion. Research is still being done on this form of nuclear energy. At some time in the future, the almost limitless energy of the sun and stars may be available for our use. Reliable estimates indicate that obtaining energy from fusion will not become practical for at least 50 years.

15-29.

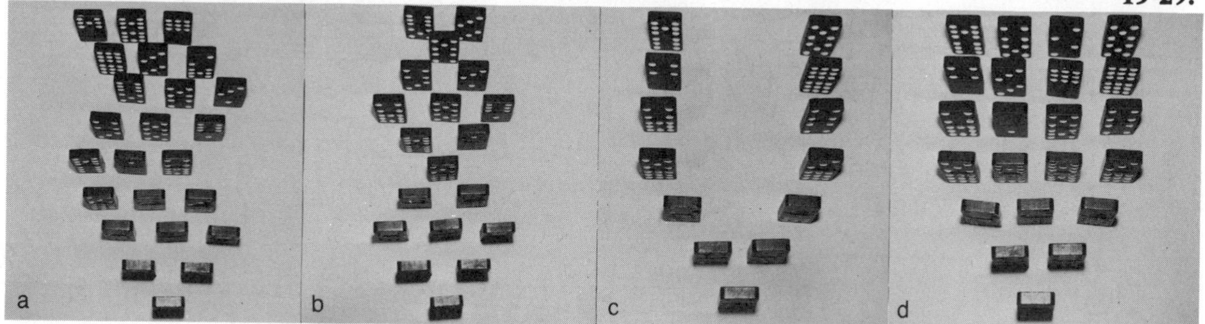

15-30.

A Controlled Chain Reaction

The control methods used in a nuclear reactor can be represented by dominoes.

A. Look at Fig. 15-29.

1. Which photo represents a chain reaction that would continue without increasing or decreasing the number of dominoes?

2. Does this photo represent a controlled or an uncontrolled chain reaction?

B. Now look at Fig. 15-30.

3. Which arrangement of dominoes would give the lowest constant level of chain reaction?

4. Which arrangement would give the greatest constant level of chain reaction?

C. Look at the arrangement of dominoes in Fig. 15-31.

5. Does this arrangement represent a controlled reaction?

6. How many dominoes would be falling at any one time?

7. What effect would removing the barrier have on the level of the reaction?

8. What part of a nuclear reactor does the barrier represent?

Materials

paper pencil

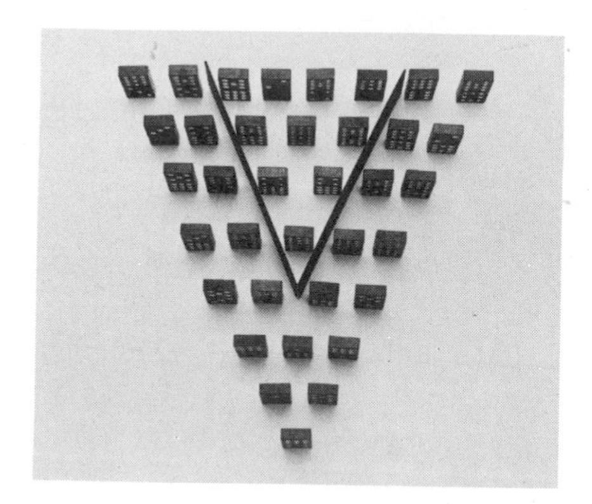

15-31.

SUMMARY

Some carbon fuels may be used up in your lifetime. We may have to turn to controlled nuclear fission to provide most of our energy. Nuclear fission reactors produce dangerous radioactive wastes. If scientists learn to control nuclear fusion, nuclear fusion reactors may someday provide a clean, safer source of energy.

QUESTIONS

Use complete sentences to write your answers.

1. How is the chain reaction controlled in a nuclear reactor?
2. Explain some of the problems arising from the use of nuclear fission to produce energy.
3. What are two advantages of using a fusion reactor to produce energy?
4. Look at Fig. 15-27. Describe how electricity is generated by a nuclear reactor.

VOCABULARY REVIEW

Match the number of the word with the letter of the phrase that best explains it.

1. binding energy
2. radioactive atom
3. half-life
4. alpha decay
5. beta decay
6. gamma decay
7. tracer
8. fission
9. fusion
10. nuclear reactor
11. chain reaction

a. The time taken for half of a given number of radioactive atoms to decay.
b. A radioactive substance used to follow atoms in chemical reactions.
c. Two small nuclei join to form a larger one.
d. A nucleus gives off electromagnetic radiation.
e. A nucleus shoots out particles made up of 2 protons and 2 neutrons.
f. A powerful force that holds the nucleus together.
g. An atom whose nucleus is changing to become more stable.
h. A large unstable nucleus splits into smaller nuclei.
i. A machine used to carry on a controlled nuclear chain reaction.
j. A nuclear change in which an electron comes from the nucleus.
k. Each step involves the same reaction.

REVIEW QUESTIONS

Complete each statement by choosing the best word or phrase, or by filling in the blank.

1. An atom is radioactive when **a.** it has an excess of electrons **b.** its nucleus is unstable and it has a small supply of binding energy **c.** it has a need for electrons **d.** its nucleus has a surplus of protons.
2. The half-life of a certain radioactive element is 5,000 years. In 15,000 years, how many of the original atoms will be unchanged? **a.** one-third **b.** one-fourth **c.** one-sixth **d.** one-eighth.
3. The age of animal remains can be estimated by radioactivity because all animals contain radioactive **a.** oxygen that increases after death **b.** carbon that decreases after death **c.** oxygen that decreases after death **d.** carbon that increases after death.

4. Look at Fig. 15-25, which shows the breakdown of energy consumption. The use that accounts for the greatest percent of consumption is __________. The percent of energy consumption that would result from combining industrial and home uses is __________.
5. Radioactivity can be used to fight some kinds of cancer because when it is aimed properly, it can **a.** cause harm to all parts of the body **b.** destroy cancer cells without harm to other parts of the body **c.** cure many illnesses **d.** all the above answers are correct.
6. Uranium has a half-life of 4.5 billion years. This means that in 4.5 billion years the earth will have __________ the amount of uranium it now has.
7. A chain reaction is a reaction in which each step **a.** is a series of links **b.** involves the same reaction as the preceding step **c.** is different from the preceding reaction **d.** is unpredictable.
8. Fusion reactions only happen in special places because **a.** very high temperatures are needed **b.** very low temperatures are needed **c.** only a few atoms are available **d.** only heavy atoms will work in this reaction.
9. Which of the following are problems arising from the use of nuclear fission to produce energy? **a.** The energy it is possible to produce is not great. **b.** The radioactive materials must be kept out of the environment. **c.** Radioactive wastes must be disposed of. **d.** The supply of uranium is limited.
10. In addition to fission reactors, another type of nuclear reactor that may someday be used to produce energy is called a __________.

REVIEW EXERCISES

Give complete but brief answers to each of the following. Use complete sentences to write your answers.

1. Explain why some atoms are radioactive.
2. Give the name and mass of two radioactive atoms.
3. If the half-life of a radioactive element is 2,000 years, how long will it be before three fourths of a given sample of its atoms decay?
4. Describe two ways in which radioactivity can be harmful to living things.
5. What are radioactive tracers and how are they used?
6. Give an example to show how energy can be released by fission of a large atomic nucleus.
7. Explain what is meant by a chain reaction.
8. How does fission differ from fusion?

9. Describe how a nuclear reactor can be used to control a fission reaction.
10. Explain some problems arising from the use of nuclear fission to produce energy.

EXTENSIONS

1. Write a report on some of the particles currently believed to exist in the nuclei of atoms. Describe the experiments that led to the discovery of these particles.
2. Write a paper discussing the "pros" and "cons" of nuclear energy. Try not to be biased in your report.
3. Research the current status of fusion research. Are the arguments for and against the use of fusion power the same as those for nuclear fission power? Describe some of the experimental fusion reactors being tested.

Throughout this book, you have been reading and learning about the discoveries and processes of science. In addition, you have seen how science has affected your daily life. As we approach the end of the twentieth century and move into the twenty-first, science and its applications will continue to play an ever-increasing and important role in our lives. Some examples of areas in which science and technology may change our way of living in the future are shown on these pages.

1. Is this the shape of the future? This piston-driven train was built using modern, lightweight materials and a futuristic, aerodynamic design.
2. Travel in the twenty-first century may be by means of trains similar to this one. Trains like the two-story electromagnetic "White Lady" could reach speeds of 500–700 km/hr.
3. The airplanes of the future may copy the hydrodynamic shapes of sea creatures. This 1,000 passenger plane was designed on the shark model.
4. Machines that can feel, hear, and talk? The key to the development of sophisticated machines such as these may be these tiny computer chips. Each chip can perform all of the functions of the room-

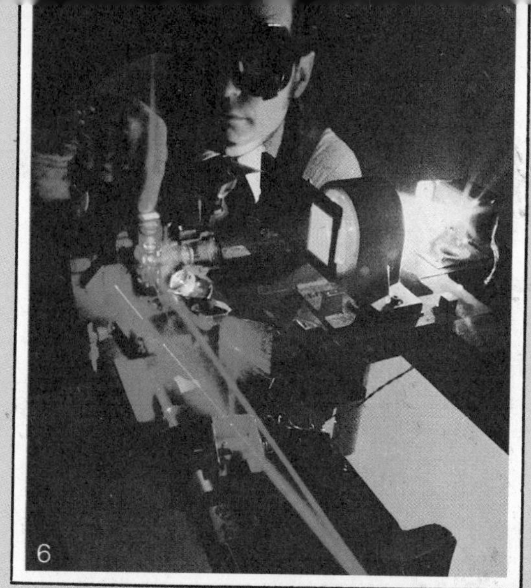

sized computers of 25 years ago.

5. The city of the future may resemble this model of the Marin Solar Village. This community was designed to use renewable energy sources, to preserve the environmental balance, and to provide a stable, low-pollution life style for its inhabitants.
6. The principle of the laser was discovered in 1960. Lasers are used to transmit radio and telephone messages, to perform delicate surgery, and to create three-dimensional works of art called holograms. In the twenty-first century, lasers may be used to help predict earthquakes, to produce controlled fusion reactions, and to propel powerful rockets.
7. Solar power satellites located 36,290 km above the equator could generate electricity from sunlight, convert this electricity into microwaves, and then beam this energy to an antenna on the earth where it could be converted back into electricity.

As you read this book, spaceships from the earth, guided by the laws of physics, are preparing to leave the solar system and travel to the stars. Who knows what wonders the scientific discoveries of the past may lead to in the future?

GLOSSARY

A

absolute zero. The temperature (−273°C) at which particles of matter stop moving.

acceleration (ik-sel-uh-**ray**-shun). The change in velocity during a given time interval; either speed or direction (or both) may change.

accelerator. A machine used to speed up atomic particles for use in research.

acidic solution. A solution containing more H^+ ions than pure water.

alcohol. A kind of organic molecule having at least one OH group.

alkali (**al**-kuh-lie) **metals.** A group of elements whose atoms all have one electron more than the stable number.

alloy. A substance that is made up of two or more metals (for example, bronze is an alloy of copper and tin).

alpha decay. A radioactive change that takes place when a nucleus shoots out an alpha particle made up of two protons and two neutrons.

alternating (**awl**-tur-nay-teeng) **current.** An electric current that changes direction in an electric circuit.

amino acid. A kind of organic molecule having both NH_2 and COOH groups.

ampere (**am**-pir). A measure of the amount of current moving past a point in an electric circuit in 1 sec.

area. The amount of surface within a given set of lines.

atom. The smallest particle of an element.

atomic mass unit. A unit used to express the masses of atomic particles and atoms; $1 \text{ amu} = 1.657 \times 10^{-24}$ g.

atomic number. The number of protons in the nucleus of an atom.

atomic particles. The basic building blocks of atoms.

aurora (uh-**rore**-uh). The northern or southern lights.

B

base. A compound that forms a solution with more OH^- ions than pure water.

beta decay. A radioactive change that causes an electron (beta particle) to shoot out of a nucleus.

binding energy. The energy that holds the nucleus of an atom together. It is also the energy required to break the nucleus apart.

boiling point. The temperature (at ordinary air pressure) at which the particles of a liquid have enough energy to become a gas.

C

calorie (**cal**-uh-ree). An amount of heat equal to that needed to raise the temperature of 1 g of water 1°C.

calorimeter. A device used to measure heat energy.

carbohydrates (kahr-boe-**hie**-drates). Compounds of carbon, hydrogen, and oxygen whose molecules contain two atoms of hydrogen for every one atom of oxygen.

catalyst (**kat**-'l-ust). A substance that changes the speed of a chemical reaction but remains the same after the reaction.

Celsius (**sel**-see-us). The name of a commonly used temperature scale. The Celsius (C) scale is generally used in science.

centimeter (**sent**-uh-meet-ur). One one-hundredth (0.01) of a meter.

chain reaction. A reaction in which each step involves the same reaction as in the previous step.

chemical activity. Describes the way in which an atom reacts with other kinds of atoms.

chemical bond. A force that joins atoms together.

chemical equation (ih-**kway**-zhun). A description of a chemical reaction using chemical formulas for the substances used and produced.

chemical family. A group of elements that are alike in their chemical behavior.

chemical reaction. A reaction in which a chemical change takes place.

compound. A substance, such as water, that cannot be broken down into simpler parts by a physical change. A compound contains only one kind of molecule.

concave. A lens shape in which the edges are thicker than the center. (The center is *caved* in.)

concentration (kon-sun-**tray**-shun). A description of the number of molecules found in a given space.

conduction (kun-**duk**-shun). Transfer of heat by direct contact.

conductor (kun-**duk**-tur). A material that can carry an electric current because electrons can move through it easily.

convection (kun-**vek**-shun). Transfer of heat by movement of a heated gas or liquid.

convex. A lens shape in which the edges are thinner than the center.

corrosion (kuh-**roe**-zhun). The eating away of the surface of a metal by chemical action.

covalent (koe-**vay**-lunt) **bond.** A chemical bond formed when atoms share two or more electrons.

crystal. A solid whose orderly arrangement of particles gives it a regular shape.

D

decibel (dB). The unit used to measure the loudness of sound.

diatomic (die-uh-**tom**-ik) **molecule.** A molecule consisting of only two like atoms, for example, H_2.

diffraction (dif-**rak**-shun). The ability of waves to bend around an obstacle in their path.

direct current. An electric current that flows in one direction in an electric circuit.

Doppler effect. An apparent change in the frequency of waves caused by the fact that the observer or the source of the waves is moving.

double bond. The bond formed when atoms are joined by sharing two pairs of electrons.

E

echo. Sound waves that are reflected by a barrier.

electric circuit. A complete path that allows electrons to move from a place rich in electrons to a place poor in electrons.

electric current. The result of electrons moving from one place to another.

electric field. A region of space around an electrically charged object in which electric forces on other charged objects are noticeable.

electric force. The force that causes two like-charged objects to repel each other or two unlike-charged objects to attract each other.

electrolysis. The process by which electric energy is used to break down water into its parts.

electrolyte (ih-**lek**-truh-lite). A substance that forms a conducting solution when dissolved in water.

electromagnet. A temporary magnet made when an electric current flows through a coil of wire wrapped around a piece of iron.

electromagnetic (ih-**lek**-troe-mag-**net**-ik) **induction** (in-**duk**-shun). Production of an electric current by motion in a magnetic field.

electromagnetic spectrum. The series of waves with many properties similar to those of light.

electromagnetic waves. A form of energy capable of moving through empty space at 3×10^8 m/sec.

electron (ih-**lek**-tron). A negatively charged particle of matter so small as to be invisible.

electron dot model. A way of picturing the outer electrons of an atom in which dots representing the outer electrons are placed around the atomic symbol.

electron shell. A region around an atomic nucleus in which electrons move.

electroscope. A device used to observe the presence of static electric charges.

element. The simplest form of matter.

endothermic (en-duh-**thur**-mik) **reaction.** A chemical reaction that absorbs energy.

end point. The point at which an acid-base indicator shows the first sign of change in a neutralization reaction.

energy. The property of matter that enables it to do work.

enzymes. Catalysts that control the rate of chemical reactions in living things.

ester. A kind of molecule formed by combining an organic acid and an alcohol.

exothermic (ek-soe-**thur**-mik) **reaction.** A chemical reaction that gives off energy in the form of heat.

experiment (ik-**sper**-uh-ment). The testing of a hypothesis.

F

fats. Esters formed by combining a particular kind of organic acid and alcohol.

fission. A nuclear reaction in which a large unstable nucleus splits into smaller nuclei.

focal length. The distance from the center of a lens to the focal point.

focal point. The point at which parallel light rays meet after being refracted.

force. Any push or pull that causes an object to move or to change its speed or direction of motion.

fossil fuels. Fuels that were formed from the remains of dead plants and animals.

fractional distillation. The process by which crude oil is separated into its parts on the basis of their different boiling points.

frequency (**free**-quen-see). The number of complete waves that pass by a point each second.

friction. The force that slows the motion between two objects in contact with each other.

fuel. A substance used as a source of chemical energy.

fuse. A part of an electric circuit that prevents too much current from flowing in the circuit.

fusion. A nuclear reaction in which two small nuclei join to form a larger nucleus.

G

gamma rays. Rays similar to X rays that cause a particular type of radioactive decay.

Geiger counter. An instrument used to detect radioactivity.

gram. A small unit of weight in the metric system. One pound contains 454 grams (g).

gravity (**grav**-it-tee). The force that pulls an object toward the center of the earth.

greenhouse effect. An increase in temperature caused by excess carbon dioxide in the atmosphere.

H

half-life. The length of time necessary for one half of a given number of atoms to change.

halogens (**hal**-uh-junz). A group of elements whose atoms all have one electron less than the stable number.

heat engine. A machine that changes heat energy into mechanical energy.

heat of fusion. The amount of heat required to change 1 g of a solid to a liquid at the same temperature.

heat of vaporization. The amount of heat required to change 1 g of a liquid to a gas at the same temperature.

heliograph. A device used to send messages by means of the sun's rays reflected from mirrors.

hertz. A unit used to measure the frequency of a wave. A frequency of 1 Hz means that one complete wave passes a point each second.

horsepower. The unit of power used in the customary (English) system of measurement.

hydrocarbon. A compound containing only carbon and hydrogen.

hypothesis (hie-**poth**-ih-sis). A prediction or intelligent guess based on patterns in observations.

I

image. The picture formed by a lens.

index of refraction. The ratio of the speed of light in air to the speed of light in another material.

indicator. A substance that changes color when put into an acid solution.

inert. A description of an atom that does not react with other atoms.

inertia. The resistance of objects to any change in motion.

insulator (**in**-suh-late-ur). A material that does not allow electrons to flow through it easily.

interference (int-er-**fir**-unts). The effect two or more waves have on each other if they overlap.

ion. An atom or molecule with an electric charge.

ionic bond. A chemical bond formed when atoms transfer electrons from one to another.

isomers (**ie**-suh-merz). Two or more molecules with the same formula but different arrangements of their atoms.

isotopes. Atoms whose nuclei contain the same number of protons but a different number of neutrons.

K

Kelvin temperature scale. A scale of temperature on which zero degrees is equal to absolute zero.

kilowatt-hour. The amount of energy supplied in one hour by one kilowatt of power. It is used to measure how much electric energy is consumed.

kinetic energy. Energy that moving objects have as a result of their motion.

kinetic theory of matter. The scientific principle that says that all matter is made of particles whose motion determines whether the matter is solid, liquid, or gas.

L

law of conservation (kon-ser-**vay**-shun) **of energy.** A natural law that says that energy cannot be created or destroyed but may be changed from one form to another.

law of conservation of matter. The law that says the same number of atoms exist after a chemical reaction as before the reaction.

lens. A piece of transparent material with curved surfaces that refract light passing through it.

lever. A rigid bar that turns on a fixed point and changes the direction and size of a force applied to it.

liter. A commonly used unit of volume in the metric system; a little less than one quart.

loudness. The effect that the energy of sound waves has on the ear.

M

magnetic field. A region of space around a magnet in which magnetic forces are noticeable.

magnetic pole. The part of a magnet where the magnetic forces are strongest.

magnetic variation. The error in a compass caused by the difference in location of the earth's magnetic and geographic poles.

mass. A measure of the amount of matter contained in an object.

mass number. The sum of the protons and neutrons in the nucleus of a particular kind of atom.

measurement. An observation that is done by counting something. Often an instrument, such as a ruler, is used in measuring.

melting point. The temperature at which a solid becomes a liquid.

metallic bond. A chemical bond formed by electrons that are not tightly held by any particular atom.

metalloid (**met**-'l-oid). An element with some properties of both metals and nonmetals.

meter. The basic unit of length in the metric system; a little more than one yard in length.

millimeter (**mil**-uh-meet-ur). One one-thousandth (0.001) of a meter or one-tenth (0.1) of a centimeter.

mirage (muh-**rah**-zh). An illusion caused by the refraction of light in which distant objects are seen upside down or floating in the air.

mixture. Any matter that contains more than one kind of molecule.

molecule (**mol**-ih-kyool). The smallest particle of a substance, such as water, that can be identified as that substance.

motion. A change in position of an object when compared to a reference point.

mutations. Changes in the cells of living things caused by excess radioactivity.

N

negative charge. The electric charge given to a hard rubber rod when rubbed with fur.

neutral. The term describing an object that has neither a positive nor a negative charge. A solution or substance that is neither an acid nor a base.

neutralization (noo-truh-luh-**zay**-shun). A chemical reaction that occurs when an acid and a base are mixed.

neutron (**noo**-tron). A particle in the atom that is electrically neutral.

noble gases. The six elements whose atoms have completely filled electron shells.

nonelectrolyte. A substance that forms a nonconducting solution when dissolved in water.

nuclear reactor. A machine used to carry on a controlled nuclear chain reaction.

nucleus. The small central core of an atom where most of the mass of the atom is located. This core is made up of protons and neutrons.

O

observation (ob-ser-**vay**-shun). Anything that we can learn by using our senses, such as sight or hearing.

ohm. A measure of the amount of resistance in an electric circuit.

organic acid. A kind of organic molecule having at least one COOH group.

organic chemistry. The chemistry of carbon compounds.

overtone. A tone of higher pitch made by the short segments of a vibrating object.

oxidation. The reaction by which an element combines with oxygen.

P

parallel circuit. An electric circuit with the various parts in separate branches.

periodic chart. An arrangement of all the elements, showing the chemical families.

permanent magnet. A magnet made of a material that tends to keep its magnetism.

petroleum. A natural mixture of compounds of carbon and hydrogen.

photon. A particle of light energy.

pitch. The property that describes the highness or lowness of a sound and is determined by the frequency of the sound wave.

polar molecule. A molecule that carries small electric charges on opposite ends.

polluted. A description of water that contains harmful substances.

positive charge. The electric charge given to a glass rod when rubbed with a silk cloth.

potential energy. Energy stored in an object as a result of a change in its position.

power. The rate at which work is done; power = work/time.

precipitate (prih-**sip**-uh-tate). A solid substance that separates from a solution.

prism. A specially shaped piece of glass. A prism divides white light into its separate colors.

proteins (**proe**-teenz). Very large molecules formed when many amino acids join together.

proton (**proe**-ton). A very small particle of matter with a positive electric charge.

R

radiation (rade-ee-**ay**-shun). Transfer of heat through space by infrared rays.

radical. A group of atoms that remain together in a chemical reaction just as if they were a single atom with a single valence.

radioactive (rade-ee-oe-**ak**-tiv). An atom is radioactive when its nucleus changes by means of decay in order to become more stable.

rays. Straight lines showing the path followed by light.

reference point. A stationary object used in the observation of motion.

reflection (rih-**fleck**-shun). The process in which a wave is thrown back after striking a barrier that does not absorb the energy of the wave.

refraction (rih-**frak**-shun). The process in which a wave changes direction because its speed changes.

resistance (rih-**zis**-tunts). Any condition that limits the flow of electrons in an electric circuit; for example, a light bulb in a circuit.

S

salt. A compound that is made up of the metallic ion from a base and the nonmetallic ion from an acid.

scientific law. A theory that has been tested many times and always found to be true.

scientific model. A kind of mental picture used by scientists to describe something that cannot be seen directly.

semiconductor. A substance that will conduct an electric current only under certain conditions.

series circuit. A circuit in which all the parts are connected one after the other.

simple machine. A device that can be used to change the direction and size of forces.

solution (suh-**loo**-shun). A mixture formed when one kind of molecule, like sugar, fills the spaces between another kind of molecule, like water.

soluble. A description of a substance that can be dissolved.

solute. The part of a solution that is dissolved.

solvent. The part of a solution that does the dissolving.

sonic boom. Sound waves caused by a jet plane as it passes the speed of sound.

sound. A form of energy caused by vibrations of the particles of matter through which the sound wave passes.

speed. Distance divided by the time needed to go that distance.

stable electron arrangement. An arrangement in which all of the electron shells are filled.

static electricity. Electricity that is caused by electric charges that are stationary, or fixed in one place.

structural formula. A diagram of a molecule.

sublimation. The process by which a solid becomes a gas without changing into a liquid.

symbol. One or two letters used to represent an atom of a particular element.

T

temperature. A measurement of the movement of particles in matter.

temporary magnet. A magnet that easily loses its magnetism.

terminal speed. The greatest speed reached by an object falling through the air.

theory. An explanation of what we know about some part of nature.

thermogram. An image made by recording infrared radiation.

thermometer. An instrument used for measuring temperature.

threshold energy. The amount of energy needed to start a chemical reaction.

tracer. Radioactive substances that are used to follow atoms through chemical reactions.

transformer. A part of an electric circuit that changes the voltage of an alternating current.

U

unstable electron. An electron that has absorbed energy and moved farther away from the atomic nucleus.

V

valence (val-luns). The number of electrons gained, lost, or shared by an atom when it forms chemical bonds.

valence electrons. Electrons in the outer shell of an atom that take part in a chemical bond.

velocity. The speed and direction of a moving object.

visible spectrum. The band of colors produced when white light is divided into its separate colors.

volt. A measure of the amount of work done in moving electrons between two points in an electric circuit.

voltmeter. An instrument used to measure voltage in an electric circuit.

volume. The amount of space occupied by an object.

W

watt. A unit used to measure the rate with which electric energy is changed into other forms of energy.

wave. A disturbance caused by the movement of energy from one place to another.

wavelength. The distance between two neighboring crests or troughs of a wave.

weight. The measurement of the force of gravity. In the metric system, gravity force is measured in newtons (N). The customary unit for measuring the gravity force is the pound (lb).

work. The work done by a force is equal to the size of the force multiplied by the distance through which the force acts ($W = F \times d$).

X

xerography. A method of producing copies by means of static electricity.

APPENDIX A: Metric Units (International System of Measurements)

METRIC-ENGLISH EQUIVALENTS

1 m = 39.37 inches
30.48 cm = 1 foot
2.54 cm = 1 inch
1 cm = 0.3937 inch
1 km = 0.621 mile
1 km = 3,280 feet
1.61 km = 1 mile
1 kg = 2.2046 pounds
0.4536 kg = 1 pound
453.6 g = 1 pound
28.35 g = 1 ounce
1 L = 1.06 quarts
943 mL = 1 quart
1 mL = 0.00106 quart

PREFIXES OF METRIC UNITS

Greater than 1:

tera (T) = 1,000,000,000,000
giga (G) = 1,000,000,000
mega (M) = 1,000,000
kilo (k) = 1,000
hecto (h) = 100
deka (da) = 10

Less than 1:

deci (d) = 0.1
centi (c) = 0.01
milli (m) = 0.001
micro (μ) = 0.000 001
nano (n) = 0.000 000 001
pico (p) = 0.000 000 000 001

COMMONLY USED METRIC UNITS

Length: The basic unit of length in the metric system is the meter (m), which is slightly more than the height of a kitchen counter top or of a doorknob.

Examples: 1 kilometer (km) = 1,000 m
1 centimeter (cm) = 0.01 m
1 meter (m) = 100 cm
1 millimeter (mm) = 0.001 m
1 meter (m) = 1,000 mm

Mass: The basic unit of mass in the metric system is the gram (g), which is equal to the amount needed to balance a standard paper clip.

Examples: 1 kilogram (kg) = 1,000 g
1 milligram (mg) = 0.001 g
1 gram (g) = 1,000 mg

Volume: The basic unit of volume in the metric system is the liter (L), which is slightly more than four times the volume of one cup.

Examples: 1 milliliter (mL) = 0.001 L
1 liter (L) = 1,000 mL

APPENDIX B: Using Decimal Numbers

1. Addition

When adding numbers, always line up the decimal points correctly.
Examples:

a. $153.6 + 6.2 + 12.4$

$$\begin{array}{r} 153.6 \\ 6.2 \\ \underline{12.4} \\ 172.2 \end{array}$$

b. $31.24 + 5.9 + 6.154$

$$\begin{array}{r} 31.24 \\ 5.9 \\ \underline{6.154} \\ 43.294 \end{array}$$

2. Subtraction

When subtracting numbers, always line up the decimal points correctly.
Examples:

a. $346.15 - 26.21$

$$\begin{array}{r} 346.15 \\ \underline{-\ \ 26.21} \\ 319.94 \end{array}$$

b. $49.20 - 15.43$

$$\begin{array}{r} 49.20 \\ \underline{-15.53} \\ 33.77 \end{array}$$

3. Multiplication

When multiplying numbers, it is not necessary to line up decimal points. The number of decimal positions (numbers to the right of the decimal point) in the product is the sum of the number of decimal positions of the numbers being multiplied (factors).
Examples:

a. 31.46×2.3

$$\begin{array}{rl} 31.46 & \text{(2 decimals)} \\ \underline{\times\ 2.3} & \text{(1 decimal)} \\ 9438 & \\ 6292 & \\ \hline 72.358 & \text{(3 decimals)} \end{array}$$

b. 415.10×0.156

$$\begin{array}{rl} 415.10 & \text{(2 decimals)} \\ \underline{\times\ 0.156} & \text{(3 decimals)} \\ 249060 & \\ 207550 & \\ 41510 & \\ \hline 64.75560 & \text{(5 decimals)} \end{array}$$

4. Division

Division expresses how many times one number goes into another number. Division may be shown as

$4.05 \div 0.5$, as $\frac{4.05}{0.5}$, or as $0.5\overline{)4.05}$

When dividing numbers, only whole numbers can be used as divisors. This means that 0.5 must be multiplied by 10 to make it a whole number. In order not to change the value, 4.05 must also be multiplied by 10. Thus the division is

$$\begin{array}{rl} & 8.1 \text{ (quotient)} \\ \text{(divisor) } 5 & \overline{)40.5} \text{ (dividend)} \\ & \underline{40.0} \\ & 5 \\ & \underline{5} \end{array}$$

Notice that the decimal in the quotient is located directly over the decimal in the dividend.
Examples:

a. $32.4 \div 0.40$

$$\begin{array}{rl} & 81 \\ 0.40. & \overline{)32.40.} \\ & \underline{32\,0} \\ & 40 \\ & \underline{40} \end{array}$$

b. $\frac{4.4748}{3.6}$

$$\begin{array}{rl} & 1.243 \\ 3.6. & \overline{)4.4.748} \\ & \underline{3\,6} \\ & 8\,7 \\ & \underline{7\,2} \\ & 1\,54 \\ & \underline{1\,44} \\ & 108 \\ & \underline{108} \end{array}$$

5. Finding An Average

To average several numbers, find their sum and then divide by the number of values that were added to give the sum. For example, find the average of the four measurements 2.3 cm, 1.8 cm, 2.1 cm, and 2.2 cm. First add:

2.3 cm
1.8 cm
2.1 cm
2.2 cm
———
8.4 cm

Then divide by 4:

$$4\overline{)8.4\text{ cm}} = 2.1\text{ cm}$$

```
    2.1 cm
 4)8.4 cm
   8.0
   ----
     4
     4
    ---
```

6. Metric Conversions

One of the advantages of using metric units is the ease of converting one unit of measurement to another. For example, a measurement is 2.5 cm. Convert this measurement to meters. Since 1 cm = 0.01 m, 2.5 cm = 0.025 m. (See Appendix A for metric prefixes.) Calculation:

$$2.5\ \cancel{\text{cm}} \times \frac{0.01\text{ m}}{1\ \cancel{\text{cm}}} + 0.025\text{ m}$$

or

$$2.5\ \cancel{\text{cm}} \times 0.01\text{ m} \div 1\ \cancel{\text{cm}} = 0.025\text{ m}$$

or

```
  2.5 cm
× 0.01 m/cm
-----------
 0.025 m
```

All conversions in the metric system result in multiplying by a decimal fraction or a multiple of 10. Remember, every measurement must consist of a number and a metric unit. The original unit disappears, leaving the unit to which you are converting.

APPENDIX C: Volume

Solids

To find the volume of a regular solid, first measure the two dimensions to find the area: length and width. Then measure the height of the solid. For example, suppose you want to know the volume of a pack of index cards. Follow these rules:

1. Measure the area of a card: the length is 12.9 cm and the width is 7.1 cm. Therefore, the area is $12.9 \text{ cm} \times 7.1 \text{ cm} = 91.59 \text{ cm}^2$.

2. You now know the area of one index card. To find the volume of a pack of index cards, multiply the area by the height of the pack, which is 1.5 cm.

$$\underset{\text{area}}{91.59 \text{ cm}^2} \times \underset{\text{height}}{1.5 \text{ cm}} = \underset{\text{volume}}{137.39 \text{ cm}^3}$$

(Remember: $\text{cm}^{(1)} \times \text{cm}^{(1)} = \text{cm}^2$ and $\text{cm}^2 \times \text{cm}^{(1)} = \text{cm}^3$.) Volume of a solid is always measured in cubed units, in this case, cubic centimeters (cm^3). Thus,

$$\text{volume} = \text{length} \times \text{width} \times \text{height}$$

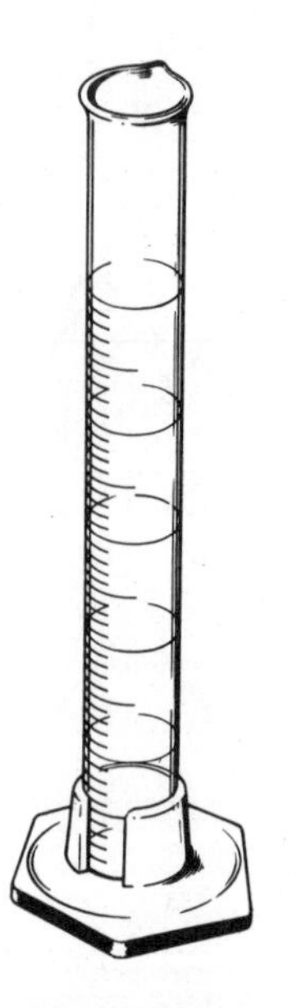

C-1

Liquids

The volume of a liquid is measured with a graduated cylinder. See Fig. C-1. When water and most other liquids are put into a glass graduated cylinder, the surface of the liquid is curved. In making volume measurements, it is important to read the mark closest to the bottom of the curve. See Fig. C-2. Today, some graduated cylinders are made of plastic and do not show this curve. When making a volume measurement, the graduated cylinder should be held so that the liquid surface is level with your eye.

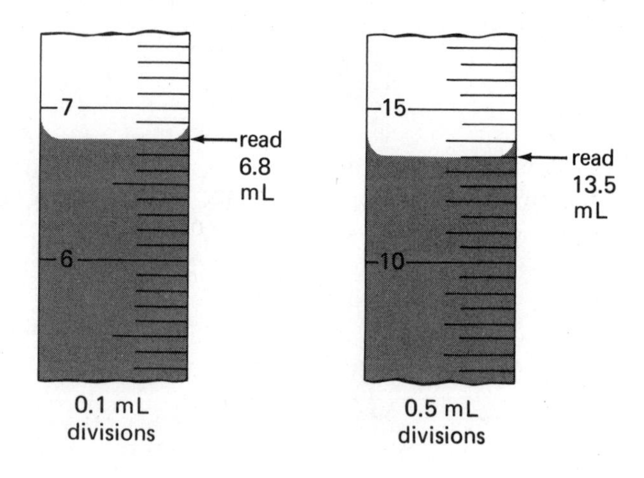

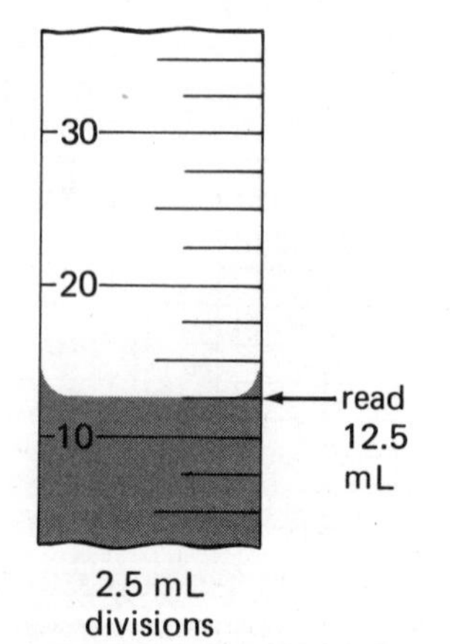

C-2

Gases

The volume of a gas varies with temperature. This variation in volume is described in two laws. The first law was formulated by the English scientist Robert Boyle and the second by the French scientist Jacques Charles.

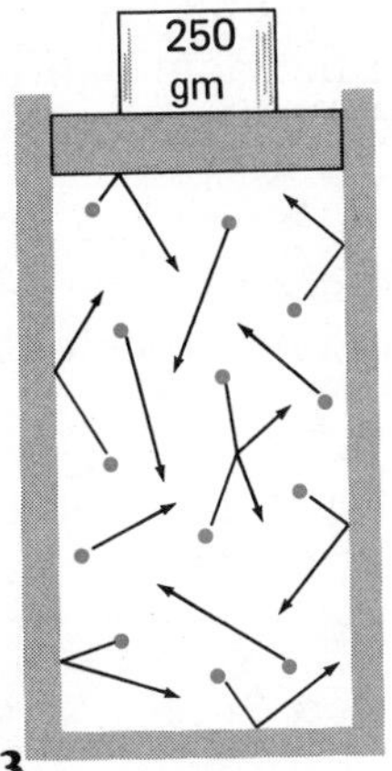

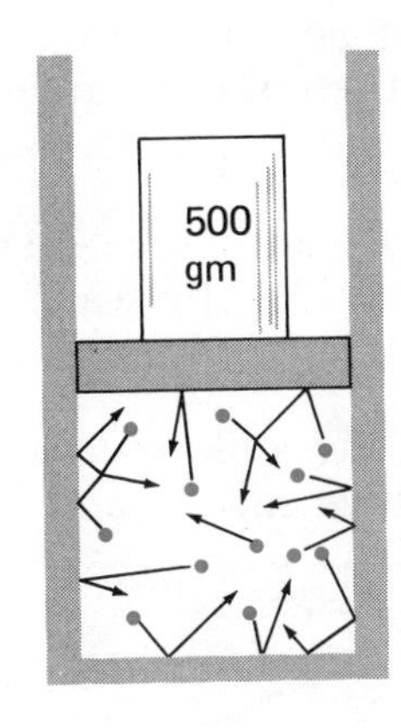

C-3

A. *Boyle's Law*

As the pressure increases, the volume of a gas decreases if the temperature does not change. See Fig. C-3. The equation for this law is

$$PV = P'V'$$

where P is the original pressure, V is the original volume, P' is the new pressure, and V' is the new volume. Boyle's original pressure-volume data are shown in the graph in Fig. C-4. From this graph you can see that the volume is decreasing as the pressure increases.

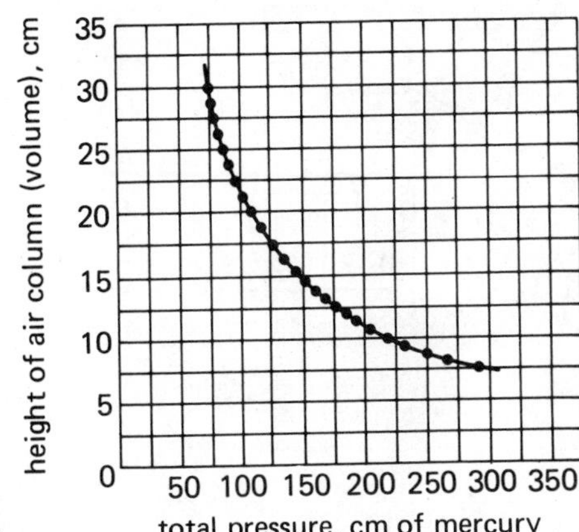

C-4

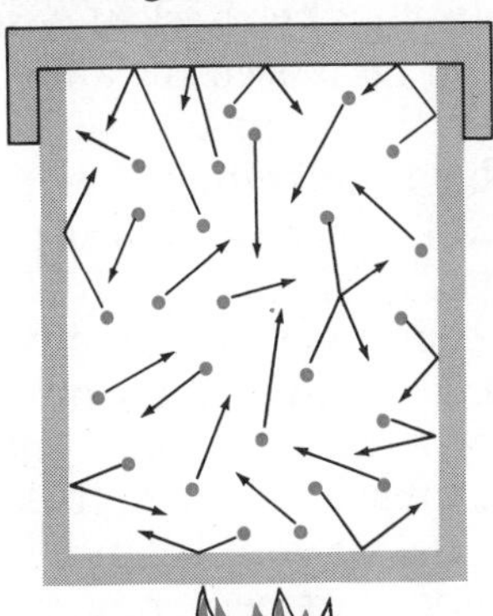

C-5

B. *Charles' Law*

As the temperature increases, the volume of a gas increases if the pressure does not change. See Fig. C-5. (The temperature must be measured in degrees Kelvin.) The equation for this law is

$$\frac{V}{T} = \frac{V'}{T'}$$

where V is the original volume, T is the original temperature, V' is the new volume, and T' is the new temperature. A graph of this relationship is shown in Fig. C-6.

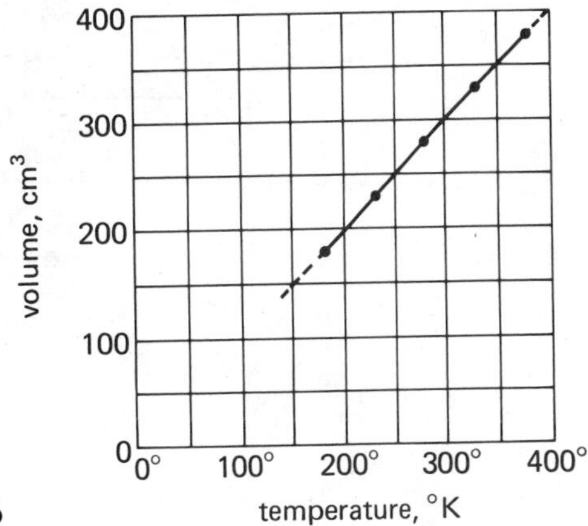

C-6

APPENDIX D: Using a Laboratory Balance

A balance is an instrument used to determine the mass of an object. There are many kinds of balances; only two of these are discussed here.

A balance must first be checked with no mass on it. It should be placed on a level surface. Some balances have an indicator such as a moving bubble to tell when they are level. If there is a level indicator, adjust the balance until the indicator shows that it is level.

The balance should always be moving a little when it is being used. When the pointer moves equally on both sides of the center of the pointer scale, the balance is "in balance." See Figs. D-1 and D-2. If your balance is level and the pointer does not center on the pointer scale, have your teacher adjust it until it does.

An equal arm balance, one kind of which is shown in Fig. D-1, uses the fact that the right and left sides of the instrument are the same size and shape. To use it, follow these steps:

1. Place the object whose mass you wish to find on the left pan of the balance.

2. Place known masses on the right pan until the pointer moves equally on both sides of the pointer scale.

3. If your balance has a numbered scale with a rider on the scale, use the rider to help balance the unknown mass.

4. The unknown mass is the total of the known masses and the rider reading.

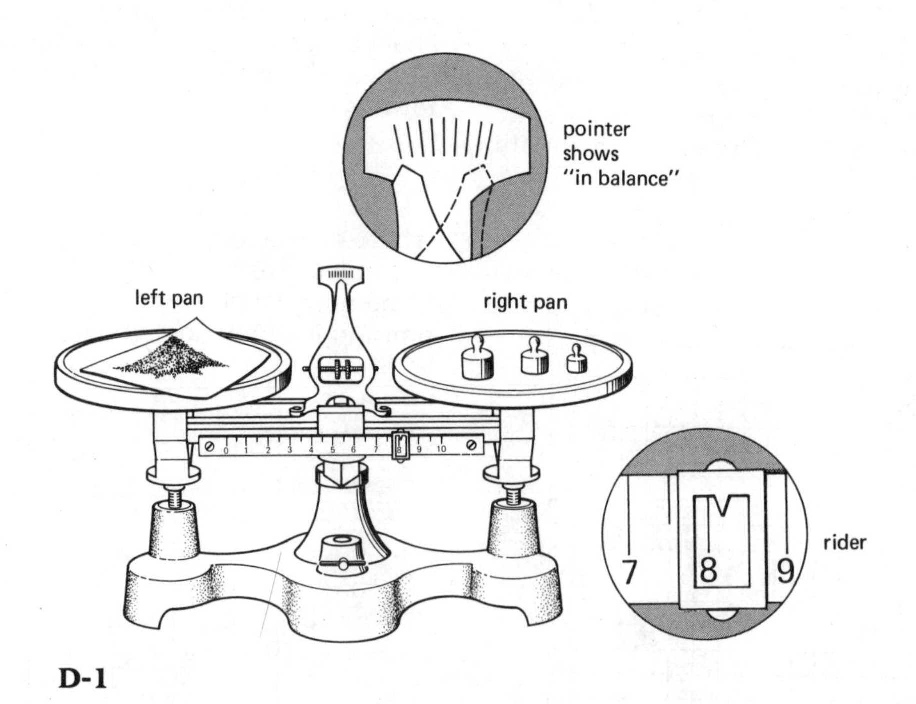

D-1

Another kind of balance is the triple beam balance shown in Fig. D-2. It has only one pan and uses three riders which are moved along the beams to balance an unknown mass placed on the pan. To use a triple beam balance, follow these steps:

1. Place the object whose mass you wish to find on the pan.

2. Move the riders until the pointer moves equally on both sides of the center of the pointer scale.

3. The unknown mass is the total of the readings of the three riders.

If you have trouble with the balance, ask your teacher for help. Do not try to adjust it yourself.

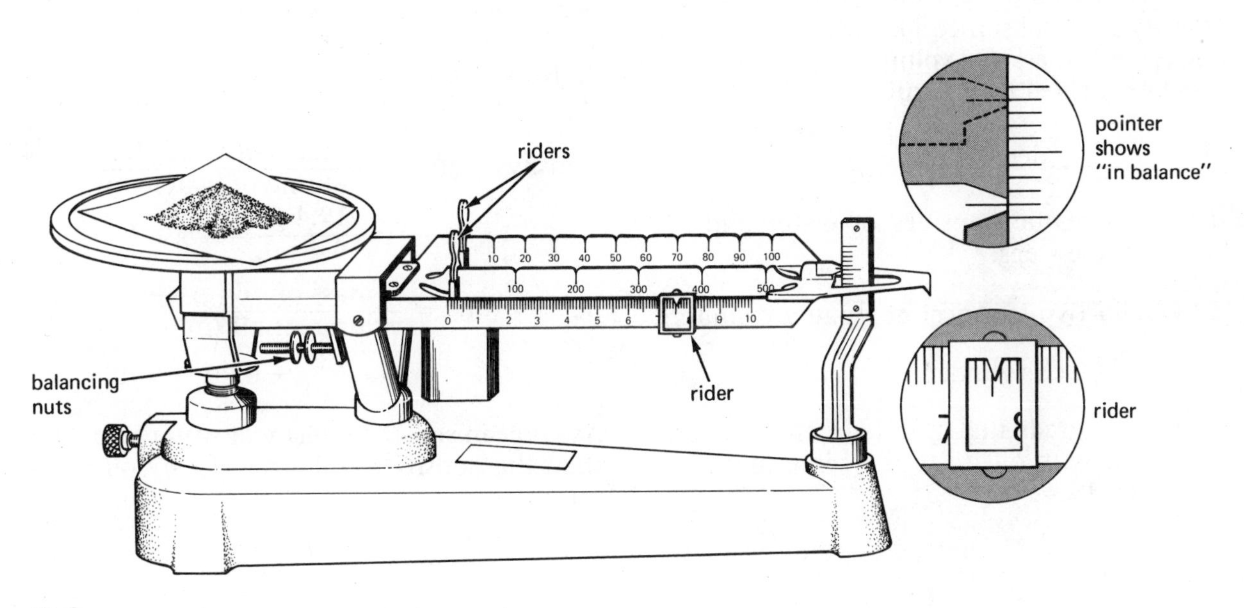

D-2

APPENDIX E: Density

A simple experiment can show what density means. Take two identical graduated cylinders. Fill one with water and the other with oil up to the same mark. Then put the two cylinders on a balance. The side of the balance with the water will swing down. At the same time, the side with the oil goes up. What does this experiment tell you? You have shown that, volume for volume, water has greater mass than oil. A scientist would say that the *density* of water is greater than that of oil. Density expresses how much mass is contained in a given volume of a substance. To find the density of a substance like water, you must measure its mass and volume. Density is equal to mass divided by volume:

$$\text{density} = \frac{\text{mass}}{\text{volume}}$$

To compare the densities of water and oil, follow these steps:

1. Obtain two identical graduated cylinders.

2. Weigh each cylinder.

3. Fill one graduated cylinder with water and the other with oil up to the same mark, for example, 19.4 mL.

4. Weigh the cylinder with the water and the one with the oil separately. Subtract to find the mass of the oil and water:

mass of graduated cylinder + water	−	mass of graduated cylinder	=	mass of water
94.7 g	−	75.3 g	=	19.4 g

mass of graduated cylinder + oil	−	mass of graduated cylinder	=	mass of oil
92.7 g	−	75.3 g	=	17.4 g

5. Now use the equation for density to find the densities of water and oil:

$$\text{density (water)} = \frac{\text{mass of water (g)}}{\text{volume of water (mL)}} = \frac{19.4\ \text{g}}{19.4\ \text{mL}} = 1\ \text{g/mL}$$

$$\text{density (oil)} = \frac{\text{mass of oil (g)}}{\text{volume of oil (mL)}} = \frac{17.4\ \text{g}}{19.4\ \text{mL}} = 0.897\ \text{g/mL}$$

As you can see, the density of water is greater than the density of oil. This is why oil floats on water.

APPENDIX F: Using the Laboratory Burner

A. Structure
There are several types of laboratory gas burners. All are alike in the way they work. The Bunsen burner is the most common kind. See Fig. F-1.

B. How to Use a Bunsen Burner
1. Connect the Bunsen burner to a gas source.

2. To light the burner, bring a lighted match up the side of the barrel. If the lighted match is held directly over the barrel, the escaping gas will often blow out the match.

C. The Flame
Figure F-1 shows what the flame should look like.

1. If the flame is large and bright yellow, there is too much gas. Use the air adjustment to allow more air to enter.

2. A gap between the flame and the top of the burner means too much air is being admitted. Use the air adjustment to decrease the amount of air entering at the base.

3. If the flame seems to be burning inside the barrel, turn off the gas. Decrease the amount of air and light the burner again. CAUTION: The barrel may be hot.

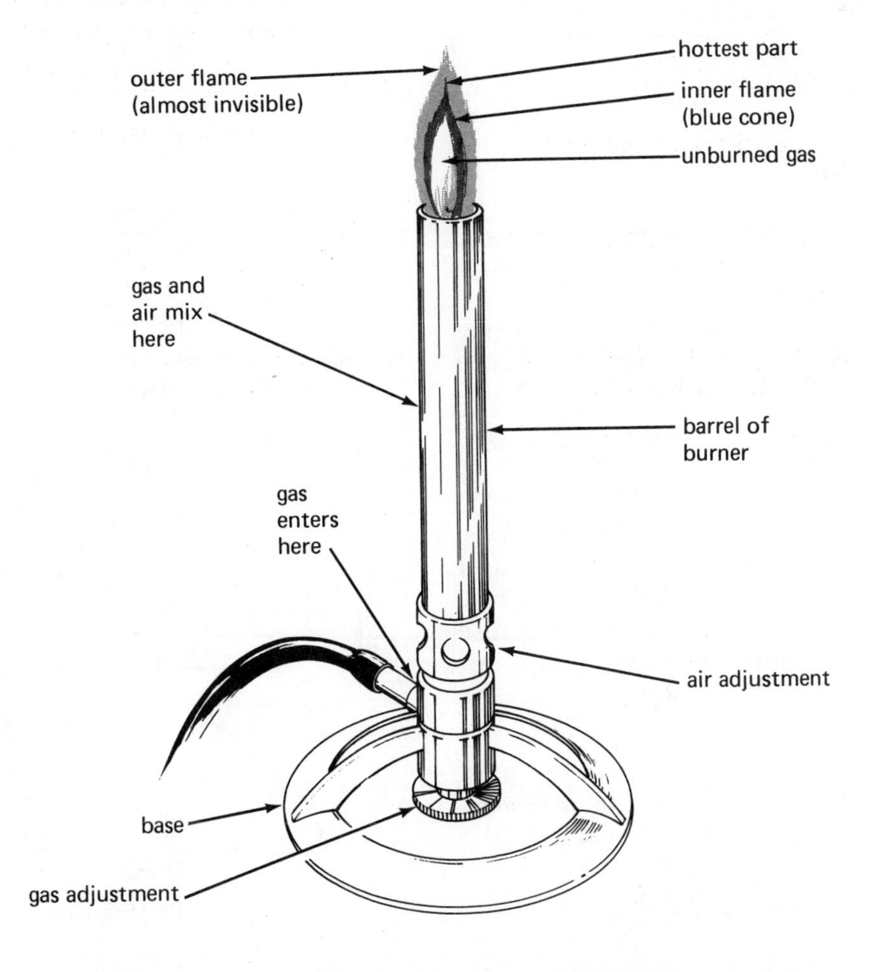

F-1

APPENDIX G: Graphing

A graph is a diagram that shows how two or more things are related. There are several kinds of graphs. These include line graphs, bar graphs, and circle graphs. The most common type of graph is the line graph. Figure G-1 shows the parts of a line graph.

A. Line graphs often represent the relationship between two things that change. For instance, as the boy shown on page 2 stays on the skateboard for a greater number of seconds, he passes a greater number of pickets. Suppose that the following observations were made.

Number of seconds	*Number of pickets*
0	0
1	2
2	4
3	6
4	9
5	12
6	15

From this table, you can see that the boy gained speed. You can now draw a graph of these results by following these steps:

1. Draw a horizontal line and label it *x* axis.

2. Draw a vertical line and label it *y* axis.

3. Mark the number of seconds on the *x* axis. The numbers of seconds are the *independent* variables. They are called variables because they change. They are independent because they are not controlled by anything else. Independent variables are always placed on the *x* axis.

4. Mark the number of pickets on the *y* axis. The numbers of pickets are the *dependent* variables. They are dependent because the number of pickets passed depends on the number of seconds that the boy stays on the skateboard. Dependent variables are always placed on the *y* axis.

5. Select a scale of numbers for each axis. The numbers on each axis should be large enough to include the highest value. For example, the largest numbers of seconds is 6 and the largest number of pickets is 15.

6. Write the numbers along the axis. To give an accurate relationship, the numbers should be spaced equally. The spacing is easier to do if you use graph paper.

7. Locate the points on the graph. To plot a point, keep in mind the relationship of pickets to seconds. The first point is (0,0). Mark a dot on (0,0). Continue to plot the points in the

G-1

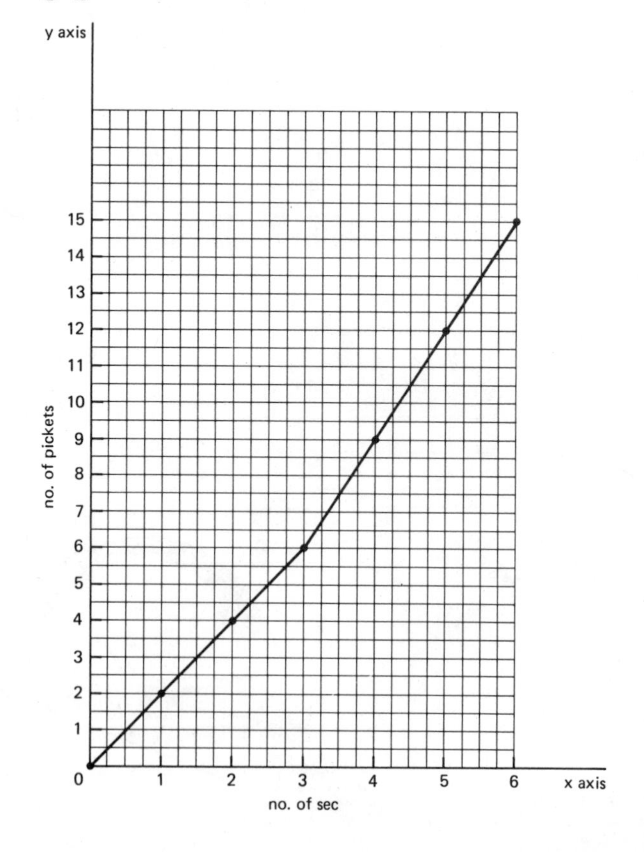

table. First read across on the x axis and then up on the y axis. Mark a dot at the proper position.

8. When you have plotted the last dot, draw a curve to connect the dots.
In the case of the skateboard data, the curve shows an increase in speed.

B. Bar graphs are used to organize data and to compare them. They can be useful in organizing the results of an experiment done by the entire class. For example, if the heights of all the students in a class are measured, the data can be organized in a bar graph. An example of the data collected from a class of 30 is shown in Figure. G-2.

C. Circle graphs are used to compare data that make up a whole or 100%. The data in Table 8-1 list the elements by percent of weight that make up 100% of the earth's crust. The circle graph in Fig. G-3 shows the same data in a different fashion.

G-2

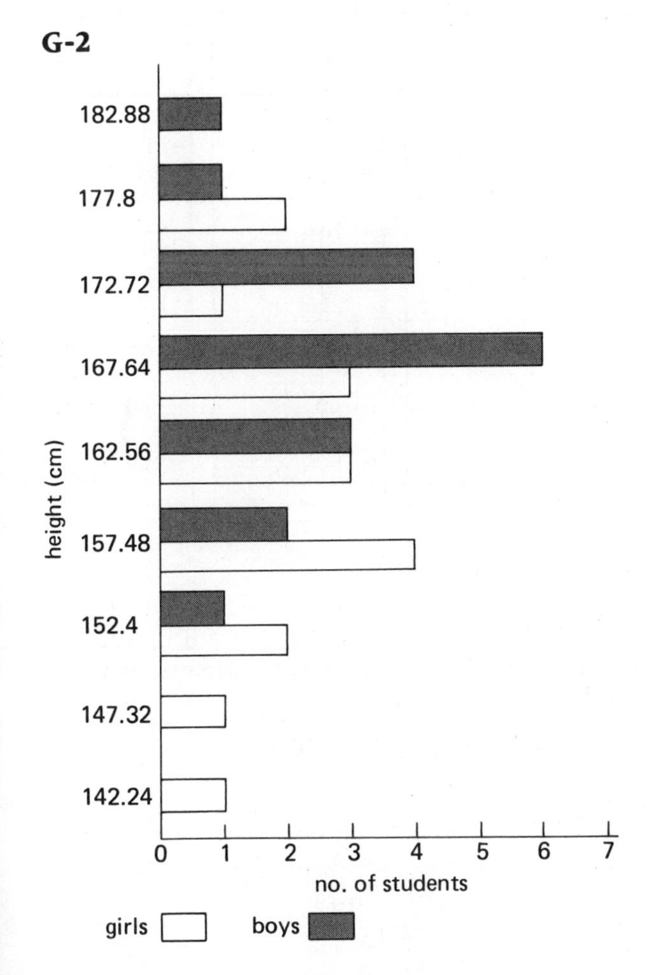

G-3

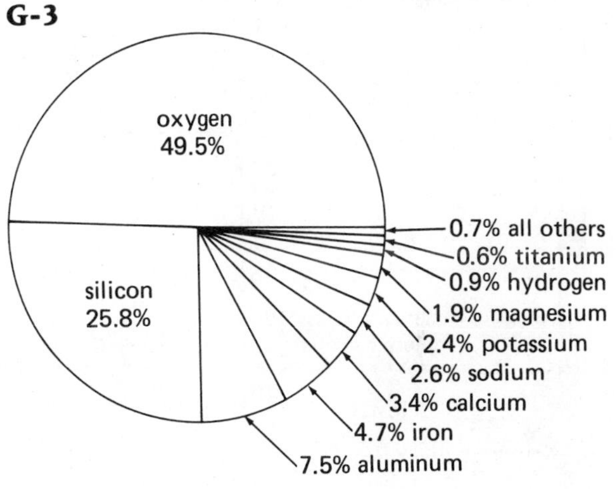

APPENDIX H: Measuring Temperature

Temperature is measured with an instrument called a thermometer. Figure H-1 shows the most common type of thermometer. When the bulb is warmed, either the mercury or alcohol (or some other fluid) in the tube expands. This expansion causes the fluid to rise in the tube. When the bulb is cooled, the fluid contracts, or shrinks, causing the level in the tube to fall. The temperature is read directly on a numbered scale along the glass tube. See Fig. H-2.

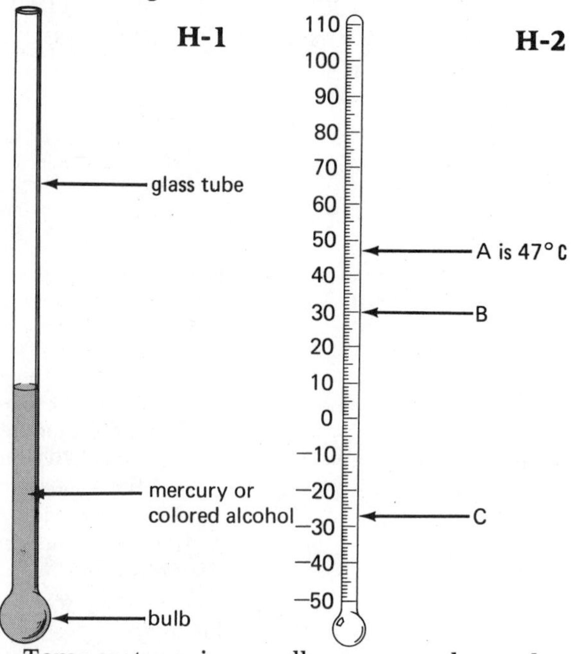

Temperature is usually measured on the Celsius scale. In the United States, the Fahrenheit scale is also used. Figure H-3 shows a comparison between the Fahrenheit and Celsius scales. To convert a temperature reading on the Celsius scale to Fahrenheit, use the following formula:

$$°F = 9/5\ (°C) + 32$$

Example:

Convert 15°C to degrees Fahrenheit.

$$°F = 9/5\ (15°) + 32$$
$$°F = 27 + 32$$
$$°F = 59°$$

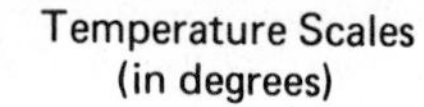

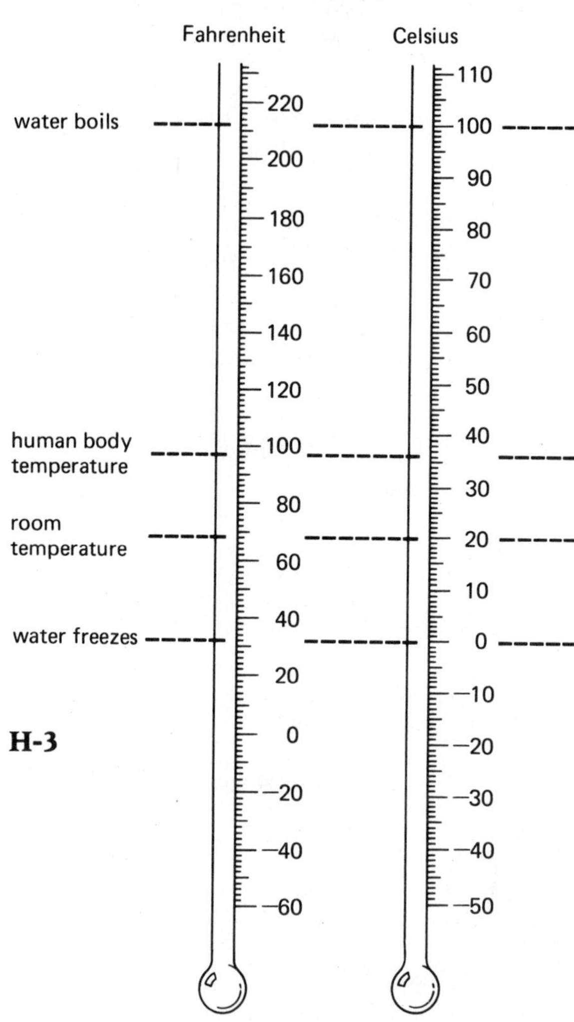

To convert a temperature reading on the Fahrenheit scale to degrees Celsius, use the following formula:

$$°C = 5/9\ (°F - 32)$$

Example:

Convert 59°F to degrees Celsius.

$$°C = 5/9\ (59° - 32)$$
$$°C = 5/9\ (27)$$
$$°C = 15°$$

PICTURE CREDITS

HRW photos by Sally Anderson-Bruce appear on pages: 27, 51, 57, 66, 70, 89, 118, 166, 172, 186, 217, 219, 232, 287, 290, 417, 425, 428, 430

HRW photos by Michael G. Brown appear on pages: 2, 35, 41, 42, 44, 45, 50, 53, 59, 61, 67, 72, 76, 83, 84, 86, 90, 106, 107, 123, 140, 141, 143, 144, 157, 173, 179, 182, 200, 202, 203, 206, 208, 214, 230, 238, 239, 249, 252, 261, 262, 270, 278, 280, 282, 283, 284, 344, 381, 398, 402, 404, 409, 411, 412, 413, 421, 438, 439, 441, 449, 463, 469

HRW photos by Russell Dian appear on pages: 52, 94, 95, 97, 115, 135, 146, 147, 162, 174, 187, 188, 211, 215, 272, 342, 348, 361, 365, 376, 396, 397, 401, 430, 431, 434, 436, 437, 469 (b)

HRW photos by William Hubbell appear on pages: 385, 428

First Steps p. 1 Wally McNamee/Woodfin Camp; p. 7 John Zimmerman/Alpha

Chapter 1 p. 17 John Zimmerman/Alpha; p. 18 (both) NASA; p. 20 Tom McHugh/Photo Researchers; p. 22 Dwight Ellefsen/FPG; p. 25 NASA; p. 26 New York Public Library; pp. 30, 31, 33 NASA; p. 37 (t) Eunice Harris/Photo Researchers; p. 37 (b) German Information Center; p. 38 (t) The Granger Collection

Chapter 2 p. 47 Eberhard Otto/FPG; p. 55 Ronald Rossi/FPG; p. 56 Mathias Oppersdorff; p. 59 (tl) & (tr) Fundamental Photographs; p. 60 Betty Medsgar; p. 63 Paramount Pictures; p. 64 Albert Schoenfield, *Swimming World*; p. 65 (t) Charles Harbutt/Magnum; p. 65 (b) Raimondo Borea/Photo Researchers; p. 69 Gary Braasch; p. 71 Spectra-Physics; p. 73 Fred Ward/Black Star; p. 74 U.S. Department of Energy

Chapter 3 p. 81 Dennis Halliman/Alpha; p. 82 (t) Fundamental Photographs; p. 88 (t) Warren Bols/Sports Illustrated; p. 88 (b) Charles Rotkin/Photography for Industry; pp. 89, 91, 92, 93, Education Development Center, Newton, Mass.; p. 96 NASA; p. 100 M. Feinberg/FPG; p. 102 Joel Gordon

Chapter 4 p. 111 Bell Labs; p. 113 NASA; p. 118 Brown Brothers; p. 119 Craig Aurness/Woodfin Camp; p. 120 Life Science Library, *Light and Vision*, Time-Life, 1966; p. 125 (t) Brian Seed/Black Star; p. 125 (b) Stephen Krasemann/Peter Arnold; p. 126 (t) William Ramsey; p. 126 (l) New York Public Library; p. 134 George Eastman House; p. 135 (m) Townsend Dickinson/Photo Researchers; p. 136 (t) Bausch and Lomb; p. 147 Vivian Fenster

Chapter 5 p. 151 Science Photo Library, London, England; p. 154 The Granger Collection; p. 156 Research Cottrell; p. 159 American Museum of Science and Energy, Oak Ridge, Tennessee; p. 167 (t) General Electric Co.; p. 176 Jack Zehrt/FPG; p. 185 Ted Cowell/Black Star; p. 186 (b) American Telephone and Telegraph; p. 190 Con Edison; p. 191 Warren Weil Communications Counselors

Chapter 6 p. 199 Manfred Kage/Peter Arnold; p. 205 Bibliotheque Nationale, Paris; p. 210 J. H. Pickerell/Black Star; p. 210 (r) David Barnes/Photo Researchers; p. 212 Al Belon/Geophysical Institute, University of Alaska; p. 219 German Information Center

Chapter 7 p. 225 Silver Image/The Image Bank; p. 226 The Granger Collection; p. 233 Arthur Grace/Stock/Boston; p. 234 Mark Godfrey/Magnum; p. 235 Carl Purcell/Photo Researchers; p. 237 (t) Copper Development Association; p. 237 (b) Owens Corning Fiberglas; p. 238 Howard Sochurek/Woodfin Camp; p. 241 Stan Osolinski/Alpha; p. 243 Weksler Instrument Corp.; p. 254 (t) Photri; p. 254 (b) Carl Zeiss Inc.; p. 255 (t) Ward's Natural Science Establishment; p. 255 (b) David Scharf/Peter Arnold; p. 255 (br) Albert Satterwaite/The Image Bank; p. 256 National Park Service

Chapter 8 p. 265 Manfred Kage/Peter Arnold; p. 266 Jerry Irwin/Sports Illustrated; p. 268 (b) Barry O'Rourke/The Image Bank; p. 269 Dennis Stock/Magnum; p. 274 Cornell Capa/Magnum; p. 277 The Granger Collection

Chapter 9 p. 287 (l) John Ross/Photo Researchers; p. 288 The Granger Collection; p. 292 William Rivelli/The Image Bank; p. 297 NASA; p. 298 Brown Brothers; p. 303 John Zimmerman/Alpha

Chapter 10 p. 310 Pete Turner/The Image Bank; p. 312 Sargent Welch Scientific Co.; p. 313 Mount Wilson and Palomar Observatories; p. 314 Sandra Corp., Western Electric; p. 318 Wide World; p. 324 Harvey Stein; p. 325 New York Public Library

Chapter 11 p. 334 Diamond Promotion; p. 353 NASA; p. 355 The Granger Collection; p. 359 Sepp Seitz/Woodfin Camp; p. 360 Erich Hartman/Magnum; p. 362 American Museum of Natural History; p. 363 Mobil Oil Corp.; p. 372 (l) E. I. du Pont de Nemours and Co.; p. 372 (r) Marcia Keegan/Peter Arnold; p. 373 U.S. Department of Energy

Chapter 12 p. 374 Carl Frank/Photo Researchers; p. 375 Joel Gordon; p. 378 (l) & (m) Ward's Scientific Establishment; p. 378 (r) Charles Rotkin/PFI; p. 380 John Blaustein/Woodfin Camp; p. 382 Dan Budnik/Woodfin Camp

Chapter 13 p. 395 Fundamental Photographs; p. 406 Mathias Oppersdorf; p. 408 Loren McIntyre/Woodfin Camp

Chapter 14 p. 417 (l) Tom Nebbia/Woodfin Camp; p. 422 (t) Michael Sullivan/Northern Exposure; p. 422 (l) Panhandle Eastern Pipe Line Co.; p. 422 (r) American Gas Assoc.; p. 425 Charles Rotkin/PFI

Chapter 15 p. 455 U.S. Department of Energy; p. 447 (t) Brown Brothers; p. 451 Science Photo Library, London; p. 454 (t) Brookhaven Labs; p. 454 (b) Dan McCoy/Rainbow; p. 457 U.S. Atomic Energy Commission; p. 458 (b) Stanford Linear Accelerator Center; p. 461 Brookhaven Labs; p. 462 (t) Lawrence Livermore Lab.; p. 462 (b) Union Carbide Corp.; p. 465 Tom McHugh/Photo Researchers; p. 467 (t) Lowell Georgia/Photo Researchers; p. 474-475 Photos 1-3 Malcolm S. Kirk, Photo 4 © 1978 Dan McCoy/Black Star, Photo 5 Van Der Ryn, Calthorpe and Partners, Photos 6-7 NASA.

ART CREDITS

Illustrated by Caliber Design Planning, Eulala Conner, Michael Goodman, Kathie Kelleher, and Vantage Studios.

INDEX

(Note: Page numbers in **boldface** type refer to illustrations and those in *italic* refer to definitions.)

A

absolute zero, *251*
accelerated motion, and speed, 18–22
acceleration, *20*–21; and force (laboratory), 44–45; and gravity, **21, 22;** and mass, **33**–34; *See also* accelerated motion
accelerators, particle, **458**–459
acetic acid, 430, **431**
acetylene molecule, 425
acetylene torch, **425**
acid(s), amino, *431;* chemical reaction with base, 406–407; neutralization of, 403, 407; organic, *431,* 432; properties of, 397–399; as proton donor, 398; test for, 398, 403; weak and strong, 398; *See also* individual acids
acid rain, 399
acidic solution(s), *397,* **398,** 399
Activity: acceleration, 34; acidic reactions, 399; a controlled chain reaction, 469; a mental model, 290; a simple electromagnet, 215; average speed, 23; catalysts in a chemical reaction, 356; chain reactions, 463; classifying acids and bases, 404; classifying elements, 321; diagrams of atoms, 306; electron dot models, 340; exothermic and endothermic reactions, 365; force gauge, 41; forces, 28; forming molecules, 346; generating an electric current, 221; half-life, 455; hard water, 383; interference, 104; infrared waves, 116; lenses, 140; levers, 61; making a compass, 208; making an air thermometer, 252; making an electric circuit, 175; measuring electric forces, 162; mixtures and compounds, 275; models of hydrocarbons, 426; motion of particles, 230; nuclear stability, 449; observation, hypothesis, experiment, 4; observing crystals, 258; observing electric charges, 156; observing spectra, 316; observing the magnetic force, 203; physical and chemical change, 270; potential and kinetic energy, 67; predicting the characteristics of an element, 327; producing oxygen from hydrogen peroxide, 280; properties of carbon compounds, 420; properties of metals and nonmetals, 378; properties of waves, 92; refraction of light, 122; solubility, 388; solubility of organic compounds, 438; substituted hydrocarbons, 432; temperature change in a chemical reaction, 351; testing insulators and conductors, 170; the amount of current in a circuit, 183; the cost of energy use, 192; the difference between temperature and heat, 245; the end point in a neutralization reaction, 409; the structure of an atom, 295; the symbols of the elements, 301; the visible spectrum, 132; three ways to transfer heat, 238; uses of energy, 74; using the metric system, 10; wave motion, 86; words about waves, 98; work and energy, 53
air, as insulator, 169; as solution, 274
air pressure, and boiling point, 257
air thermometer, 250
airships, 318
alcohol(s), *429,* 430; properties of, 268; reaction with organic acid, 432
alcohol molecule, 268
alkali metals (table), *319*
alloys, *375,* (table), 376
alpha decay, *447,* 448
alpha particle, **447**
alternating current, *175;* generation of, 218, **219,** 220
alternators, 220
aluminum atom, 375
aluminum oxide, formula for, 345
amine group, 431
amino acids, 431; in proteins, 438
ammeter, 179
ammonia, as basic solution, 403–404
ampere, *179*
amplitude, wave, *85*
analysis, *282*
angle of incidence, *144*
angle of reflection, *144*
antifreeze, **430**
area, *9*
argon, 320
Aristotle, and falling objects, 32
artificial atoms, **462,** 463
atom(s), *279,* artificial, **462,** 463; chemical behavior of, 335; chemically active, 319; inert, 319; ionization of, 314, **315;** mass of (*See* atomic mass);

nucleus of (*See* atomic nucleus); structure of, 288, **289,** (table), 290, 292, **293, 294;** *See also* atomic particles; elements; individual atoms
atomic bomb, **457,** 459
atomic fission, *459*
atomic fusion, **460,** 461, 468
atomic mass, *304*–305
atomic mass unit, *289*
atomic models, 290; Bohr's, 298
atomic nucleus, *293,* **294;** binding energy of, 446
atomic number, of elements, *294,* (table), 295
atomic particles, 288; acceleration of, 457, **458,** 459; structure of, *288,* **289,** (table), 290
atomic theory, 279
auroras, *212*
automobile engines, 228–229
average speed, *19*–20

B

baking powder, 385
balanced chemical equation, **349, 350**
balanced forces, 37–40
bar magnet, **200,** 201
base(s), *402;* chemical reaction with salt(s), 407; as household products, 402; properties of, 401–403; strong and weak, 402; test for, 403
basic solutions, 403–404
battery, **173**
Becquerel, Henri, and radioactivity, 446–447
bellows, **355**
benzene, 426
Berzelius, and chemical symbols, 299–300
beta decay, *448,* 449
beta particle, 448
bicycle pump, **249**
binding energy, *446*
biodegradable materials, 382
Bohr, Niels, and atomic model, **298**
boiling point, *257*
bond(s), see chemical bonds
bright-line spectrum, 313
butane, 364
butane molecule, 423–424

C

calcium carbonate, eggshell as, **397**
calcium chloride, formula for, 345
calorie, *244,* 361
Calorie, *245,* 361
calorimeter, **245, 361**
camera lens, **135,** 139
capacitor, 178
carbohydrates, *435*–436
carbon, properties of, **418, 419,** 420
carbon atom, bonding in, **418**
carbon compounds, 418, **419,** 420
carbon dioxide, effect on body, 418–419
carbon dioxide molecule, **419**
carbon fuels, **362,** 363, 364
carbon monoxide, effect on body, 418, **419;** formation of, 418
carbon-14 atom, **448,** 449
carbon-14 dating, 451, **452**
catalyst(s), *356*
catalytic converters, 356
Celsius, *242*
Celsius temperature scale, 242–*243,* **251**
centigrade, *243*
centimeter, *8*
chain reaction, *459,* **460**
charge(s), electric, *See* electric charge(s)
chemical activity, *319*
chemical bond, *335*–336; double, *418;* kinds of, 342, **343, 344,** 345
chemical change, *269*
chemical energy, 70
chemical equation, *348,* **349, 350**
chemical family(ies), *324*–327
chemical reaction(s), *348*–350; between acid and base, 406–407; endothermic, *361*–362; energy in, 359–364; exothermic, 359, **360,** 361; neutralization, *403,* 407; (laboratory), 411–413; speed of, 353–356; zinc with hydrogen chloride, **397**
chemical tests, for organic compounds (laboratory), 440–441
chlorine, properties of, 342
chlorine atom, 342–343
chlorine molecule, formation of, **343**
chloroform, **428,** 429
circuit breakers, **188,** *189*
coal, **362,** 364
coke, 364
color(s), 126–132; primary, **131;** *See also* light
colorblindness, 132
compass, magnet as, 205, **206**
compound(s), *274;* breakdown of, **278, 279;** carbon, 418, **419,** 420
concave lens, **140**
concentration, *354*
conduction, *232,* **233**
conductor, *169;* of heat, 233
conservation, of energy, 63–66, and electromagnetic induction, 221; of matter, law of, 349
continuous spectrum, 313
controlled experiment, *3*
convection, *233,* 234, **235**
convex lens, *135,* **136, 137, 138,** 139, **140**
corrosion, *376,* **377**
Coulomb's law, 163

covalent bond, *343*
critical mass, *459*
crystal, *254;* ice, **254, 255**
crystalline substances, melting and boiling points of, 255
cube, volume of, **9**
Curie, Marie, and discovery of radium, 445, **447**
current(s), *See* electric current(s)
cyclotron, **314**

D

Dalton, John, and atomic theory, 279, 288
dark-line spectrum, **313**
decay, *See* radioactive decay
decibels, *102*
decomposition, *282*
Democritus, atomic theory of, 288
diagnosis, radioactive substances in, **451**
diamond, **418;** cutting of, **254,** 255
diatomic molecule, *344*
diesel engine, **70**
diffraction, wave, *102,* **103**
direct current, *175*
distillation, **363,** *364*
Doppler effect, *91,* **92**
double bonds, *418*
double sugar, 435
double-ring molecule, 435
dry cell, **173**

E

Earth, as magnet, 205, **206, 207, 208;** satellites of, **40;** elements in crust (table), 280
echo, *97*
eggshell, as calcium carbonate, **397**
Einstein, Albert, and matter-energy relation, 462
electric charge(s), of atomic particles, 289, (table), 290; kinds of, 152, **153, 154,** 155, 156; like and unlike, **153, 154,** 155; negative, *155*–156; neutral, *155;* positive, *155*–156
electric circuit, *173,* **174;** parallel, *182;* series, *181*
electric current, *167;* breakdown of water by, 278–279; (laboratory), 194–195
electric energy, 70
electric field, *161*–162
electric force, *159,* **160, 161,** 162
electric generator, **219**
electric light bulb, **188**
electric machine(s), 218, **219,** 220, 221
electric meter, **190**
electric motor, **214, 215**
electric power, 185–191
electrical appliances, cost of operating, **190, 191**
electrical cell, **173**
electrical charge(s), uses of, 155, 156
electrical conductivity of solutions (laboratory), 390–391
electrical energy, cost of use, **190, 191;** measurement of, **52**
electrical transformer, **220**–221
electrical shocks, 189
electricity, and magnetism, 212–215; measurement of 178–182; static, 160; *See also* electric charge(s); electric force; electrical energy
electrolysis, *282*
electrolyte, *385*–386
electromagnet, *213,* **214;** superconducting, 212
electromagnetic energy, 71
electromagnetic induction, *218*–221
electromagnetic spectrum, *113,* (table), 114, 115, 116
electromagnetic waves, *113;* dangers of, 115
electron(s), *166*–167, *288,* 290; unstable, 311–*312; See also* electron shell(s)
electron dot model, *337*
electron shell(s), *298,* **299, 300**
electronic machines, 185, **186**
electronic thermometer, **243**
electronics, careers in, **210, 211**
electroscope, 161, **162**
element(s), **279,** (table), 280; arrangement of (table), 336; artificial, **462,** 463; atomic numbers of (table), 295; classification of, **325, 326;** common (table), 280; 327; electron shells of (table), 299; mass number of, 305; periodic chart of, **326;** radioactive, 446, **447, 448,** 449; spectra of, **312,** 313; symbols for, 299, 300, (table), 301; valence numbers of (table), *338; See also* atoms(s); individual elements
end point, *407*
endothermic reaction, *361,* 362
energy, *51;* binding, *446,* changes in, **65,** 66; chemical, 70; conservation of, 63–66, 221; consumption by people, 465, **466;** conversion of, **70;** electrical, 52, 70; electromagnetic, 71; heat, 71; kinetic, *64,* **65;** magnetic, 70; mechanical, 69, 227–229; nuclear, 70, 457, **458, 459, 460;** potential, *63,* **64, 65;** as scientific concept, 69; solar, **73;** sources of, 72–74; uses of, 69–74; wave, 83; and work, 50–51; *See also* heat
energy label, **191**
engine(s), automobile, 228; heat, *229;* steam, 228, **229, 230**
enzymes, 356

equilibrium, state of solution, 273
ester, *432*
ethane molecule, **422,** 423, 424
ethanol, 430
ethylene glycol, **430**
ethylene molecule, 425
exothermic reaction, 359, **360,** 361
experiment, controlled, *3*
eye, lens of, **135, 138**

F

falling objects, and gravity, 32, **33**
Faraday, Michael, and electromagnetic induction, 217–218
Fahrenheit temperature scale, **242**
farsightedness, **140**
fat(s), **436, 437;** polyunsaturated and saturated, 425
fat molecule, **437**
Fermi, Enrico, and nuclear fission, 459
fertilizers, 377
fiber optics, *121*
fireworks, **360**
first class lever, *56,* **57, 58**
first law of motion, 26–27
fission, *459*
fluorine atom, 319, **320;** mass of, **304**
focal length, *136*
focal point, 136
food, chemical changes in, 354–355; energy content of, 361; *See also* human diet
force(s), and acceleration (laboratory), 44–45; action and reaction, 38; balanced, 37–40; electric (*See* electric force); and friction, *25*–26; *See also* gravity; gravitational force
formic acid, 431
fossil fuels, 72–73, **363, 364,** 365
fractional distillation, **363,** 364
fractions, petroleum, **363,** *364*
Franklin, Benjamin, and French stove, 235; study of electricity, **154,** 155
freezing, of water, 269
freon, 429
frequency, wave, *85*
friction, *25, 26*
fructose, **434,** 435
fusion, **460,** *461;* heat of (laboratory), 260–262
fusion reactors, 468
fuel(s), *362,* **363;** *See also* energy
fulcrum, *56,* **57**
fundamental tones, 101, **102**
fuse, **188,** *189*

G

Galileo, and air thermometer, 250; falling object experiment, 32–33; and speed of light, 113
gamma decay, 449
gamma rays, 116, *448,* 449
gas(es), behavior of, 248, **249, 250, 251;** kinetic theory of, **249, 250, 251;** noble, *315,* 320; *See also* individual gases
gas fuels, 364
gear wheel, **59**
geiger counter, **449**
generator, electric, **219**
Gilbert, William, and magnetism of earth, 206
glucose, **434,** 435
gold, properties of, 375
gold atom, 297–298
grams, *9*
graphite, 418, 419
gravitational force, 38–40; *See also* gravity
gravitional potential energy, calculation of, 64
gravity, *27;* and acceleration, **21, 22;** and falling objects, 32–**33**
greenhouse effect, 419–*420*
guitar strings, vibration of, **95,** 96, **97**

H

half-life, *452,* **453**
halogens, (table), *320*
hard water, 383
heat, of chemical reaction, 360, **361;** conduction of, *232,* **233, 235;** convection of, *233,* 234, **235;** as energy, 71, **226, 227, 228, 229;** of fusion (laboratory), 260–262; production by nuclear reactor, 466–467; radiation of, *234,* **235;** and temperature, 241–245; transfer of, 232–238
heat energy, measurement of, 243, **244, 245**
heat engines, *229*
heat of fusion, *256*
heat of vaporization, *257*
heliographs, *118,* 119
helium, in airships, 318
helium atom, 318–319, 447
helium welding, 320
Henry, Joseph, and electromagnetic induction, 216–217
hertz, *85*
high-frequency sounds, 97
Hindenburg, **318**
home, heating of, 233, 234, **235, 236, 237**
home heating, careers in, **210, 211**
home insulation, **237**
home wiring system, **187, 188, 189**
horseshoe magnet, **200**
hot air heating system, **236**
human body, elements in (table), 280; water in, 381

human diet, 408–409, **434,** 435, **436, 437;** *See also* food
hydraulic press, 60
hydrocarbon(s), *422;* substituted, 428–432; *See also* individual hydrocarbons
hydrocarbon molecules, 429–432
hydrocarbon molecule sequence, 423, (table), 424
hydrochloric acid, *398*
hydrogen, in airships, 318; atomic number of, **294;** as element, 279; in water, **278,** 279
hydrogen atom, 293, **294,** 304, 318–319
hydrogen chloride, chemical reaction with zinc, **397**
hydrogen ion, 398
hydrogen isotopes, 304, **305**
hydrogen molecule, formation of **335,** 336, 337
hydrogen peroxide, chemical changes in, 355, **356**
hydronium ion, 398
hydroxide radical, 429
hypothesis, 3

I

ice, 269; heat of fusion of, 257
ice crystal, **254,** 255
ignition coil, 220
image, *134*
inclined planes, **60**
index of refraction, *120,* (table), 121
indicator, *398*
inert atom, *319*
infrared radiation, 234, **235**
infrared waves, 115
insulation, home, **237**
insulator, *169, 233*
interference, wave, *100,* **101, 104**
iodine, radioactive, **451,** 454
ion(s), *314*–315; solubilities of (table), 387; *See also* electrolyte(s)
ionic bond, *344*–345
iron, on periodic chart, **378**
isomers, *424*
isotopes, *304*

J

jet engine, **38,** 228–229

K

Kelvin temperature scale, *251*
kilocalorie, *245*
kilogram, *9*
kilowatt, *52*
kilowatt-hour, *190*–191
kinetic energy, *64,* **65**
kinetic theory of matter, *248,* **249, 250**

L

laboratory, characteristics of sound, 106–107; chemical tests for organic compounds, 440–441; electric current, 194–195; electrical conductivity of solutions, 390–391; heat of fusion, 260–262; force and acceleration, 44–45; neutralization reaction, 411–413; power, 76–77; reflection and refraction, 143–145; voltaic cell, 330–331
laser beam, **71,** 129
law of conservation of energy, 66
law of conservation of matter, 349
lens, *134,* **135;** camera, **135;** concave, *139,* **140;** convex, *135,* **136, 137, 138,** 139, **140;** of human eye, **135, 138,** 139
lever(s), 56, **57, 58**
Lewis, Gilbert, and electron dot model, 336–337
light, properties of, **112,** 113; reflection of **130,** *131;* refraction of, 119, **120, 121, 122, 127,** 128; speed of, 113; *See also* color(s); spectrum(a)
light bulb, **188**
light rays, *119*
light waves, 112–116; movement of, 118–122
lightning, **167,** 168, 169; as electricity, **154,** 155
lightning rods, **168**
lime, 403
linear accelerator, **458,** 459
liquid(s), and kinetic theory, **249;** pH of (table), 403; properties of, 257
liter, *9*
lithium atoms, **319;** mass number of, 305
litmus, 398
longitudinal waves, 83, **84**

M

machine, efficiency of, *58;* simple, 55–60
magnesium hydroxide, as base, 403
magnesium radical, 341
magnet(s), bar, **200,** 201; as a compass, **206;** creating of, **201, 202;** earth as, 205, **206, 207, 208;** as energy source, 70; in electric motor, **214,** 215; electromagnet, *213,* **214;** horseshoe, **200;** temporary and permanent, 202
magnetic bottle, 461

magnetic energy, 70
magnetic field, *202;* of earth, 207, **208**
magnetic poles, *200,* **201, 202;** of earth, 206, **207**
magnetic variation, *207*
magnetism, and electricity, 212–215
magnifying glass, **136**
marsh gas, 423
mass, *31;* and acceleration, **33,** 34; atomic (*See* atomic mass); of atomic particles, 289, (table), 290; and weight, 30–34
mass number, *305*
matter, *267;* ancient theory of, 288; changes in, 269; conservation of, 349; kinetic theory of, *248,* **249, 250;** phases of, 248, **249,** 269
measurement, *7;* making of, 3; metric system of, 7–9; units of, 7–9
mechanical advantage, of lever, 57
mechanical energy, 69; from heat, 227–229; *See also* engine; machines
melting, of ice, 269
melting point(s), *255,* **256,** (table), 257
Mendeleev, Dmitri and classification of elements, **325,** 326, 327
mercury, spectrum of, **312**
metal(s), alkali, (table), *319;* properties of, **375, 376**
metallic bond, *375–376*
metalloids, *377–378*
meter, *8;* electric, **52**
methane, 364
methane molecule, 423, 424
methanol, 429, **430**
metric system, 7–9
microwaves, 115
millimeter, *8*
mirage(s), *121,* **122**
mixture(s), *272,* **273**
models, atomic, 290
molecule(s), *267–269;* diatomic, *344;* double-ring, 435; formula for, 344–345; hydrocarbon, 429–432; measuring size of, 267; in solution, 273
moon, light on, *113*
motion, *19–22;* accelerated, 18–22; first law of, 26–27; second law of, 33–34; third law of, 37–38
motor(s), *See* electric motor(s)
music, and sound, 100–103
mutations, 453

N

negative electric charge, *155–156*
neon, 320
neutral electric charge, *155*
neutralization reaction, *403,* 407; (laboratory), 411–413
neutrons(s) in atomic fission, **459**
newton(s), 27
Newton, Sir Isaac, **26;** first law of motion, 26–27; and gravitational force, 38–39; second law of motion, 33–34; third law of motion, 37–**38**
neutrons, *289,* 290
nitric acid, 399
nitrogen, as fertilizer, 377
noble gas(es), *315,* 320
noise, 102–103
noise levels (table), 103
nonelectrolyte, *386*
nonmetals, properties of, 376–377
nuclear bomb, *See* atomic bomb
nuclear energy, 70; production of, 466, **467**
nuclear fusion, **460,** *461*
nuclear fusion reactors, 468
nuclear power plants, dangers of, 467–**468**
nuclear reactions, **459, 460, 461, 462,** 463; controlled and uncontrolled, 465, **466**
nuclear reactor, *466,* **467, 468**
nucleus, *See* atomic nucleus

O

observation(s), making of, *2–4*
Oersted, Hans Christian, and electromagnetism, 212
ohm, *180*
Ohm, Georg, and resistance, 180
Ohm's law, 180–181
organic acid(s), *431,* 432; reaction with alcohol, 432
organic chemistry, *420*
organic compounds, chemical tests for (laboratory), 440–441;saturated and unsaturated, 425
organic solvents, 426
overtones, 101, *102*
oxidation, *348,*
oxygen, as element, 279; in water, **278,** 279
oxygen atom, 293, **294;** atomic mass of, **304;** atomic number of, **294**

P

parallel circuit, *182*
particle(s), *See* atomic particle(s)
patterns, in observations, 3
periodic chart of elements, *326*
permanent magnet, *202*
petroleum, *363–364*
pH of common liquids (table), **403**
pH scale, 403
phases of matter, 248, **249**
phenolphthalein, 407

phosphorus, as fertilizer, 377
photon, *112*
photosynthesis, 361–362
physical change, *269*
pistons, **59,** 60
pitch, *97*
planets, and gravitational force, **39, 40**
polar molecule, *380*
pole(s), of magnet, **200,** 201, **202**
pollution, thermal, 383; water, *381,* **382**
polonium, 447
polyethylene molecule, 425
polyunsaturated fats, 425
positive electric charge, *155*–156
potential energy, *63,* **64, 65**
power, *52;* (laboratory), 76–77
precipitate, 386–*387*
primary colors, **131**
prism, 126
propane, 364
propane molecule, **422,** 423, 424
protein(s), *437*–438
protein molecule, **437**
proton(s), *167, 289,* 290
pulleys, 58, **59**

R

radar waves, 114–115
radiation, *234,* **235**
radical, *338,* (table), 339
radio waves, 114
radioactive decay, **447, 448,** 449
radioactive elements, 446, **447, 448,** 449; half-life of, **452, 453**
radioactive iodine, **451,** 454
radioactive tracers, 452, **453**
radioactive wastes, disposal of, 467, **468**
radioactivity, *446;* dangers of, 453; in treatment of disease, **451, 454;** uses of, **451, 452, 453, 454**
radium, 445, 447
radium atom, 447, 451; half-life of, **453**
radon, **447**
rainbow, **127, 128**
rays, *119*
reactor, *See* nuclear reactor
real image, *137*–138
reference point, *19*
reflection (laboratory), 143–145; wave, *90;* refraction (laboratory), 143–145, light, 119, **120, 121, 122;** wave, **88,** *89,* **92**
resistance, *179, 180*
resources, renewable and nonrenewable, 72
ripple tank, waves in, **89**
Roemer, and speed of light, 113
roller coaster, and energy changes, **65**
Rumford, Count, and heat as energy, **226,** 227
Rutherford, Ernest, and atomic structure, 292–293

S

salt, *407;* in diet, 408–409; uses of, 408–409
salt crystal, **255;** dissolving of, **381**
salt solution, 397; as electrolyte, 386, **387**
salt water solution, **274**
salts, production of, 407, **408**
satellites, earth, **40**
saturated fats, 425
saturated organic compound, 425
saturated solution, 273
scientific law, *3–4*
scientific method, 2–4
scientific model, 165–*166;* of atom, 290
screw, making of, **60**
scrubbing, *156*
sea salt, 408
sea water, dissolved substances in (table), 408
second class lever, *56,* **57, 58**
second law of motion, 33–34
semiconductor, *377*–378
series circuit, *181*
sewage, treatment of, 382–383
silicon, on periodic chart of elements, **378**
silver, **376**
simple machine, *56*
simple sugars, 435, 436
sky, color of, **129**
smokestack gases, removal of particles from, **156**
sodium, properties of, 342
sodium atom, 342, **343,** 344
sodium chloride, 342, 406; *See also* salt
sodium chloride crystals, **344**
sodium chloride molecule, formation of, **344,** 345
salt water, 383
sodium hydroxide, as base, 407–408
solar cells, 73, 172
solar energy, **73**
solar heating system, **236, 237**
solar system, 297
solid(s), and kinetic theory, **249;** melting of, **255,** 256; structure of, 254–256
solid fuels, 364
soluble substance, *387*
solute, *381*
solution(s), *273;* acidic, 397–399; electrical conductivity of (laboratory), 390–391; hydrogen chloride, 397; salt, 397; saturated, 273; *See also* individual solutions

solvent, *381*
sonic booms, *96*
sound, *94;* characteristics of (laboratory), 106–107; high frequency, 97; and music, 100–103; *See also* sound waves
sound waves, 94–97
spectrum(a), kinds of, 313; visible, *126,* **127**
spectrum tube, **312**
speed, and accelerated motion, 18–22; average, *19*–20; calculation of, *2*–3; of light, 113; wave, 84–**85**
spring, and energy, **63,** 64
stable electron arrangement, *315*
standing wave(s), 83
star, spectrum of, 313
starch(es), 435, **436**
starch molecule, 436
state of solution equilibrium, 273
static electricity, 160
steam carriage, **38**
steam engines, 227, **228, 229**
steam turbine, **229**
structural formula, *423*
substituted hydrocarbons, 428–432
sublimation, *256*
sucrose, 435
sugar(s), **434,** 435, **436;** burning of, 278; as compound, **278**
sugar solution, 273; as nonelectrolyte, **386**
sulfate radical, 338
sulfur, on periodic chart of elements, **378**
sulfuric acid, 398
sulfurous acid, 399
sun, fusion reactions in, **461**
sunburn, 116
superconducting electromagnets, 212
symbols, chemical, *299,* 300, (table), 301

T

taste, 396–397
television color tube, **131, 132**
temperature, *241;* and chemical reactions, 354–355; of gases, **250, 251;** and heat, 241–245; measurement of, 241–245
temporary magnet, *202*
terminal speed, *22*
theory, *3*
thermogram, 237, **238**
thermometer(s), **242, 243**
third class lever, **57, 58**
third law of motion, 37, **38**
Thompson, Benjamin, *See* Rumford, Count
Thomson, J. J., atomic theory of, 288
threshold energy, *354*
thyroid gland, radioactive tracers in, 451, 454
tone(s), musical, 101, **102**
tongue, and tasting, 396–397
tracers, *453,* **454**
transformer, *187;* electrical, **220,** 221
transverse wave(s), *83,* **84**
turntable, force on, 37

U

unsaturated organic compound, 425
unstable electron, 311–*312*
uranium-235 atom, fission of, **459**

V

valence, *337*
valence electrons, *338,* (table), 339
velocity, *20*
vibration, and sound, **95,** 96, **97**
vinegar, 430, **431**
vinegar solution, properties of, 398
visible light waves, 115
visible spectrum, *126,* **127**
volt, *178*–179
Volta, Alessandro, and electrical cell, 172–173
voltaic cell (laboratory), 330–331
voltmeter, 179
volume, *9*
Van de Graaf generator, **160,** 161

W

water, boiling point of, 257; analysis of, **282, 283,** 284; as compound, 274; as energy source, 74; equation for formation of, **349;** hard and soft, 383; heat of vaporization of, 257; heating of, **227;** in human body, 381; hydrogen and oxygen production from, **278,** 279; as neutral solution, 403, 407; pollution of, 381, **382;** properties of, 269; as solvent, 381; and sugar solution, 273
water gas, 364
water molecule, 267–269, **380;** formation of, 343
water waves, **82,** 83
watt, *52,* 190
Watt, James, and steam engine, 228, **229**
watt-hour, *190*
wave(s), *82;* amplitude, *85;* diffraction, *102,* **103;** electromagnetic, *113;* and energy, 83; gamma rays, 116; infra-

red, 115; interference, *100,* **101, 104;** light, 112–116; longitudinal, *83,* **84;** microwaves, 115; motion of, 88–91; properties of, 83–85; radar, 114–115; radio, 114; reflection of, *90;* refraction of, **88,** *89,* **92;** sound, 94–97; speed of, 84, **85;** standing, 83; transverse, *83,* **84;** ultraviolet, 115–116; water, **82,** 83; X rays, 116; *See also* light
wavelength, *84*
wedge, 60
weight, *27;* and mass, 30–34
weightlessness, **30, 31**
wind, as energy source, 73, **74**
windmills, **74**
wood alcohol, 430
work, *50;* and energy, 50–52

X

X rays, 116
xerography, 155, 156

Z

zinc, chemical reaction with hydrogen chloride, 397

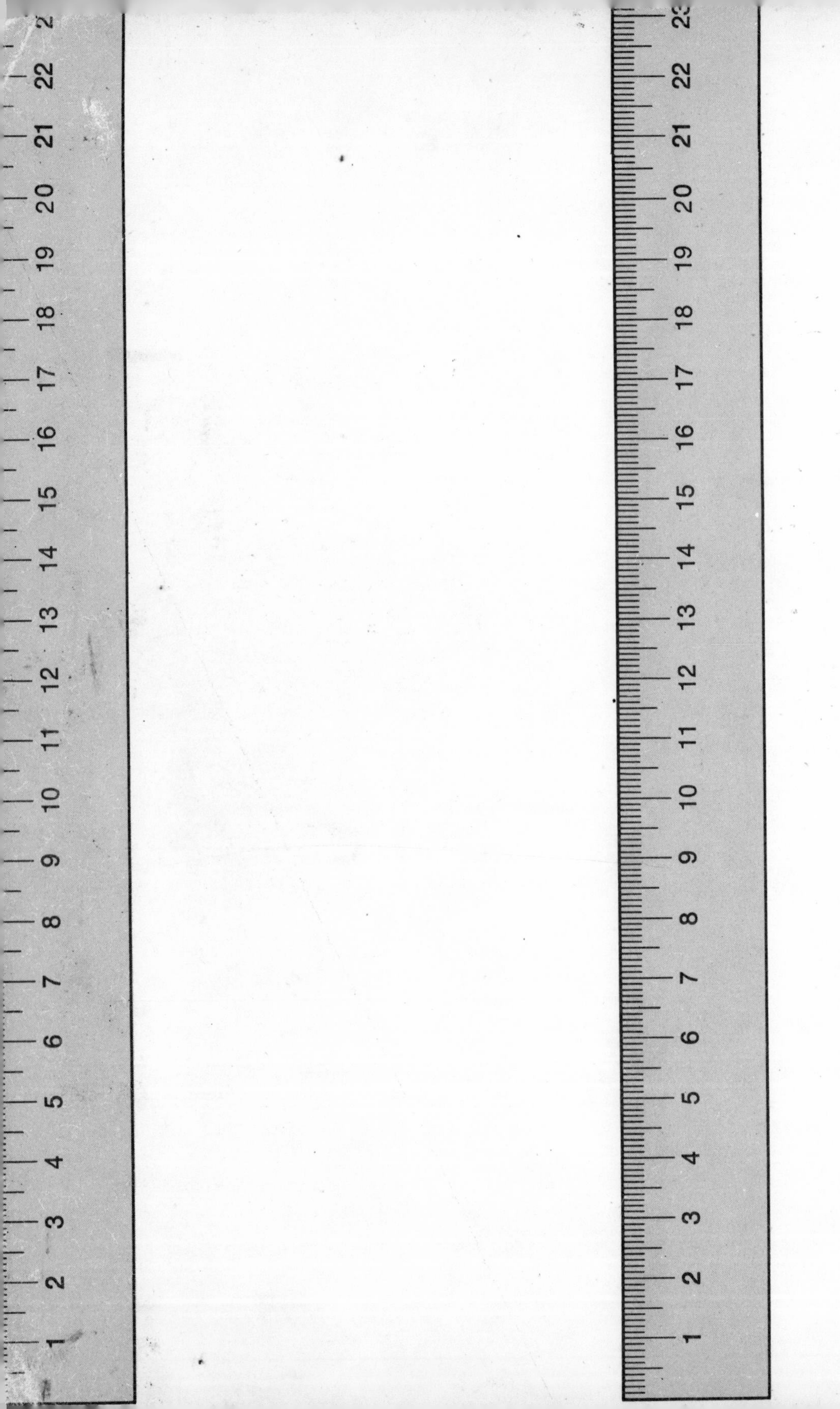